IMMUNOLOGICAL RECOGNITION *of* PEPTIDES *in* MEDICINE *and* BIOLOGY

Edited by

Netty D. Zegers
Wim J.A. Boersma
Eric Claassen

CRC Press
Boca Raton New York London Tokyo

Library of Congress Cataloging-in-Publication Data

Immunological recognition of peptides in medicine and biology / Netty D. Zegers, Wim J. A. Boersma, and Eric Claassen, editors
p. cm.
Includes bibliographical references and index.
ISBN 0-8493-8967-4 (alk. paper)
1. Peptides--Immunology. 2. Immune recognition. 3. Synthetic vaccines. I. Zegers, Netty D. II. Boersma, Wim J. A.
III. Claassen, Eric
[DNLM: 1. Peptides--analysis. 2. Peptides--immunology. 3. Peptides--therapeutic use. QU 68 I336 1995]
QR186.6.P76I495 1995
616.07′9--dc20
DNLM/DLC
for Library of Congress 95-3480
CIP

International Standard Book Number 0-8493-8967-4
Library of Congress Card Number 95-3480
Printed in the United States of America 1 2 3 4 5 6 7 8 9 0
Printed on acid-free paper

AFFILIATIONS

Netty D. Zegers was a member of the section of peptide-based immunochemistry in the department of Immunology in the TNO-Medical Biological Laboratory in Rijswijk. She has been involved in the production of monoclonal antibodies for immunodiagnostic and immunopathologic applications mainly. Presently, she leads a group using similar approaches for the development of rapid antibody- and probe-based diagnostics for infectious agents and toxins in the Division of Immunological and Infectious Diseases at the TNO Institute for Prevention and Health.

Wim J.A. Boersma, PhD, was head of the section of peptide-based immunochemistry in the department of Immunology in the TNO-Medical Biological Laboratory. Research topics were: anti-human immunoglobulin antibodies (IgA, IgG subclasses) and the development of antibodies to a number of proteins based on a known DNA base sequence only. Presently, he is heading the section of vaccine development based on live oral vaccine carriers in the division of Immunological and Infectious Diseases of the TNO Institute for Prevention and Health in Leiden, the Netherlands.

Prof. Eric Claassen, PhD, in the department of Immunology and Medical Microbiology at the TNO-Medical Biological Laboratory in Rijswijk developed immunoenzyme cytochemistry methodology to detect the presence and production of cytokines simultaneously with antibody-forming cells, antigen specificity of antibody-forming cells and other immune cell-related characteristics. In his group the development of peptide enzyme conjugates was initiated. Presently, he is both a professor of cellular immunology at the Erasmus University of Rotterdam and the deputy head of the division of Immunological and Infectious Diseases of the TNO Institute for Prevention and Health in Leiden, the Netherlands.

TNO Prevention and Health
P.O.Box 2215
2301 CE Leiden
The Netherlands

Phone +31 71 18 1433
Fax +31 71 18 1901

E-mail:
nd.zegers@pg.tno.nl
wja.boersma@pg.tno.nl
ehjhm.claassen@pg.tno.nl

AUTHORS

Prof. Dr. R.C. Aalberse
Central Laboratory of The Netherlands Red Cross
Blood Transfusion Service
Amsterdam, The Netherlands

Dr. A. van Amerongen
ATO-DLO
Wageningen, The Netherlands

Dr. J.R. Appel
Torrey Pines Institute for Molecular Studies
San Diego, CA 92121

Prof. Dr. R. Arnon
Department of Chemical Immunology
Weizmann Institute of Science
Rehovot, Israel, 76100

M. van den Berg
Central Laboratory of The Netherlands Red Cross
Blood Transfusion Service
Amsterdam, The Netherlands

Prof. Dr. W. G. Bessler
Institut für Immunbiologie der Universität
Freiburg, Germany

Dr. W.J.A. Boersma
Division of Immunological and Infectious Diseases
TNO Prevention and Health
Leiden, The Netherlands

Ing. R.M.P. Brandt
Dept. of Immunohaematology and Blood Bank
University Hospital Leiden
Leiden, The Netherlands

Dr. S. Buus
Institute for Medical Microbiology and Immunology
University of Copenhagen
DK-2200 Copenhagen N., Denmark

Dr. L. Castagnoli
Dipartimento di Biologia
Universitá di Roma Tor Vergata
Rome, Italy

Dr. G. Cesareni
Dipartimento di Biologia
Universitá di Roma Tor Vergata
Rome, Italy

Prof. Dr. E. Claassen
Division of Immunological and Infectious Diseases
TNO Prevention and Health
Leiden, The Netherlands

Dr. R. Cortese
Istituto di Ricerche di Biologia Molecolare
P. Angeletti, Pomezia
Rome, Italy

Dr. L. Dente
Dipartimento di Biologia
Universitá di Roma Tor Vergata
Rome, Italy

Dr. H.J. Dyson
Department of Molecular Biology
The Scripps Research Institute
La Jolla, CA 92037

Dr. F. Felici
Istituto di Ricerche di Biologia Molecolare
P. Angeletti, Pomezia
Rome, Italy

Dr. K. Gerritse
TNO Prins Maurits Laboratory
Rijswijk, The Netherlands

Dr. L. Gorbics
The Wistar Institute of Anatomy and Biology
Philadelphia, PA 19104

Dr. J. Hammer
Roche Milano Ricerche
I-20132 Milano, Italy

Dr. W. van 't Hof
Oral Biochemistry ACTA
Vrije Universiteit
Amsterdam, The Netherlands

Dr. R.A. Houghten
Torrey Pines Institute for Molecular Studies
San Diego, CA 92121

Dr. G. Iannolo
Dipartimento di Biologia
Universitá di Roma Tor Vergata
Rome, Italy

Dr. R. Jemmerson
Department of Microbiology
University of Minnesota Medical School
Minneapolis, MN 55455

Prof. Dr. G. Jung
Institut für Organische Chemie der Universität
D-72076 Tübingen, Germany

Dr. W.M. Kast
Dept. of Immunohaematology and Blood Bank
University Hospital Leiden
Leiden, The Netherlands

Dr. J.P.M. Langedijk
Central Veterinary Institute (ID-DLO)
Lelystad, The Netherlands

Dr. J.P.M. Langeveld
Central Veterinary Institute (ID-DLO)
Lelystad, The Netherlands

Dr. A. Luzzago
Istituto di Ricerche di Biologia Molecolare
P. Angeletti, Pomezia
Rome, Italy

Prof. Dr. C.J.M. Melief
Dept. of Immunohaematology and Blood Bank
University Hospital Leiden
Leiden, The Netherlands

Prof. Dr. R.H. Meloen
Central Veterinary Institute (ID-DLO)
Lelystad, The Netherlands

Dr. P. Monaci
Istituto di Ricerche di Biologia Molecolare
P. Angeletti, Pomezia
Rome, Italy

Dr. S. Modrow
Institut für Medizinische Mikrobiologie und Hygiene
Universität Regensburg
D-93042 Regensburg, Germany

Dr. A. Nicosia
Istituto di Ricerche di Biologia Molecolare
P. Angeletti, Pomezia
Rome, Italy

Dr. J.M. van Noort
Division of Immunological and Infectious Diseases
TNO Prevention and Health
Leiden, The Netherlands

Dr. L. Otvos
The Wistar Institute of Anatomy and Biology
Philadelphia, PA 19104

Dr. P. Panina-Bordignon
Roche Milano Ricerche
I-20132 Milano, Italy

Dr. L.Ø. Pedersen
Institute for Medical Microbiology and Immunology
University of Copenhagen
DK-2200 Copenhagen N., Denmark

Dr. C. Pinilla
Torrey Pines Institute for Molecular Studies
San Diego, CA 92121

Dr. W.C. Puijk
Central Veterinary Institute (ID-DLO)
Lelystad, The Netherlands

Dr. W.M.M. Schaaper
Central Veterinary Institute (ID-DLO)
Lelystad, The Netherlands

Prof. Dr. M. Sela
Department of Chemical Immunology,
Weizmann Institute of Science,
Rehovot,Israel, 76100

Dr. F. Sinigaglia
Roche Milano Ricerche
I-20132 Milano, Italy

Dr. A. Stryhn
Institute for Medical Microbiology and Immunology
University of Copenhagen
DK-2200 Copenhagen N., Denmark

Dr. G.I. Szendrei
The Wistar Institute of Anatomy and Biology
Philadelphia, PA 19104

Dr. P. Valsasnini
Roche Milano Ricerche
I-20132 Milano, Italy

Dr. C. Vetriani
Dipartimento di Biologia
Universitá di Roma Tor Vergata
Rome, Italy

Dr. L. Urge
The Wistar Institute of Anatomy and Biology
Philadelphia, PA 19104

Dr. H. Wolf
Institut für Medizinische Mikrobiologie und Hygiene
Universität Regensburg
D-93042 Regensburg, Germany

Dr. N.D. Zegers
Division of Immunological and Infectious Diseases
TNO Prevention and Health
Leiden, The Netherlands

IMMUNOLOGICAL RECOGNITION OF PEPTIDES IN MEDICINE AND BIOLOGY

PREFACE

In this book selected specialists give an insiders view on the state of the art in their sector of expertise in the various fields of applications of peptides in modern medicine and biology. From experimental preclinical methodology and applications, to the basics and hands on methodology, you will get an overview of specific new technologies which may generate your enthusiasm for new developments in your own field of medical biological interest. Examples are derived from immunology, but the results are applicable in a wide field of biology and medicine. The various chapters are composed so that medical biological students as well as scientists that want to enter the field will find important clues to set out on an expedition in peptidology.

This book consists of four parts: Part I: Modalities of Functional Peptide Characterization; Part II: Structure Function Relations and Synthesis of Special, Derivatized Peptides; Part III: Immunological Recognition of Peptides; and Part IV: Preclinical and Experimental Clinical Applications of Peptides.

In Part I, modalities of functional peptide characterization are discussed. To understand how ligands or parts of ligands bind to receptors, it is of utmost importance to develop methods to determine the sequence and structure of these ligands. Two major types of approaches can be distinguished: (i). random synthetic peptide libraries or specific selected synthetic peptides and (ii). random peptide libraries expressed in micro-organisms or filamentous bacteriophages.

Ad. i. When using a synthetic peptide library, the sequence and identity of the protein which binds to the receptor may be unknown (hormone receptors, antibodies, substrate-binding sites). *Clemencia Pinilla et al. (Chapter 1)* give examples of the organization of a synthetic peptide library and demonstrate the feasibility of this approach. New putative applications are proposed and discussed.

When the sequence of a protein is known or predicted from the DNA base sequence, the binding of receptors or antibodies to the protein may be investigated using the PEPSCAN methodology. In this method peptides overlapping in amino acid sequence are synthesized. In a specific PEPSCAN ELISA-like system the binding of proteins (receptors) to the peptides can be demonstrated, thereby identifying the binding site. *R. H. Meloen et al. (Chapter 2)*, who were among the inventors of this methodology, give an overview of the present state of the art and demonstrate the power of this method not only in identification of antigenic sites of infectious agents (virus) but also in the optimization of artificial ligands which can be applied for the generation of highly immunogenic peptide vaccines. A variant of the PEPSCAN method is applicable to identify T cell epitopes. A simple modification of the PEPSCAN method was developed by *Wim van 't Hof et al. (Chapter 3)*. They replaced the pins, which are used as the solid phase on which peptides are synthesized in the PEPSCAN method, by paper discs. It is demonstrated that in using this relatively simple methodology a "poor men's" PEPSCAN is easily performed.

Ad ii. Micro-organisms can be made to synthesize "random" peptides and to expose these peptides, mainly as part of a fusion protein, on their surface. Vaccine models have been widely based on this approach. Filamentous phages can be made to express a random library of peptides which are expressed on their surface. Phages expressing specific peptide sequences will bind to proteins or receptors. The selected phages in this way indirectly establish the binding site. The phage, which is bound, is selected to produce the binding peptide in large quantities. *G. Cesareni et al. (Chapter 4)* discuss the various methods and limitations of phage library techniques.

Functional identification of peptide sequences can be based on ligand-receptor interactions. A specific receptor decisive for immune interaction and the integrity of the immune system is the major histocompatibility complex (MHC) molecule which binds T cell epitopes. Consequently, the isolation of peptides from peptide MHC complexes is a direct approach to identify T cell epitopes as described by *Søren Buus et al. (Chapter 5)*. The role and specificity of T cell epitopes is a matter of debate. Originally, T cell epitopes were considered highly specific for certain MHC-T cell receptor combinations. Now it has become clear that a certain degeneration of this system can be observed under various circumstances. This promiscuity of T cell epitopes is shown by *F. Sinigaglia et al. (Chapter 6)*. The sequences selected by the immune system are generated by enzymatic processing of antigens by antigen-presenting cells. Peptides during this process are bound to MHC molecules and as such presented on the cell surface. To generate T cell epitopes without going through the procedure of MHC isolation and peptide elution, a method was developed to perform *in vitro* processing based on the use of enzymes abundantly present in antigen-presenting cells. *J.M. van Noort (Chapter 7)* describes the methodology, experimental results and possible applications and limitations of the technique. The way T cell epitopes should be presented to the immune system especially in relation to their structural association to B cell epitopes is described for a few typical examples by *Netty D. Zegers and Wim J.A. Boersma (Chapter 8)*.

Part II gives a representative view on how peptide sequences can be identified in experimental systems as described above. Alternatively, specific epitopes or binding sites can be theoretically predicted based on the properties of individual amino acids, the properties of a specific sequence of amino acids and the knowledge with respect to protein structure and behavior at a given known amino acid sequence. Based on these data, algorithms have been developed which to a large extend predict B cell and T cell epitopes or ligand-binding sites. *Susan Modrow and Hans Wolf (Chapter 9)* who were among the first to develop this methodology discuss basics and application of prediction algorithms.

To understand peptide-protein and protein-protein interactions it is important to investigate the structure of these molecules. *H. Jane Dyson (Chapter 10)* shows which methods can be applied to get detailed information on the structure of protein elements under experimental and physiological conditions.

Synthesis, application and results based on modified peptides are described by *L.Otvos et al. (Chapter 11)* who present their results using glycopeptides and phosphopeptides which may mimic the natural structure (secondary, tertiary) of peptides better than is possible with a synthetic amino acid sequence alone. Peptides in general have a lower immunogenicity than observed with intact proteins. *Günther Jung and Wolfgang G. Bessler (Chapter 12)* demonstrate clearly the advantages and possibilities of peptides with intrinsic adjuvanticity which can be regarded as model vaccines.

In Part III, in the chapter by *Wim J.A. Boersma and Eric Claassen (Chapter 13)*, various adjuvants and immunological applications of peptides with respect to the elicitation of specific antibodies are described. A set of practical methods is included.

The chapter by *Eric Claassen and Wim J.A. Boersma (Chapter 14)* introduces the presently available methodology for immunoenzyme histochemical detection of epitopes in tissues of experimental and clinical origin for the determination of the specificity of ligand-receptor interactions. The examples given are mainly derived from immunological experimentation, but they will find, and in part already have found, application in the whole field of medical biology and clinical medicine.

For efficient antigen presentation *in vivo* as well as *in vitro/in assay*, complex synthetic structures, which basically consist of multipeptide structures, can efficiently be applied. *Netty D. Zegers et al. (Chapter 15)* give a short overview and recent results of these immunogenic structures of which diagnostic as well as vaccine applications can be envisaged. *Ronald Jemmerson (Chapter 16)* demonstrates how recognition of proteins and antibodies is based on the primary and secondary structure of peptides as well as on the way these peptides are presented to the immune system where the recognition at the level of proteins and peptides plays an important role in response initiation and regulation. Minor modifications in amino acid sequences of proteins can lead to dramatic changes in the function of these proteins, resulting in the development of pathology. Based on the use of carefully selected synthetic peptide sequences and effective immunization protocols, antibodies have been developed which allow detection of single amino acid genetic variants of proteins for elegant immunodetection. Some examples as well as the strategy of production are discussed by *Netty D. Zegers et al. (Chapter 17)*.

In Part IV, we have gathered mainly preclinical and even some experimental clinical applications of peptides. In these chapters the present state of potential applications is demonstrated. *Ruth Arnon and Michael Sela (Chapter 18)* describe the various ways in which the immunogenicity of otherwise low immunogenic peptides is enhanced with the main aim of using them as vaccines for prevention of infectious diseases. The adjuvant activity applied is mainly restricted to the natural adjuvanticity of proteins of viral and bacterial origin. The key issues in development of peptide vaccines such as genetic compatibility and support of humoral as well as class I and class II MHC-restricted cellular responses are addressed. *W. Martin Kast et al. (Chapter 19)*, experts in the field of cellular immunology, demonstrate that by using selected peptide sequences and effective immunization it is possible to use synthetic peptide vaccines to induce class I MHC-restricted cellular anti-tumor responses. By application of this most promising technique, cure of tumor-bearing animals was obtained. *Koen Gerritse (Chapter 20)* gives an overview of the various existing animal models for autoimmune diseases. There, peptide determinants play a role as disease-inducing epitopes of the self antigens to which the destructive capacity of the immune system is aberrantly directed. The models discussed are used for analysis of the pathogenicity of disease as well as for experimental therapeutic regimens based on application of peptides as disease modulators. These disease modulators may either act as tolerizing or as suppressive epitopes or competitors for inductive peptides.

All together we have tried to give you a view on the present possibilities of the use of (synthetic) peptides from the most advanced applications to the basics of materials and methods. Albeit that such an overview is never complete, we hope this will stimulate you to develop new ideas and applications in the specific fields you are all working in. We will closely follow the results.

Netty D. Zegers, Wim J.A. Boersma and Eric Claassen

TABLE OF CONTENTS

PART III: Immunological Recognition of Peptides

PART IV: Preclinical and Experimental Clinical Applications of Peptides

PART V: Index

Chapter 1

SYNTHETIC PEPTIDE LIBRARIES FOR RESEARCH AND DRUG DISCOVERY

Clemencia Pinilla, Jon R. Appel and Richard A. Houghten

TABLE OF CONTENTS

I. INTRODUCTION

The increasing use of peptides for immunological research, diagnostics, pharmaceuticals, and vaccines has led to the development of new immunodiagnostic and therapeutic agents. One of the advances that has led to this increase in peptide research is the improvements made in synthetic peptide chemistry (Geysen et al., 1984; Houghten, 1985; Jung et al., 1992). Over the past decade, new methods developed for the preparation of large numbers of individual synthetic peptides have greatly simplified and/or decreased the time needed for synthesis with earlier techniques (Merrifield, 1963). It is now possible, in modestly equipped laboratories employing a variety of synthetic techniques, to synthesize hundreds to thousands of individual peptides per year in appropriate quantities at a fraction of the cost per peptide when compared to previous methods. Approaches utilizing simultaneous multiple peptide synthetic methods have become essential for laboratories whose research requires large numbers of synthetic peptides.

An increasing need exists for new, pharmacologically active compounds, as well as the need for more rapid and less expensive screening methods. In the past, drug screening has been accomplished through the synthesis and testing of large numbers of individual analogs of a compound having an initial activity of interest. Thus, an initial compound, termed a "lead sequence", which may have poor initial biological activity, is manipulated through laborious individual chemical modifications and further refinement studies to arrive at the property profile desired (low toxicity, enhanced bioavailability, half-life, etc.). This classical approach to drug discovery, which involves the development of structure activity relationships (SARs), is ultimately limited to the number of individual compounds that can be synthesized and screened. Although peptides often tend to be poor final drug candidates because they do not typically have good oral or central activity, in many instances they can be modified to improve their activity. Of equal importance, peptides as small molecules possess sufficient chemical information to assist in the design of pharmacologically relevant compounds. In spite of the advances in the synthesis of large numbers of individual peptides over the past 8–10 years, the task of individually synthesizing and screening all combinations of the 20 natural

0-8493-8967-4/95/$0.00+$.50

L-amino acids using any of the existing approaches for even short peptides is impractical [64,000,000 six-residue peptides (20^6), 1,280,000,000 seven-residue peptides (20^7), etc.].

The recent introduction of peptide libraries composed of tens to hundreds of millions of sequences offers a fundamental, practical advance in the study and use of peptides for basic research in immunology, as well as for the potential development of therapeutic drugs (Birnbaum et al., 1992). The earlier use of simple mixtures of peptides containing hundreds of peptides has facilitated antigen-antibody interaction studies and the development of improved enzyme inhibitors and sequences that bind to proteins. Peptide mixtures can thus be readily prepared and screened for specific cases in which prior sequence information is available about a peptide or protein antigen. Peptide libraries, defined as containing tens to hundreds of millions of peptides, provide a more general and much broader approach for the identification of novel peptides when using existing diagnostically or pharmacologically relevant assay systems. Synthetic peptide library approaches are more rapid than traditional approaches by many orders of magnitude, and they represent a significant advance in all areas of drug screening and discovery. This chapter reviews the generation and use of synthetic peptide libraries for the study of immunological recognition, particularly antigen-antibody interactions. The advantages and disadvantages of the various peptide library formats will also be discussed. Finally, new strategies for the development of differently formatted peptide libraries will be presented.

II. PEPTIDE SYNTHESIS

II.1. Multiple peptide synthesis

Within the past decade, major improvements in synthetic peptide chemistry have greatly increased the number of peptides available for immunological research. Two of the major advances in solid-phase peptide chemistry responsible for this increased availability are the pin technology (Geysen et al., 1984) and the simultaneous multiple peptide synthesis approach (SMPS), also known as the "tea bag" method (Houghten, 1985). Both methods take advantage of the common washing steps used for the entire set of peptides, with only the specific amino acid being added requiring individual treatment. The pin method generates microgram quantities of 100 or more individual peptides simultaneously on the end of polyethylene rods. Peptides thus prepared can be assayed while attached to the pins. One hundred or more different peptides can also be synthesized simultaneously using the tea bag method, but this approach allows multiple milligram quantities to be prepared following cleavage from their individual resins using a multiple cleavage apparatus (Houghten et al., 1986). Nonsupport-bound peptides generated in this manner can be further characterized by reversed-phase high performance liquid chromatography (RP-HPLC), mass spectrometry, and other analytical procedures. Peptides generated by either method have proven particularly useful for the precise identification of the amino acids making up antigenic determinants recognized by antibodies. Overlapping peptides representing a protein sequence are now routinely used to identify antigenic determinants (Barakat et al., 1991; Cheng et al., 1991; Radford et al., 1990; Geysen et al., 1987; Selvey et al., 1990; Trifilieff et al., 1991). Also, omission and substitution analogs of a particular peptide sequence have proven useful in mapping the specificity of antibodies that recognize peptides (Appel et al., 1990; Pinilla et al., 1993b). More recently, the knowledge that T cell receptors recognize peptides presented on the surface of MHC class I and II molecules enables the use of large numbers of peptides or peptide analogs for this area of research.

A new approach for the simultaneous synthesis of thousands of peptides combines the technologies of photolithography with solid-phase chemistry (Fodor et al., 1991). Peptide synthesis is carried out on a derivatized silica substrate, then photolabile protected amino acids are coupled to the solid-phase amino terminus. Deprotection and coupling cycles are controlled using photolithographic methods, which enables miniaturization and precise spatial localization. This method appears to be very useful for the study of immunological systems in which a large number of analogs are necessary. However, this approach is technically sophisticated and not currently available for general use.

II.2. Peptide Libraries

The development of synthetic peptide libraries offers a significant increase in the number and diversity of peptides capable of being screened for drug discovery and various biological studies, as compared to the synthetic methods described earlier. By definition, synthetic peptide libraries should only refer to a set of peptides that consist in total of all, or close to all, of the possible combinations of amino acids expected for a given peptide length. The peptide mixtures making up a library can have both defined and mixture positions (mixtures are made up of hundreds to millions of peptides). The use of amino acid mixtures at individual positions of known peptide sequences was found in earlier studies to facilitate antigen-antibody interactions, and more recently a number of groups have used mixtures at one or more positions in peptide sequences for a variety of applications (Birkett et al., 1991; Owens et al., 1991; Petithory et al., 1991; Zuckermann

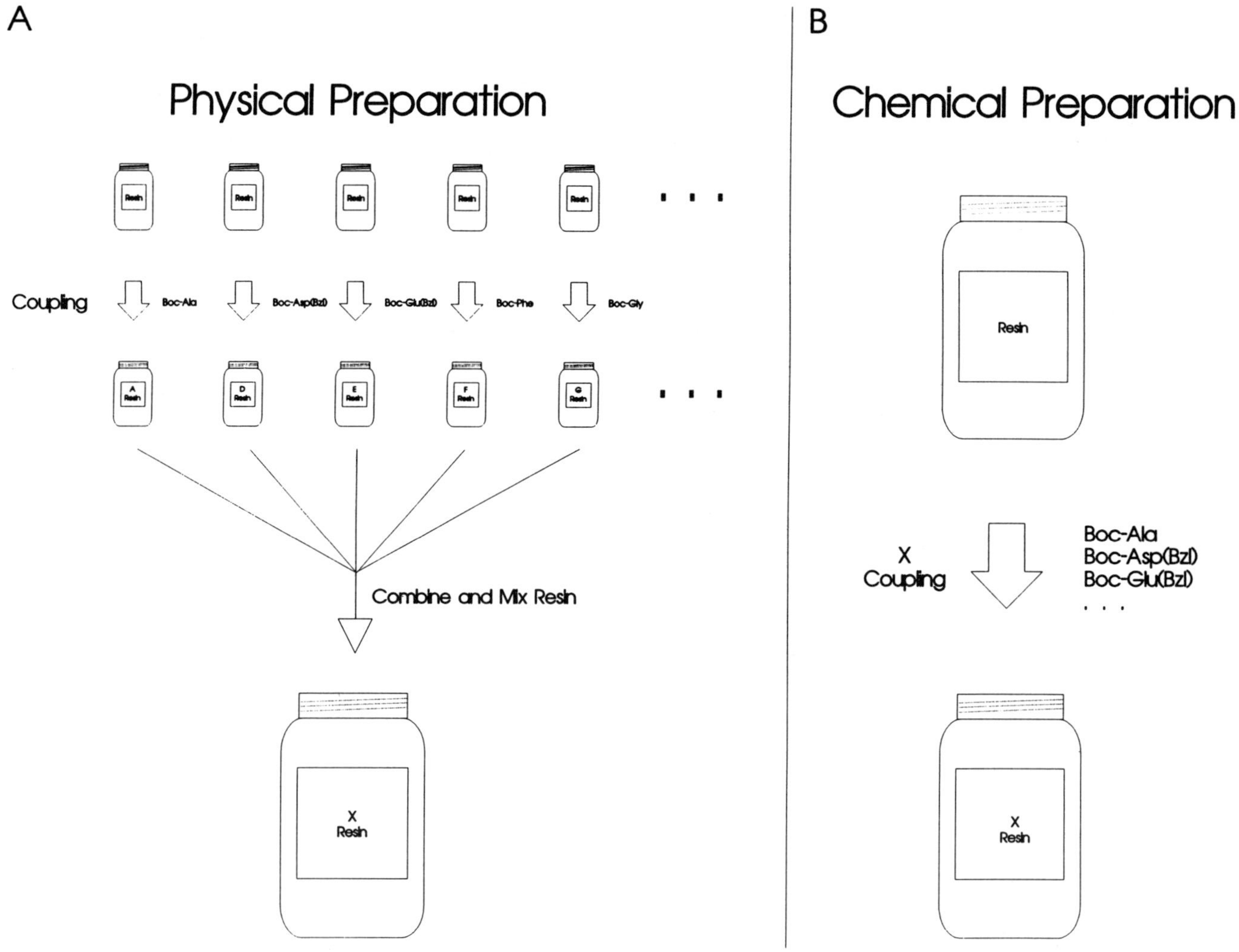

Fig. 1. Preparation of peptide libraries using "physical" and chemical approaches.

et al., 1992). While a variety of different peptide library approaches have been presented, all rely upon the inherent selectivity of biological receptor systems. Each method is more rapid than traditional methods of drug discovery by many orders of magnitude, and each represents a true revolution over earlier approaches.

II.2.a. Peptide libraries prepared with the resin-mixing, "physical" approach

The preparation of synthetic peptide libraries can be divided into two different approaches: physical and chemical. The first method can generally be termed the divide, couple, and recombine (DCR) physical method because it involves physically dividing and recombining the peptide resins after each coupling step. Briefly, equal portions of resin (and therefore, equal numbers of milliequivalents) are individually coupled to each of the protected amino acids. Initially described peptide libraries were made up of 18 to 19 amino acids (cysteine is often omitted). It is essential that coupling of each amino acid to the resin goes to completion, which can be easily monitored using either Kaiser's ninhydrin test (Kaiser et al., 1970), Gisin's picric acid procedure (Gisin, 1972), or Lebl's bromophenyl blue procedures (Krchnak et al., 1988). Following coupling completion, the resins are combined and thoroughly mixed. The resulting mixture [termed a one "X" resin and represented as X-resin (Fig. 1A)] is then divided into equal aliquots (corresponding to the number of amino acids used), and the α-amino protecting group is then removed. The resulting free α-amino X-resin is then taken through another coupling step. The process is repeated until the desired length of peptide mixture is obtained. This approach yields very close to equimolar peptide mixtures. The physical method for the preparation of complex peptide mixtures (hundreds of thousands to millions) was presented independently by two different groups (Houghten et al., 1991; Lam et al., 1991). A description of this approach for the preparation of very limited mixtures (<200 tetrapeptides) was also presented (Furka et al., 1991).

The physical resin mixture approach presented by Lam et al. was used to prepare a resin-bound (i.e., not available in solution as free peptide) pentapeptide library, in which each bead contained only a single peptide (one bead/one peptide) (Lam et al., 1991). Each bead was found to contain 50–200 pmol of peptide. Using the natural L-amino acids, excluding cysteine, the number of peptides prepared for this pentapeptide library was expected to total 2,476,099 (19^5). The screening results of this peptide library will be discussed later in this chapter.

Synthetic peptide combinatorial libraries (SPCLs) prepared using the same physical resin mixture approach, as presented by Houghten et al., consist of hexapeptide libraries which were grouped into 400 different peptide mixtures (Houghten et al., 1991; Houghten et al., 1992a). This peptide library consisted of mixtures in which the first two positions of each hexapeptide mixture were defined with a single L-amino acid (represented as O), while the remaining four positions were present as mixtures (represented as X). Nineteen of the 20 natural L-amino acids were incorporated (cysteine was excluded to avoid disulfide aggregates). The N-acetylated, C-terminal amidated form of this library can generally be represented by the sequence Ac-O_1O_2XXXX-NH_2, where O_1 and O_2 are specific amino acids (i.e., O_1O_2 = AA, AC, AD, etc., through YV, YW, YY, for a total of 400 combinations). The XXXX-resin was prepared using the DCR process in combination with SMPS (Houghten, 1985). Each X position represents a very close to equimolar mixture of the 19 amino acids, and four mixture positions yield 130,321 combinations. Since each of the 400 different peptide mixtures consists of 130,321 (19^4) different peptides, the peptide library is made up of 52,321,400 different hexapeptides. The peptide mixtures prepared by this method were cleaved from their respective resins. Libraries prepared in this manner offer the significant advantage of being able to be screened in solution assays. It should also be noted that at a typical concentration of 5 mg/ml, each of the peptides present in each peptide mixture has a concentration of 50 nM.

II.2.b. Chemical synthetic methods

The second method used to generate synthetic peptide libraries can be termed the chemical ratio method (Fig. 1B). In this approach, the peptide mixture resins making up the peptide library are prepared using a specific ratio of amino acids, empirically defined to give an equimolar incorporation of each amino acid at each coupling step. The degree of equimolarity of the resulting peptide mixtures does not approach that of the physical method due to the difference in amino acid coupling rates, but the ratio of amino acids used at each coupling step must be adjusted to give approximately equimolar representation, which can then be confirmed by amino acid analysis.

A peptide library prepared using the chemical synthetic method was presented by Geysen and co-workers (Geysen et al., 1986). This peptide library was prepared on pins in which the relative concentrations of the 20 L-amino acids in a reaction mixture were adjusted to yield approximately equimolar representation, as determined by amino acid analysis. This peptide library consisted of octapeptides in which the third and fourth position were defined with each of the 20 L-amino acids, while the remaining six positions were present as mixtures. This library can be represented by the formula Ac-XXO_1O_2XXXX-S_s, where S_s represents the solid support on which the synthesis is carried out. This peptide library contains more than 25 billion individual peptide sequences divided among 400 different peptide mixtures. The use of this peptide library will be discussed later in this chapter.

A differently formatted peptide library, termed a positional scanning SPCL (PS-SPCL) and prepared using the chemical synthetic method, was recently presented (Pinilla et al., 1992). Using a specific chemical ratio of 18 of the 20 natural L-amino acids (cysteine and tryptophan were omitted), approximate equimolar representation was found for each amino acid within the peptide mixtures as determined by amino acid analysis (Dooley et al., 1993a; Ostresh et al., 1994a). This PS-SPCL is made of six individual, single-position defined peptide libraries. Each position peptide library consists of hexapeptides in which one position is defined with one of the 18 L-amino acids, while the remaining five positions are mixtures of the same 18 amino acids. The six peptide libraries can be represented as Ac-O_1XXXXX-NH_2, Ac-XO_2XXXX-NH_2, Ac-XXO_3XXX-NH_2, Ac-XXXO_4XX-NH_2, Ac-XXXXO_5X-NH_2, and Ac-XXXXXO_6-NH_2, totalling 108 peptide mixtures. Each peptide mixture represents 1,889,568 (18^5) individual sequences; the entire peptide library thus contains 34,012,224 (18 x 1,889,568) different hexamers. This peptide library was shown to rapidly identify antigenic determinants recognized by monoclonal antibodies. The screening results of this PS-SPCL will be discussed later in this chapter.

II.2.c. Format of peptide library

One of the most important aspects in synthesizing peptide libraries is the format or configuration of each peptide library. Although the initial synthetic peptide libraries prepared by the physical method consisted of penta- and hexapeptides, their formats were different in that the "one bead, one peptide" approach (Lam et al., 1991) was a single mixture of more than 2 million individual bead-bound peptides, while the SPCL approach (Houghten et al., 1991) consisted of peptide mixtures with two positions individually defined with amino acids totalling 400 peptide mixtures (52 million hexapeptides). It should be noted that while the physical method of peptide library synthesis yields very close to equimolar concentrations, this approach is limited to peptides four or five residues in length if the 20 natural L-amino acids are used, since increasing the number of mixture positions requires 20 times more starting resin for each position added to ensure the presence of a statistical number of beads for every peptide. A variety of such libraries in different formats has been prepared, and their synthesis and use will be discussed later in this chapter.

III. IDENTIFICATION OF ANTIGENIC DETERMINANTS USING PEPTIDE LIBRARIES

To date, the utility of synthetic peptide library approaches has been illustrated for the identification of peptides recognized by antibodies. Table 1 summarizes the antigenic determinants identified using several different peptide library approaches. Peptide libraries used for the identification of antigenic determinants have been assayed by ELISA or by methods in which the antibody has been labeled with a fluorochrome or other readily detectable reagent. A protocol for the preparation and screening of SPCLs for the identification of antigenic determinants has been presented (Pinilla et al., 1994). Here we will review the immunological work carried out with soluble SPCLs, followed by a discussion of the work presented using pin- and bead-bound synthetic peptide library approaches.

III.1 Synthetic Peptide Combinatorial Libraries (SPCL)

III.1.a. SPCL screening methods using competitive ELISA

A number of monoclonal antibodies (MAb) have been screened using an SPCL composed of 400 different peptide mixtures, which is represented by the formula $Ac\text{-}O_1O_2XXXX\text{-}NH_2$ (Houghten et al., 1991; Houghten et al., 1992a). Typically, we have used competitive ELISA in which an antigenic peptide or protein, recognized by the target monoclonal antibody, is adsorbed to microtiter plates. For the initial screening of the peptide library, each peptide mixture is assayed at a concentration of 5 mg/ml for inhibition of antibody binding to the antigen on the plate. The known antigen is used in solution and serves as a control for the sensitivity of the assay. By testing serial dilutions of the peptide mixtures that were found to have significant inhibition activity, IC_{50} values (the concentration necessary to inhibit 50% of MAb binding to the antigen on the plate) can be determined, which permit the selection of the most effective peptide mixture from the initial screening.

An iterative process is then carried out in which the subsequent X positions of the peptide mixtures are individually defined with each of the 20 natural L-amino acids. This process involves ranking, selecting, and reducing the number of peptide sequences while synthetically defining one more position at each step. It is sometimes necessary to move forward with several peptide mixtures that have similar activities upon screening the peptide library. Often, the difference between these peptide mixtures is due to a conservative substitution at the same position (i.e., a valine residue substituted for an isoleucine residue). Since each iteration requires the synthesis of 20 new peptide mixtures, it is important to select only the most effective peptide mixtures to keep the total number synthesized to a minimum. As noted previously, a typical synthesis using the tea bag approach (Houghten, 1985) consists of 100–150 peptide mixtures, enough for five to seven iterations. If there are more than five to seven peptide mixtures with relatively good activity, then those peptide mixtures having amino acid side chains that have significant chemical differences are pursued. Thus, both DVXXXX and DIXXXX would not be moved forward, but if DVXXXX and AKXXXX were both found to be effective, they would both be followed, since they are quite different in chemical character. Those cases not pursued initially, however, can always be moved forward at a later date.

The screening of SPCLs, along with the iterative selection and enhancement process, can be likened to finding the proverbial needle in the haystack. The initial "haystack" is divided into 400 groups of 160,000 "straws" each, with only one of the groups containing the needle. The metal detector can be likened to the target antibody that will be assayed. If the correct peptide sequence (needle) recognized by the antibody is present in the peptide mixture, a strong signal will be found. To identify the active peptide within the peptide mixture, the next position is defined with the 20 natural L-amino acids (20 different groups or haystacks). Upon assaying the 20 groups (which now contain 8000 peptides, or straws, per mixture) with the target antibody and identifying the active peptide mixture, this iterative process is continued until a final sequence is identified.

III.1.b. Use of SPCLs against three different MAbs

In the first application of the SPCL approach for antigenic determinant mapping, a known peptide-MAb interaction was used as a model system (Houghten et al., 1991). This peptide-MAb interaction had been extensively characterized by ELISA using individual

TABLE 1
Summary of synthetic peptide library approaches used for the identification of antigenic determinants recognized by monoclonal antibodies

Synthetic method	Peptide library	Total # of sequences	MAb	Immunogen	Antigenic determinant	Peptide(s) identified from library	IC_{50} (nM)	References
SPCL (free peptide)	Ac-O_1O_2XXXX-NH_2	34 million	19B10	[1]HA1:75–110	DVPDYA	Ac-DVPDYA-NH_2	30	Houghten et al., 1991
		52 million	17D09	HA1:75–110	DVPDYA	Ac-DVPDYA-NH_2	2	Shown here
			125-10F3	v-fes 703–732	PYPNLS	Ac-PYPNLP-NH_2	4	Houghten et al., 1992a
						Ac-PYPNLS-NH_2	6	"
						Ac-YPYPNL-NH_2	2	Appel et al., 1992
						Ac-YPNLSS-NH_2	140	"
	O_1O_2XXXX-NH_2	52 million	3E7	β-endorphin	YGGFMT	YGGFMT-NH_2	3	Pinilla et al., 1993a
PS-SPCL (free peptide)	Ac-O_1XXXXX-NH_2	34 million	17D09	HA1:75–110	DVPDYA	Ac-DVPDYA-NH_2	2	Pinilla et al., 1992
	Ac-XO_2XXXX-NH_2					Ac-DIPDYA-NH_2	2	"
	Ac-XXO_3XXX-NH_2					Ac-DVEDYA-NH_2	2	"
	Ac-XXXO_4XX-NH_2					Ac-DVPDYS-NH_2	4	"
	Ac-XXXXO_5X-NH_2							
	Ac-XXXXXO_6-NH_2		125-10F3	v-fes 703–732	PYPNLS	Ac-PYPNLR-NH_2	13	"
Mimotope (pins)	Ac-XXO_1O_2XXXX	25 billion	15	[2]FMDV	Unknown	WQMRHS	[3]N.D.	Geysen et al., 1986
	XXO_1O_2XX	64 million	B13F2	Myohemerythrin (69–87)	DFLEKI	DFLEKI	N.D.	Geysen and Mason, 1993
Split synthesis (one bead/ one peptide)	XXXXX	2 million	3E7	β-endorphin	YGGFMT	YGGFQ	15	Lam et al., 1991
						YGGFA	33	"
						YGGFT	37	"
						YGAPM	52	Lam et al., 1993

[1]influenza virus hemagglutinin; [2]foot-and-mouth disease virus; [3]not determined

omission and substitution peptide analogs (Appel et al., 1990). This prior knowledge of the specificity of the peptide-antibody interaction provided a clear model, which could be used to test the practical utility of the SPCL approach. In the first library prepared, 18 of the 20 natural L-amino acids were used in both the defined and mixture positions (for ease of synthesis, cysteine and tryptophan were excluded). This initial library consisted of 324 different peptide mixtures for a total library containing more than 34 million hexapeptides (324 x 18^4). Upon screening each peptide mixture of the SPCL (Ac-O_1O_2XXXX-NH_2) and the subsequent iterations by competitive ELISA, a single peptide (Ac-DVPDYA-NH_2) was found to effectively inhibit MAb 19B10 interacting with a 13-residue peptide (Ac-YPYDVPDYASLRS-NH_2 from the influenza virus hemagglutinin, HA1:98–110). This sequence exactly matched the antigenic determinant sequence known to be recognized by this MAb.

More recently, another MAb, which was generated against the same peptide immunogen as MAb 19B10 (Wilson et al., 1984), was also extensively characterized by competitive ELISA using substitution analogs of Ac-YPYDVPDYASLRS-NH_2 (Pinilla et al., 1993b) and by structural means using x-ray crystallography (Rini et al., 1992). This MAb (17D09) was screened against an SPCL (Ac-O_1O_2XXXX-NH_2) which contained tryptophan in the four mixture positions (19^4 = 130,321) and cysteine and tryptophan in the defined positions (20^2 = 400), resulting in 400 different peptide mixtures for a peptide library totalling more than 52 million peptide sequences. Upon screening each of the 400 different peptide mixtures, the peptide mixture Ac-DVXXXX-NH_2 was found to have the greatest inhibition of antibody binding (IC_{50} = 58,000 nM). Twenty new peptide mixtures were then synthesized in which the third position of Ac-DVXXXX-NH_2 was defined with the 20 L-amino acids (Ac-DVOXXX-NH_2). The most effective inhibiting peptide mixture was Ac-DVPXXX-NH_2 with an IC_{50} = 15,000 nM. The iterative process was repeated in order to define the fourth position of Ac-DVPXXX-NH_2 by synthesizing another set of 20 peptide mixtures (Ac-DVPOXX-NH_2). Only one peptide mixture (Ac-DVPDXX-NH_2, IC_{50} = 1,100 nM) was found to have better activity than Ac-DVPXXX-NH_2 from the previous iterative step. The same degree of specificity was found when the fifth position of Ac-DVPDXX-NH_2 was defined, in that only one peptide mixture, Ac-DVPDYX-NH_2, IC_{50} = 40 nM, was found to be effective. This clearly illustrates the importance of the presence of specific peptide sequences in each peptide mixture throughout the iterative process. Upon defining the sixth and final position of the sequence, Ac-DVPDYA-NH_2 with an IC_{50} = 2 nM was found to be the most effective peptide recognized by this monoclonal antibody. The sequence -DVPDYA- exactly matches the antigenic determinant found in earlier studies to be recognized by this monoclonal antibody. This precise identification of an antigenic determinant proves the concept of the SPCL approach for its use in mapping antigenic determinants, as well as its usefulness for the identification of peptides in a variety of other immunological assays. It should also be noted that no information about the sequence of the antigen or antibody is required to carry out determinations of this kind. Prior knowledge of the characteristics of such interactions, however, can be helpful.

Table 2 summarizes the SPCL screening and iterative process that was used to identify the antigenic determinant recognized by MAb 17D09. Note that upon screening the 400 different peptide mixtures of the SPCL for the most effective peptide mixture, the number of peptide sequences is reduced from 52 million to 130,321. In four iterative steps a total of 80 peptide mixtures or peptides were screened to identify one optimal peptide sequence. This reduction in the number of active sequences found through the iterative process is reflected in the increase in inhibition activity at each step.

In another example, the same peptide library was screened against an MAb, which was made against a synthetic peptide representing residues 703–732 of the oncogene-related protein from v-fes (Houghten et al., 1992a; Appel et al., 1992). In this example, the antigenic determinant that was recognized by the MAb was previously characterized using single omission and substitution analogs (Appel et al., 1990; Pinilla et al., 1993b) and served as another test case to verify the SPCL approach using a different peptide-antibody interaction. Each of the 400 peptide mixtures was assayed to determine its ability to inhibit the interaction of MAb 125-10F3 with the antigenic peptide Ac-GASPYPNLSNQQT-NH_2 by competitive ELISA. Several peptide mixtures were found to cause significant inhibition of MAb binding, namely Ac-PYXXXX-NH_2, Ac-CYXXXX-NH_2, Ac-VYXXXX-NH_2, Ac-IYXXXX-NH_2, Ac-AYXXXX-NH_2, and Ac-YPXXXX-NH_2. The IC_{50}s of these peptide mixtures were determined, and Ac-PYXXXX-NH_2 (IC_{50} = 20 μM) and Ac-YPXXXX-NH_2 (IC_{50} = 200 μM) were found to be the most effective. An iterative process was carried out for Ac-PYXXXX-NH_2, in which 20 new peptide mixtures were synthesized with the third position defined (Ac-PYOXXX-NH_2) (Houghten et al., 1992a). The most effective inhibiting peptide mixture was found to be Ac-PYPXXX-NH_2 (IC_{50} = 860 nM). The peptide mixture Ac-PYGXXX-NH_2 (IC_{50} = 12,056 nM) was the only other peptide mixture found to have a lower IC_{50} than Ac-PYXXXX-NH_2. On defining the fourth position of Ac-PYPOXX-NH_2, the peptide mixture Ac-PYPNXX-NH_2 was found to be the only

TABLE 2
SPCL screening and iterative process against MAb 17D09

Step	Peptide	Defined sequences x mixture combinations			Total # of sequences	IC_{50} (nM)
Screening	Ac-OOXXXX-NH_2	400	x	130,321	52,128,400	
Selection	**Ac-DVXXXX-NH_2**	1	x	130,321	130,321	58,000
Synthesis/screening	Ac-DVOXXX-NH_2	20	x	6,859	137,180	
Selection	**Ac-DVPXXX-NH_2**	1	x	6,859	6,859	15,000
Synthesis/screening	Ac-DVPOXX-NH_2	20	x	361	7,220	
Selection	**Ac-DVPDXX-NH_2**	1	x	361	361	1,100
Synthesis/screening	Ac-DVPDOX-NH_2	20	x	19	380	
Selection	**Ac-DVPDYX-NH_2**	1	x	19	19	40
Synthesis/screening	Ac-DVPDYO-NH_2	20	x	1	20	
Peptide	**Ac-DVPDYA-NH_2**	1			1	2

mixture that significantly inhibited antibody binding. Its IC_{50} was more than 1400 times lower than the most effective peptide mixture in this set. The relative importance found for this position as recognized by this MAb corresponds with the results found from earlier studies using a complete set of substitution analogs of the antigenic peptide. Upon defining the fifth position, Ac-PYPNLX-NH_2 (IC_{50} = 6 nM) was found to inhibit antibody binding 10 times better than the next best peptide mixture. The final position was defined, resulting in 20 peptides (Ac-PYPNLO-NH_2), in which 13 peptides, including the known antigenic determinant sequence (-PYPNLS-), had IC_{50} values within threefold of the most effective peptide (Ac-PYPNLP-NH_2, IC_{50} = 4 nM). Since no significant increase in activity was found, the sixth position of the antigenic determinant can be considered redundant, which was also found in previous mapping studies.

We have also carried out the iterative process for the peptide mixture Ac-YPXXXX-NH_2 (IC_{50} = 200,000 nM), which was found to be the second most effective peptide mixture from the initial screening of the SPCL (Appel et al., 1992). It was initially believed that the activity of Ac-YPXXXX-NH_2 was due simply to the sequence homology that it has with the second and third positions of the known antigenic determinant -PYPNLS-. We chose to pursue the identification of specific sequences from Ac-YPXXXX-NH_2 through the iterative process to test this hypothesis, as well as to accumulate a base of information about the utility of SPCLs in such studies.

Thus, the third position of Ac-YPXXXX-NH_2 was defined, resulting in 20 new peptide mixtures (Ac-YPOXXX-NH_2). Assayed by competitive ELISA, only Ac-YPNXXX-NH_2 (IC_{50} = 7329) and Ac-YPYXXX-NH_2 (IC_{50} = 12,780) were found to inhibit antibody binding at a lower concentration than Ac-YPXXXX-NH_2. Since these peptide mixtures were different in chemical composition and were found to have similar activities, the iterative process was carried out for both sequences. Upon defining the fourth position of Ac-YPNOXX-NH_2, the peptide mixture Ac-YPNLXX-NH_2 was found to be the most effective inhibitor (IC_{50} = 376 nM). The iterative process was continued, in which 20 peptide mixtures for Ac-YPNLOX-NH_2 were assayed; Ac-YPNLPX-NH_2 (IC_{50} = 164 nM) and Ac-YPNLSX-NH_2 (IC_{50} = 257 nM) were found to be the most effective peptide mixtures. The serine residue was anticipated at this position from the known antigenic determinant sequence. Also, the small increase in activity found for the peptide mixtures upon defining the fifth position is consistent with the fact that serine at this position in the known antigenic determinant is relatively redundant. As expected, no significant increase in activity was found upon defining the final position of either Ac-YPNLPO-NH_2 or Ac-YPNLSO-NH_2.

The same iterative process was carried out for Ac-YPYXXX-NH_2. A substantial increase in activity was found for Ac-YPYPXX-NH_2 (IC_{50} = 408 nM). Another significant increase was found upon assaying the peptide mixtures with the fifth position defined, resulting in Ac-YPYPNX-NH_2 (IC_{50} = 37 nM). The next best peptide mixture, Ac-YPYPHX-NH_2, was found to be nearly 1500-fold less effective, illustrating the relative importance of the asparagine residue in this peptide-MAb interaction. In the final iterative step, Ac-YPYPNL-NH_2 (IC_{50} = 2 nM) was found to inhibit antibody binding 7- to 10-fold better than the next best peptide(s).

From the final results of the iterative process for both Ac-PYXXXX-NH_2 and Ac-YPXXXX-NH_2, it is evident that Ac-PYXXXX-NH_2 was initially more effective primarily because at least 20 peptides representing Ac-PYPNLO-NH_2 within Ac-PYXXXX-NH_2 were equally effective in inhibiting antibody binding. In contrast, only one peptide, namely Ac-YPYPNL-NH_2, which was found from the iterative process of Ac-YPXXXX-NH_2, inhibited antibody binding at a similar concentration. The differences in inhibitory activity of the two peptide mixtures, Ac-PYXXXX-NH_2 and Ac-YPXXXX-NH_2, can be attributed to the number of high affinity peptide sequences (i.e., 20 vs. 1) within each peptide mixture. Also, whereas Ac-PYPNLS-NH_2 and Ac-YPYPNL-NH_2 were found to have similar inhibitory activities, the 40 peptides representing Ac-YPNLSO-NH_2 and Ac-YPNLPO-NH_2 were found to have 100-fold less activity, which indicates the importance of the presence of at least one additional residue at the amino terminal of the -YPNL- motif.

In a final example of the utility of our standard SPCL approach, we screened a nonacetylated SPCL having the same format (O_1O_2XXXX-NH_2) against MAb 3E7 (Pinilla et al., 1993a) which was raised against β-endorphin (Gramsch et al., 1983) in order to identify the antigenic determinant as well as compare these results with those found by others investigating MAb 3E7 using different peptide library approaches (Cwirla et al., 1990; Lam et al., 1991). The 400 peptide mixtures of the nonacetylated SPCL were assayed by competitive ELISA in a similar manner as the previous examples. Only peptide mixtures YGXXXX-NH_2 (IC_{50} = 2900 nM) and YAXXXX-NH_2 (IC_{50} = 10,000 nM) were found to cause significant inhibition of MAb 3E7 binding to β-endorphin. An iterative process was carried out for YGXXXX-NH_2 in order to define the remaining positions of this peptide mixture. The two most effective peptide mixtures found when defining the third position were YGFXXX-NH_2 (IC_{50} = 400 nM) and YGGXXX-NH_2 (IC_{50} = 1000 nM). The fourth position was defined for each peptide mixture, namely YGFOXX-NH_2 and YGGOXX-NH_2. However, no significant improvement in antibody-inhibiting activity was found for YGFOXX-NH_2; YGFLXX-NH_2 and YGFWXX-NH_2 were found to have IC_{50}s similar to YGFXXX-NH_2, and the range of activities from the most effective to the least effective peptide mixture was approximately 30-fold. In contrast, after assaying YGGOXX-NH_2, only YGGFXX-NH_2 was found to have significant activity over the next best peptide mixture (20-fold) and the least effective peptide mixture (35,000-fold), indicating the relative importance of the phenylalanine at the fourth position of this peptide-MAb interaction. We chose to further pursue only YGGFXX-NH_2 because of its greater activity over YGFLXX-NH_2. Upon defining the fifth position (YGGFOX-NH_2), peptide mixtures having methionine, leucine and isoleucine at these positions were found to have significant antibody-inhibiting activity when compared to YGGFXX-NH_2. Upon completion of the iterative process for the most effective peptide mixture, YGGFMO-NH_2, YGGFMT-NH_2 was found to have the greatest antibody-inhibiting activity, even though only an eightfold difference in activity was found for a majority of the defined peptides. The peptide YGGFMT-NH_2 exactly matches the first six amino acids of β-endorphin.

III.2. Positional Scanning SPCL (PS-SPCL)

Recently, we have presented an approach that reduces the time necessary for screening and identifying the active peptide sequence to a single day (Pinilla et al., 1992), thus eliminating the need to select and synthesize new peptide mixtures at each iterative step. As first presented, each of the six peptide libraries prepared using the chemical ratio method is composed of 18 peptide mixtures in which a single position is defined (represented as O) with one of 18 of the 20 natural L-amino acids (cysteine and tryptophan were omitted), while the remaining five positions are mixtures (represented as X) of the same 18 amino acids. The six peptide libraries differ only in the location of the defined position. The first PS-SPCL presented is represented as Ac-O_1XXXXX-NH_2, Ac-XO_2XXXX-NH_2, Ac-XXO_3XXX-NH_2, Ac-XXXO_4XX-NH_2, Ac-XXXXO_5X-NH_2, and Ac-XXXXXO_6-NH_2, totalling 108 peptide mixtures. Each peptide mixture represents 1,889,568 (18^5) individual sequences; the entire peptide library thus contains 34,012,224 (18 x 1,889,568) different hexamers.

Upon a single screening of the 108 peptide mixtures of this PS-SPCL, the antigenic determinants recognized by two different monoclonal antibodies were correctly identified. In the first example, competitive ELISA was used, in which each peptide mixture of this PS-SPCL was assayed for inhibition of MAb 17D09 binding to the antigenic peptide Ac-YPYDVPDYASLRS-NH_2. Upon a single screening, the most effective amino acid residue for a position, as well as the relative specificity of that position, can be determined. In this example, only one amino acid residue was effective (threshold level >40%) for position 1 (aspartic acid), position 4 (aspartic acid), and position 5 (tyrosine). More than one amino acid residue was effective at position 2 (isoleucine and valine), position 3 (glutamic acid, proline, and glutamine), and position 6 (alanine and serine). To confirm the results of the screening, 12 individual peptides representing the combinations of the most effective amino acid residues at each position were synthesized. The peptide Ac-DVPDYA-NH_2, having one of the highest affinities for the MAb of the 12 individual peptides synthesized, corresponded exactly to the known antigenic determinant for this peptide-MAb interaction. Five peptides, representing single or double substitutions at the second, third, and sixth positions of this antigenic determinant were also found to have affinities to the MAb similar to Ac-DVPDYA-NH_2.

In a similar manner, the PS-SPCL was also screened for inhibition of MAb 125–10F3 binding to the antigenic peptide Ac-GASPYPNLSNQQT-NH_2. Again, 12 individual peptides were synthesized that represent the combinations of the most effective amino acid residues for a position. The most effective peptides were Ac-PYPNLL-NH_2 and Ac-PYPNLR-NH_2. The first five positions corresponded to the sequence of the known antigenic determinant (PYPNLS). None of the peptide mixtures having the sixth position defined inhibited antibody binding more than 50%; therefore, this position can be considered redundant, which is in agreement with previous studies.

The major advantage of PS-SPCLs over previously described SPCLs is that no iterative synthesis and selection process is required for the identification of optimal peptides. However, since the binding threshold of the PS-SPCL is lower than the dual fixed position SPCL due to the number of peptides making up each peptide mixture (1,889,568 vs. 130,321, respectively), it is likely that only peptides with high affinity to antibodies, receptors, or other acceptor molecules will be identified. It should be noted that since the PS-SPCL is composed of six separate positional SPCLs, each one can be considered independent of the others. Therefore, each positional SPCL can be independently screened and pursued using the iterative synthesis and selection process described earlier (Houghten et al., 1991; Houghten et al., 1992a).

III.3. Mimotope

The generation of peptide mixtures on pins using chemical ratio methods was presented by Geysen et al. and termed the mimotope approach (Geysen et al., 1986). As presented, each pin contained thousands of octapeptides, each with two positions defined and six positions as mixtures. This peptide library can be represented by the formula Ac-XXO_1O_2XXXX-S_s. The 400 different pins, each having a different peptide mixture, when assayed for their ability to bind to a monoclonal antibody, were found to yield various levels of antibody binding. The peptide mixture having the highest antibody binding, as determined by optical density using direct binding ELISA, was then enhanced by synthesizing a new set of peptide mixtures with a third position defined and the others maintained as mixtures. Assaying this set of peptide mixtures for direct antibody binding resulted in a sequence that could be further defined by repeating this process. In the first presentation of this method, the authors' aim was to identify a peptide (mimotope) that mimicked a discontinuous determinant. The monoclonal antibody used in this study was believed to recognize a discontinuous determinant on the foot-and-mouth-disease virus (FMDV), since it did not recognize any of the individual overlapping peptides representing the protein used to produce the antibody. The final sequence that was identified through the iterative process was the hexapeptide Ac-WQMRHSXX-S_s, since no significant improvement was found for the final two positions. Since the peptides used in this approach were covalently attached to the pins, no specific, quantitative information in regard to the affinity of the peptide-antibody interaction was obtained. Among the limitations of this approach is the low peptide capacity of each pin (1 μmol); thus, if six positions of the peptide are composed of mixtures, the amount of each individual peptide present on each pin is about 0.015 pmol, which is an insufficient amount of peptide for a majority of antigen-antibody interactions. Also, no analytical information was given in this study as to the relative amounts of amino acids present in the peptide mixtures. There also may be significant differences in peptide recognition due to the immobilized peptide vs. a free form of the same peptide.

A control study using this approach against a monoclonal antibody known to recognize a specific antigenic determinant was recently presented using a smaller peptide library which can be represented as XXO_1O_2XX (Geysen et al., 1993). The reported amount of peptide on each pin was 100 nmol, and since the four positions as mixtures total 160,000 peptides per pin, the amount of each individual peptide was less than 1 pmol. The set of 400 peptide mixtures was tested against a monoclonal antibody that recognized a six-residue determinant of myohemerythrin using direct binding ELISA. The best peptide mixture was found to be XXLEXX. An iterative process of synthesis and selection was carried out to define the remaining positions of the peptide mixture. The final peptide sequence deduced in this manner was found to be DFLEKI, which corresponds to the antigenic determinant previously determined using individual peptide analogs (Fieser et al., 1987).

III.4. Split synthesis

Another synthetic peptide library approach was presented in which a pentapeptide library was synthesized on beads using the physical synthesis method, which the authors refer to as the "split synthesis" approach (Lam et al., 1991). Each bead contains a single peptide, and the peptide library totals approximately more than 2 million peptides. The mixture of the bead-bound peptides was assayed with monoclonal antibody 3E7, whose specificity was known to recognize the amino terminal of β-endorphin (YGGFMT) (Gramsch et al., 1983). The peptide library was mixed with enzyme-conjugated MAb; upon detection with a color reagent, several beads were separated from the mixture. The bead-bound peptides were then sequenced to identify peptides that were recognized by MAb 3E7. The peptide YGGFQ was found to have the highest binding affinity of the isolated bead-bound peptides, even though the expected determinant sequence was not found. This may be due to the possible absence of the sequence YGGFM from the library, or the bead may not have been identified or sequenced or the peptide may have been conformationally unavailable to the MAb due to the bead support.

In an extension of the first study using MAb 3E7, several different peptide libraries were screened in order to identify a variety of peptides (Lam et al., 1993). It is not clear how many peptide libraries were screened or how they differed from one another. It is clear, however, that only those peptides identified which had the structural motif YG(G/A)(F/L/M/W) (15 out of 41 peptides) were found to have affinities lower than 10 μM. For comparison, the affinities for this anti-β-endorphin MAb 3E7 have been reported for YGGFL-OH (IC_{50} = 27 nM) (Lam et al., 1993), which is naturally occurring leucine-enkephalin, and YGGFMT-NH_2 (IC_{50} = 3 nM) (Pinilla et al., 1993a), which represents the N-terminal of β-endorphin.

The bead-bound peptide library initially assayed against the β-endorphin antibody was also assayed against streptavidin, which is known to bind to biotin with very high affinity. Of 75 reactive beads retrieved, only 28 were sequenced; the consensus sequence -HPQ- was found in 23 of the sequenced peptides. This consensus sequence was also found using a peptide library expressed on phage (Devlin et al., 1990). This same pentapeptide library was also screened against avidin, and a similar consensus sequence -HP(Y/F)P- was detected. In a separate screening, a peptide library without histidine was used in order to identify specific non-HPQ sequence motifs that interact with streptavidin (Lam et al., 1992). A higher concentration of streptavidin-alkaline phosphatase conjugate was required to detect approximately 200 beads. Again, only a fraction (14/200 or 7%) of the reactive bead-bound peptides was sequenced, and the peptides that were identified showed little sequence homology with those found from the histidine-containing library.

Recently, a method has been described in which peptide libraries prepared by the split synthesis approach are able to be selectively cleaved from their beads (Lebl et al., 1993). This allows, in some cases, the screening of the peptide library in solution as a result of selectively and independently cleaving the peptides from the bead. This approach, termed orthogonal linkage-cleavage chemistry, uses different amino acid protecting groups to ensure completely independent cleavage of peptide under various conditions. The successful cleavage and screening of a peptide library using this approach has yet to be reported.

IV. NEW DEVELOPMENTS USING SYNTHETIC PEPTIDE LIBRARIES

IV.1. Potential uses of peptide libraries for other immunological studies

The majority of the work reviewed here has shown that peptide libraries can effectively be used for antigenic determinant mapping studies. The use of peptide libraries should not, however, be limited to only epitope mapping. Peptide libraries can also be used to identify new, unanticipated peptide ligands that effectively inhibit or block antibody binding. In fact, peptide libraries, specifically SPCLs, have been utilized in other areas of drug discovery research such as radioreceptor assays for the identification of a range of opioid peptides (Houghten et al., 1992a; Houghten et al., 1993; Dooley et al., 1993b), and in microdilution assays for the development of novel antimicrobial peptides (Houghten et al., 1991; Houghten et al., 1992b; Houghten et al., 1992a) and the development of trypsin inhibitors (Eichler et al., 1993). Although the screening of peptide libraries for the purpose of mapping discontinuous determinants recognized by antibodies represents a basic challenge in the field of immunochemistry, a more realistic and versatile application of peptide libraries would include the identification of peptides that specifically inhibit or disrupt an antigen-antibody interaction. For drug discovery and development studies, this objective would appear to be more relevant since the initial lead compound identified can then be further improved using more traditional drug design methods.

Although the use of synthetic peptide libraries to map antigenic determinants is still in its infancy, one can envision the use of peptide libraries to identify specific sequences that bind to MHC class I and II molecules as well as other immunologically related proteins, cell surface receptors, and growth factors such as interleukins, interferons, cytokines, etc. Peptide libraries can also be envisioned to be of value for mapping antigenic T cell sites for those systems in which specific hybridomas are used. For studies such as these, peptide libraries in different formats and composition can be prepared to better suit the interactions that are studied.

IV. 2. New, differently formatted peptide libraries

Although synthetic peptide library technology is in its nascent stage, a wide range of different peptide libraries can be synthesized using current peptide chemistry strategies. It appears unlikely, however, that there will be a single "all-purpose" peptide library that is optimally useful for all studies. Certainly, the same peptide library can be used for screening against an antibody, receptor, enzyme, or any other high affinity binding protein. However, there are many other fields of study in which peptides affect a certain function that could be examined with peptide libraries. The advantage of synthetic peptide libraries is that different formats can be "custom made" for a particular application in a short amount of time, whether it is in immunochemistry or other fields of medicinal research.

The most obvious of these strategies is to generate peptide libraries consisting entirely of D-amino acids. Such a library can be envisioned to have direct applications in many fields of study, including microbiology and enzymology, in addition to immunology. To date, only SPCLs consisting of all D-amino acids have been used for the identification of novel, potent antimicrobial peptides; for the development of opioid peptide inhibitors; and for the development of D-amino acid peptides that bind to antibodies (Dooley et al., 1994). Another SPCL that has been prepared consists of a tetramer sequence in which one position is defined, with the remaining three positions as mixtures using all L-, D-, and the 19 unnatural amino acids. This peptide library has also been shown to be effective for the identification of antimicrobial peptides and opioid peptide inhibitors (Blondelle et al., 1994; Dooley et al., 1995). SPCLs have also been prepared with different peptide lengths. For example, a positional scanning SPCL of 10 residues in length (i.e., Ac-O_1XXXXXXXXX-NH_2, Ac-XO_2XXXXXXXX-NH_2, etc.) has been prepared. For this PS-SPCL, one position is defined with 18 amino acids and the remaining nine positions as mixtures of the 18 amino acids. Each peptide mixture contains approximately 198 billion different peptides for a peptide library totalling more than 3 trillion sequences. This PS-SPCL has been successfully screened using competitive ELISA against MAb 17D09 for the identification of the antigenic determinant (Pinilla et al., 1994). The same antigenic determinant sequence (-DVPDYA-) was found in the context of the decapeptide sequence, with some positional overlapping. Current solid-phase chemistries can effectively synthesize peptides up to 30 residues. Undoubtedly, a peptide library consisting of 30 residues would include an enormous diversity of individual peptides and would inherently have potential problems arising from aggregation, secondary structure conformations, and solubility. However, in time, the synthetic problems of such diverse peptide libraries may be overcome. Also, one can envision peptide libraries prepared to have structural constraints built in. A majority of the peptide libraries have avoided the use of cysteine simply to control the linearity of the sequences. By introducing cysteine into the synthesis of peptide libraries and carefully controlling the oxidation of disulfide formation, peptide libraries can be designed in a number of conformational arrays for specific applications. A library derived from the chemical transformation of an existing N-acetylated L-amino acid PS-SPCL was recently prepared and used to identify new, permethylated compounds which exhibit potent antimicrobial activity against gram-positive bacteria (Ostresh et al., 1994b). In a similar manner, the same starting PS-SPCL was exhaustively reduced to identify polyamines with opioid and antimicrobial activity (Cuervo et al., 1995).

V. CONCLUSIONS

The generation and use of synthetic peptide libraries will have a significant impact on how immunological research is carried out in the future. In the past, the availability of individual synthetic peptides was a limiting factor in the majority of immunochemical studies. Peptide synthesis is becoming more routine and less expensive, and the use of synthetic peptides is continually growing. It is anticipated that the availability of synthetic peptide libraries will also increase. The future use of peptide libraries will certainly complement existing rational drug design methods currently used to study antigen-antibody interactions and MHC-peptide-T cell receptor interactions such as x-ray crystallography, nuclear magnetic resonance, and computer modeling, since large numbers of optimal peptide sequences can be rapidly identified.

ACKNOWLEDGEMENTS

The authors wish to thank Eileen Silva for her assistance in preparing this manuscript. This work was funded, in part, by Houghten Pharmaceuticals, Inc., San Diego, CA.

ABBREVIATIONS

MHC Major histocompatibility complex
MAb Monoclonal antibody
SPCL Synthetic Peptide Combinatorial Library
ELISA Enzyme-linked immunosorbent assay

VI. REFERENCES

Appel JR, Pinilla C, Niman H et al. Elucidation of discontinuous linear determinants in peptides. J Immunol 144:976, 1990

Appel JR, Pinilla C and Houghten RA. Identification of related peptides recognized by a monoclonal antibody using a synthetic peptide combinatorial library. Immunomethods 1:17, 1992

Barakat S, Briand JP, Van Regenmortel MHV et al. Mapping of epitopes on U1 snRNP polypeptide A with synthetic peptides and autoimmune sera. Clin Exp Immunol 86:71, 1991

Birkett AJ, Soler DF, Wolz RL et al. Determination of enzyme specificity in a complex mixture of peptide substrates by N-terminal sequence analysis. Biochemistry 196:137, 1991

Birnbaum S and Mosbach K. Peptide screening. Curr Opin Biotechnol 3:49, 1992

Blondelle SE, Takahashi E, Weber PA et al. Identification of antimicrobial peptides using combinatorial libraries made up of unatural amino acids. Antimicrob Agents Chemother, 38:2280, 1994

Cheng H, Foong Y, Sam C et al. Epstein-barr virus nuclear antigen 1 linear epitopes that are reactive with immunoglobulin A (Iga) or IgG in sera from nasopharyngeal carcinoma patients or from healthy donors. J Clin Microbiol 29:2180, 1991

Cwirla SE, Peters EA, Barrett RW et al. Peptides on phage: A vast library of peptides for identifying ligands. Proc Natl Acad Sci USA 87:6378, 1990

Cuervo JH, Weitl F, Ostresh JM et al. Polyalkylamine chemical combinatorial libraries. In: Maia HLS ed:Peptides 1994: Proceedings of the 23rd European Peptide Symposium, Escom, Leiden, in press, 1995

Devlin JJ, Panganiban LC and Devlin PE. Random peptide libraries: A source of specific protein binding molecules. Science 249:404, 1990

Dooley CT and Houghten RA. The use of positional scanning synthetic peptide combinatorial libraries for the rapid identification of opioid receptor ligands. Life Sciences 52:1509, 1993a

Dooley CT, Chung NN, Schiller PW et al. Acetalins: New opioid receptor antagonists determined through the use of Synthetic Peptide Combinatorial Libraries. Proc Natl Acad Sci USA 90:10811, 1993b

Dooley CT, Chung NN, Wilkes BC et al. An all D-amino opioid peptide with central analgesic activity from a combinatorial library. Science 266:2019, 1994

Dooley CT, Hope SK and Houghten RA. Identification of tetrameric opioid peptides from a combinatorial library composed of L-, D- and non-proteinogenic amino acids. In: Maia HLS ed:Peptides 1994: Proceedings of the 23rd European Peptide Symposium, Escom, Leiden, in press, 1995

Eichler J and Houghten RA. Identification of substrate-analog trypsin inhibitors through the screening of synthetic peptide combinatorial libraries. Biochemistry 32:11035, 1993

Fieser TM, Tainer JA, Geysen HM et al. Influence of protein flexibility and peptide conformation on reactivity of monoclonal anti-peptide antibodies with a protein α-helix. Proc Natl Acad Sci USA 84:8568, 1987

Fodor SPA, Read JL, Pirrung MC et al. Light-directed, spatially addressable parallel chemical synthesis. Science 251:767, 1991

Furka A, Sebestyen F, Asgedom M et al. General method for rapid synthesis of multicomponent peptide mixtures. Int J Pep Prot Res 37:487, 1991

Geysen HM, Meloen RH and Barteling SJ. Use of a peptide synthesis to probe viral antigens for epitopes to a resolution of a single amino acid. Proc Natl Acad Sci USA 81:3998, 1984

Geysen HM, Rodda SJ and Mason TJ. *A priori* delineation of a peptide which mimics a discontinuous antigenic determinant. Mol Immunol 23:709, 1986

Geysen HM, Rodda SJ, Mason TJ et al. Strategies for epitope analysis using peptide synthesis. J Immunol Meth 102:259, 1987

Geysen HM and Mason TJ. Screening chemically synthesized peptide libraries for biologically-relevant molecules. BioMed Chem Lett 3:397, 1993

Gisin BF. The monitoring of reactions in solid-phase peptide synthesis with picric acid. Anal Chim Acta 58:248, 1972

Gramsch C, Meo T, Riethmüller G et al. Binding characteristics of a monoclonal β-endorphin antibody recognizing the N-terminus of opioid peptides. J Neurochem 40:1220, 1983

Houghten RA. General method for the rapid solid-phase synthesis of large numbers of peptides: Specificity of antigen-antibody interaction at the level of individual amino acids. Proc Natl Acad Sci USA 82:5131, 1985

Houghten RA, Bray MK, De Graw ST et al. Simplified procedure for carrying out simultaneous multiple hydrogen fluoride cleavages of protected peptide resins. Int J Pep Prot Res 27:673, 1986

Houghten RA, Pinilla C, Blondelle SE et al. Generation and use of synthetic peptide combinatorial libraries for basic research and drug discovery. Nature 354:84, 1991

Houghten RA, Appel JR, Blondelle SE et al. The use of synthetic peptide combinatorial libraries for the identification of bioactive peptides. BioTechniques 13:412, 1992a

Houghten RA, Blondelle SE and Cuervo JH. Development of new antimicrobial agents using a synthetic peptide combinatorial library involving more than 34 million hexamers. In: Epton R ed: Innovation and Perspectives in Solid Phase Synthesis: Peptides, Polypeptides and Oligonucleotides Intercept Ltd, Andover, MA, p237, 1992b

Houghten RA and Dooley CT. The use of synthetic peptide combinatorial libraries for the determination of peptide ligand in radio-receptor assays: Opioid peptides. BioMed Chem Lett 3:405, 1993

Jung G and Beck-Sickinger A. Multiple peptide synthesis methods and their applications. Angew Chem 31:367, 1992

Kaiser ET, Colescott RL, Blossinger CD et al. Color test for detection of free terminal amino groups in the solid-phase synthesis of peptides. Anal Biochem 34:595, 1970

Krchnak V, Vagner J, Safar P et al. Noninvasive continuous monitoring of solid-phase peptide synthesis by acid-base indicator. Coll Czech Chem Comm 53:2542, 1988

Lam KS, Salmon SE, Hersh EM et al. A new type of synthetic peptide library for identifying ligand-binding activity. Nature 354:82, 1991

Lam KS and Lebl M. Streptavidin and avidin recognize peptide ligands with different motifs. Immunomethods 1:11, 1992

Lam KS, Hruby VJ, Lebl M et al. The chemical synthesis of large random peptide libraries and their use for the discovery of ligands for macromolecular acceptors. BioMed Chem Lett 3:419, 1993

Lebl M, Pátek M, Kocis P et al. Multiple release of equimolar amounts of peptides from a polymeric carrier using orthogonal linkage-cleavage chemistry. Int J Pept Prot Res 41:201, 1993

Merrifield RB. Peptide synthesis. I. The synthesis of a tetrapeptide. J Am Chem Soc 85:2149, 1963

Ostresh JM, Winkle JH, Hamashin VT et al. Peptide libraries: Determination of relative reaction rates of protected amino acids in competitive couplings. Biopolymers, 34:1681, 1994a

Ostresh JM, Husar GM, Blondelle SE et al. "Libraries from libraries": Chemical transformation of combinatorial libraries to extend the range and repertoire of chemical diversity. Proc Natl Acad Sci USA 91:11138, 1994b

Owens RA, Gesellchen PD, Houchins BJ et al. The rapid identification of HIV protease inhibitors through the synthesis and screening of defined peptide mixtures. Biochem Biophys Res Comm 181:402, 1991

Petithory JR, Masiarz FR, Kirsch JF et al. A rapid method for determination of endoproteinase substrate specificity: Specificity of the 3C proteinase from hepatitis A virus. Proc Natl Acad Sci USA 88:11510, 1991

Pinilla C, Appel JR, Blanc P et al. Rapid identification of high affinity peptide ligands using positional scanning synthetic peptide combinatorial libraries. Biotechniques 13:901, 1992

Pinilla C, Appel JR and Houghten RA. Synthetic peptide combinatorial libraries (SPCLs): Identification of the antigenic determinant of β-endorphin recognized by monoclonal antibody 3E7. Gene 128:71, 1993a

Pinilla C, Appel JR and Houghten RA. Functional importance of amino acid residues making up peptide antigenic determinants. Mol Immunol 30:577, 1993b

Pinilla C, Appel JR and Houghten RA. Investigation of antigen-antibody interactions using a soluable nonsupport-bound synthetic decapeptide library composed of four trillion sequences. Biochem J 301:847, 1994

Radford AJ, Wood PR, Billman-Jacobe H et al. Epitope mapping of the *Mycobacterium bovis* secretory protein MPB70 using overlapping peptide analysis. J Gen Microbiol 136:265, 1990

Rini JM, Schulze-Gahmen U and Wilson IA. Structural evidence for induced fit as a mechanism for antibody-antigen recognition. Science 255:959, 1992

Selvey LA, Tindle RW, Geysen HM et al. Identification of B-epitopes in the human papillomavirus 18 E7 open reading frame protein. J Immunol 145:3105, 1990

Trifilieff E, Dubs MC and Van Regenmortel MHV. Antigenic cross-reactivity potential of synthetic peptides immobilized on polyethylene rods. Mol Immunol 28:889, 1991

Wilson IA, Niman HL, Houghten RA et al. The structure of an antigenic determinant in a protein. Cell 37:767, 1984

Zuckermann RN, Kerr JM, Siani MA et al. Identification of highest-affinity ligands by affinity selection from equimolar peptide mixtures generated by robotic synthesis. Proc Natl Acad Sci USA 89:4505, 1992

Chapter 2

PEPSCAN TO DETERMINE T AND B CELL EPITOPES

R.H. Meloen, W.C. Puijk, J.P.M. Langeveld, J.P.M. Langedijk, A. van Amerongen, and W.M.M. Schaaper

TABLE OF CONTENTS

I. INTRODUCTION

Mapping of B cell epitopes (also called antigenic determinants, antigenic sites, antibody combining sites, etc.) and T cell epitopes of a given protein is the essential part of the design of synthetic vaccines or diagnostics and an essential element for the understanding of the immune response and the pathology of major diseases. Thus, it is not surprising that a vast array of methods has evolved to map and define B and T cell epitopes in much detail. Methods which are for the most part equally applicable to those receptor-ligand interactions in which the ligand is a peptide or a protein.

B cell epitopes of proteins have been studied in detail using both monoclonal and polyclonal antibodies. Especially the study of the spatial structure of a number of antibody-antigen complexes by X-ray analysis has produced a detailed description of the nature of B cell epitopes (Amit et al., 1986; Bentley et al., 1989, 1990; Colman et al., 1987; Tulip et al., 1989; Sheriff et al., 1987; Padlan et al., 1989). The results show that B cell epitopes consist of 12 to 20 amino acids usually located on different protein strands which are spatially near to each other on the intact protein.

Because most of the protein surface can serve as the B cell epitope, the amino acid make-up of B cell epitopes does not seem to vary much from that of the "normal" protein surface (Janin and Chothia, 1990). The size of the interface of a B cell epitope and its paratope (i.e., antigen combining site of the antibody) varies between 600 and 1000 Å^2. (If the interface is regarded as a circle, this circle would have a diameter of 28 to 36 Å.)

The interaction between antibody and B cell epitope does not necessarily require the B cell epitope to be fully intact. Parts of the B cell epitope as linear peptides may bind the antibody, although at lower affinity.

We found that on the average, one out of three monoclonal antibodies could be mapped with linear peptides, while normally each polyclonal antiserum contains antibodies which can bind with linear peptides. Thus, although linear peptides will rarely represent full epitopes (Meloen et al., 1988), they can be used to map B cell epitopes.

0-8493-8967-4/95/$0.00+$.50

From a practical point of view, B cell epitopes can be divided in three groups (Fig. 1):

1. Real linear epitopes. These epitopes are easily mapped using peptides, but probably do not occur very often.
2. Discontinuous epitopes with relative large linear parts. The linear parts of these epitopes are easily mapped using peptides. Epitopes of this group probably form a large minority existing of all epitopes.
3. Discontinuous scattered epitopes. These epitopes probably form the majority of all existing epitopes. They are difficult to map with peptide-based methods. One has to rely on methods like "site-directed mutagenesis" or, in case of replicating agents, on "escape mutations". Often, the data will only produce a useful definition of the epitope if the spatial structure of the protein carrying this epitope is known.

Although systematic methods to map scattered discontinuous epitopes are still being developed, it is sometimes possible to define peptides which partially mimic scattered discontinuous epitopes. This will be discussed in more detail below.

Epitopes can be mapped using numerous approaches: they can be "predicted" from the amino acid sequence of a protein, they can be inferred from the location of amino acid mutations which disrupt antibody binding, they can be assessed by amino acid sequencing of antibody-binding clones of genomic expression libraries and finally they can be assessed using peptides. The latter approach can be subdivided in two categories: one is based on peptide expression libraries (Lenstra et al., 1992; Cwirla et al., 1990; Devlin et al., 1990; Scott and Smith, 1990), and the other is based on synthetic peptide libraries (Geysen et al., 1984, 1985b; Meloen et al., 1991; Houghten et al., 1991; Lam et al., 1991; Fodor et al., 1991). Both categories of methods can be used to provide sets of overlapping peptides of known linear amino acid protein sequences and sets of random peptides. So far the method which provides sets of overlapping peptides of known amino acid protein sequences: the PEPSCAN (Geysen et al., 1984, 1985; Meloen et al., 1991) has been most useful in the systematic mapping of numerous B and, as we will discuss below, T cell epitopes.

Using the PEPSCAN, B cell epitopes have been mapped for proteins of viruses, bacteria, parasites, hormones and allergens.

The viruses include foot-and-mouth disease virus (Geysen et al., 1984, 1985a; Meloen and Barteling, 1986a; Meloen et al., 1986b, 1987, 1991; Posthumus et al., 1991b; McCahon et al., 1989), parvoviruses (Brown et al., 1992; Langeveld et al., 1993, 1994), herpes viruses (Middeldorp and Meloen, 1988; Jacobs et al.,

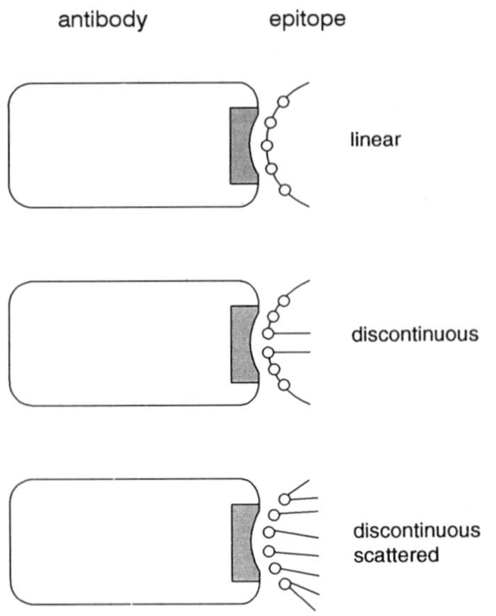

Fig. 1. Schematic representation of the three theoretically possible epitope configurations (epitope: antigenic determinant of a monoclonal antibody). The antibody is represented as a rough outline of the Fab fragment. The hatched area indicates the part of the antibody in contact with the amino acids of the epitope. The amino acids of the epitope are represented by small circles. If the amino acids of the epitope form a single protein strand the epitope is called linear; if the amino acids are located on two or more protein strands which are spatial next to each other due to the conformation of the antigen molecule, the epitope is called discontinuous. The most extreme case has been called discontinuous scattered. (from Meloen et al., 1991)

1990), retroviruses (Goudsmit et al., 1988a, 1988b; Krone et al., 1988; Goudsmit et al., 1989a, 1989b, 1989c; Meloen et al., 1989; Boucher et al., 1990; Brake et al., 1990; Goudsmit et al., 1990; Langedijk et al., 1990, 1991; Wolf et al., 1991; Zwart et al., 1991; Javaherian et al., 1992; Langedijk et al., 1992a; 1992b; D'Souza et al., 1991; Weijer et al., 1993), corona viruses (Kusters et al., 1989; Luytjes et al., 1989; Bredenbeek et al., 1990; Enjuanes et al., 1990; Posthumus et al., 1990, 1991c; Gebauer et al., 1991, and others (Versteeg-v.Oosten et al., 1989; Coursaget et al., 1991; Snijders et al., 1991; Chassot et al., 1993). The bacteria include Mycobacteria (Radford et al., 1990). The parasites include Malaria, trypanosoma and chlamydia (v. Amerongen et al., 1989, 1990, 1992; Stüber et al., 1990; Clarke et al., 1990; Ahlborg et al., 1991; Miles et al., 1989). The hormones include hCG and inhibin (Grootenhuis et al., 1989; v. Amerongen et al., 1994). The allergens include house dust mite major allergen der p II (Van 't Hof et al., 1991) and cat allergen Fel d I (Van 't Hof et al., 1993). Other proteins include alpha-bungarotoxins (Tzartos and Remoundos, 1990), fat globules (Xing et al., 1991) and tumor necrosis factor-alpha (Stiemer et al., 1992).

Using PEPSCAN, T cell epitopes are also readily mapped. Both Th cell and CTL epitopes have been

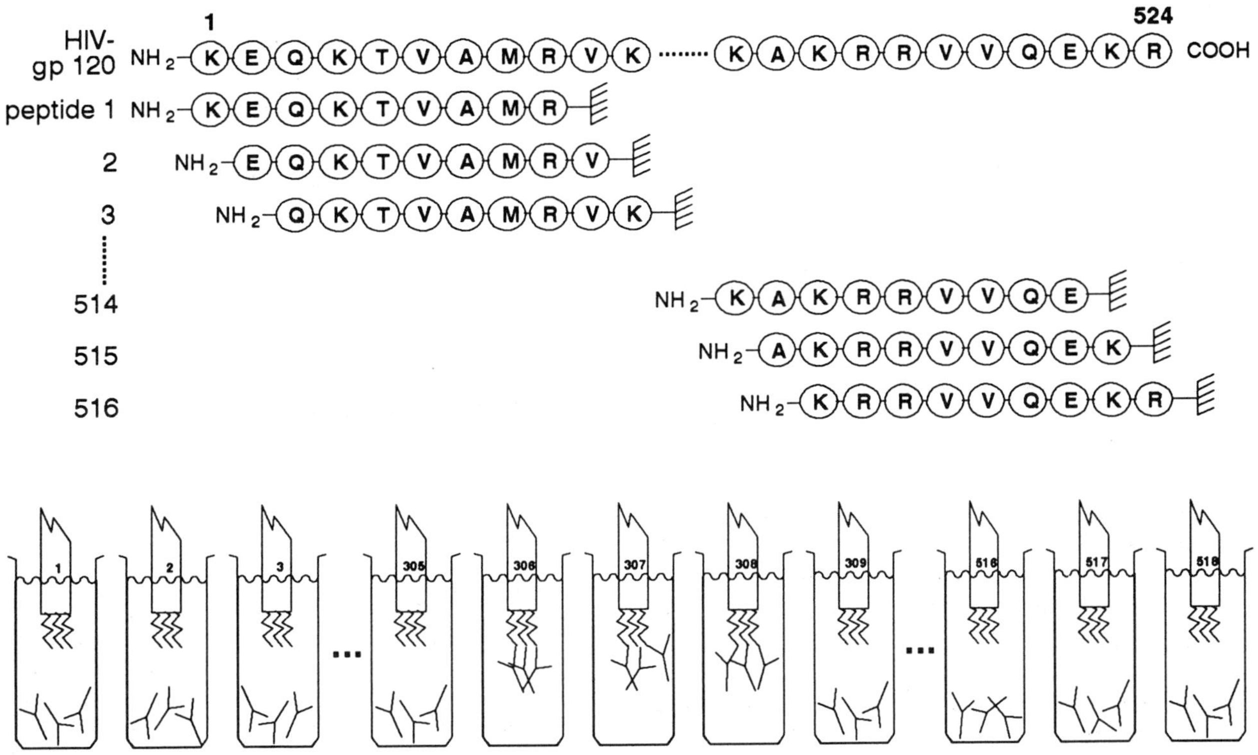

Fig. 2. Schematic representation of the PEPSCAN method. On top is shown the primary amino acid sequence of gp120 of HIV-1, the major glycoprotein of HIV-1. Each circle represents an amino acid. For the amino acids the single lettercode is used (i.e., A: alanine, C: cysteine, D: aspartic acid, E: glutamic acid, F: phenyl alanine, G: glycine, H: histidine, I: isoleucine, K: lysine, L: leucine, M: methionine, N: asparagine, P: proline, Q: glutamine, R: arginine, S: serine, T: threonine, V: valine, W: tryptophan, Y: tyrosine).
The amino acid sequence of gp120 of HIV-1 is then divided into overlapping peptides, as indicated. Peptide nr 1 is the peptide which starts with amino acid nr 1 and ends with amino acid nr 9; peptide nr 2 is the peptide starting at amino acid nr 2 to amino acid nr 10, etc. The peptides are synthesized on polyethylene rods, as shown at the bottom of this figure; the peptides are indicated as /\/\/\ . (The rods are in such a configuration that they fit into microtiter wells; this greatly simplifies handling of the rods and data manupulation.) All rods with peptides are then contacted with the same antibody (indicated as -<). Some peptides will bind this antibody. After the rods are taken out of the antibody solution and washed, the antibody still present on the rods (bound to the peptide) can then be tested with anti-antibody conjugate for the presence of antibody. This directly produces the sequence of the peptide which binds the antibody. After this process the antibody can be removed from the peptides and the peptides can be reused. One technician can easily test 2000 different peptides daily for reactivity with a given antibody. Over 2000 different peptides are being synthesized on a routine basis at our laboratory each month. (From Meloen et al., 1991.)

mapped for proteins of numerous agents including viruses (van Lierop et al., 1992; Kast et al., 1989, 1991; Rimmelzwaan et al., 1990), bacteria (Geluk et al., 1992; Zee et al., 1989; Wauben et al., 1993; van Eden et al., 1989; Vordermeier et al., 1992), parasites (Süss et al., 1993; Guttinger et al., 1991) and allergens (Carballido et al., 1993).

Thus, PEPSCAN methods offer a fast and systematic initial approach to the delineation of B and T cell epitopes and may further help to define epitopes in detail at the level of single amino acids. It has been successfully applied to map B cell epitopes during the last 10 years and T cell epitopes during the last 4 years.

Its possibilities and its limitations will be illustrated with results obtained with foot-and-mouth disease virus (FMDV), parvovirus, HIV-1 and malaria, while later we will summarize the state of the art of useful synthetic peptides for peptide vaccines. Furthermore, we will discuss how PEPSCAN-based epitope mapping relates to epitope mapping using other approaches.

II. PEPSCAN METHOD

The PEPSCAN method itself has been amply described (Geysen et al., 1984, 1985a, 1987; Geysen, 1990; Maeji et al., 1991; Bray et al., 1990; van der Zee et al., 1989).

The principle of the method is outlined in Fig. 2. Synthetic peptides of a given length (e.g., 9- or 12-mers are synthesized on solid supports. Usually a complete set of overlapping peptides covering the whole sequence of a given protein is prepared. The peptides are covalently bound to solid supports and can be used to test their ability to bind antibodies. Bound antibodies are then located using anti-antibody marked with peroxidase using established ELISA procedures. After the assay has been performed, the peptides are cleaned from antibodies in SDS and mercapto-ethanol containing solutions and can be reused. At the present, one technician can test 2000 peptides daily in ELISA, while a batch of 3000 pin-bound peptides can be produced in 4–6 weeks, involving at least two technicians. (Alternatives are being developed to increase this peptide number by at least a factor of 10 applying the same manpower.) PEPSCAN peptides used for B cell epitope mapping are always acetylated at the N-terminus. Otherwise the interaction with the antibodies may suffer from the presence of positive charged amino groups, which do not occur "normally" in the original protein.

From an operational point of view, peptides of 8–9 amino acids long were much more useful for mapping than peptides of 6–7 amino acids long [the original assumption that all linear epitopes were 6 to 7 amino acids long (Twining et al., 1980; Geysen et al., 1984) turned out not to be the case].

PEPSCAN-generated peptides can be equally well used for T cell epitope assays. To do so the peptides must be cleaved off from the solid support. After the first method was published (van der Zee et al., 1989), numerous more efficient alternatives were developed (Schaaper et al., 1992a, 1992b; Maeji et al., 1990). The methods are still being refined today.

Peptides used for T cell epitope mapping differ in a few aspects from those used for B cell epitope mapping. These peptides should not be acetylated, in compliance with the "natural" occurrence of free NH_3^+ groups of peptides in the groove of the MHC molecules. Often, depending on the method, unnatural linker molecules are applied at the C-terminal. These linker molecules may in principle be a disadvantage; however, in practice they do not appear to interfere dramatically. The length of the peptides seems to be especially important for mapping of CTL epitopes; this should be between 9 and 11 amino acids (Falk et al., 1992; Rötzschke et al., 1992).

A general problem applying both to peptides for B and T cell epitope mapping is that some amino acid sequences are diffucult to synthesize, due to their particular amino acid make-up. Using PEPSCAN with cleavable peptides, purity can be checked; however, it is difficult to check all individual peptides and one has to assume that once in a while antibody-binding

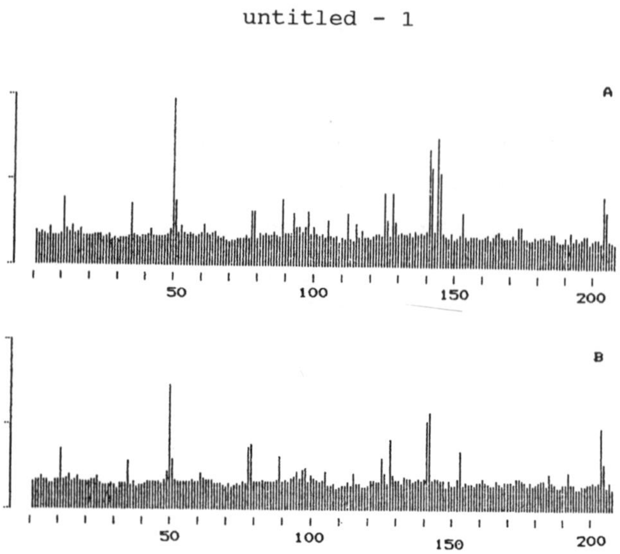

Fig. 3. (A) PEPSCAN of polyclonal sera raised with whole foot-and-mouth disease virus (FMDV) in rabbits. The overlapping peptides are six amino acids long and cover the total length of the amino acid sequence of VP1 of FMD virus type A10. The numbers on the horizontal axis correspond with the N-terminal amino acid of the hexapeptides. Vertically, the adsorbancies at 450 nm obtained with each peptide in an ELISA are plotted. Maximum adsorbancy on the vertical axis is 2.0. (B) The same peptides as in (A) tested with the same sera used in (A) after it was adsorbed with intact FMD virus. (From Meloen and Barteling, 1986.)

peptides may be overlooked due to this problem. However, when using overlapping peptides stepped up by one amino acid, chances are small that a whole area of antibody-binding peptides will be missed. We estimate that most "linear" B cell and T cell epitopes (which are linear by their very nature) can be mapped using PEPSCAN.

III. PEPTIDES REACTING WITH POLYCLONAL ANTIBODIES

Polyclonal antisera raised against the foot-and-mouth disease virus were the first sera studied with the PEPSCAN, i.e., with all six amino acid long overlapping peptides of the viral capsid protein VP1 (Geysen et al., 1984). The results showed (Fig. 3) that multiple peptides of VP1 bound antibodies. Because we were mainly interested in surface exposed reactive peptides, while it was expected that not all reactive peptides were exposed on the intact virion, the polyclonal sera were adsorbed with whole virus. The adsorbed sera (which were not supposed to contain antibodies against external peptides) were retested with the same set of overlapping peptides. The results showed that antibodies which bound with one or two overlapping peptides between amino acids 140–160 had disappeared. This showed that these peptides were exposed on the intact virion in contrast to the other peptides. This result was in accordance with earlier observations (Strohmaier et al., 1982; Bittle et al., 1982; Pfaff et al., 1982) and was confirmed later (Acharya et al., 1989). As a matter of fact this amino acid sequence forms the tip of a loop which extends from the FMDV surface and carries the cell attachment site of the virus (Logan et al., 1993).

Interestingly, based on results with replacement sets of peptides (see below), we predicted (Geysen et al., 1984) that part of this loop would form an alpha-helix. Recent structural information confirmed this prediction (Logan et al., 1993).

Thus, (1) polyclonal antisera raised against complex antigens may contain antibodies directed to internal and external peptides. (2) Adsorption of these with sera with intact antigen and retesting will directly help to differentiate between the internal and external location of reactive peptides.

Although these earlier experiments were done using hexapeptides, it turned out that nonapeptides are in general more suitable (as discussed previously).

IV. PEPTIDES REACTIVE WITH MONOCLONAL ANTIBODIES

In our experience, monoclonal antibodies (MAbs) tested in PEPSCAN exhibit one or more of the following aspects:

1. They react only with peptides which are part of the epitope.
2. They react with additional peptides unrelated to the epitope.
3. They do not react at all or react weakly with multiple peptides.

These aspects are best illustrated with a set of MAbs raised against FMDV and selected for their ability to neutralize the virus *in vitro*. When tested against all nine amino acid long overlapping peptides of VP1, VP2 and VP3 of FMDV (all three capsid proteins are exposed on the virus surface: Acharya et al., 1989), the MAbs could be grouped on the basis of their reactivity pattern (Fig. 4). In group I, MAb 4 only reacted with an amino acid sequence of the 140–160 area [this was the loop area earlier identified (Bittle et al., 1982; Pfaff et al., 1982; Strohmayer et al., 1982)]. However, this MAb appeared to be an exception, because the other MAbs reacted with additional peptides. A study with the aid of the three-dimensional model of FMDV of the location of these additional reactive peptides showed that they were unrelated to the epitope and probably reacted because of some similarities in amino acid sequence. An extreme example is the reactivity pattern of MAb 18 of group II. This MAb reacted strongly with at least three different peptides, which have no obvious sequence similarities. Study of the location of these peptides with the three-dimensional model of FMDV revealed that only the C-terminal peptide of VP1 was located on the viral surface. Additional experiments with these peptides showed that only the surface located peptide induced neutralizing activity when used to vaccinate animals and competed with whole virus for binding to MAb 18 (Meloen et al., 1987). The reactivity of the nonepitope peptides with the MAbs suggests that these antibody-binding peptides, in some way, resemble the real epitope. For instance, they could share a few properly spaced amino acids. Often, this has been observed; however, cases did occur also in which no obvious relationship was found between the amino acid sequence of the real epitope and additional antibody-binding peptides. MAbs of a third group (results not shown) showed weak reactivity patterns with peptides of VP1, VP2 and VP3. Although none of these patterns were identical, they showed that similar sets of peptides were bound by multiple MAbs. When these weakly reacting peptides were located on the viral surface they appeared to be in close proximity (Posthumus, 1991a) and could theoretically form a scattered discontinuous antigenic site (i.e., a set of scattered discontinuous epitopes). This was in accordance with competition studies and escape mutations (McCahon et al., 1989; Thomas et al., 1988); however, direct proof could not be obtained due to our inability to reconstruct this (important) antigenic site in the form of a linear peptide. Reconstruction of discontinuous antigenic sites is still the main stumbling block for the design of synthetic vaccines in particular and peplater in this chap,ter.) drugs in general. (This will be discussed further later in this chapter.)

Thus, three different sorts of reactivity patterns may be obtained when MAbs are tested in PEPSCAN. To determine which one of the reactive peptides forms the epitope or part of it, the availability of the three-dimensional structure is a major advantage, if not a necessity. In other words, the role of promising peptides found with PEPSCAN should always be confirmed using unrelated methods like the induction of anti-peptide sera with the appropriate activities, competition experiments, site-directed mutagenesis, and a study of the molecular structure of the target protein (if available).

V. PRECISE ROLE OF THE INDIVIDUAL AMINO ACIDS OF AN ANTIBODY-BINDING PEPTIDE

Using different sets of peptides allows the precise definition of the epitopes at the level of single amino acids, i.e., the core, the size and the role of individual amino acids of antibody-binding peptides.

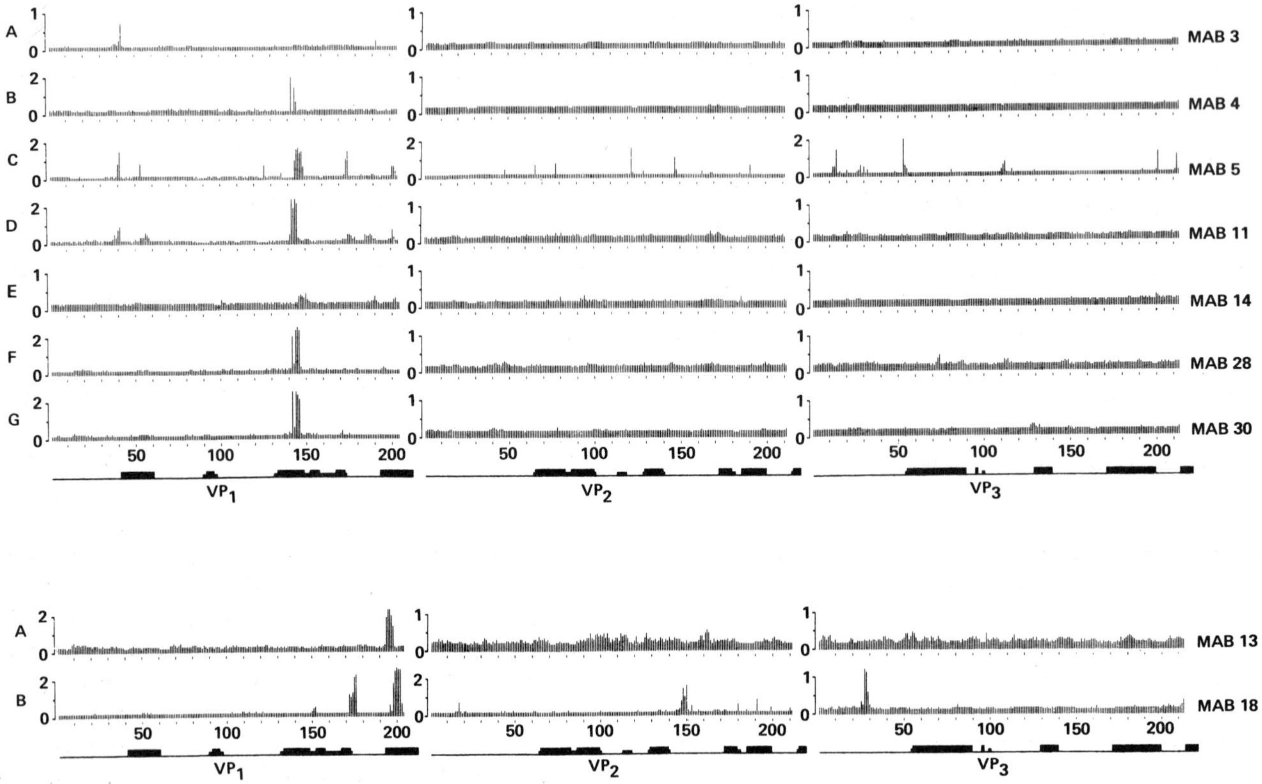

Fig. 4. PEPSCAN results with monoclonal antibodies tested on nonapeptides derived from VP1, VP2 and VP3 of FMDV. (Upper graph) MAb 3, 4, 5, 11, 14, 28 and 30 reaction pattern with overlapping nonapeptides of VP1, VP2 and VP3 of FMDV-A10. Ascitic fluid dilutions are $2x10^{-3}$, 10^{-3}, $2x10^{-3}$, $2x10^{-3}$, $6.6x10^{-5}$, $2x10^{-5}$, $5x10^{-3}$ and $5x10^{-3}$ respectively. (Lower graph) Reaction pattern of MAb 13 and MAb 18. Dilutions: 10^{-3} and 10^{-4}. The numbers on the horizontal axis are according to the sequence and correspond with the N-terminal amino acid of the peptides. On the vertical axis the extinction at 405 nm is indicated. (From Posthumus 1991a.)

Currently we use three approaches.

1. The "*core*" (i.e., the continuous sequence of amino acids of an antibody-binding peptide which cannot be deleted) of the antibody-binding peptide can be defined using a *set of deletion peptides*. This set is based on the previously determined amino acid sequence of the antibody-binding peptide. Peptides are produced in which systematically amino acids one at a time are deleted. This approach allowed us for instance to determine the "cores" of neutralizing MAbs raised against gp120 which reacted with the V3 loop (Fig. 5).
2. The *size* of the antibody-binding peptide, the shortest antibody-binding peptide and the contribution to antibody binding of the amino acids flanking the core can be determined using sets of *overlapping peptides with different lengths*. Such sets cover that part of the amino acid sequence of the protein which contains the antibody-binding peptide. Because peptides of different length are used, the increase in binding can be somewhat related to the addition of single amino acids to the shortest antibody-binding peptide. For instance, this approach allowed us to further delineate the antibody-binding peptide of a neu-

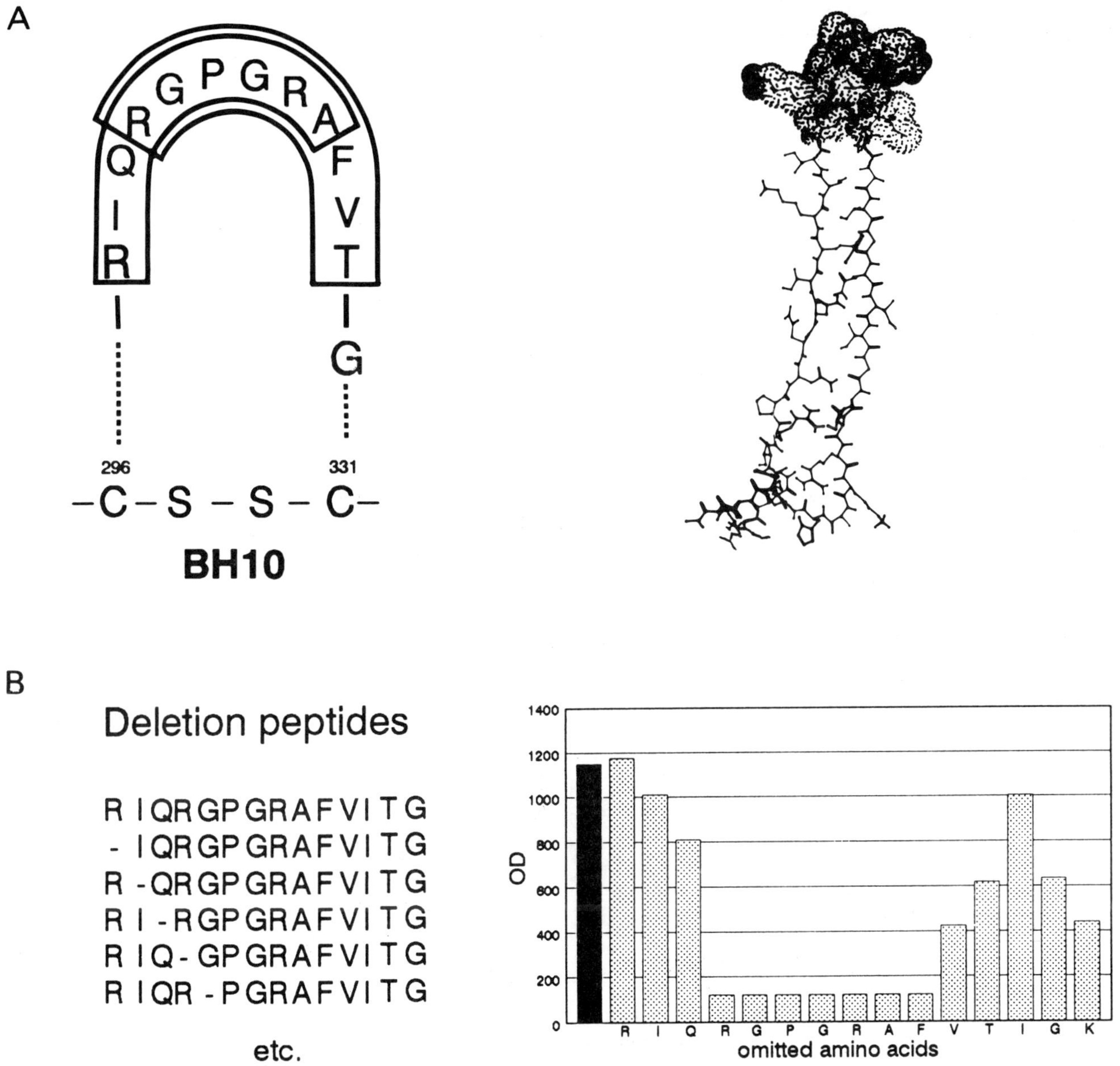

Fig. 5. (A) Major immunodominant antigenic site of HIV-1: the V3 loop. To the left, the V3 loop of strain BH10 is schematically shown. To the right, a theoretical structure of the V3 loop is shown. The antibody-binding peptide which forms the antigenic determinant is located on the top. (B) Definition of the core of an antigenic determinant using a deletion set of peptides. The amino acids of a reactive peptide are deleted, one at a time; the resulting peptides are tested with the antibody. If the deleted amino acid belongs to the core of the antigenic determinant, the reactivity of the peptide with the antibody is lost. To the left is an example of a set of deletion peptides. -= deleted amino acid. To the right is a typical result with the peptides of the V3 loop of HIV-1 and a neutralizing monoclonal antibody. The binding activity is shown as a bar proportional to the optical density in the ELISA. The black bar represents the activity of the parent peptide RIQRGPGRAF-VTIGK. The crossed bars represent the reactivities of the deleted peptides. The amino acid below each bar represents the amino acid deleted from the parent sequence. The core of this antigenic determinant is RGPGRAF. In this example, the results are shown of a MAb which neutralizes HIV-1 and is directed against the V3 loop. (From Langedijk et al., 1991.)

tralizing MAb raised against gp120 of HIV-1 which reacted with the V3 loop (Fig. 6).

3. The *role* of the individual amino acids can be examined using *replacement sets of peptides*. When the size of the antibody-binding peptide is properly assessed (using approach 2), all amino acids (one at a time) of the antibody-binding peptide are systematically replaced by all other amino acids. This set of peptides is then tested for their antibody-binding properties. The results immedi-

a

schematically: K - S - R - I - Q - R - G - R - A - F - V - T - I - G

3 long
K - S - R
S - R - I
R - I - Q
etc.

4 long
K - S - R - I
S - R - I - Q
R - I - Q - R
etc.

b

example:

R - A - F absorbance at 450 nm is 0.500
G - R - A - F " 1.500

c

result:

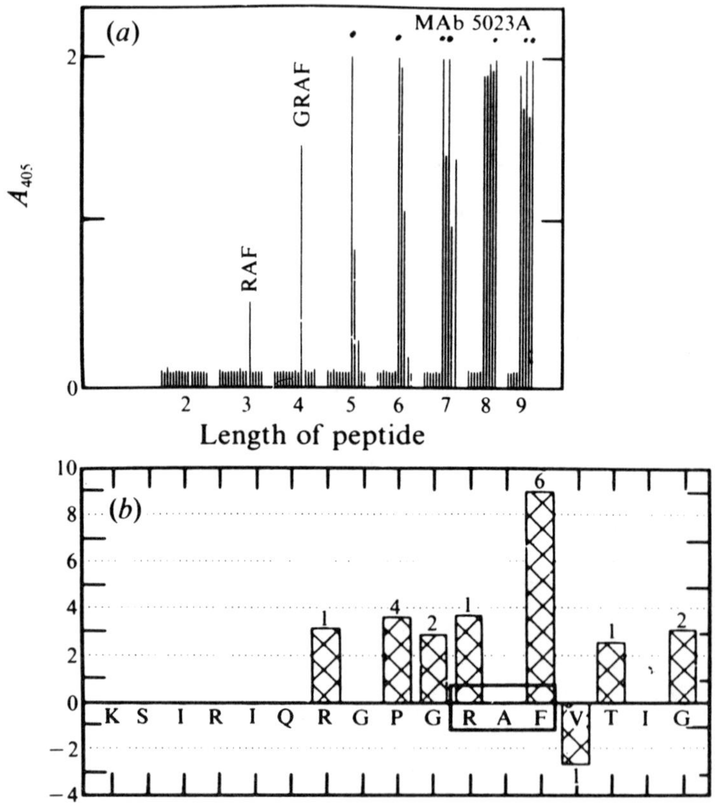

Fig. 6. Sets of peptides of multiple length are used to determine the smallest antibody-binding peptide and the contribution of the flanking amino acid. The principle of the approach is indicated schematically in (a) and (b). In (a) the set of peptides is shown. It consists of all overlapping 3 amino acid long, 4 amino acid long, 5 amino acid long, up to 9 amino acid long peptides. In (b) it is shown that after these peptides are tested for antibody binding, peptides that differ one amino acid in length and also differ in antibody binding allow an assessment of the contribution of the extra amino acid in the longer peptide. In this example, G from GRAF apparently contributes to antibody binding. In (c) (Upper graph) the results are shown of the whole set. In (c) (Lower graph) the results are analyzed for all subsets of peptides that differ one amino acid. (From Langedijk et al., 1991.)

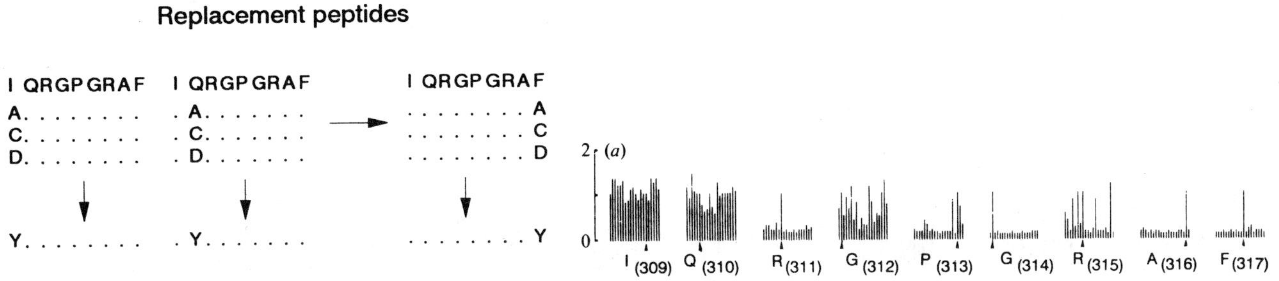

Fig. 7. Definition of the irreplaceable amino acids of an antigenic determinant by using a set of replacement peptides. (Left part) Replacement peptides are obtained by replacing each amino acid with all other amino acids, one at a time. For a nonapeptide, this gives a set of 180 different peptide analogs. In this example, the replacement set is shown for the peptide of the V3 loop of HIV-1: IQRGPGRAF. (Right part) The set of peptides is tested against an MAb; in this example, results obtained with the serum of an individual infected with HIV are shown. The binding activity is shown as a vertical line proportional to the optical density in the ELISA. Each group of 20 lines corresponds to the complete set for one of the amino acids in the peptide. Within each group, the left-hand line corresponds to substitution of the original amino acid by S, and the following lines to its substitution by amino acids GDNEQKRYHWFMLIVPAT. This grouping accounts for the size and polarity of the side chain and is derived from the circle arrangement described by Doolittle (1985) which places amino acids with similar properties near each other. To facilitate presentation we have "cut" the circle between T and S; thus some amino acids with similar properties (i.e., small side chains) are shown at the left- and right-hand sides. The sequence of the parent peptide is indicated. Four of the amino acids cannot be replaced. In this example, the results are shown of the same MAb used in Fig. 5.

ately show which amino acids can be or cannot be replaced. A typical example obtained with a neutralizing MAb raised against gp120 of HIV-1 is shown in Fig. 7. A rather intriguing example is further shown in Fig. 8. In this case, not an MAb but a polyclonal serum of an individual, infected with HIV, was used. In contrast to patterns normally encountered with polyclonal antisera, an unusual number of amino acids are irreplaceable. Because the V3 loop is a major neutralizing antigenic site, this illustrates that HIV-1 has apparently developed an efficient way to circumvent the immune surveillance: a single amino acid change is sufficient to dissolve antibody binding. At the present time, this phenomenon forms one of the major problems that needs to be solved in order to develop an efficient vaccine against HIV-1.

VI. PEPTIDES WHICH MIMIC DISCONTINUOUS EPITOPES

One of the major goals of the pharmaceutical industry is to find faster and more systematic ways to define peptides or other low cost, low molecular products which mimic functional parts of ligands or receptors. From an operational point of view, a similar problem is to define peptides which mimic epitopes. The assumption is that linear peptides exist which mimic discontinuous epitopes. Although this assumption is still being debated, a number of approaches have emerged to define "surrogate" epitopes.

VI. 1. Surface simulation peptides

The earliest approaches to do so were based on the design of relative small peptides which "simulated" or mimicked part of a known three-dimensional protein structure (Atassi, 1978). This approach produced peptides which were able to compete with a native antigen for antibody. Although at present many more three-dimensional structures of proteins are known, "surface-simulation" peptides, useful as drugs or vaccines, are not known to us.

VI. 2. "Surrogate" antibody-binding peptides derived from the universal dipeptide set

Assuming that discontinuous epitopes can be mimicked by linear peptides, the universal dipeptide set of peptides has been applied to find antibody-binding dipeptides (the set comprises just 400 different dipeptides, readily accessible by PEPSCAN). If an antibody-binding dipeptide was found, it could be elongated at both ends with all existing amino acids. The best antibody-binding peptides were elongated again, etc. In this way, longer peptides derived of dipeptides can be defined which bind antibody, i.e., mimotopes (Geysen, 1985; Geysen et al., 1986a, 1986b, 1987). However, so far no reports have become available of mimotopes which induce antibodies cross-reactive with the original antigen.

The reason for these failures is unknown, but they could be related to the absence of the native conformations among the major conformations of the "surrogate" peptide in solution. The antibody-binding properties of the surrogate peptide could be explained

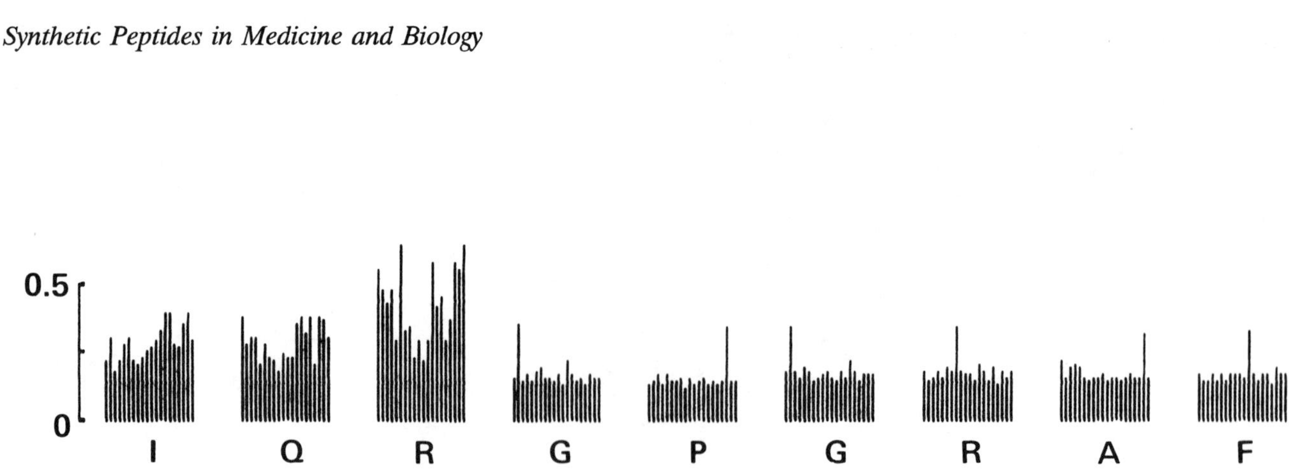

Fig. 8. Polyclonal serum of an individual infected with HIV-1 tested against a replacement set of peptides of the amino acid sequence which forms the top of the V3 loop. Note the irreplaceability of the six C-terminal amino acids.

by induced fit: the antibody induces the native conformation of the peptide upon binding. Because the surrogate peptide itself does not possess the proper conformation, it will not induce antibodies with the proper paratope (i.e., mirror image of the antigen).

VI. 3. Antibody-binding peptides found with the aid of peptide libraries

This approach is applicable to each antibody, but it should be especially useful if the native antigen is unknown. Numerous large peptide libraries have been described, being either expression libraries (Lenstra et al., 1992; Devlin et al., 1990; Scott ans Smith, 1990; Cwirla et al., 1990) or based on sets of synthetic peptides (Lam et al., 1991; Houghton et al., 1991; Fodor et al., 1991; Barbas et al., 1992). However, success with these libraries has been limited, probably due to two inherent disadvantages.

First, although the libraries may contain millions of peptides, they still contain only a fraction of all possible amino acid combinations. For instance, a complete library of all eight-long amino acids (which we feel is the minimum size) should contain 26.10^9 different peptides; the available libraries contain at best 4.10^6 peptides: only 0.015%.

Second, screening of the library with just a single antibody may often require an extraordinary amount of time and reagents.

In our hands, only on rare occasions have expression libraries yielded additional antibody-binding peptides to those previously defined using the much faster and efficient PEPSCAN method (Lenstra et al., 1992; Langeveld et al., in prep). A typical example is shown in Table 1.

Nevertheless, the search for alternative approaches to define discontinuous epitopes is still going on and sometimes results in rather unexpected approaches like the one of Vlatakis et al., 1993, who uses a synthetic molecular mold to define a three-dimensional molecular structure.

VI.4. Definition of high affinity antibody-binding peptides from low affinity antibody-binding peptides using small peptide libraries

In two cases (van Amerongen et al., 1989, 1990, 1994), high affinity antibody-binding peptides were defined from low affinity ones. The low affinity peptides were obtained using a PEPSCAN of the native protein. By using a replacement set of peptides for this particular peptide, amino acid replacements were introduced which yielded a peptide with 100 to 1000 times higher affinity (Fig. 9). These results suggest that if a low affinity antibody-binding peptide can be defined, it is possible to design a high affinity peptide using the systematic PEPSCAN methods. To find the initial low affinity peptides, especially in case of discontinuous epitopes, is still a matter of luck. Systematic approaches are not yet available.

VII. DETERMINATION OF T CELL EPITOPES

Peptides bound to solid supports cannot be used in T cell assays; however, when cleaved from solid supports, they can be readily used. Multiple methods to this end have been developed. This was done successfully for the first time by introducing an acid labile sequence between the peptide and the solid support (van der Zee et al., 1989) (Fig. 10A). This approach produced sufficient peptide free in solution to stimulate T cell clones (approximately ~0.1 to 1 μg per PEPSCAN pin). It has been applied successfully to numerous systems (Kast et al., 1991; van der Zee et al., 1989; Süss et al., 1993; Suttinger et al., 1991; Carballido et al., 1993; Rimmelzwaan et al., 1990). However, when peripheral blood lymphocytes or less sensitive T cell clones need to be stimulated or MHC binding needs to

TABLE 1
Comparison of PEPSCAN and a random-sequence expression library

MAb 1	
PEPSCAN*	-P-S-N-S-E-A-
Library**	-L-L-P-S-N-S-E-A-C
	-S-P-N-S-E-A-C
	-P-V-T-P-I-N-S-E-A-C
	-P-G-T-V-G-G-V-N-S-E-A-C
	-P-V-C-C-Q-G-H-H-Q-P-A-H-S-E-A-C
	-P-G-R-P-S-H-S-E-A-C
	-P-S-H-S-D-H
MAb 2	
PEPSCAN*	-S-F-F-S-Y-G-E-I-
Library**	-P-G-Q-P-F-S-F-G-S-L-L-I-D
MAb 3	
PEPSCAN*	-D-T-S-N-P-V-K-T-
Library**	-N-Y-S-D-P-V-R-K-
	-D-T-S-D-P-T-L-K-
MAb 4***	
PEPSCAN*	---
Library**	---

* These antibody-binding peptides were found using standard PEPSCAN on primairy sequences of native protein.

** These antibody-binding peptides were found screening a random peptide library. (From Lenstra et al., 1992; Langeveld et al., in prep.)

*** This is to indicate that almost always, in our hands, MAbs that did not bind in PEPSCAN, using the native primairy amino acid sequence, did not bind any random sequences of the libraries used either.

be measured, much larger amounts of free peptides are needed. This was achieved using the PEPSCAN system, improved polymer grafts and alternative cleavage methods (Bray et al., 1990; Maeiji et al., 1991). Applying these alternatives, milligram amounts of free peptides can be prepared per pin, thus allowing the simultaneous synthesis of hundreds if not thousands of peptides.

To define the T cell epitopes at the level of single amino acids, essentially similar approaches can be taken, as in the case of B cell epitopes. For instance, in the case of mycobacterial HSP65, a replacement set of peptides was prepared and tested (Wauben et al., 1993) (Fig. 10B) In contrast to the replacement set results of B cell epitopes, the amino acids of this T cell epitope only allowed very limited replacements, which is in accordance with the results of others (Brown et al., 1991). This may reflect the different roles of B and T cell epitopes; in contrast to B cell epitopes, T cell epitopes interact with two proteins (i.e., the MHC molecule and the T cell receptor), allowing much less tolerance to changes in the amino acid sequence of the T cell epitope.

The properties of peptides used for T cell epitope assays differ from those for B cell epitopes in a number of aspects. T cell epitopes should preferably have free amino termini, especially in case of MHC class I, as discussed previously. In contrast, B cell epitopes must have acetylated N-termini, because otherwise the charge at the N-terminal will interfere with antibody binding; most B cell epitopes of a protein will not have a free N-terminal under "normal" conditions. Ideally,

a

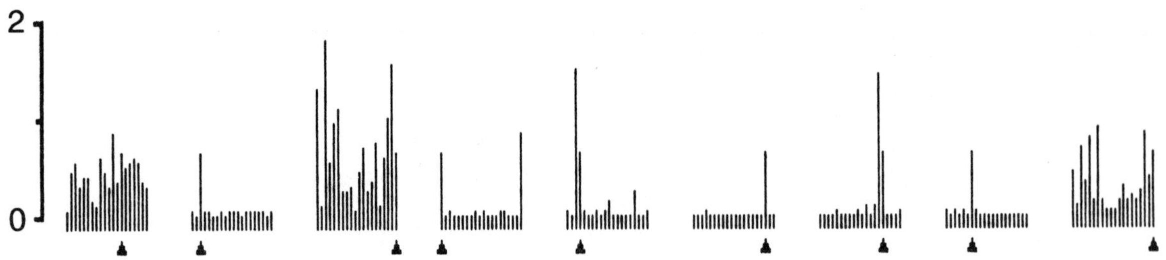

b

		elisa reactivity	affinity
original:	L - D - T - S- N - P - V - K - T	1200	1.3×10^5
improved:	F - D - D - T- D - P - I - K - K	1100000	2.6×10^8

Fig. 9. Systematic improvement of a peptide of the malaria parasite, *Plasmodium falciparum*, which reacts with a transmission blocking monoclonal antibody. This peptide, LDTSNPVKT, was obtained applying PEPSCAN to the gametocyte protein pfs 25. The reactivity of LDTSNPVKT with the monoclonal antibody is low. Using a replacement of peptides, we were able to improve its reactivity in an ELISA 100 to 1000-fold. The improved peptide, FDDTDPIKK, also had a 1000-fold higher affinity for the antibody compared to the original peptide. (From van Amerongen et al., 1989, 1992.)

the C-terminal of T cell epitopes should be a carboxy group; unfortunately, this forms a problem from a technical point of view and has as yet not been solved for the pin system. Nonetheless, peptides with a C-terminal amino acid like β-alanine or diketero-piperazine have been very useful both in Th cell and CTL bioassays and MHC binding assays.

Finally, the length of T cell epitopes for MHC class I molecules is probably limited between 9 to 11 amino acids (Falk et al., 1992; Rötzschke, 1992); longer peptides may not fit into the MHC class I groove and react considerably less well. For MHC class II, the length of the peptide is much less important. In our hands, a useful length is approximately 15 amino acids.

VIII. STATE OF THE ART FOR SYNTHETIC VACCINES

Often PEPSCAN or other peptide-based methods have been used to define candidates for peptide vaccines. In Table 2 a list is given of the more important ones. These peptides were selected for their ability to induce neutralizing antibodies. Of these peptides three also produced partial protection in the target animal. Because partial protection (i.e., not all animals are protected) is unacceptable, these peptides have not become commercially acceptable vaccines. The only exception could be a canine parvovirus peptide. We have recently shown that this peptide, from the N-terminal part of the canine parvovirus capsid protein (Fig. 11), when used to vaccinate dogs, is able to protect animals from infection. This could become the first example of a peptide vaccine which satisfies the rigorous criteria for a commercial vaccine.

IX. CONCLUSION

Linear and simple discontinuous B cell epitopes can be readily mapped and defined at the level of single amino acids using systematic approaches like PEPSCAN. Discontinuous scattered epitopes cannot be solved yet using a systematic general strategy. However, special approaches are sometimes possible, especially if a low affinity antibody-binding peptide can be found. Such antibody-binding peptides can be systematically changed into high affinity ones using PEPSCAN. Because T cell epitopes are linear amino acid sequences, they can be mapped systematically and defined at the level of single amino acids with the aid of PEPSCAN, just as in case of "linear" B cell epitopes.

As expected, these approaches can be used as well to map the ligands of receptors as was recently shown (Richter et al., 1992). Thus, the approaches described are not only useful in case of B and T cell mapping, but they can be just as useful in drug design in general.

X. METHODOLOGY

The procedures of the PEPSCAN technique have been extensively described (Geysen et al., 1984, 1985a, 1987)

A

170 180 190 200

EGVITVEESNTFGLQLELTEGMRFDKGYISGYFVTD	T cell response (SI) A2b	A2c
EGVITVEES	<2	<2
EGVITVEESN	<2	<2
GVITVEESNT	<2	<2
VITVEESNTF	<2	<2
ITVEESNTFG	<2	<2
TVEESNTFGL	<2	<2
VEESNTFGLQ	<2	<2
EESNTFGLQL	63	14
ESNTFGLQLE	773	440
SNTFGLQLEL	812	413
NTFGLQLELT	616	395
TFGLQLELTE	<2	<2
FGLQLELTEG	<2	<2
GLQLELTEGM	<2	<2
LQLELTEGMR	<2	<2
QLELTEGMRF	<2	<2
LELTEGMRFD	<2	<2
ELTEGMRFDK	<2	<2
LTEGMRFDKG	<2	<2
TEGMRFDKGY	<2	<2
EGMRFDKGYI	<2	<2
GMRFDKGYIS	<2	<2
MRFDKGYISG	<2	<2
RFDKGYISGY	<2	<2
FDKGYISGYF	<2	<2
DKGYISGYFV	<2	<2
KGYISGYFVT	<2	<2
GYISGYFVTD	<2	<2
M. bovis 65-kDa protein (5 μg/ml)	589	557

B

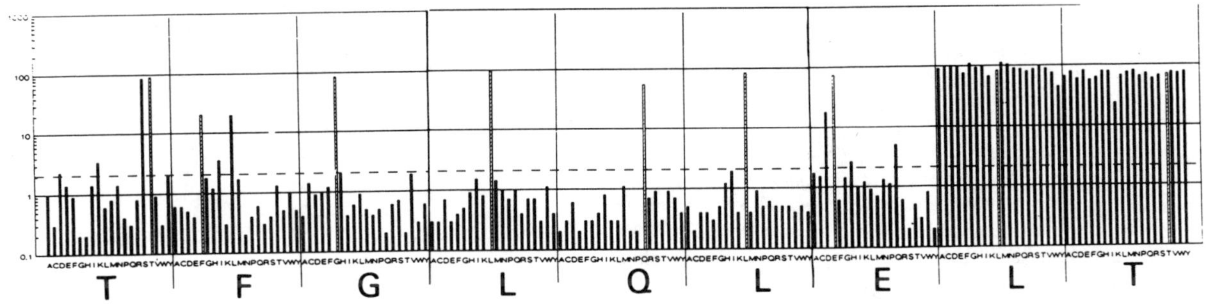

Fig. 10. (A) Epitope mapping of a T cell epitope. Overlapping peptides were tested for their ability to stimulate a T cell clone to proliferate. (From van der Zee et al., 1989.) (B) Set of replacement peptides based upon the reactive sequence of (A) were tested for their ability to stimulate a T cell clone to proliferate. The replaced amino acid is indicated on the horizontal axis for each amino acid position. The open bars are peptides with the parent sequence. (From Wauben et al., 1993.)

and can be carried out in laboratories suitably equipped for peptide chemistry and preferably located within a larger organic chemical environment. Peptides can be split off the solid supports to be used in bioassays as described by van der Zee et al. (1989).

If required, PEPSCAN can be performed or sets of peptides can be provided as a service by PEPSCAN SYSTEMS BV, P.O. Box 8323, 3503 RH Utrecht (phone: 31.30.941.456; fax: 31.30.941.526).

ACKNOWLEDGEMENTS

We would like to thank Janke de Jager for her help with the preparation of the manuscript and Joop Iedema and Fred van Welie for the preparation of the artwork.

ABBREVIATIONS

CPV	Canine parvovirus
CTL	Cytotoxic T cell
ELISA	Enzyme-linked immunosorbent assay
FMDV	Foot-and-mouth disease virus
HIV	Human immunodeficiency virus
MAb	Monoclonal antibody
MHC	Major histocompatibility complex
SDS	Sodium dodecyl
Th	T-helper

TABLE 2
State of the art of
synthetic peptide vaccines

	NT	Laboratory animals	Target animals	References
HSV-1	+	+	?	Kochen et al., 1988
TGE-virus	+	+	?	Posthumus et al., 1990
HIV-1	+	+	?	Goudsmit et al., 1988a
Semliki Forest Virus	+	+	+/-	Snijders et al., 1991
Rabies	+	+	+/-	Dietzschold et al., 1990
Foot-and-Mouth Disease	+	+	+/-	Bittle et al., 1982
Parvovirus	+	+	+	Langeveld et al., 1994

XI. REFERENCES

Acharya R, Fry E, Stuart D et al. The three-dimensional structure of foot-and-mouth disease virus at 2.9 Å resolution. Nature 337:709, 1989

Ahlborg N, Berzins K and Perlmann P. Definition of the epitope recognized by the Plasmodium falciparum-reactive human monoclonal antibody 33G2. Mol Biochem Parasitol 46:89, 1991

van Amerongen A, Sauerwein RW, Beckers PJA et al. Identification of a peptide sequence of the 25 kD surface protein of Plasmodium falciparum recognized by transmission-blocking monoclonal antibodies: Implications for synthetic vaccine development. Parasite Immunol 11:425, 1989

van Amerongen A, Meloen RH, Beckers PJA et al. PEPSCAN based amino acid modifications of a *P. falciparum* pfs25 peptide sequence substantially increase reactivity with transmission blocking monoclonal antibodies. In: Vaccines 90. Cold Spring Harbor Laboratory Press, Cold Spring Harbor, New York, p 461, 1990

van Amerongen A, Beckers PJA, Plasman HH et al. Peptides reactive with a transmission blocking monoclonal antibody against *Plasmodium falciparum* Pfs25: 2000-fold affinity increase by PEPSCAN-based amino acid substitutions. Peptide Res 5:269, 1992

van Amerongen A, Plasman HH, Kuperus D et al. Characterization of peptide sequences from ß-human chorionic gonadotropin reactive with monoclonal antibodies: Improvement of peptide affinities by specific amino acid replacements. Peptide Res 7:83, 1994

Amit AG, Mariuzza RA, Phillips SE et al. Three-dimensional structure of an antigen-antibody complex at 2.8 A resolution. Science 233:747, 1986

Atassi MZ. Precise determination of the entire antigenic structure of lysozyme. Immunochemistry 15:909, 1978

Barbas CF III, Bain JD, Hoekstra DM et al. Semisynthetic combinatorial antibody libraries: A chemical solution to the diversity problem. Proc Natl Acad Sci USA 89:4457, 1992

Bentley GA, Bhat TN, Boulot G et al. Immunochemical and crystallographic studies of antibody D1.3 in its free, antigen-liganded, and idiotope-bound states. Cold Spring Harb Symp Quant Biol 54:239, 1989

Bentley GA, Boulot G, Riottot MM et al. Three-dimensional structure of an idiotope-anti-idiotope complex. Nature 348:254, 1990

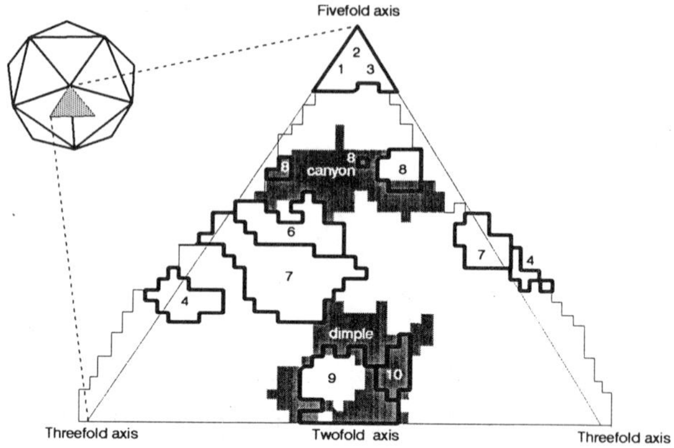

Fig. 11. Distribution of antigenic sites mapped to the surface of CPV capsids. The triangular blowup represents 1 of 60 mathematical repeating units from which the icosahedral virus surface is composed. The deep domains - canyon and dimple - are presented in grey. With PEPSCAN, 10 antigenic sites were mapped (Langeveld et al., 1993). Based on X-ray structural data (Tsao et al., 1991), nine of these sites could be assigned to the viral surface as indicated with the numbered areas enclosed by heavy solid lines. Sites 1-3 may serve as candidates for a peptide-based vaccine.

Bittle JL, Houghten RA, Alexander H et al. Protection against foot-and-mouth disease by immunization with a chemically synthesized peptide predicted from the viral nucleotide sequence. Nature 298:30, 1982

Boucher CA, Krone WJ, Goudsmit J et al. Immune response and epitope mapping of a candidate HIV-1 p17 vaccine HGP30. J Clinic Lab Anal 4:43, 1990

Brake DA, Goudsmit J, Krone WJA et al. Characterization of murine monoclonal antibodies to the tat protein from human immunodeficiency virus type 1. J Virol 64:962, 1990
Bray AM, Maeji NJ, and Geysen HM. The simultaneous multiple production of solution phase peptides; Assessment of the Geysen method of simultaneous peptide synthesis. Tetrahedron Letters 31:5811, 1990
Bredenbeek PJ, Snijder EJ, Noten AFH et al. The polymerase gene of corona- and toroviruses: Evidence for an evolutionary relationship. In: Cavanagh D and Brown TDK eds: Pathogenesis and molecular biology of coronaviruses. Plenum Press, London -Adv Exp Med Biol 276:307, 1990
Brown LE, Jackson DC, White DO et al. Extension of a minimal T cell determinant allows relaxation of the requirement for particular residues within the determinant. Int Immunol 3:1307, 1991
Brown CS, Jensen T, Meloen RH et al. Localization of an immunodominant domain on baculovirus-produced parvovirus B19 capsids: Correlation to a major surface region on the native virus particle. J Virol 66:6989, 1992
Carballido JM, Carballido-Perrig N, Kägi M et al. T cell epitope specificity in human allergic and nonallergic subjects to bee venom phospholipase A2. J Immunol 150:3582, 1993
Chassot S, Lambert V, Kay A et al. Fine mapping of neutralization epitopes on duck hepatitis B virus (DHBV) pre-S protein using monoclonal antibodies and overlapping peptides. Virology 192:217, 1993
Clarke JL, Drake AF, Vallace GR et al. PEPSCAN and circular-dishroism analysis of a repetitive antigen of Trypanosoma cruzi. Biochem Soc Trans 18:868, 1990
Colman PM, Laver WG, Varghese JN et al. Three-dimensional structure of a complex of antibody with influenza virus neuraminidase. Nature 326:358, 1987
Coursaget P, Lesage G, Le Cann P et al. Mapping of linear B-cell epitopes of hepatitis B surface antigen. Virology 142:461, 1991
Cwirla SE, Peters EA, Barrett RW et al. Peptides on phage: A vast library of peptides for identifying ligands. Proc Natl Acad Sci USA 87:6378, 1990
D'Souza MP, Durda P, Hanson CV et al. Evaluation of monoclonal antibodies to HIV-1 by neutralization and serological assays: An international collaboration. AIDS 5:1061, 1991
Devlin JJ, Panganiban LC and Devlin PE. Random peptide libraries: A source of specific protein binding molecules. Science 249:404, 1990
Dietzschold B, Gore M, Marchadier D et al. Structural and immunological characterization of a linear virus-neutralizing epitope of the rabies virus glycoprotein and its possible use in a synthetic vaccine. J Virol 64:3804, 1990
Doolittle RF. Proteins. Scientific American 253:74, 1985
van Eden W, Noordzij A, Hensen EJ et al. A modified PEPSCAN method for rapid identification and characterization of T-cell epitopes in protein antigens. In: Vaccines 89, Cold Spring Harbor Laboratory Press, Cold Spring Harbor, New York, USA, p 33, 1989
Enjuanes L, Gebauer F, Correa I et al. Location of antigenic sites of the S-glycoprotein of Transmissible Gastroenteritis Virus and their conservation in Coronaviruses. In: Cavanagh D and Brown TDK eds: Pathogenesis and Molecular Biology of Coronaviruses, Plenum Press, New York/London - Adv Exp Med Biol 276:159, 1990
Falk K, Rötzschke O, Stevanovic S et al. Allele-specific motifs revealed by sequencing of self-peptides eluted from MHC molecules. Nature 351:29, 1992
Fodor SPA, Read JL, Pirrung MC et al. Light-directed, spatially adressable parallel chemical synthesis. Science 251:767, 1991
Gebauer F, Posthumus WPA, Correa I et al. Residues involved in the antigenic sites of Transmissible Gastroenteritis coronavirus S glycoprotein. Virology 183:225, 1991
van Geluk A, van Meijgaarden KE, Janson AAM et al. Functional analysis of DR17(DR3)-restricted mycobacterial T cell epitopes reveals DR17-binding motif and enables the design of allele-specefic competitor peptides. J Immunol 149:2864, 1992
Geysen HM, Meloen RH and Barteling SJ. Use of peptide synthesis to probe viral antigens for epitopes to a resolution of a single amino acid. Proc Natl Acad Sci USA 81:3998, 1984
Geysen HM, Barteling SJ and Meloen RH. Small peptides induce antibodies with a sequence and structural requirements for binding antigen comparable to antibodies raised against the native protein. Proc Natl Acad Sci USA 82:178, 1985a
Geysen HM. Antigen-antibody interactions at the molecular level: Adventures in peptide synthesis. Immunology Today 6:364, 1985b
Geysen HM, Rodda SJ and Mason TM. The delineation of peptides able to mimic assembled epitopes. In: Porter R, and Whelan J, Eds: Synthetic Peptides as Antigens, John Wiley Ltd, Chichester, Ciba Foundation Symposium 119, p 130, 1986a
Geysen HM, Rodda SJ and Mason TM. A priori delineation of a peptide which mimics a discontinuous antigenic determinant. Molec Immunol 23:709, 1986b
Geysen HM, Rodda SJ, Mason TJ et al. Strategies for epitope analysis using peptide synthesis. J Immunol Meth 102:259, 1987
Geysen HM. Molecular technology: Peptide epitope mapping and the pin technology. South East Asian J Trop Med Public Health 21:523, 1990
Goudsmit J, Debouck C, Meloen RH et al. HIV type 1 neutralization epitope with conserved architecture elicits early type-specific antibodies in experimentally infected chimpanzees. Proc Natl Acad Sci USA 85:4478, 1988a
Goudsmit J, Meloen RH, Rusche JR et al. Baculovirus expressed gp160 of HIV-1 induces antibodies with isolate-specific binding to a nine-amino acid sequence related to type-specific cell fusion inhibition. AIDS 2:231, 1988b
Goudsmit J, Bakker M, Smit L et al. Shared and cross reactive epitopes on peptide antigens binding HIV-1 neutralizing antibodies of susceptible hosts. In: Vaccines 89, Cold Spring Harbor Laboratory, Press, Cold Spring Harbor, New York, p 145, 1989a
Goudsmit J, Bakker M, Smit L et al. Specifity of chimpanzee antibodies binding a strain specific HIV-1 neutralization epitope of the external envelope. J Med Primatol 18:357, 1989b
Goudsmit J, Meloen RH, Brasseur R et al. Human B-cell epitopes of HIV-2 transmembrane protein are similarly spaced as in HIV-1. J AIDS 2:297, 1989c
Goudsmit J, Meloen RH and Brasseur R. Map of sequential B-cell epitopes of the HIV-1 transmembrane protein using human antibodies as probe. Intervirology 31:327, 1990
Grootenhuis AJ, Steenbergen J, Timmerman MA et al. Inhibin and activin-like activity in fluids from male and female gonads: Different molecular weight forms and bioactivity/immunoactivity ratios. J Endocrinology 122:293, 1989
Guttinger M, Romagnoli P, Vandel L et al. HLA polymorphism and T cell recognition of a conserved region of p190, a malaria vaccine candidate. Int Immunol 3:899, 1991
van 't Hof W, Driedijk PC, Beck-Sickinger AG et al. Epitope mapping of the *Dermatophagoides pteronyssinus* house dust mite major allergen *Der p* II using overlapping synthetic peptides. Mol Immunol 28:1225, 1991

van 't Hof W, van Milligen FJ, van der Berg M et al. Epitope mapping of the cat (*Felis domesticus*) major alergen *Fel d* I by overlapping synthetic peptides and monoclonal antibodies against native and denatured *Fel d* I. Allergy 48:255, 1993

Houghten RA, Pinilla C, Blondelle SE et al. Generation and use of synthetic peptide combinatorial libraries for basic research and drug discovery. Nature 354:84, 1991

Jacobs L, Meloen RH, Rhiza H-J et al. Epitope analysis of glycoprotein I of Pseudorabies Virus. J Gen Virol 71:881, 1990

Janin J and Chothia C. The structure of protein-protein recognition sites. J Biol Chem 265:16027, 1990

Javaherian K, Langlois AJ, Schmidt S et al. The principal neutralization determinant of simian immunodeficiency virus differs from that of human immunodeficiency virus type 1. Proc Natl Acad Sci USA 89:1418, 1992

Kast WM, Offringa R, Peters PJ et al. Eradication of adenovirus E1-induced tumors by E1A specific cytotoxic T lymphocytes. Cell 59:603, 1989

Kast WM, Roux L, Curren J et al. Protection against lethal Sendai virus infection by *in vivo* priming of virus-specific cytotoxic T lymphocytes with an unbound synthetic peptide. Proc Natl Acad Sci USA 88:2283, 1991

Kochen CH, Geerligs HJ, Bos CA et al. Immunological properties of an N-terminal fragment of herpes simplex virus type 1 glycoprotein D expressed in Escherichia coli. Arch Virol 103:267, 1988

Krone WJA, Debouck C, Epstein LG et al. Natural antibodies to HIV-tat epitopes and expression of HIV-1 genes *in vivo*. J Med Virol 26:261, 1988

Kusters JG, Jager EJ, Lenstra JA et al. Analysis of an immunodominant region of avian coronavirus IBV. J Immunol 143:2692, 1989

Lam KS, Salmon SE, Hersh EM et al. A new type of synthetic peptide library for identifying ligand-binding activity. Nature 354:82, 1991

Langedijk JPM, Schalken JJ, Tersmette M et al. Location of epitopes on the major core protein p24 of the human immunodeficiency virus (HIV). J Gen Virol 71:2609, 1990

Langedijk JPM, Back NKT, Durda PJ et al. Neutralizing activity of anti-peptide antibodies against the principal neutralization domain of HIV-1. J Gen Virol 72:2519, 1991

Langedijk JPM, Vos WFHM, Van Doornum GJJ et al. Identification of cross-reactive epitopes recognized by HIV-1 false-positive sera. AIDS 6:1547, 1992a

Langedijk JPM, Back NKT, Kinney-Thomas E et al. Comparison and fine mapping of both high and low neutralizing monoclonal antibodies against the principal neutralization domain of HIV-1. Arch Virol 126:129, 1992b

Langeveld JPM, Casal JI, Vela C et al. B-cell epitopes of canine parvovirus: Distribution on the primary structure and exposure on the viral surface. J Virol 67:765, 1993

Langeveld JPM, Casal JI, Osterhaus ADME et al., First peptide vaccine providing protection against viral infection in target animal: Studies on canine parvovirus in dogs. J Virol 68:4506, 1994

Lenstra JA, Erkens JHF, Langeveld JPM et al. Isolation of sequences from a random-sequence expression library that mimic viral epitopes. J Immunol Meth 152:149, 1992

van Lierop M-JC, van Maanen K, Meloen RH et al. Proliferative lymphocyte responses to foot-and-mouth disease virus and three FMDV-peptides after vaccination or immunization with these peptides in cattle. Immunology 75:406, 1992

Logan D, Abu-Ghazaleh R, Blakemore W et al. Structure of a major immunogenic site on foot-and-mouth disease virus. Nature 362:566, 1993

Luytjes W, Geerts D, Posthumus W et al. Amino acid sequence of a conserved neutralizing epitope of murine coronaviruses. J Virol 63:1408, 1989

Maeji NJ, Bray AM and Geysen HM. Multi-rod peptide synthesis strategy for T-cell determinant analysis. J Immunol Meth 134:23, 1990

Maeji NJ, Bray AM, Valerio RM et al. Systematic screening for bioactive peptides. Peptide Res 4:142, 1991

McCahon D, Crowther JR, Belsham GJ et al. Evidence for at least four antigenic sites on type 0 foot-and-mouth disease virus involved in neutralisation; Identification by single and multiple site monoclonal antibody resistant mutants. J Gen Virol 70:639, 1989

Meloen RH and Barteling SJ. Epitope mapping of the outer structural protein VP1 of three different serotypes of FMDV. Virology 149:55, 1986a

Meloen RH, Puyk WC, Meijer DJA et al. The use of peptides to locate epitopes on the surface of foot-and-mouth disease. In: Peeters H Ed: Protides of the Biological Fluids 34, Pergamon Press, New York, p 103, 1986b

Meloen RH, Puyk WC, Meijer DJA et al. Antigenicity and immunogenicity of synthetic peptides of foot-and-mouth disease virus. J Gen Virol 68:305, 1987

Meloen RH, Puyk WC, Meijer DJA et al. Use of peptides to locate and characterize epitopes. In: Ginsberg H, Brown F, Lesner RA et al. Eds: Vaccines 88, Cold Spring Harbor Laboratory, Press, Cold Spring Harbor, New York, p 35, 1988

Meloen RH, Liskamp R and Goudsmit J. Specificity and function of the individual amino acids of a major determinant of HIV-1 that induces neutralization activity. J Gen Virol 70:1505, 1989

Meloen RH, van Amerongen A, Hage-van Noort M et al. The use of peptides to reconstruct conformational determinants. A brief review. Annales de Biologie Clinique 49:231, 1991

Middeldorp JM and Meloen RH. Epitope-mapping on the Epstein-Barr virus major capsid protein using systematic synthesis of overlapping oligopeptides. J Virol Meth 21:147, 1988

Miles MA, Wallace GR and Clarke JL. Multiple peptide synthesis (PEPSCAN Method) for the systematic analysis of B- and T-cell epitopes: Application to parasite proteins. Parasitology Today 5:397, 1989

Padlan EA, Silverton EW, Sheriff S et al. Structure of an antibody-antigen complex: Crystal structure of the HyHEL-10 Fab-lysozyme complex. Proc Nat Acad Sci USA 86:5938, 1989

Pfaff E, Mussgay HO, Schulz GE et al. Antibodies against a preselected peptide recognize and neutralize foot-and-mouth disease virus. EMBO Journal 1:869, 1982

Posthumus WPA, Lenstra JA, Schaaper WMM et al. Analysis and simulation of a neutralizing epitope of Transmissible Gastroenteritis Virus. J Virol 64:3304, 1990

Posthumus WPA. Localization of epitopes on viral surface proteins. In: Localization of Epitopes on Viral Surface Proteins: Thesis Rijksuniversiteit Utrecht, p 16, 1991a

Posthumus WPA, Puijk WC, Thomas AAM et al. Surface locations of foot-and-mouth disease virus type A10 epitopes of neutralizing monoclonal antibodies. In: Localization of Epitopes on Viral Surface Proteins: Thesis University of Utrecht, Utrecht, p 21, 1991b

Posthumus WPA, Meloen RH, Enjuanes L et al. Linear neutralizing epitopes on the peplomer protein of coronaviruses. In: Cavanagh D and Brown TDK Eds: Pathogenesis and Molecular Biology of Coronaviruses. Plenum Press, New York/London - Adv Exp Med Biol 276:181, 1991c

Radford AJ, Wood PR, Billman-Jacobse H et al. Epitope mapping of the *Mycobacterium bovis* secretory protein MPB70 using overlapping peptide analysis. J Gen Microbiol 136:265, 1990

Richter A, Conlan JW, Ward ME et al. Multidomain binding of transforming growth factor α to the epidermal growth factor receptor. Biochemistry 31:9546, 1992

Rimmelzwaan GF, Poelen MCM, Meloen RH et al. Delineation of canine parvovirus T cell epitopes with peripheral blood mononuclear cells and T cell clones from immunized dogs. J Gen Virol 71:2321, 1990

Rötzschke O, Falk K, Stevanovic S et al. Peptide motifs of closely related HLA class I molecules encompass substantial differences. Eur J Immunol 22:2453, 1992

Schaaper WMM, Beekman NJCM, Van Lierop M-JC et al. Improvements in the synthesis of cleavable peptides on pins. In: Schneider CH and Eberle AN Eds: Peptides 1992, Proceedings of the Twenty Second European Peptide Symposium, p 312, 1992a

Schaaper WMM, Beekman NJCM, Hage-van Noort M et al. Synthesis of large numbers of peptides for rapid screening of bioactive sequences. In: Smith JA and Rivier JE Eds: Peptides: Chemistry and Biology: Proceedings of the 12th American Peptide Symposium, ESCOM Science Publishers, Leiden, The Netherlands, p 651, 1992b

Scott JK and Smith GP. Searching for peptide ligands with an epitope library. Science 249:386, 1990

Sheriff S, Silverton EW, Padlan EA et al. Three-dimensional structure of an antibody-antigen complex. Proc Natl Acad Sci USA 84:8075, 1987

Snijders A, Benaissa-Trouw BJ, Oosterlaken TAM et al. Identification of linear epitopes on Semliki Forest virus E2 membrane protein and their effectiveness as a synthetic peptide vaccine. J Gen Virol 72:557, 1991

Stiemer RH, Gausepohl H, Mirshahi M et al. Immunological characterization of an immunomodulatory epitope in S-antigen/arrestin with a sequence motif common to tumor necrosis factor alpha. Immunol Lett 32:233, 1992

Strohmaier K, Franze R and Adam KH. Location and characterization of the antigenic portion of FMDV immunizing protein. J Gen Virol 59:295, 1982

Stüber D, Bannwarth W, Pink JRL et al. New B-cell epitopes in the plasmodium falciparum malaria circumsporozoite protein. Eur J Immunol 20:819, 1990

Süss G, Matile H, Meloen RH et al. Identifying polymorphic regions of the p190 protein from different *Plasmodium falciparum* strains by using specific T cells. Parasite Immunol 15:127, 1993

Thomas AAM, Woortmeijer WJ, Puyk WC et al. Antigenic sites on foot-and-mouth disease virus type A10. J Virol 62:2782, 1988

Tsao J, Chapman MS, Agbandje M et al. The three-dimensional structure of canine parvovirus and its functional implications. Science 251:1456, 1991

Tulip WR, Varghese JN, Webster RG et al. Crystal structures of neuraminidase-antibody complexes. Cold Spring Harb Symp Quant Biol 54:257, 1989

Twining SS, Lehmann H and Atassi M. The antibody response to myoglobin is independent of the immunized species. Biochem J 191:681, 1980

Tzartos SJ and Remoundos MS. Fine localization of the major alpha-bungarotoxin binding site to residues alpha 189–195 of the Torpedo acetylcholine receptor. Residues 189, 190 and 195 are indispensable for binding. J Biol Chem 265:21462, 1990

Versteeg-van Oosten J, Langeveld S, Voorma HO et al. Epitope Mapping of the fusion protein of measles virus. In: Vaccines 89, Cold Spring Harbor Laboratory, Press, Cold Spring Harbor, New York, p 521, 1989

Vlatakis G, Andersson LI, Müller R et al. Drug assay using antibody mimics made by molecular imprinting. Nature 361:645, 1993

Vordermeier HM, Harris DP, Friscia G et al. T cell repertoire in tuberculosis: Selective anergy to an immunodominant epitope of the 38-kDa antigen in patients with active disease. Eur J Immunol 22:2631, 1992

Wauben MHM, van der Zee R, Joosten I et al. Peptide variant of an arthritis related T-cell epitope induces T cells that recognise this epitope as a synthetic peptide but not in its naturally processed form. J Immunol 150:5722, 1993

Weijer K, Pfauth A, van Herwijnen JR et al. Induction of feline leukemia virus neutralizing antibodies by immunization with synthetic peptides of FeLV-env gene. Vaccine, 11:946, 1993

de Wolf F, Meloen RH, Bakker M et al. Characterization of human antibody binding sites on the external envelope of human immunodeficency virus type 2. J Gen Virol 72:1261, 1991

Xing PX, Reynolds K, Pietersz GA et al. Effect of variations in peptide sequence on anti-human milk fat globule membrane antibody reactions. Immunology 72:304, 1991

van der Zee R, van Eden W, Meloen RH et al. Efficient mapping and characterization of a T-cell epitope by the simultaneous synthesis of multiple peptides. Eur J Immunol 19:43, 1989

Zwart G, Langedijk H, Van der Hoek L et al. Immunodominance and antigenic variation of the principal neutralization domain of HIV-1. Virology 181:481, 1991

Chapter 3

EPITOPE MAPPING STUDIES WITH PAPER DISC-BOUND PEPTIDES

Wim van 't Hof, Marjan van den Berg, and Rob C. Aalberse

TABLE OF CONTENTS

I. INTRODUCTION

For the epitope mapping studies with synthetic peptides and large numbers of antisera it is essential that the used method allows the parallel synthesis of large numbers of peptides and offers a convenient procedure for large-scale immunologic testing. The PEPSCAN method (Geysen et al., 1987) does allow the parallel synthesis of large numbers of peptides; however, the parallel testing of large numbers of antisera is not feasible: the peptides cannot be cleaved from the polyethylene pins and only one antiserum at a time can be tested with the complete panel of peptides. In order to test 100 pin-bound peptides with 100 antisera within a reasonable time, the synthesis of 10,000 pin-bound peptides would be necessary, i.e., without regeneration of the pins. Several methods of simultaneous multiple peptide synthesis have been described that permit the parallel solid-phase synthesis of large numbers of peptides (Houghten, 1985; Ruggeri et al., 1986; Schnorrenberg and Gerhardt, 1989; Appel et al., 1990; Gausepohl et al., 1991). We found that with the peptides made by simultaneous multiple peptide synthesis with the T-bag (tea bag) method and subsequently coupled to Sepharose-particles large numbers of sera could be tested simultaneously (Van 't Hof et al., 1991, 1993a; Van Milligen et al., 1994). However, we felt that for large numbers of peptides a method in which the solid phase synthesis matrix would be the same as the solid phase in the immunologic testing procedure would be more practical. Methods for simultaneous multiple peptide synthesis have been reported in which simple materials such as paper (Frank and Döring, 1988; Eichler et al., 1989) or cotton (Lebl and Eichler, 1989; Berg et al., 1989; Daniels et al., 1989) were used as the solid-phase synthesis matrix. Frank and co-workers described the "spot-synthesis" on paper sheets, allowing a PEPSCAN-like approach with the parallel synthesis of large numbers of peptides on a paper sheet (Frank et al., 1991). However, as in the PEPSCAN, only one antiserum at a time can be tested with each sheet of paper. For our purposes of epitope mapping of allergens, peptide synthesis on paper is a particularly interesting technique, since paper discs are often used as a solid phase in the *in vitro* diagnosis of allergy (Pharmacia, Phadebas-RAST, Uppsala, Sweden). Krchňák and co-workers described a simple method of multiple peptide synthesis in which paper discs were used as the solid-phase synthesis matrix in polypropylene reactor columns, coupled to a manually operated peptide synthesizer (Krchňák et al., 1990).

0-8493-8967-4/95/$0.00+$.50

After completion of the synthesis the paper disc-bound peptides were incubated with antiserum in a paper disc-ELISA in plastic syringes equipped with polypropylene sinters. After development of the color, these syringes were emptied in ELISA plates and the absorbance was measured. The authors described the synthesis of four peptides with 10 paper discs each, which were cut into four pieces for subsequent use as the solid phase in an ELISA. However, both the method of synthesis as well as the determination of antibody binding are not practical for large numbers of peptides or antisera. We performed a T-bag synthesis with paper discs as the solid-phase matrix and compared the results obtained with these paper disc-bound peptides with those obtained with the PEPSCAN method to demonstrate that a PEPSCAN-like approach with paper disc-bound peptides is possible. Using the T-bag method of simultaneous multiple peptide synthesis combined with a paper disc-RIA, both the synthesis and subsequent immunologic testing of large numbers of peptides is feasible. For the cat major allergen *Fel d* I 146 T-bags were made with 15,000 paper discs, thereby permitting 100 simultaneous PEPSCAN-like epitope mapping tests. This synthesis and subsequent testing can be performed within a few working weeks.

II. RESULTS

II.1. PEPSCAN with paper discs and Epitope Mapping Kit

The cat major allergen *Fel d* I is a homodimer, of which each monomer consists of two polypeptide chains. Amino acid sequence analysis and cDNA cloning revealed that chain 1 consists of 70 and chain 2 of 92 amino acid residues (Morgenstern et al., 1991). Of both chains, all nonapeptides, overlapping by eight residues and spanning the entire sequences of both chains, were synthesized, i.e., 62 peptides for chain 1 and 84 peptides for chain 2. Peptides 141–146 and 143–148 of interleukin-6, the latter peptide representing a linear epitope for monoclonal antibody MAb 7 raised against denatured IL-6 (Brakenhoff et al., 1990), were also synthesized. A monoclonal antibody against denatured sperm whale myoglobin (Geysen, 1985) has been included in the Epitope Mapping Kit (Cambridge Research Biochemicals, Cambridge, UK) as control antibody. It binds to the peptide with the sequence PLAQ, but not to the peptide with the sequence GLAQ. Both these peptides were synthesized as well.

The 150 peptides were synthesized on paper discs as well as on polyethylene pins of the Epitope Mapping Kit using the Fmoc/*t*Bu strategy with BOP/HObt activation. Paper discs were activated with 1,1'-carbonyldiimidazole (Bethell et al., 1981), after which they

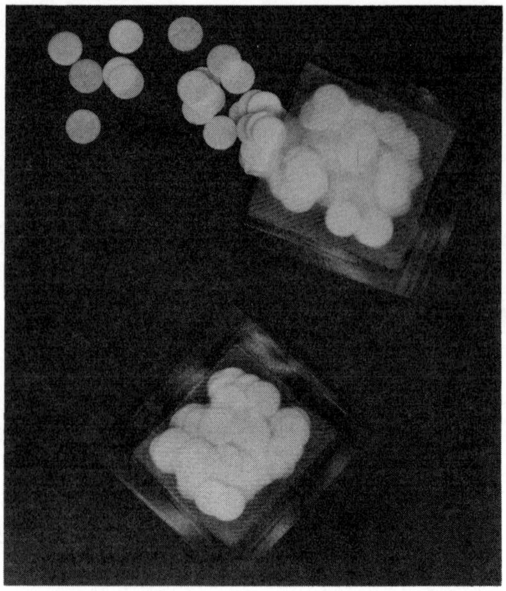

Fig. 1. T-bags containing derivatized paper discs. The T-bag was indelibly labeled by inserting numbers from a transfer sheet between the bottom seals.

were treated with an excess of 1,2-diamino-ethane to obtain free amino groups on the discs, onto which the peptides were synthesized. Of each peptide about 100 copies were made. During the synthesis each group of 100 paper discs was separated from the rest of the paper discs by inclusion in polypropylene T-bags (Fig. 1).

In previous studies epitopes had been found with longer Sepharose-coupled peptides of *Fel d* I with monoclonal antibodies against native and reduced/alkylated *Fel d* I (Van 't Hof et al., 1993a) and IgE antibodies from a number of cat-allergic patients (Van Milligen et al., 1994). A number of these antibodies were also tested with the paper disc-bound and pin-bound nonapeptides of *Fel d* I. The monoclonal antibodies against native *Fel d* I did not bind to any of the paper disc-bound or pin-bound peptides (Van 't Hof et al., 1993b). With monoclonal antibodies against reduced/alkylated *Fel d* I we generally found binding in the same regions of the molecule as with the Sepharose-coupled peptides (Van 't Hof et al., 1993a, 1993b). Monoclonal antibody FdR-1a against reduced/alkylated chain 1 of *Fel d* I bound to peptide 13–21 in the paper disc-RIA and the PEPSCAN, a region that had not been detected with the Sepharose-coupled peptides (Fig. 2). IgE antibodies against *Fel d* I also bound to the same regions of the molecule that we had found with the Sepharose-coupled peptides.

II.2. Signal-background ratios in paper disc-RIA and pin-ELISA

In order to compare the sensitivity of our epitope mapping method with the PEPSCAN we compared the signal-background ratios for both methods. The results are given in Table 1. In the PEPSCAN we compared binding of the control monoclonal antibody to the pin-bound peptides PLAQ and GLAQ with that to the control pins included in the Epitope Mapping Kit (E^{405} 1.328 and 0.138 in the pin-ELISA for the pin-bound peptides that we had synthesized ourselves, compared to 1.655 and 0.119 for the control pins). This indicates that the efficiency of BOP/HObt activation is comparable to that of the reaction with activated esters, which are included in the Epitope Mapping Kit. The major advantage of our coupling procedure is that the reaction time has been reduced from 16 to 1 h.

II.3. Reproducibility of antibody binding

The reproducibility of antibody binding to paper disc- and pin-bound peptides PLAQ and GLAQ was checked by performing the paper disc-RIA and the pin-ELISA in six-fold replicates. In the paper disc-RIA the mean percentage binding to PLAQ was 22.5 (standard deviation 0.9) and to GLAQ 2.6 (standard deviation 0.4). In the pin-ELISA the mean absorbance at 405 nm was 1.44 for PLAQ (standard deviation 0.19) and 0.13 for GLAQ (standard deviation 0.01). In the paper disc-RIA with MAb 7 we found in nine-fold replicates a mean percentage binding to peptide 143–148 of IL-6 of 11.1 (standard deviation 0.5) and to peptide 141–146 of 1.0 (standard deviation 0.3). In the regeneration experiments (Section II.4) we also found good reproducibility. However, one aberrant disc was found, underlining the importance of duplicate testing.

II.4. Reusability/regeneration of paper disc-bound peptides

One of the advantages of the PEPSCAN is that after the ELISA the antibodies can be eluted from the pins, after which the pin-bound peptides are ready for another ELISA. This procedure can be repeated several times, but eventually the pins cannot be cleaned efficiently enough and the signal-background ratio drops. We investigated whether the same principle holds for the paper disc-RIA by eluting MAb 7 from 28 paper disc replicates with covalently bound peptide 143–148 of IL-6 with 0.5 M glycine/HCl, pH 2.5, containing 10% dioxane, followed by another incubation, for several times. As we can see in Table 2, the residual binding after each elution rapidly increased. However, this acid buffer is a rather mild reagent, often used to elute antibodies from the affinity purification matrix without disturbing the binding capacity of these antibodies. We did not investigate whether other elution conditions are more efficient, i.e., giving less residual binding after each elution. However, using the acid glycine/HCl buffer, the paper disc-bound peptides can be regenerated and used at least several times.

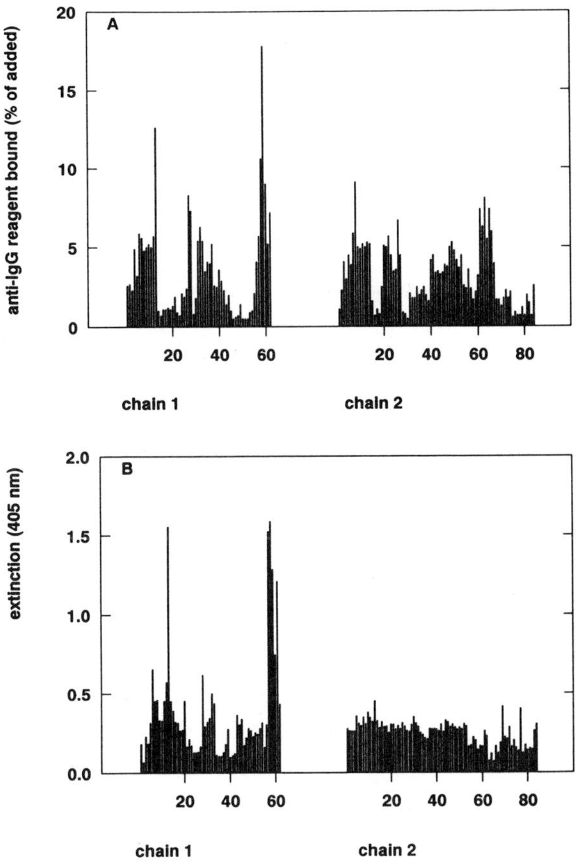

Fig. 2. Results of paper disc-RIA (A) and pin-ELISA (B) with monoclonal antibody FdR-1a against reduced/alkylated chain 1 of *Fel d* I for both chains of this allergen. In the paper disc-RIA it bound to peptides 13-21 and 58-66 to 60-68 of chain 1 and in the pin-ELISA to peptides 13-21 and 57-65 to 61-69. In our epitope mapping studies with Sepharose-coupled peptides we had previously shown binding to the overlapping peptides 53-66 and 60-70 of chain 1 (Van 't Hof et al., 1993a).

II.5. Specificity control of antibody binding to paper disc-bound peptides

Antibodies can bind to synthetic peptides nonspecifically, and therefore specificity control is extremely important. The specificity of binding can be tested in different ways (Van 't Hof et al., 1993c).

In the paper disc-RIA with FdR-1a we found high binding to a large number of paper disc-bound peptides (Fig. 2A). When we compared the binding of nonrelated isotype and concentration-matched monoclonal antibodies to the paper disc-bound peptides of *Fel d* I, we found almost the same nonrandom binding pattern as the pattern found for FdR-1a. These control monoclonal antibodies, however, did not bind strongly to peptides 13–21 and 59–67 of chain 1, contrary to FdR-1a. This may be regarded as an indication that FdR-1a binds to these peptides specifically. In Fig. 3 we have subtracted for each paper disc-bound peptide the average binding value of these three nonrelated

TABLE 1
Signal-background ratios in paper disc-RIA and PEPSCAN

Monoclonal antibody	Peptide	Paper disc-RIA	PEPSCAN
Control antibody	PLAQ	8.8	10.9
CRB-pins			13.9
MAb 7	IL-6: 143–148	10.7	15.9
FdR-1a	Ch.1: 13–21	4.1	5.8
	Ch.1: 59–67	5.8	6.0
FdR-2a	Ch.2: 43–51	13.2	9.5
FdR-2b	Ch.1: 12–20	3.3	
	Ch.1: 13–21		4.4
	Ch.1: 60–68	2.9	4.6
	Ch.2: 50–58	8.2	9.9

As background the binding to IL-6: 141–146 was taken for MAb 7, the binding to GLAQ for the control monoclonal antibody included in the Epitope Mapping Kit and the average binding to the low binding peptides for the monoclonal antibodies against reduced/alkylated *Fel d* I.

TABLE 2
Results of the regeneration experiments using 28 paper discs with covalently bound peptide 143–148 of IL-6 and MAb 7

	Mean	Standard deviation
After first elution	0.7	0.3
After second incubation	11.1	1.0
After second elution	1.7	0.4
After third incubation	11.8	0.6
After third elution	3.5	0.3

An increase in the residual binding is observed after each elution.

monoclonal antibodies from the binding value of FdR-1a. Binding to peptides other than peptides 13–21 and 59–67 of chain 1 was then considered as background, giving a cut-off value for specificity of average binding plus three times the standard deviation. We think that additional, more reliable, methods for establishing specificity are required, particularly inhibition of antibody binding by the immunizing protein or by related peptides.

Binding of FdR-1a to paper disc-bound peptides 13–21 and 59–67 of chain 1 was inhibited by reduced/alkylated *Fel d* I. It was also, to a lesser extent, inhibited by native *Fel d* I (Fig. 4), which may be caused by small contaminations of denatured *Fel d* I in the cat extract used (Van Milligen et al., 1992). Binding to peptide 13–21 was not inhibited by the related peptide 9–24 (Fig. 4A). Binding to peptide 59–67 was inhibited by the related peptide 53–66, but not by the related peptide 60–70 (Fig. 4B). Binding of MAb 7 to paper disc-bound peptide 143–148 of IL-6 was inhibited by unfolded IL-6. Unfolding was established by adding up to 1% SDS to the incubation buffer. Binding was also inhibited by the related peptide 141–155, but not by the related peptide 133–148 (Fig. 4C).

III. DISCUSSION/EVALUATION

The PEPSCAN-like approach with paper disc-bound peptides proved to be a good alternative for the PEPSCAN itself. The T-bag method with paper discs as the solid phase allowed the parallel synthesis of large numbers of peptides and the use of the same discs, without cleavage of the peptides, in a paper disc-RIA allowed the parallel testing of large numbers of antisera. This possibility is a large advantage over the PEPSCAN, with which only one serum can be tested at a time. The paper discs could be regenerated by elution of the bound antibodies, which makes the number of antibodies that can be tested even larger. The reproducibility of antibody binding was generally good. The signal-background ratios were good as well and comparable to those in the PEPSCAN. As we demonstrated, it was quite easy to synthesize 15,000 paper disc-bound peptides and test a considerable fraction of these simultaneously with several antisera.

Whenever we find antibody binding to synthetic peptides, it is extremely important to establish whether this binding is specific. Several specificity criteria can be used (Van 't Hof et al., 1993c). For FdR-1a and MAb 7 binding was compared with binding to control peptides and with binding of nonrelated monoclonal antibodies. For FdR-1a this led to an operational cut-off value for specificity as depicted in Fig. 3. We think that a more reliable method of investigating specificity of antibody binding is formed by inhibition studies with the immunizing protein: in the case of FdR-1a, reduced/alkylated *Fel d* I, and in the case of MAb 7, unfolded IL-6. Inhibition by the identical peptide or related peptides is sometimes considered a good method of specificity control (Savoca et al., 1991). However, we found that these kind of studies may be severely complicated by the effects of different length, pH and ionic strength of the related peptides. Binding of FdR-1a to paper disc-bound peptide 13–21 of chain 1 was not inhibited by peptide 9–24, although it contains all residues of the shorter peptide (Fig. 4A). This may be caused by strong folding of peptide 9–24 or by adoption of another conformation than in the native protein. Binding of the same monoclonal antibody to

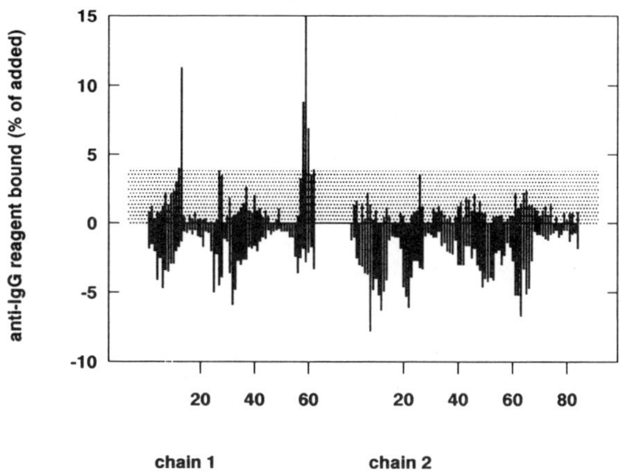

Fig. 3. Results of paper disc-RIA with monoclonal antibody FdR-1a. For each paper disc-bound peptide the average binding of three nonrelated isotype and concentration-matched monoclonal antibodies, given on the negative axis, was substracted. These control antibodies did not bind strongly to peptides 13-21 and 59-67 of chain 1. Then for FdR-1a, binding to all *Fel d* I peptides, except these peptides 13-21 and 59-67, was used to calculate a cut-off value for specificity (mean plus three times standard deviation), represented by the shaded area.

paper disc-bound peptide 59–67 was inhibited by peptide 53–66, which lacks the C-terminal residue of the shorter peptide, but not by peptide 60–70, which lacks the N-terminal residue of the shorter peptide (Fig. 4B). At least some inhibition by peptide 60–70 was expected as FdR-1a also bound to paper disc-bound peptides 58–66 and 60–69 of chain 1. The positively charged protonated α-amino group of residue 60 may obstruct antibody binding to peptide 60–70 by electronic repulsion. This positive charge is neither present in the native *Fel d* I or the reduced/alkylated chain 1, nor in the acetylated paper disc-bound peptide 60–69. When peptide 60–70 is coupled to Sepharose a fraction of the peptide molecules will be coupled *via* this N-terminal amino-group, thus neutralizing the positive charge and possibly allowing antibody binding again. This theory may explain why binding of FdR-1a to this Sepharose-coupled peptide was not inhibited by a 1 mM solution of the free peptide itself (Van 't Hof et al., 1993a). Electronic repulsion by the protonated ϵ-amino group of the extra C-terminal lysine, added to peptides 53–66 and 60–70 to facilitate coupling to Sepharose in earlier studies (Van 't Hof et al., 1993a, Van Milligen et al., 1994), appeared to be negligible. Peptide 53–66, in which the extra C-terminal lysine is located even closer to the binding residues of paper disc-bound peptide 59–67 than in peptide 60–70, did inhibit binding.

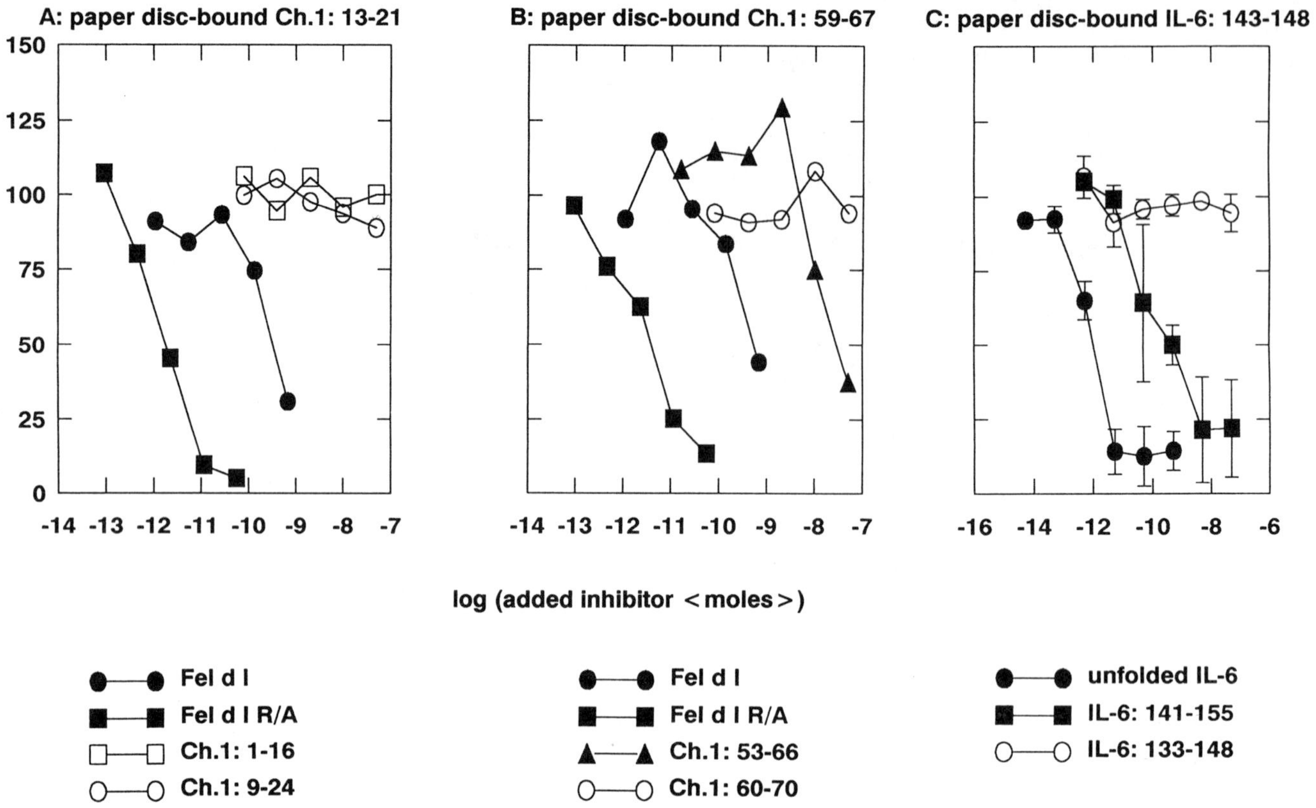

Fig. 4. Inhibition assays. Binding of monoclonal antibody FdR-1a, against reduced/alkylated chain 1 of *Fel d* I, to paper disc-bound peptide 13-21 of chain 1 (A) was inhibited by reduced/alkylated *Fel d* I (filled squares) and native *Fel d* I (filled circles), but not by the related peptide 9-24 (open circles). Binding to the paper disc-bound peptide 59-67 of chain 1 (B) was also inhibited by reduced/alkylated *Fel d* I (filled squares) and native *Fel d* I (filled circles), but also by the related peptide 53-66 (closed triangles). It was, however, not inhibited by the related peptide 60-70 (open circles). Binding of monoclonal antibody MAb 7, against denatured Il-6, to paper disc-bound peptide 143-148 of IL-6 (C) was inhibited by unfolded IL-6 (filled circles) and the related peptide 141-155 (filled squares), but not by the related peptide 133-148 (open circles).

Binding of MAb 7 to paper disc-bound peptide 143–148 of IL-6 was inhibited by peptide 141–155. Obviously, in this case the extra positive charge of the protonated α-amino group of amino acid residue 141, compared to its neutral charge in the native or unfolded protein, has little effect on antibody binding. However, binding to paper disc-bound peptide 143–148 was not inhibited by peptide 133–148. Possibly, in this case the negative charge of the α-carboxylic acid group of amino acid 148 residue obstructs antibody binding by polar interactions. This hypothesis could be tested by synthesis of peptide 133–148 as an carboxamide or by testing a longer peptide in which the free C-terminus is located further away from the binding region 143–148. This negative charge does not exist in the native or unfolded protein, nor in the pin-bound peptide 143–148. Thus, in general, the addition of extra amino acid residues or coupling to a solid phase may have a noticeable impact on antibody binding to synthetic peptides. In our opinion, these effects of different length, pH and ionic strength make inhibition by related peptides an unpredictable specificity criterion.

IV. GENERAL METHODOLOGY

IV.1. Functionalization of paper discs

Discs of filter paper (Schleicher & Schüll, Dassel, FRG: no. 0859, medium, 60 g/m^2), with a diameter of 0.6 cm, were punched out with a standard perforator. To 30 g (±15,000) of paper discs, 15 g of 1,1'-carbonyldiimidazole (Bethell et al., 1981) were added in 750 ml DMF in a 2 liter erlenmeyer flask. After 1 h of stirring at room temperature, the paper discs were washed quickly, twice with DMF and twice with water (±300 ml). A solution of 112 g of ethylenediammoniumdichloride and 36 g of NaOH in 500 ml, pH 8.5, was added immediately. After 2 h of stirring at room temperature, the reaction mixture was decanted, after which 500 ml 0.5 M glycine, pH 8.5, was added to block any remaining activated carboxy groups. After 1 h of stirring at room temperature, the discs were alternately washed with 0.1 M acetate buffer, pH 4.0, and 0.1 M borate buffer, pH 8.0, three times, twice with water, twice with methanol (all ±300 ml) and air dried.

IV.2. Determination of the number of free amino groups on paper discs and pins

The number of free amino groups on the paper disc was determined by acylation by SPDP, followed by spectrophotometric determination of the subsequently released pyridine-2-thione upon treatment with DTT (Ngo, 1986). Of four pins from the Epitope Mapping Kit, the Fmoc-β-alanine spacers were deprotected with 20% piperidine in DMF. They were washed twice with DMF, once with methanol and air dried. As control, we took four pins of which the Fmoc-protecting group was not removed. Each pin was acylated for 30 min with a mixture of 155 μl phosphate buffer (0.1 M, pH 7.5, with 0.1 M NaCl) and 20 μl of a 20 mM (6.25 mg/ml) solution of SPDP in absolute ethanol. The pins were washed twice with the phosphate buffer and twice with the reduction buffer, a 0.1 M sodium citrate buffer, pH 4.5, with 0.1 M NaCl. Each pin was reduced with 175 μl of a 50 mM (7.7 mg/ml) solution of DTT in this reduction buffer for 30 min. For each pin the reaction mixture and three washings with 175 μl reduction buffer were combined, the total volume was made up to 1400 μl with reduction buffer, and the extinction at 343 nm was measured (0.185–0.210) and corrected for the extinction value of the unprotected pins (0.089–0.098). The concentration of pyridine-2-thione (ϵ_{343}= 8080 l mol^{-1} cm^{-1}), and from this value the number of free amino groups on each pin, was calculated with the formula: (A^{343}- blank) x dilution factor x volume/(ϵ_{343} x length of cell) = number of free amino groups/pin, e.g., (0.185–0.098) x 8 x 0.0014/(8080 x 0.5) = 241 x 10^{-9} mol/pin. The number of free amino groups ranged from 240–340 nmol/pin, which is very well in accordance with the manufacturer's specification of 150–300 nmol/pin.

In the same way the number of free amino groups on the derivatized paper discs was determined to be 45 nmol/disc. As background, nonderivatized paper discs were used. The kind of paper was not optimized; other types of filter paper may yield higher numbers of free amino groups per disc.

IV.3. Peptide synthesis on paper discs

T-bags were made of polypropylene mesh (Spectrum Medical Industries, Los Angeles, CA), mesh opening 75 μm, with a standard impulse sealer (TISH-300, San Diego Bag, San Diego, CA). They were indelibly labeled by insertion of a number from a transfer sheet (Aldrich, Brussels, B) in between the bottom seals. Each T-bag was filled with 200 mg (approximately 100) derivatized paper discs (Fig. 1).

Before each coupling step the T-bags were sorted on the amino acid prescribed by the desired sequence and they were put into 20 polypropylene flasks (Rofa-Mavi, Beverwijk, the Netherlands) of 50, 100, 250 and 500 ml, depending on the number of T-bags, that served as reaction vessels. Per T-bag 2.5 ml of a solution containing 0.2 M of the desired Fmoc-protected amino acid (Novabiochem, Läufelfingen, Switserland), 0.2 M BOP (Richelieu Biotechnologies Inc., Quebec, Canada), 0.2 M HObt (Fluka, Buchs, Switserland) and 0.26 M DIPEA (Aldrich, Brussels, Belgium) in DMF were added. Dimethylamine, a decomposition product of DMF that leads to premature Fmoc-deprotection, had been removed by purging each freshly openened bottle of PA-grade DMF (Aldrich, Brussels, Belgium) with nitrogen gas until no more amine-like smell could be detected. Semi-permanent side chain protection of arginine was achieved with the Pmc group (Ramage and Green, 1987), of histidine (Sieber and Riniker, 1987) and cysteine (McCurdy, 1989) with the Trt group and of the other amino acids with the *t*Boc or *t*Bu group. The reaction time for the coupling steps was 1 h. After each coupling step all T-bags were collected in a 500 ml polypropylene flask (Rofa-Mavi, Beverwijk, the Netherlands) for joint washing and deprotection. After each coupling step the T-bags were washed extensively: twice with 200 ml DMF (2 min), once with 200 ml dichloromethane (2 min), and twice with 200 ml DMF (2 min) again. Deprotection was performed with 30% piperidine in DMF (200 ml) for 15 min. After each deprotection step the T-bags were washed extensively: three times with 200 ml DMF (2 min), once with 200 ml 2-propanol (5 min), once with 200 ml dichloromethane (2 min), once with 200 ml 2-propanol (5 min), twice with 200 ml dichloromethane (2 min), and twice with 200 ml DMF (2 min). All washing solvents were of LAB quality (Merck, Darmstad, FRG). After completion of the synthesis, the peptides were acetylated with 200 ml of a solution of 5% acetic anhydride and 1% DIPEA in DMF. Collective deprotection of the side chains was performed with 200 ml TFA containing 5% thioanisole (Aldrich, Brussels, Belgium) at room temperature for 4 h.

IV.4. PEPSCAN with modified Fmoc-protocol

In the coupling steps each pin was incubated with 150 μl of the same reaction mixtures as described in Section IV.3 in the poyethylene microtiter plates included in the Epitope Mapping Kit for 1 h. The deprotection with 20% piperidine in DMF and the extensive washings after each coupling and deprotection step were performed in standard polypropylene boxes. In order to immerse the tops of the pins, approximately 250 ml of solvent was necessary. The same washing protocols as in Section IV.3 were used. After completion of the synthesis, the peptides were acetylated with 150 μl/pin of a solution of 5% acetic anhydride and 1% DIPEA in DMF. The amount of TFA used in the final side chain deprotection was reduced to 0.5 ml/pin by performing this reaction in polypropylene deep-well ELISA plates (Beckman Instruments Inc., Fullerton, CA), onto which the polyethylene reaction blocks can be snapped as a lid, rather than in polypropylene boxes. After the final side chain deprotection, the pins were washed with dichloromethane, dichloromethane with 5% DIPEA and with dichloromethane again. After air drying they were washed with water and immersed in methanol for 16 h. After air drying the pins were ready for ELISA testing.

IV.5. Radioimmunoassays with paper disc-bound peptides

After opening the T-bags, each paper disc-bound peptide was incubated separately end over end with 50 μl of diluted antiserum and 500 μl PBS-AT at room temperature for 16 h. After washing the paper discs five times with PBS-T to remove unbound antibodies, a detection reagent, 1 ng, radiolabeled with 400 Bq ^{125}I, was added in 500 μl PBS-ATS. After a 16 h incubation the discs were washed four times with PBS-T to remove excess radiolabeled antibody. The amount of detection reagent bound was measured by counting the γ-radiation and results were expressed as percentage of added radiolabel.

In the regeneration experiments at this stage the antibody binding was disrupted by incubation with 500 μl/disc 0.5 M glycine/HCl, pH 2.5, with 10% dioxane. After washing five times washing with PBS-T the discs were ready for another paper disc-RIA.

IV.6. ELISA with pin-bound peptides

The pins were precoated at room temperature in an ELISA plate with 200 μl/pin of PBS-T containing 1% (w/v) bovine serum albumin (demineralized BoSerAl powder, Organon Technika, Turnhout, Belgium) and 1% (w/v) ovalbumin (Sigma, St Louis, MO) for 1 h. The pins were each incubated with 150 μl of a 1/1000 dilution of the monoclonal antibody at 4°C for 16 h. After washing four times washing with PBS-T for 10 min to remove unbound antibodies, the pins were incubated with 150 μl of a 1/2000 dilution of a HRP-labeled goat-anti-mouse Ig reagent (Central Laboratory of The Netherlands Red Cross Blood Transfusion Service, Amsterdam, The Netherlands) at room temperature for 1 h. The pins were washed four times with PBS-T for 10 min to remove the excess HRP-labeled detection reagent. They were incubated in the dark with 150 μl of a 100:2:1 mixture of 0.1 M citrate buffer, pH 4.0, a fresh-made 40 mM solution of ABTS and a stock solution of 0.6% hydrogen peroxide at room temperature. After sufficient green color had developed in some of the wells (generally after 15–30 min), the reaction was terminated by removing the pins from the ELISA plates. The extinction at 405 nm was measured directly. The pins were immediately immersed in PBS-T to prevent drying. The antibodies bound were removed from the pins by extensive washing with 0.1 M glycine/HCl, pH 2.5, for 15 min. After washing with PBS, water and methanol and air drying, the pins were generally ready for the next ELISA test. In one case this was not enough to remove all antibodies. After 1 h sonication at 60°C in PBS with 1% SDS and 0.1% thioglycolic acid in an ultrasonic cleaning bath all antibodies were eluted from the pins.

IV.7. Inhibition studies with paper disc-bound peptides

Antibodies, diluted in PBS-AT (50 μl), were preincubated at room temperature with a serial dilution series of inhibitor (50 μl) for a minimal of 2 h. After addition of 500 μl PBS-AT and a paper disc-bound peptide, the complete procedure described in Section IV.5 was followed.

ABBREVIATIONS

BOP	Benzotriazole-1-yl-oxy-tris(-dimethylamino) phosphonium hexafluorophosphate
DIPEA	*N,N*-Diisopropyl-ethylamine
DMF	*N,N*-Dimethylformamide
DTT	Dithiothreitol
Fmoc	9-Fluorenylmethoxycarbonyl
HObt	1-Hydroxybenzotriazole
PBS	Phosphate buffered saline
PBS-T	PBS with 0.1% Tween 20
PBS-AT	PBS-T with 0.3% bovine serum albumin
PBS-ATS	PBS-AT with 4.5% normal bovine serum and 0.5% normal sheep serum
Pmc	2, 2, 5, 7, 8-pentamethylchroman-6-sulfonyl
SPDP	*N*-Succinimidyl-3-(2-pyridyldithio) propionate
*t*Boc	*tert.* Butoxycarbonyl
*t*Bu	*tert.* Butyl
TFA	Trifluoroacetic acid
Trt	Trityl

V. REFERENCES

Appel JR, Pinilla C, Niman H et al. Elucidation of discontinuous linear determinants in peptides. J Immunol 144:976, 1990

Berg RH, Almdal K, Pedersen WB et al. Long-chain polystyrene-grafted polyethylene film matrix: A new support for solid phase peptide synthesis. J Am Chem Soc 111:8024, 1989.

Bethell GS, Ayers JS, Hearn MTW et al. Investigations of the activation of various insoluble polysaccharides with 1,1'-carbonyldiimidazole and the properties of the activated matrices. J Chromatography 219:361, 1981

Brakenhoff JPJ.Molecular cloning and structure function analysis of human interleukin-6. Thesis, Amsterdam, 1990

Daniels SB, Bernatowicz MS, Coull JM et al. Membranes as solid supports for peptide synthesis. Tetrahed Lett 30:4345, 1989

Eichler J, Beyerman M and Bienert M. Application of cellulose paper as support material in simultaneous peptide synthesis. Collect Czech Chem Commun 54:1746, 1989

Frank R and Döring R. Simultaneous multiple peptide synthesis under continuous flow conditions on cellulose paper discs as segmental solid supports. Tetrahedron 44:6031, 1988

Frank R, Güler S, Krause S et al. Facile and rapid "spot-synthesis" of large numbers of peptides on membrane sheets. In: Giralt E and Andreu D Eds: Peptides 1990 (Proceedings of the 21st European Peptide Symposium), ESCOM, Leiden, p 151, 1991

Gausepohl H, Kraft M, Boulin C et al. A multiple reaction system for automated simultaneous peptide synthesis. In: Girault E and Andreu D Eds: Peptides 1990 (Proceedings of the 21st European Peptide Symposium), ESCOM, Leiden, p 206, 1991

Geysen HM. Antigen-antibody interaction at the molecular level. Adventures in peptide synthesis. Immunol Today 6:364, 1985

Geysen HM, Rodda SJ, Mason TJ et al. Strategies for epitope analysis using peptide synthesis. J Immunol Methods 102:259, 1987

Houghten RA. General method for the rapid solid-phase synthesis of large numbers of peptides: Specificity of antigen-antibody interaction at the level of individual amino acids. Proc Natl Acad Sci USA 82:5131, 1985

Krchňák V, Vágner J, Novák J et al. A general procedure for evaluation of immunological relevance of synthetic peptides: Peptides synthesized on paper in enzyme-linked immunosorbent assay. Anal Biochem 189:80, 1990

Lebl M and Eichler J. Simulation of continuous solid phase synthesis: Synthesis of methionine Enkephalin and its analogs. Pept Res 2:297, 1989

McCurdy SN. The investigation of Fmoc-cysteine derivatives in solid phase peptide synthesis. Pept Res 2:147, 1989

Morgenstern JP, Griffith IJ, Brauer AW et al. Amino acid sequence of Fel d I, the major allergen of the domestic cat: Protein sequence analysis and cDNA cloning. Proc Natl Acad Sci USA 88:9690, 1991

Ngo TT. A simple spectrophometric determination of solid supported amino groups. J Biochem Biophys Methods 12:349, 1986

Ramage R, and Green J. N^G-2,2,5,7,8-pentamethylchroman-6-sulphonyl-L-arginine: A new acid labile derivative for peptide synthesis. Tetrahed Lett 28:2287, 1987

Ruggeri Z, Houghten RA, Russell SR et al. Inhibition of platelet function with synthetic peptides designed to be high-affinity antagonist of fibrinogen binding to platelets. Proc Natl Acad Sci USA 83:5708, 1986

Savoca R, Schwab C, and Bosshard HR. Epitope mapping employing immobilized synthetic peptides. How specific is the reactivity of these peptides with antiserum raised against the parent protein? J Immunol Methods 141:245, 1991

Schnorrenberg G and Gerhardt H. Fully automatic simultaneous multiple peptide synthesis in micro-molar scale - rapid synthesis of series of peptides for screening in biological assays. Tetrahedron 45:7759, 1989

Sieber P and Riniker B. Protection of histidine in peptide synthesis: A reassessment of the trityl group. Tetrahed Lett 28:6031, 1987

Van 't Hof W, Driedijk PC, Van den Berg M et al. Epitope mapping of the Dermatophagoides pteronyssinus house dust mite major allergen Der p II using overlapping synthetic peptides. Molec Immunol 28:1225, 1991

Van 't Hof W, Van Milligen FJ, Van den Berg M et al. Epitope mapping of the cat (Felis domesticus) major allergen Fel d I using overlapping synthetic peptides and monoclonal antibodies against native and denatured Fel d I. Allergy 48:255, 1993a

Van 't Hof W, Van den Berg M, and Aalberse RC. The use of T-bag synthesis with paper discs as the solid phase in epitope mapping studies. J Immunol Methods 161:177, 1993b

Van 't Hof W, Van den Berg M, and Aalberse RC. How to demonstrate specificity of anti-body binding to synthetic peptides? Letter to the editors. J Immunol Methods 161:273, 1993c

Van Milligen FJ, Van Swieten P, and Aalberse RC. Structure of the major cat allergen Fel d I in different allergen sources: An immunoblotting analysis with monoclonal antibodies against denatured Fel d I and human IgE. Int Archs Allergy Immunol 99:63, 1992

Van Milligen FJ, Van 't Hof W, Van den Berg M et al. IgE epitopes of the cat (Felis domesticus) major allergen Fel d I: A study using overlapping synthetic peptides. J Allergy Clin Immunol 93:34, 1994

Chapter 4

CONSTRUCTION AND UTILIZATION OF PEPTIDE LIBRARIES DISPLAYED BY FILAMENTOUS BACTERIOPHAGE

G. Cesareni, L. Castagnoli, L. Dente, G. Iannolo, C. Vetriani, F. Felici, A. Luzzago, P. Monaci, A. Nicosia and R. Cortese

TABLE OF CONTENTS

0-8493-8967-4/95/$0.00+$.50

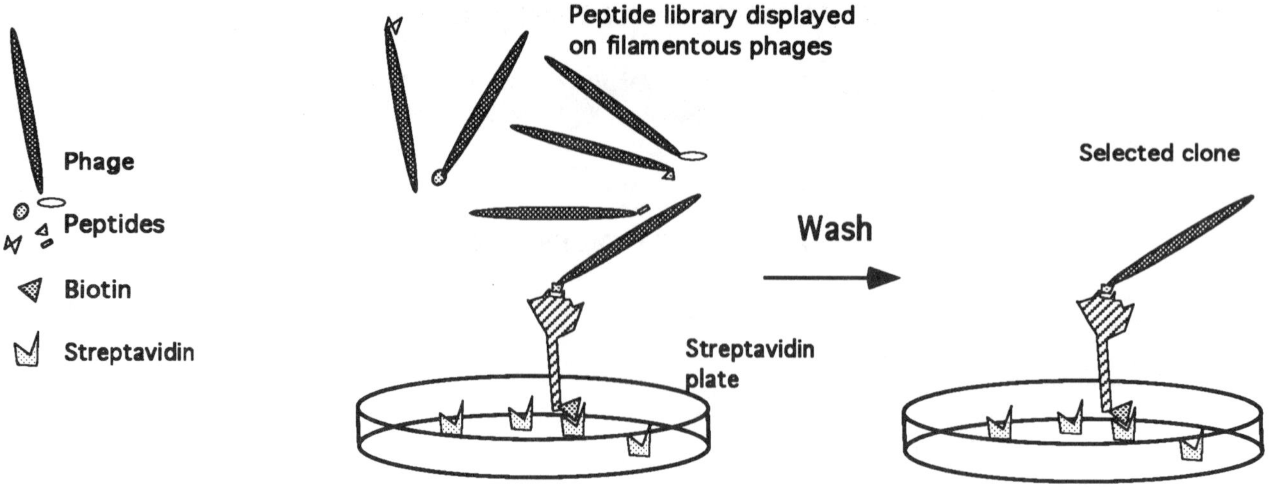

Fig. 1. Selection from an epitope library of a filamentous phage displaying a peptide that binds an Fab fragment of immunoglobulin. The Fab fragment is linked to a plastic Petri dish via a biotin-streptavidin bridge.

I. OVERVIEW

Peptide display technology permits the construction of vast collections (libraries) of peptides of random amino acid sequences and the selection of those elements that bind to a given target molecule. Since this approach may prove useful for problems in both applied and fundamental immunology, it will be described in some detail in this chapter.

The essential feature of the method, as illustrated in Fig. 1, is that each peptide of the library is synthesized *in vivo* and linked to its coding sequence. The target molecule of interest, for instance, a biological receptor, can then be used to affinity purify those peptides that stick to it and, most importantly, to select the peptide-bound coding sequences for subsequent cloning and amplification. Although various schemes for the practical realization of this idea have been thought of, and some of them accomplished, this is most efficiently achieved by genetically fusing the peptide coding sequences to the gene for a phage capsid protein (Smith, 1985; Parmley and Smith, 1988). The insertion point is judiciously chosen in order not to disturb phage assembly and to achieve the display of the fused peptide on the surface of the phage capsid available for interaction with the target. Other systems include the display of peptides directly on the surface of *Escherichia coli* (Agterberg et al., 1987; Charbit et al., 1988; Freudl, 1989; Hedegaard and Klemm, 1989; Wu et al., 1989); the fusion to a DNA binding protein, like for instance the lac repressor (Cull et al., 1992); or, at least in principle, the direct selection of nascent peptides on polysomes (Tuerk and Gold, 1990).

Smith (1985) was the first to show that a minor capsid protein (pIII) of filamentous bacteriophage (f1, fd, M13) can be effectively used for this purpose. Fusion of peptides to its amino terminus results in the production of infective viral particles that display the peptide. The peptide-tagged phage can then be affinity purified, over a large excess of wild-type phage, by soaking the phage mixture in the presence of a peptide-specific antibody linked to a solid support. The antibody-bound bacteriophages are then released and used to start a new homogeneous phage culture where each phage displays the peptide. An important characteristic of filamentous phage is its ability to survive exposure to extreme conditions (95°C, pH 2), thereby allowing recovery of the bound phages by breakage of the noncovalent bonds between displayed and target molecules without eliminating phage infectivity.

This experiment has been generalized by utilizing a variety of display methods and by constructing large collections of bacteriophages, each encoding and presenting on the virion capsid a peptide of different amino acid sequence (Scott and Smith, 1990; Cwirla et al., 1990; Devlin et al., 1990; Felici et al., 1991). This library can then be exploited as a source of peptide structures that can be challenged with practically any target molecule (ligate) to search for ligands. The selected phages are then characterized and the amino acid sequences of the displayed peptides are deduced by sequencing the DNA of the single stranded phage chromosome. The structural information extracted from the consensus derived from the selected peptides could then be used either to elucidate the binding mechanism or, from an applied point of view, to develop nonpeptide drugs of therapeutical interest.

Although this review is restricted to the discussion of small peptide libraries, it is worth mentioning that phage display technology has been successfully applied to large proteins or protein domains (for a review, see Cesareni, 1992). Perhaps the most successful achievement is the construction of phage libraries where each element displays a different immunoglobulin variable domain and the demonstration that it is possible to select from such libraries antigen-binding sites that bind practically to any antigen (Marks et al., 1991).

II. BIOLOGY OF FILAMENTOUS BACTERIOPHAGE

II.1. The growth cycle

Filamentous bacteriophages are a class of ssDNA phage that can only infect male bacteria, either F^+ or Hfr (for reviews about phage biology and morphogenesis, see Rasched and Oberer, 1986; Model and Russell, 1988; Russel 1991). When observed under the electron microscope, the appearance of an M13 bacteriophage is that of a flexible filament 6–10 nm thick and 900 nm long. The phage receptor on the bacterial surface is the tip of a thread-like structure, the sex pilus encoded by the F episome in male strains. At the beginning of the infective process, receptor recognition triggers the retraction of the pilus, the release of the DNA from the coat structure and its injection into the cytoplasm through a mechanism which at present is poorly characterized. The single stranded phage chromosome is immediately made double stranded by host enzymes and then, with the involvement of the phage encoded protein pII, replicated several times to build a pool of double stranded replicative form (RF) DNA. In the meantime, the remaining phage genes are transcribed and translated. Later in the infection cycle, the product of gene V (pV) determines the switch from double stranded to single stranded DNA replication. The newly synthesized ssDNA is complexed with pV, and finally phage assembly takes place at the inner membrane where newly synthesized coat proteins and other proteins necessary for phage assembly are stored. Although the details of the assembly process are not fully understood, it is known that two noncoat proteins, the product of gene IV and gene I, are required to produce phage particles.

From the point of view of phage display technology, an important characteristic of the filamentous phage life cycle is that infection does not lead to bacterial lysis or cell death. Plaques are observed because infected cells have a longer doubling time and can be identified as a less turbid area on a confluent lawn of noninfected bacteria. If the phage chromosome carries a drug resistance marker, infected cells can be selected and propagated on antibiotic plates in the same way as any plasmid transformed cell. The infected cell goes on dividing and releasing virus, thus producing approximately 300 phage particles per bacterium per generation.

II.2. Virus architecture

When a suspension of phage particles is denatured in SDS and analyzed by standard polyacrylamide gel electrophoresis, a very prominent band can be identified in the low molecular weight range. This corresponds to the product of gene VIII. Only one other, less distinct band (the product of gene III, pIII, can be visualized in these conditions in the region corresponding to an apparent molecular weight of approximately 60 kilodaltons (Pratt et al., 1969). By growing the phage in the presence of radiolabelled amino acids, three more coat proteins can be identified: the products of genes VI, VII and IX. The abundance of the minor coat proteins (pIII, pVI, pVII, and pIX) is estimated in approximately five copies/phage particle, while pVIII is represented approximately 2700 times.

X-ray diffraction and NMR experiments have led to the conclusion that most of the pVIII peptide is a single, slightly distorted, α-helix with an amino terminal extension of five disordered residues (Marvin, 1990; Glucksman et al., 1992; Opella et al., 1987). The 2700 subunits form a helical tube surrounding a core of single stranded circular DNA.

The ends of the viral filament are capped by two structures formed by the four minor coat proteins the products of gene VII and IX, at the side that emerges first from the infected bacterium, and VI and III, at the opposite end (Lopez and Webster 1983). Although no direct information is available about the three-dimensional structures of these terminal complexes, the observation of amino acid sequence similarities between the minor proteins and pVIII has led to a tentative structural model of the virus termini (Makowski, 1992) (see Fig. 2).

III. SELECTION OF SITES FOR PEPTIDE INSERTION

If the three-dimensional structure of filamentous bacteriophage was known at high resolution, sites tolerant to peptide display could be identified by simple inspection of the phage model. However, sufficient details are only known in the case of the major coat protein, and we do not have a molecular model for the less abundant minor coat proteins. In this latter case the missing structural information can be supplemented by biochemical, immunological or genetic data.

III.1. pIII

pIII is the largest coat protein, 430 amino acids long as deduced from the DNA sequence of the gene, and plays an important role both structurally and functionally. pIII is synthesized with an amino terminal signal peptide that is cleaved off by signal peptidase after translocation through the membrane. Eventually, it ends up in the inner membrane, anchored by a single membrane spanning domain, with the amino terminus facing the periplasmic space. Subsequently, the protein is transferred from the membrane to the tip of the virion that leaves the cell last. Subtilisin treatment splits the protein into two domains of similar size (Amstrong et al., 1981); the amino terminal domain mediates attachment to the bacterial receptor, the F pilus, while the carboxy terminal domain anchors the protein to the rest of the capsid and is important for phage morphogenesis and stability (Stengele et al., 1990). There is also evidence that pIII is essential for translocating the phage chromosome across the bacterial membranes during infection, possibly by forming a hydrophilic pore through the membrane (Wuttke et al., 1989).

Smith (1985) was the first to show that pIII could be used to display peptides on the coat surface. He utilized a naturally occurring *Bam*HI restriction endonuclease site to insert a DNA sequence encoding a fragment of the *Eco*RI endonuclease, approximately in the middle of gene III, and to demonstrate that the peptide could be incorporated into the phage capsid in a conformation available for interaction with a peptide-specific antibody. More recently, vectors have been constructed that carry unique restriction endonuclease recognition sites in the DNA sequence encoding the amino terminus of pIII. Although no detailed study has yet been published, evidence is accumulating that peptide inserts are tolerated in this region, with minor, if any, sequence limitations, and that they do not dramatically affect phage stability or infectivity (Parmley and Smith, 1988; Scott and Smith 1990; Devlin et al., 1990; Cwirla et al., 1990).

Given the relatively large dimensions of pIII, it is possible that a better understanding of the fold of this protein would lead to the identification of other regions of the protein surface that are suitable for displaying peptides in a constrained conformation.

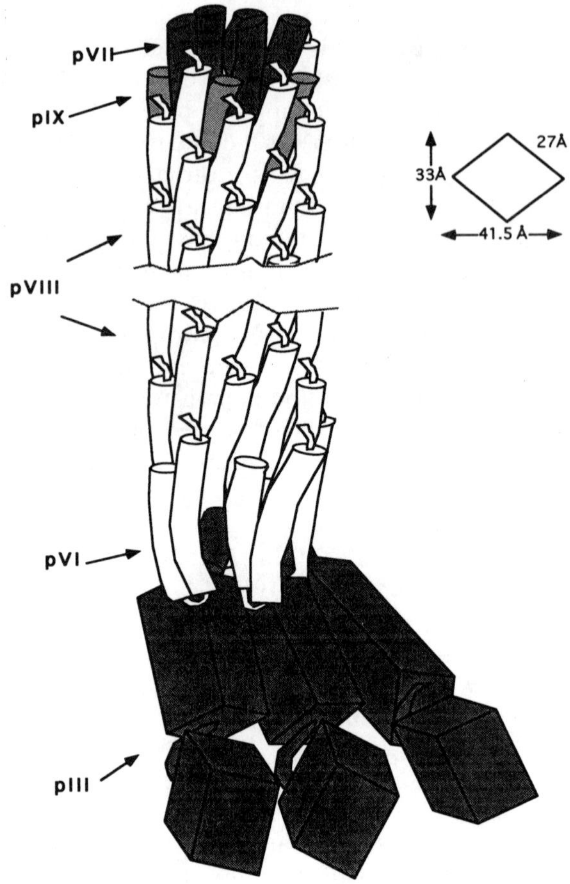

Fig. 2. Schematic representation of filamentous phage structures. This has been designed freely on the basis of experiments and models of Banner et al. (1981), Makowski (1992), Amstrong et al. (1981), Glucksman et al. (1992).

III.2. The major coat protein

Like pIII, pVIII is synthesized with an amino terminal extension cleaved by leader peptidase. Through a poorly understood assembly mechanism, 2700 molecules eventually end up forming a helical tubular array that covers the single stranded chromosome and forms a structure somewhat reminiscent of the scales of a fish. In the virion, the positively charged residues at the carboxy terminus of pVIII neutralize the negative phosphate in the DNA, while the amino terminus is exposed to the solvent.

The very simple structure of the major coat protein — single α-helix starting with a disordered pentapeptide — does not leave many options about the insertion site. The amino terminus that constitutes the major antigenic determinant of the virus is the target for insertion of peptides of various sizes.

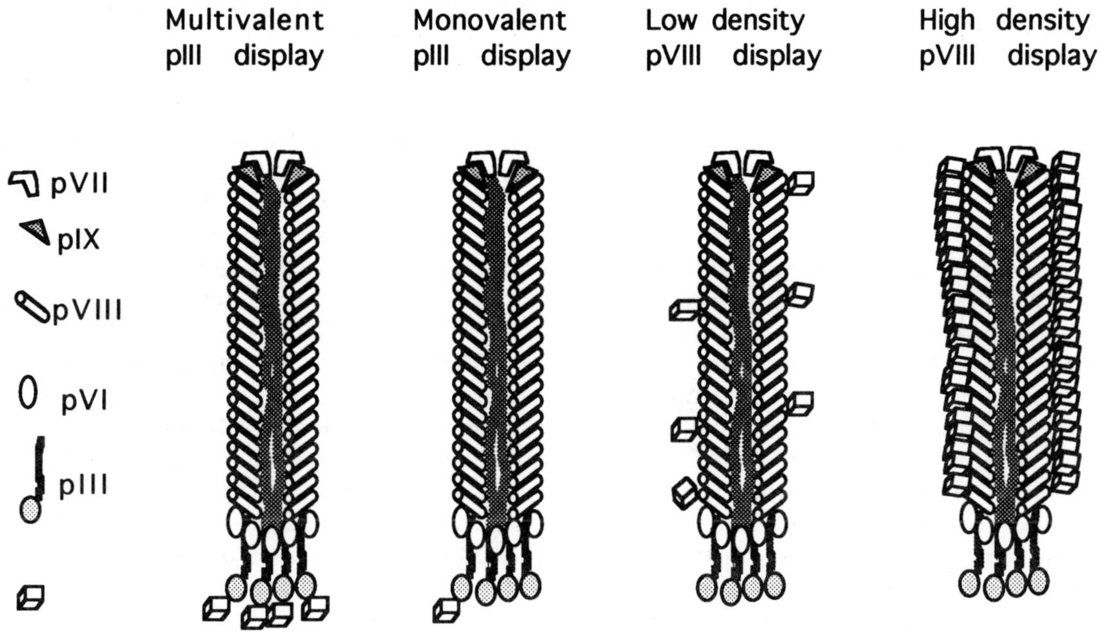

Fig. 3. The figure illustrates the hybrid phage particles that are obtained by utilizing four different strategies for peptide display. Peptides are represented as small cubes.

Different groups have shown that peptides as long as five residues are tolerated by the phage structure when inserted in different positions of the amino terminal region of pVIII (Ilíchev et al., 1989; Greenwood et al., 1991; Felici et al., 1991). Attempts to insert specific peptides longer than six amino acids have failed: the reasons for this failure are not understood. To overcome this problem that prevents the display of a large number of relatively long peptides, a phagemid system was developed (Greenwood et al., 1991; Felici et al., 1991). This vector system permits construction of a chimeric phage where some "recombinant" pVIII are assembled interspersed into otherwise wild-type capsids (see below and Fig. 3).

At least in principle, a little more sophisticated engineering could permit pVIII to be exploited to build libraries of α-helical surfaces or, possibly, to make peptide constrained libraries. The latter could be achieved by making insertions into the α-helical region where a natural loop is found in the related filamentous bacteriophage Pf1 (Nambudripad et al., 1991).

III.3. Minor coat proteins

The remaining three minor coat proteins (pVI, pVII and pIX) have not been used yet for peptide display. They are all relatively short proteins that contribute to form the structures that terminate the regular helical capsid. Although no direct information is available about their conformation, on the basis of amino acid sequence alignment with pVIII, it was recently proposed that they might also be α-helical (Makowski, 1992). The resulting model that identifies the regions to be exposed to the solvent could be used as a working hypothesis for the development of these proteins as vectors for peptide display.

IV. VECTOR CONSTRUCTION

The vectors that have been developed for the purpose of peptide display can be grouped into two major classes. To the first one belong M13 (fd) derivatives that have been engineered to introduce suitable unique restriction endonuclease recognition sites in the genetic loci that code for the protein regions that display foreign peptides, i.e., the amino termini of either pIII or pVIII. The second class of vectors includes phagemids that contain the engineered version (with unique restriction endonuclease recognition sites) of the phage genes encoding either pIII or pVIII.

IV.1. Phage vectors

Vectors belonging to this class permit the construction of hybrid phage where each copy of the relevant capsid protein displays the inserted peptide. As a consequence, pIII phage vectors will display approximately three to five copies of the foreign peptide while, in the equivalent pVIII vectors, the entire surface of the filament will be saturated by 2700 copies. It is important to recognize that in either vector two copies of the displayed peptides are either sufficiently close (approximately 30 Å in the case of pVIII) or linked to a flexible support (in the case of pIII) to be able to bind

Fig. 4. (Right) Vector maps. fUSE 5 (Scott and Smith, 1990) is a phage vector suitable for displaying peptides at the amino terminus of pIII. The product of the TnA gene confers resistance to the antibiotic tetracycline. PC89 (Felici et al., 1991) and PC107 (unpublished) are phagemids that direct the synthesis of pVIII and pIII, respectively. *bla* is the gene encoding β-lactamase. pORI is the origin of pMB1 plasmid replication.

a multivalent target molecule, for instance, an antibody, at the same time.

IV.1.a. Fusion to pIII

The pIII display system has been developed by Smith and collaborators who have recently built a versatile collection of vectors, derived from fd Tet, each containing a different restriction endonuclease recognition site in the sequence encoding the amino terminus of pIII (Parmley and Smith 1988; Scott and Smith 1990; G. Smith, personal communication). One of these is illustrated in Fig. 4. A convenient property of these vectors is that they confer resistance to tetracycline on the infected host and can thus be propagated like a plasmid, independently of phage function. Furthermore, fd Tet derivatives do not kill their host even in the absence of phage assembly, contrary to what has been observed in other strains of filamentous phage (Smith, 1988). This feature has permitted the construction and propagation, as plasmids, of phage vectors that do not synthesize an active pIII, because a restriction endonuclease site was inserted out of frame. Insertions spliced into this site, however, can restore the correct reading frame, thereby producing infective particles. In library construction (see below) this offers a convenient way of selecting clones containing productive inserts.

IV.1.b. Fusion to pVIII

Ilíchev et al. (1989) first showed that pVIII proteins carrying short amino terminal extensions can assemble into functional viral capsids. Later, we and others found that a substantial number of hybrid pVIII carrying peptides longer than six residues can be incorporated into phage only when interspersed in an otherwise wild-type capsid (Grenwood et al., 1991; Felici et al., 1991). Although, as we have recently demonstrated, this size limit is not absolute (Iannolo and Cesareni, unpublished), pVIII libraries are more conveniently assembled in phagemid vectors, since most peptide inserts would not result in productive phage in phage vectors.

IV.2. Phagemid vectors

Phagemids are plasmid vectors that carry the replication origin of filamentous bacteriophage (Dente et al., 1983; Cesareni and Murray, 1987). They normally replicate under the control of the plasmid replication origin. However, when the plasmid containing bacteria are superinfected by a wild-type bacteriophage, the phage replication origin is activated by the product of gene II. Single stranded phagemid chromosomes are then produced and incapsidated into phage particles. If the phagemid contains a phage gene that directs the synthesis of a modified form of a capsid protein, this will coassemble with the native product synthesized by the helper phage and will determine the formation of chimeric phage (Fig. 3).

This "two gene system" permits the display of peptides that interfere with the function of the carrier protein either pIII or pVIII. The ratio of hybrid to native protein can be modulated, within a certain range, by engineering the modified capsid gene, in the phagemid, under the control of an inducible promoter (Castagnoli and Cesareni, unpublished).

IV.2.a. Fusion to pIII

Since peptides of 6–20 amino acids do not normally abolish pIII function and phage infectivity, pIII phagemid display vectors have been exploited mostly for the purpose of presenting relatively large peptides (Bass et al., 1990; McCafferty et al., 1990). A phagemid vector suitable for this scope is illustrated in Fig. 4. Phage particles that are assembled after superinfection by a wild-type phage of a strain harboring a pIII phagemid, on average, display less than one hybrid protein per phage. These phages are monovalent with respect to peptide presentation, and since they eliminate avidity effects they could be used for experiments of affinity maturation (Lowman et al., 1991).

Phagemids, and phage vectors that like fd Tet can be propagated as plasmids, can better maintain library representativity by eliminating the growth disadvantage of inserts that impair phage infectivity to different extents during amplification.

IV.2.b. Fusion to pVIII

pVIII phagemid vectors have been commonly used since it was discovered that some inserts larger than six residues severely hinder pVIII from forming an infective phage capsid. On the other hand, most peptides can be incorporated into chimeric phage mixed with a larger proportion of wild-type coat protein. Felici et al. (1991) have used plasmid pC89 (Fig. 4) to construct a large library of nonapeptides inserted near the amino terminus of pVIII. Most of the hybrid pVIII genes that form the library direct the synthesis of a chimeric pVIII protein that can be incorporated into phage particles by superinfection with a helper phage. However, the proportion of hybrid pVIII in chimeric phage varies from clone to clone, ranging from less than 5% up to 30%, depending on the sequence of the peptide

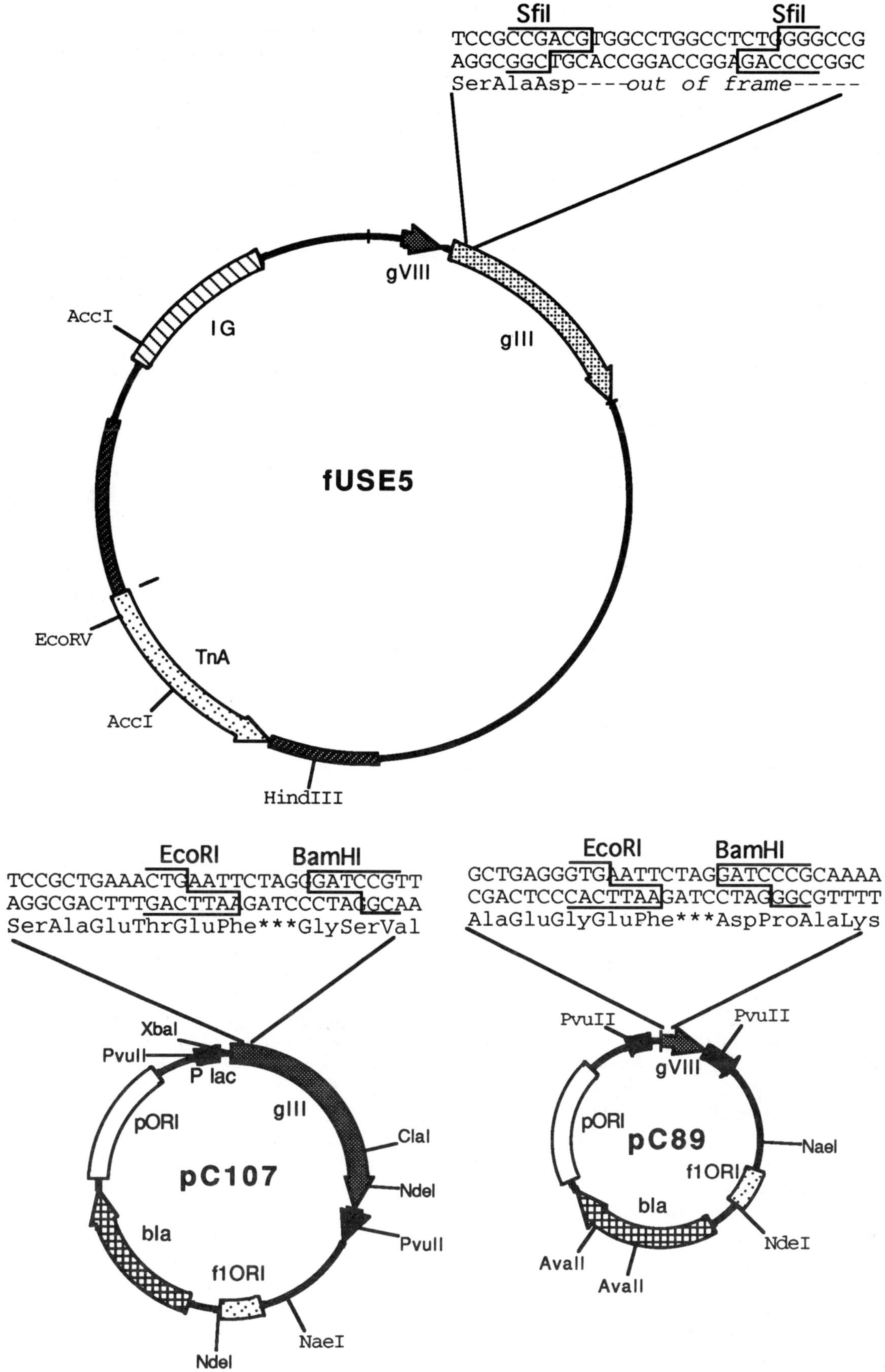
SfiI
SfiI
TCCGCCGACGTGGCCTGGCCTCTGGGGCCG
AGGCGGCTGCACCGGACCGGAGACCCCGGC
SerAlaAsp----out of frame-----
gVIII
AccI
IG
gIII
fUSE5
EcoRV
TnA
AccI
HindIII
EcoRI
BamHI
TCCGCTGAAACTGAATTCTAGGGATCCGTT
AGGCGACTTTGACTTAAGATCCCTAGGCAA
SerAlaGluThrGluPhe***GlySerVal
XbaI
PvuII
P lac
pORI
gIII
ClaI
pC107
NdeI
PvuII
bla
f1ORI
NdeI
NaeI
EcoRI
BamHI
GCTGAGGGTGAATTCTAGGATCCCGCAAAA
CGACTCCCACTTAAGATCCTAGGGCGTTTT
AlaGluGlyGluPhe***AspProAlaLys
PvuII
PvuII
gVIII
pORI
pC89
NaeI
f1ORI
bla
NdeI
AvaII
AvaII

that needs to be displayed (Cesareni, unpublished observations).

V. PEPTIDE LIBRARIES

A peptide library is made up of a collection of phages, each displaying a different peptide fused to one of the phage coat proteins (for a review see Smith and Scott, 1993). For storage purposes, however, libraries originally containing 10^7–10^8 independent clones are normally amplified by growing them for a limited number of generations. Thus, peptide libraries consist of a phage suspension (10^{11}–10^{14} phages/ml) containing several copies of each original clone. Since clones displaying different peptides vary in their growth performance, they will multiply at different rates during library amplification, and their abundance in the final cell suspension may differ from what is predicted by straightforward statistical considerations.

V.1. Library construction

The diversity of the various clones in a peptide library is obtained by cloning, into suitable vectors, degenerate oligonucleotides synthesized by incorporating mixtures of the nucleotide precursors at selected positions. Upon transformation, this oligonucleotide library will be translated into an oligopeptide library where each amino acid is represented proportionally to the number of triplets that encode the specific amino acid in the genetic code. For instance, an amino acid like methionine (1 codon) will be found six times less frequently than serine (6 codons). This bias can be somewhat diminished by incorporating only the nucleotides G and T (or C) in the third codon positions of the degenerate oligonucleotide. By doing so, one further eliminates the *ochre* and the *opal* triplets, which carry the extra bonus of decreasing the fraction of inserts that encode nonsense sequences. The strategy used to obtain a double stranded DNA fragment containing a degenerate sequence flanked by appropriate restriction endonuclease recognition sites may vary in different experiments. Two of these are illustrated in Fig. 5. The digested oligonucleotide is then ligated to plasmid DNA and transformed by electroporation into competent bacteria. Since the vector DNA is prepared by digestion with appropriate restriction endonucleases (or by a single endonuclease that like *Sfi*I may cut in two different sites leaving noncompatible ends) and liberated from the short intervening fragment by selective precipitation, it cannot recircularize by itself and most of the transformed bacteria contain a plasmid circularized through the insertion of one of the oligonucleotides of degenerate sequence. In a standard experiment one should expect 10^6–10^7 transformants per microgram per gram of plasmid DNA. Thus, several electroporations are necessary to obtain a library of 10^7–10^8 independent clones.

V.2. Evaluation of genetic diversity

In theory, as stated above, each residue should be represented in the peptide library at a frequency proportional to the number of the corresponding codons in the genetic code. In practice, the library could be biased if conditions ensuring that the four nucleotide precursors are incorporated at the same frequency at each degenerate position are not used during oligonucleotide synthesis. The genetic diversity of a library is further affected by the different growth proficiency of the various elements during the amplification process. It is then important to be able to rapidly assess if a library to be used is not grossly biased, since this would decrease its effective size and as a consequence the probability of a specific clone being isolated during screening. For instance, DNA extracted from the library could be sequenced to ensure that each nucleotide is represented equally in the degenerate positions. Alternatively, as proposed by Cwirla et al. (1990), a DNA sequence could be carried out on a number of different clones, and the occurrence of the various amino acids in the translated peptides could be compared with the frequency expected on the basis of the genetic code. Christian et al. (1992) have proposed an alternative method based on a colony hybridization experiment that use probes synthesized *ad hoc* to test diversity.

VI. SELECTION AND SCREENING METHODS

The power of phage display technology is that of allowing one to look, at the same time in a small volume, at a number of peptides so large (10^8) that one would never conceive of screening as isolated clones. The selection method is such that often several rounds of affinity purification are needed to eliminate most background clones. We should realize, however, that during each growth cycle which follows affinity selection different clones are amplified at different rates. A growth selection is then superimposed on an affinity selection. As soon as sufficient rounds of affinity selection (usually one is sufficient) have been performed to raise the proportion of binding clones to a screenable value, say 1 in 1000, it is advisable to stop further selection and to use a screening method to identify the positive clones. In this way, some good binders, but poor growers, are not lost, and the collection of selected clones is a better representation of the binders in the initial library.

VI.1. Affinity selection (biopanning)

The selection schemes that are used in most experiments are minor variations of the protocol that was originally devised by Parmley and Smith (1988). A small volume of the ligate molecule is first reacted to a solution containing a suspension of phage displaying different peptides and then linked via a biotin-strepta-

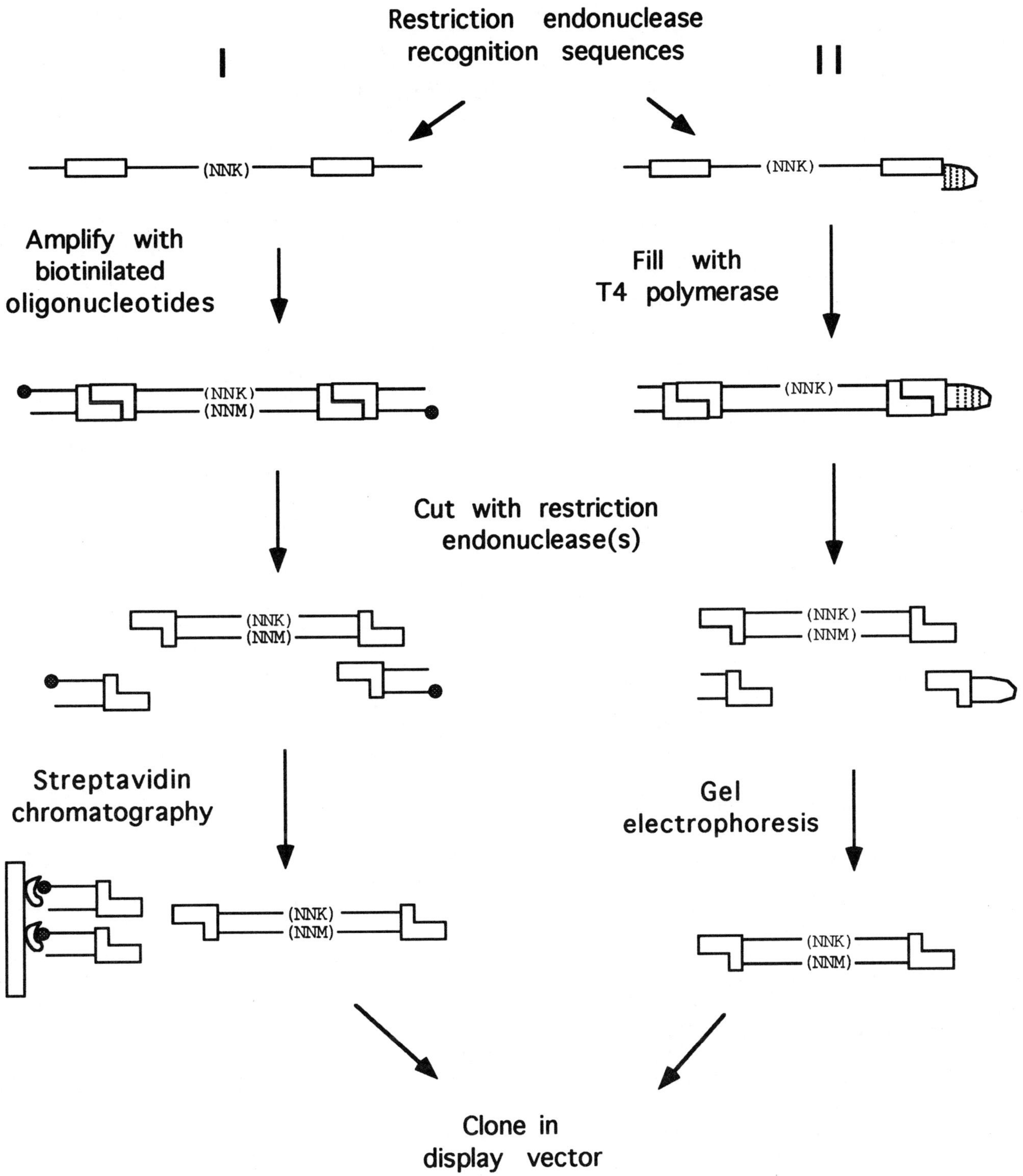

Fig. 5. Strategies for library construction. Both schemes start from an oligonucleotide containing a degenerate sequence flanked by restriction endonuclease recognition sites. N represents any nucleotide while K is either G or T. The strategy on the left (Scott and Smith, 1990) utilizes biotinilated oligonucleotides to obtain large quantities of double stranded DNA by polymerase amplification. Streptavidin chromatography is then used to remove the fragment termini or the unreacted oligos after incubation with restriction endonucleases. The strategy on the right uses a self-priming oligonucleotide which should increase the efficiency of both the annealing and priming reaction (Christian et al., 1992).

vidin bridge to a solid support: a polystyrene Petri dish, polystyrene beads, acrylamide beads or Sepharose beads (see Sections VIII.1 and VIII.2). After several washes, the bound phages are eluted at low pH, still in an infective form. Such an experiment usually permits clones that bind a ligate with affinity in the micromolar range to be enriched 10^3-fold in comparison to phage displaying random peptides. The eluted phage particles are then used to transduce an associated drug resistance marker to a host bacterium and are replicated with or without helper phage (depending on the vector that is utilized) to produce a phage suspension that can either be used for another run of biopanning or screened directly for its capacity to bind the ligate of interest.

VI.2. Colony or plaque immunological screening

After one or two selection rounds the eluted clones can be tested for their ability to specifically bind the ligate molecule by ELISA assay or by immunoscreening (see Sections VIII.3, VIII.4 and VIII.5). The clones that have survived panning are spread at high density over a plate to form either colonies or plaques on a lawn of sensitive bacteria, transferred to a nitrocellulose filter and tested for their ability to bind the target molecule. We find this immunological screening to be more sensitive and reliable when clones displaying peptides fused to pVIII are analyzed as plaque forming units. However, since pVIII libraries are only available in phagemid vectors, in order to obtain plaques, clones are plated as phagemid containning bacteria infected by a helper phage. On a lawn of a sensitive strain these bacteria will generate plaques that contain a mixture of chimeric phagemid and phage displaying the hybrid peptide encoded by the gene VIII on the phagemid.

VI.3. Affinity maturation

A number of groups have demonstrated that, by following the procedure outlined above, it is possible to isolate phage displaying peptides that bind the target macromolecule with affinities in the micromolar range. Although peptides with higher affinities may be present in the library, they have no selective advantage in the biopanning procedure. This may be due to their multivalent character (in the case, for instance, of an IgG) or to the high concentration of monovalent (for instance Fab) target molecules. In these conditions even low affinity binders are not discriminated against because of avidity effects induced by simultaneous binding to multiple targets.

Barrett et al. (1992) have shown that by controlling these factors, essentially by decreasing the concentration of a monovalent target, it is possible to discriminate between phage displaying peptides with affinities ranging from 1 μM to 10 nM. Following an alternative approach one may use a monovalent display vector. For instance, it is possible to control the promoter of a pIII-phagemid vector so that the secreted phage particles contain less then one peptide/phage. These particles can only bind one target molecule independently from its valency and concentration and do not show avidity effects.

One strategy that can be adopted to identify high affinity binders may be the following. Starting from a library in a multivalent vector, for instance a pVIII-phagemid, one should select, at low stringency, a collection of clones, as large as possible irrespective of what their affinities are, for their ability to stick to the target molecule of interest. The nucleotide sequences encoding the selected peptides are then amplified in moderately mutagenic conditions to explore new variations of the amino acid sequences that are known to bind and subsequently cloned into monovalent display vectors. The following selection round is carried out in stringent conditions to discriminate between high affinity binders and low affinity binders.

VII. APPLICATIONS

VII.1. Epitope mapping

As soon as peptide libraries became available they were searched with monoclonal antibodies raised against linear epitopes. In most experiments it was possible to recover clones, the majority of which displayed peptides identical or closely resembling the original epitope (Scott and Smith, 1990; Cwirla et al., 1990; Felici et al., 1991). This result indicates that epitope mapping through biopanning of phage peptide libraries is a viable alternative to other techniques. This was first demonstrated by Stephen and Lane (1992) who utilized this approach to identify a previously unmapped epitope that was unmasked by a conformational change induced by a point mutation in the protein p53. However, clones whose sequence is unrelated to that of the original epitope are sometimes affinity selected. Using a hexapeptide library, Scott and Smith (1990) isolated, among other related clones, a phage displaying the peptide CRFVWC which bears no resemblance to the epitope (DFLEKI) that elicited production of the anti-myoemerythrin monoclonal antibody used in affinity selection. Similarly, the phage isolated by Christian et al. (1992), utilizing an anti-gp120 monoclonal antibody, could be grouped into two classes. Neither of the hexamer motifs characterizing the two classes is related to the amino acid sequence of the original gp120 epitope.

The majority of anti-protein antibodies do not recognize linear epitopes and rather bind residues that are distant in the primary sequence and become spatially close as a consequence of the folding process. These epitopes are known as conformational or assembled epitopes. Can an antibody that recognizes such an assembled epitope find high affinity "mimotopes" among the linear peptides in the library? Smith and Scott (quoted in Smith 1991) were able to find low affinity mimotopes only in 50% of the attempts. Felici et al. (1993) searched a nonapeptide library displayed on pVIII with a monoclonal antibody raised against a conformational epitope of the *Bordetella pertussis* toxin. They were able to find mimotopes whose binding to the monoclonal antibody could be competed by the natural antigen, indicating that the selected clones were binding to the antigen-binding site.

The low success rate and the low affinity of the selected peptides could be a consequence of the displayed peptides being flexible and free to take many conformations. Although this can be considered an advantage for some applications, as it amplifies the effective size of the libraries, it nonetheless decreases the affinity of any given ligand for its ligate. Initial attempts to construct constrained peptide libraries focused on designing libraries where the peptides were forced to take a circular conformation by disulfide bond formation (O'Neil et al., 1992). Luzzago et al. (1993) constructed and utilized such a library to select mimotopes of a conformational epitope of the protein ferritin. From the consensus sequence of the selected peptides, they were able to identify in the three-dimensional structure of ferritin the residues that bind the antigen-binding site.

VII.2. Searching the libraries for ligands that mimic hormones, enzymes inhibitors or non-proteinaceous substances

It is conceivable that by screening larger and structurally more sophisticated peptide libraries, in the near future it will be possible to find elements that mimic practically any conformational epitope. It is unlikely, however, that the inspection of the amino acid sequences of the displayed peptides would allow one to identify the residues that, in the original antigen, are implicated in the interaction with the antibody. This is especially true in those cases where the three-dimensional structure of the antigen is not known.

These observations are particularly relevant to attempts to widen the applications of this technology and to extend it to the search for ligands for proteins other than antibodies (receptors, enzymes, ...). Although rumor is spreading that this has turned out to be slightly more complicated than just selecting a linear epitope, some encouraging reports are starting to come in. The first to isolate peptide ligands for a protein that is not an antibody was Devlin et al. (1990) who selected peptides that compete with biotin for binding to streptavidin. Phages bearing the consensus peptide YPY were selected by two different groups from a hexapeptide and an octapeptide library using the plant lectin concanavalin A (con A) as a ligate (Scott et al., 1992; Oldenburg et al., 1992). This tripeptide (YPY) was found to be highly selective for con A, since phage presenting these peptides do not bind closely related pea or lentil lectins.

An indication that constrained libraries might be the right place to look for high affinity binders comes from the work of O'Neil et al. (1992) who were able to find peptides that have affinities in the nanomolar range for GPIIb/IIIa in a library of hexapeptides constrained by disulfide bridges.

More recently, Smith et al. (1993) have selected, from a linear hexapeptide library, clones displaying peptides that bind the S protein (a fragment of bovine pancreatic ribonuclease) with affinities in the micromolar range. The selected peptides show a common sequence motif, and one of them, YNFEVL, synthesized chemically, was shown to behave as an antagonist of the S peptide in that it prevents it from restoring enzyme activity to S protein. One of the clearest demonstrations of the versatility of the phage library approach is given by the work of Sinigaglia and collaborators who successfully applied the technology to study the effect of human MHC class II polymorphism on peptide-MHC interaction (see Chapter 6 for details) (Hammer et al., 1992, 1993).

VII.3. Immunization

Soon after filamentous phage had been developed as carriers for immunogenic peptides, it was realized that these hybrid structures could be used for vaccine development. Compared to the conventional technique based on chemical synthesis of peptides, they present the following advantages:

1.) Antigen purification is relatively simple, and standard PEG precipitation of the secreted phage may be sufficient for several uses.

2.) The chemical purity of the molecules used as immunogens is guaranteed by the high fidelity of the translation machinery *in vivo*. This characteristic contrasts the 70–90% purity that can be obtained by standard solid phase synthesis of peptides.

3.) The phage constitutes an easily renewable and inexpensive source of antigen.

4.) Conjugating the antigen to a carrier molecule is not necessary since the antigen is already synthesized as a fusion protein.

Antigens displayed by pIII are good immunogens (de la Cruz et al., 1988). However, recent results indicate that the multiple display by fusion to pVIII is far better (Willis et al., 1993; Felici et al., 1993; Minenkova et al., 1993). Willis et al. (1993) have shown that two major antigenic determinants of the circumsporozoite protein of the malaria parasite, *Plasmodium falciparum*, are strongly immunogenic in different strains of mice in the absence of adjuvants and that the elicited antibodies are highly specific for the individual epitopes. The immune response was found to be T cell dependent and to undergo class switching from IgM to IgG.

In conclusion, the ease of preparation and purification of antigens, the strength and specificity of the response and the fact that no external adjuvants are needed makes immunization by epitopes presented by phage a simpler and cheaper way to obtain an anti-peptide response than the conventional method based on chemical synthesis and coupling to a carrier antigen.

VII.4. Perspectives

The success rate in experiments that use monoclonal antibodies to search peptide libraries suggests that the libraries already available are complete, or almost complete, as they contain elements able to bind to most immunoglobulins. The availability of larger and more sophisticated libraries is likely to further raise the success rate.

A desirable extension of the technology is the ability to utilize a complex antibody mixture to search an epitope library. This allows one to fingerprint the antibody repertoire of an individual by selecting a collection of library clones that bind to the different serum components. Although this approach presents some technical complications, preliminary results indicate that, at least in some cases, this is feasible (Felici, Monaci, Nicosia, Luzzago, and Cortese, unpublished).

If a significant proportion of the immunoglobulins in the serum is the result of the response to the challenge by an infective agent, some of the epitopes selected by the serum are mimotopes of the molecular surface of the infective agent and could be used, at least in principle, as a vaccine to induce a protective response in noninfected individuals. The success and the generality of this strategy, however, depends on an unanswered question. If one uses an immunoglobulin against a specific infective agent to select "peptide clones", how many of these are able to elicit an immune response that recognizes the original antigen and protects from infection? Felici et al. (1993) addressed this question using clones selected for their ability to bind a monoclonal antibody that neutralizes the *Bordetella pertussis* toxin. Although in that case the results were rather discouraging, recent experiments with different antibodies and polyclonal sera suggest that the difficulties are not insurmountable and that the main limitation resides in the ability to rapidly test the immune response against a large number of clones or clone mixtures (Felici, Monaci, Nicosia, Luzzago, and Cortese, unpublished observations).

VIII. METHODOLOGY

VIII.1. Selection of clones from oligopeptide libraries in pVIII

(Adapted from Parmley and Smith, 1988.)

Streptavidin-Coated Plates

- Pipette 4 ml of filter-sterilized coating buffer (see Section VIII.6) into a polystyrene plate (60 x 15 mm, Falcon 1007).
- Tilt the plate to wet the entire surface.
- Add 40 μl of streptavidin solution (1 mg/ml in coating buffer).
- Mix and leave overnight at 4°C in a humidified plastic box.

First Antibody Reaction

- Pipette 10^{10} – 10^{11} TU (Ampicillin transducing units) of the pVIII library in 1x TBS/NaN^3 into a 1.5 ml Eppendorf tube.
- Add the primary antibody (preadsorbed overnight to 4×10^{11} PFU of UV-killed* M13-KO7/μg of Ab; the preadsorption step is not necessary if you use a purified monoclonal antibody) to a final concentration of 1 μM for the first round of selection and 0.1 nM for the subsequent steps.
- Allow to react overnight at 4°C.

* Irradiate 50 μl drops of 10^{13}/ml phage suspension with Stratalinker (Stratagene; 2×10^5 joules, 8 cycles, auto-crosslink mode).

Biopanning

- Add to the reacted library the biotinylated secondary antibody (preadsorbed overnight with 4×10^{11} PFU of UV-killed M13-KO7/μg of Ab) to a final concentration of 250 μM for the first round of selection and 25 pM for the subsequent steps.
- Allow to react 3 hours at room temperature.
- Discard the streptavidin solution from the plates.
- Fill each plate with 10 ml of Biopanning Blocking Solution. Leave 1 hour at room temperature.
- Wash three times with TBS/Tween.
- Add 1 ml TBS/Tween to the reacted library and pipette onto the streptavidin-coated plate; rock at room temperature for 10 min.
- Pour and shake off the panned library and wash the plate 10 times with 10 ml TBS/Tween over a 60–90 min period.
- Pipette 800 μl of elution buffer into the plate, rock for 30–60 min.

- Remove the eluate, add to 1.5 ml Eppendorf tubes containing 48 μl of 2 M Tris Base (unadjusted pH) and titrate the TU.

Titration of TUs

- Prepare XL-1 blue TB cells.*
- Add 10-fold serial dilutions of eluted library to 0.1 ml aliquots of cell suspension.
- Leave 30 min at 37°C.
- Spread on 100 μg/ml ampicillin plates + Xgal+IPTG.

* Grow XL-1 blue cells from minimal + Tet (20 μg/ml) plate in Terrific broth + Tet (20 μg/ml) at 37°C with vigorous shaking up to an O.D.$_{600}$ between 0.125–0.250, measured on a 1/10 dilution. Incubate for a further 15 min at 37°C without shaking and then use within 1 hour.

VIII.2. Phage enrichment by antibody-coated polystyrene beads

Beads coating

- Incubate polystyrene beads (Ø 6.4mm; Precision Plastic Ball Co., Chicago, IL) with 10 μg of antibody per milliliter in coating buffer. Add enough solution to cover beads (for example coat 25 beads with ≈9 ml per glass bottle). Incubate for 12–18 hours at 4°C with rotation.
- Next day wash beads three times with PBS, one time with H_2O (10 min each time). At this stage you can store beads frozen at -20°C.
- Before use, block beads by incubating with TBS/Tween + 10 mg/ml BSA for 4 hours or more at 4°C. Wash briefly two times with TBS/Tween.

Phage adsorption

- Mix together in a 5 ml polypropylene Falcon tube:
  - 12 Ab-coated beads.
    A 10-fold excess over library phage of carrier
  - UV-killed phage particles (M13-K07).
  - ≈3.4 ml of TBS/Tween + 1 mg/ml BSA to fill the tube.
- Incubate by rotating at 4°C for at least 4 hours.
- Add the phage library (10^{11} TU) and leave rotating 12–18 hours at 4°C.

Phage elution

- Collect and store supernatant.
- Wash beads 10 times with TBS/Tween rotating at 4°C, 10 min each time.
- Transfer beads to a glass tube. Add 2.4 ml of elution buffer. Incubate with strong agitation 5 min at RT.
- Transfer eluate to a polypropylene tube and neutralize it by adding 240 μl 2M TrisHCl pH 9.0.
- Titrate input phage, nonadsorbed phage and adsorbed phage as in Section VIII.1.

VIII.3. Plaque immunoscreening

Phage amplification

- Infect ≈3ml of XL-1 blue TB cells with the enriched phage. Incubate 15 min without agitation at 37°C and a further 30 min with vigorous shaking at 37°C.
- Plate infected cells onto LB + Amp (100 μg/ml) + 1% glucose (usually 10^5–10^6 cells per 230x230 mm plate). Incubate overnight at 37°C.
- Next day, scrape cells with ≈50ml of LB + Amp (100 μg/ml) + Tet (20 μg/ml) + 10% glycerol. Store in aliquots at -20°C.

Phage plating

- Inoculate a culture from the scraped cells in LB + Amp (100 μg/ml) + Tet (20 μg/ml) at a starting O.D. = 0.05. Grow at 37°C, with vigorous shaking till O.D. ≈ 0.240. Leave a further 15 min without agitation.
- Infect 1 ml of this culture with M13-K07 helper phage. Incubate at 37°C for 15 min without agitation, and a further 45 min with vigorous shaking.
- Spin 5 min in microfuge. Discard supernatant and resuspend bacterial pellet in 1 ml of LB. Repeat three times, to remove the excess M13-K07.
- Mix:
  - $5x10^4$–$1x10^5$ infected bacteria.
  - 0.8 ml of XL1-blue TB cells.
  - 15 ml of melted top agar (48°C)). It is important to plate onto well dried 15 cm Ø LB plates.

  To measure the titre and the percentage of blue clones (see Felici et al., 1991), plate a 10^{-2}–10^{-3} dilution on XL-1 blue lawn on LB + IPTG + Xgal.
- Incubate plates for a couple of hours at 37°C, then lay a nitrocellulose filter onto the plate. Incubate at 37°C overnight.

Blocking

- Store plates with filters 30–60 min at 4°C, then mark the filter/plate correspondence with a needle and peel off the filters.

- Wash the filters in I-buffer 4 times for 30 min at room temperature with gentle agitation.

First Antibody
- Incubate filters with an appropriate MAb dilution (for example, 1 μg/ml) in I-buffer (typically use 20 ml of incubation mix/filter) and leave it overnight at 4°C with gentle agitation.

Washing
- Wash the filters 10 times for 10 min in PBS/NP40 at 4°C with agitation.

Secondary Antibody
- Incubate the filters with the AP-conjugated anti-mouse (Fc specific) antibody, diluted 1:5000 in I-buffer (typically use 20 ml of incubation mix).
- Incubate 4 hours at 4°C with gentle agitation.

Washing
- Wash filter 10 times for 10 min in PBS/NP40 with agitation.

Development
- Wash the filters in substrate buffer. Then incubate filter with developing solution at room temperature (typically 15 ml/filter).
- Stop reaction by washing with H_2O.

VIII.4. Preparation of phage supernatants for ELISA assay

Plating
- Pick a positive plaque and transfer phage into 200 μl TBS.
- Heat 20 min at 70°C to kill bacterial cells.
- Spin 5 min in microfuge, full speed and recover supernatant.
- Infect 0.2 ml of XL-1 blue TB cells with 0.5 μl of pVIII clone phage suspension.
- Incubate 30 min at 37°C.
- Plate 5 μl of the infected mix onto LB + Amp (100 μg/ml) + 1% glucose plates. Incubate overnight at 37°C.

Phage growth
- Inoculate one colony in 2 ml of LB + Amp (100 μg/ml) + Tet (20 μg/ml).
- Grow up with vigorous shaking to $O.D._{(600nm)} \approx 0.25$.
- Leave for 15 min with very gentle agitation.
- Infect cells with helper phage M13-K07 and add IPTG (0.1 mM final concentration).
- Leave 15 min at 37°C with no agitation.
- Incubate for 5 hours at 37°C with strong agitation.

Phage recovery
- Centrifuge the bacteria/phage suspension in 2.2 ml Eppendorf tubes, 15 min, full speed at 4°C. Carefully recover supernatant.
- Spin again 15 min, full speed at 4°C.

Phage sup storage
- Carefully remove cleared supernatant. Add NaN_3 up to a 0.02% final concentration.
- Aliquot and store at 4°C.
- Titrate phage supernatant concentration as Amp transducing units.

VIII.5. ELISA assay

Coating
- Aliquot in each well 100 μl of rat anti-pIII monoclonal antibody (57D1, by P. Delmastro, unpublished data) diluted in coating buffer (1 μg/ml). Leave 3 hours at 37°C or overnight at 4°C on rocking platform.

Washing
- Discard and wash by adding 250 μl of washing buffer at RT. Repeat 10 times.

Blocking
- Add 250 μl of ELISA blocking buffer. Leave 1 hour at 37°C on a rocking platform.

Phage addition
- Discard and add a mixture of:
  cleared phage supernatant 50 μl.
  ELISA blocking buffer 50 μl.
  Leave 1 hour at 37°C on rocking platform.

Washing
- Discard and wash ten times with washing buffer at room temperature (leave 1–2 min each time).

First Antibody
- Discard washing buffer and aliquot in each well 100 μl of the appropriate MAb dilution in ELISA blocking buffer. Leave overnight at 4°C on a rocking platform.

Washing
- Discard and wash ten times with cold washing buffer at 4°C (leave 1–2 min each time).

Secondary Antibody
- Add anti-mouse Ig antibody (Fc specific) alkaline-phosphatase conjugated (use a serum

preadsorbed to rat IgG; 1:5000 dilution in ELISA blocking buffer in a total volume of 100 μl).
- Leave 4 hours at 4°C.

Washing
- Discard and wash ten times with cold washing buffer at 4°C (leave 1–2 min each time).

Development
- Wash with 200 μl of substrate solution. Add 100 μl of developing solution to each well. Incubate 30–60 min at 37°C in the dark. Read O.D. at 405 nm.

VIII.6. Solutions and Buffers

Biopanning Blocking Solution
29 mg/ml BSA (SIGMA A-3912)
3 μg/ml streptavidin in filter-sterilized 0.1 M $NaHCO_3$, 0.02% NaN_3
pH adjusted to 9.2 with NaOH

BSA stock solution
100 mg/ml BSA (Sigma A-3912)
0.22 μm filter sterilize

Coating buffer
50 mM Na_2CO_3 (pH 9.6)
0.02% NaN_3

Developing solution
66 μl NBT (50 mg/ml in dimethyl-formamide)
33 μl BCIP (50 mg/ml in 70% dimethyl-formamide) dissolved in 10 ml of substrate buffer

ELISA blocking buffer
PBS
0.05% Tween 20
5% nonfat dry milk

ELISA washing buffer
PBS
0.05% Tween 20

Elution buffer
0.1N HCl
1 mg/ml BSA
pH adjusted to 2.2 with glycine

I-buffer
PBS
0.1% NP40
5% nonfat dry milk

PBS/NP40
PBS
0.1% NP40

Substrate buffer
100 mM Tris-Cl pH 9.6
100 mM NaCl
5 mM $MgCl_2$
0.05% NaN3

TBS/NaN_3
50 mM TrisHCl pH 7.5
150 mM NaCl
0.02% NaN_3

TBS/Tween
TBS 1x
0.5% Tween 20

ACKNOWLEDGEMENTS
The work done in Tor Vergata was supported by an EC Bridge grant to G.C and by C.N.R (Progetto finalizzato Ingegneria Genetica).

ABBREVIATIONS

Ab	Antibody
Amp	Ampicillin
Ap	Alkaline phosphatase
BSA	Bovine serum albumin
ConA	Concanavalin A
IPTG	Isopropyl-β-D-thiogalactopyranoside
LB	Luria broth
MAb	Monoclonal antibody
O.D.	Optical density
PEG	Polyethylene glycol
PFU	Plaque forming unit
RT	Room temperature
TB	Terrific broth
TBS	Tris borate solution
Tet	Tetracycline
UV	Ultraviolet
X-gal	5-Bromo-4-chloro-3-indolyl-β-D-galactoside

IX. REFERENCES

Agterberg M, Adriaanse H and Tommassen J. Use of outer membrane PhoE as a carrier for the transport of a foreign antigenic determinant to the cell surface of Escherichia coli K12. Gene 58:145, 1987

Amstrong J, Perham RN and Walker RN. Domain structure of bacteriophage fd adsorption protein. FEBS Letters 135:167, 1981

Banner DW, Nave C and Marvin DA. Structure of the protein and DNA in fd filamentous bacteriophages. Nature 289:814, 1981

Barret RW, Cwirla SE, Ackerman MS et al. Selective enrichment and characterization of high affinity ligands from collections of random peptides on filamentous phage. Anal Biochem 204:357, 1992

Bass S, Greene R and Wells J. Hormone phage: An enrichment method for variant proteins with altered binding properties. Proteins 8:309, 1990

Cesareni G and Murray JAH. Plasmid vectors carrying the replication origin of filamentous single stranded phages. Genetic Engineering: Principles and Methods 8:135, 1987

Cesareni G. Peptide display on filamentous phage capsids; a new powerful tool to study protein-ligand interaction. FEBS Letters 307:66, 1992

Charbit A, Molla A, Saurin A et al. Versatility of a vector for expressing foreign polypeptides at the surface of gram-negative bacteria. Gene 70:181, 1988

Christian RB, Zuckermann RN, Kerr JM et al. Simplified methods for construction, assessment and rapid screening of peptide libraries in bacteriophage. J Mol Biol 227:711, 1992

Cull MG, Miller JF and Schatz PJ. Screening for receptor ligands using large libraries of peptides linked to the C terminus of the lac repressor. Proc Natl Acad Sci USA 89:1865, 1992

Cwirla SE, Peters EA, Barret RW et al. Peptides on phage: A vast library of peptides for identifying ligands. Proc Natl Acad Sci USA 87:6378, 1990

De la Cruz VF, Lal AA and McCutchan TF. Immunogenicity and epitope mapping of foreign sequences via genetically engineered filamentous phage. J Biol Chem 263:4318, 1988

Dente L, Cesareni G and Cortese R. pEMBL: A new family of single stranded plasmids. Nucl Acids Res 11:1645, 1983

Devlin JJ, Panganiban LC and Devlin PE. Random peptide libraries: A source of specific protein binding molecules. Science 249:404, 1990

Felici F, Castagnoli L, Musacchio A et al. Selection of antibody ligands from a large library of oligopeptides expressed on a multivalent exposition vector. J Mol Biol 222:301, 1991

Felici F, Luzzago A, Folgori A et al. Mimicking of discontinuous epitopes by phage-displayed peptides. II. Selection of clones recognized by a protective monoclonal antibody against the Bordetella pertussis toxin from phage peptide libraries. Gene 128:21, 1993

Freudl R. Insertion of peptides into cell-surface exposed area of the Escherichia coli OmpA protein does not interfere with export and membrane assembly. Gene 82:229, 1989

Glucksman MJ, Bhattacharjee S and Makowski L. Three-dimensional structure of a cloning vector. X-ray diffraction studies of filamentous bacteriophage M13 at 7 Å resolution. J Mol Biol 226:455, 1992

Greenwood J, Willis AE and Perham RN. Multiple display of foreign peptides on a filamentous bacteriophage. Peptides from Plasmodium falciparum circumsporozoite protein as antigens. J Mol Biol 220:821, 1991

Hammer J, Takacs B and Sinigaglia F. Identification of a motif for HLA-DR1 binding peptides using M13 display libraries. J Exp Med 176:1007, 1992

Hammer J, Valsasnini P, Tolba K et al. Promiscuous and allele specific anchors in HLA-DR binding peptides. Cell 74:197, 1993

Hedegaard L and Klemm P. Type 1 fimbriae of Escherichia coli as carriers of heterologous antigenic sequences. Gene 85:115, 1989

Ilíchev AA, Minenkova OO, Tat´kov SL et al. Production of a viable variant of the M13 phage with a foreign peptide inserted into the basic coat protein. Dokl Akad Nauk SSSR 307:481, 1989

Lopez J and Webster RE. Morphogenesis of filamentous bacteriophage f1: orientation of extrusion and production of polyphage. Virology 127:177, 1983

Lowman HB, Bass SH, Simpson N et al. Selecting high-affinity binding proteins by monovalent phage display. Biochemistry 30:10832, 1991

Luzzago A, Felici F, Tramontano A et al. Mimicking of discontinuous epitopes by phage-displayed peptides. I. Epitope mapping of human H ferritin using a phage library of constrained peptides. Gene 128:51, 1993

Makowski L. Terminating a macromolecular helix. Structural model for the minor proteins of bacteriophage M13. J Mol Biol 228:885, 1992

Marks JD, Hoogenboom HR, Bonnert TP et al. By-passing immunization. Human antibodies from V-gene libraries displayed on phage. J Mol Biol 222:581, 1991

Marvin DA. Model building studies of inovirus: Genetic variation on a geometric theme. Int J Biol Macromol 12:125, 1990

McCafferty J, Griffiths AD, Winter G et al. Phage antibodies: Filamentous phage displaying antibody variable domains. Nature 348:552, 1990

Minenkova OO, Ilychev AA, Kishchenko GP et al. Design of specific immunogens using filamentous phage as the carrier. Gene, 128:85, 1993

Model P and Russel M. Filamentous bacteriophage. In: Calendar R ed: The Bacteriophage, Plenum Publishing Corp, New York, p 375, 1988

Nambudripad R, Stark W and Makowski L. Neutron diffraction studies of the structure of filamentous bacteriophage Pf1. Demonstration that the coat protein consists of a pair of α-helices with an intervening, non-helical surface loop. J Mol Biol 220:359, 1991

O'Neil KT, Hoess RH, Jackson SA et al. Identification of novel peptide antagonists for GPIIb/IIIa from a conformationally constrained phage peptide library. Proteins 14:509, 1992

Oldenburg KR, Loganathan D, Goldstein IJ et al. Peptide ligands for a sugar binding protein isoated from a random peptide library. Proc Natl Acad Sci USA 89:5393, 1992

Opella SJ, Stewart PL and Valentine KG. Protein structure by solid state NMR spectroscopy. Q Rev Biophys 19:7, 1987

Parmley SF and Smith GP. Antibody-selectable filamentous fd phage vectors: Affinity purification of target genes. Gene 73:305, 1988

Pratt D, Tzagaloff H and Beaudoin J. Conditional lethal mutants of the small filamentous colipahge M13. II. Two genes for coat proteins. Virology 39:42, 1969

Rasched I and Oberer E. Ff coliphages: Structural and functional relationship. Microbiol Rev 50:401, 1986

Russel M. Filamentous phage assembly. Mol Microbiol 5:1607, 1991

Scott JK and Smith GP. Searching for peptide ligands with an epitope library. Science 249:386, 1990

Scott JK, Loganathan D, Easley RB et al. Discovery of a family of concanavalin A-binding peptides from a hexapeptide epitope library. Proc Natl Acad Sci USA 89:5398, 1992

Smith GP. Filamentous fusion phage: Novel expression vectors that display cloned antigens on the virion surface. Science 228:1315, 1985

Smith GP. Filamentous phage assembly: Morphogenetically defective mutants that do not kill the host. Virology 167:156, 1988

Smith GP. Surface presentation of protein epitopes using bacteriophage expression systems. Current Opinion in Biotechnology 2:668, 1991

Smith GP and Scott JK. Libraries of peptides and proteins displayed on filamentous phage. Methods Enzymol 217:228, 1993

Smith GP, Schultz DA and Ladbury JE. A ribonuclease S-peptide antagonist discovered with a bacteriophage display library. Gene 128:37, 1993

Stengele I, Bross P, Garces X et al. Dissection of functional domains in phage fd adsorption protein. J Mol Biol 212:143, 1990

Stephen CW and Lane DP. Mutant conformation of p53. Precise epitope mapping using a filamentous phage epitope library. J Mol Biol 225:577, 1992

Sternberg MJE and Islam SA. Local protein sequence similarity does not imply a structural relationship. Protein Engng 4:125, 1990

Tuerk C and Gold L. Systematic evolution of ligands by exponential enrichment: RNA ligands to bacteriophage T4 DNA polymerase. Science 249:505, 1990

Willis AE, Perham RN and Wraith D. Immunological properties of foreign peptides in multiple display on a filamentous bacteriophage. Gene 128:79, 1993

Wu JY, Newton S, Judd A et al. Expression of immunogenic epitopes of epatitis B surface antigen with hybrid flagellin proteins by a vaccin strain of Salmonella. Proc Natl Acad Sci USA 86:4726, 1989

Wuttke GG, Keppner J and Rasched I. Pore-forming properties of the adsorption protein of filamentous phage fd. Biochim Biophys Acta 985:239, 1989

Zacher AN, Stock CA, Golden II et al. A new filamentous phage cloning vector: fd-tet. Gene 9:127, 1980

Chapter 5

THE ANALYSIS OF PEPTIDE BINDING TO MHC AND OF MHC SPECIFICITY

Søren Buus, Lars Østergaard Pedersen and Anette Stryhn

TABLE OF CONTENTS

This chapter will focus on the interaction between peptide and major histocompatibility gene complex (MHC) molecules and on how to measure this interaction. Being able to determine and characterize peptide binding to MHC allows one to examine an important intermediary step in the generation of an antigen-specific T cell response. Indirectly, it also allows one to address events occurring before and/or after peptide binding. Thus, there has been a strong and sustained interest in generating reliable peptide-MHC binding assays. A large and steadily growing number of different methods to detect peptide binding to MHC molecules have been proposed over the years, and it would be impossible to go through all of them in detail. Unfortunately, interpretations and, in particular, comparisons between laboratories are complicated by the many different assays and by the fact that most of them are nonquantitative. We will concentrate on a few, mostly quantitative, assays in the hope that those who are about to enter this field will find one of them useful. It is not the intention to give a comprehensive review on the structure and function of the MHC; however, a brief introduction to the complicated nature of MHC restricted antigen presentation is required.

I. INTRODUCTION TO MHC RESTRICTION AND ANTIGEN PRESENTATION

I.1. Antigen recognition by B and T cells

The specific immune system has established at least two different recognition systems, B cells and T cells. Several pertinent differences between the two systems exist. B cells can recognize almost any organic substance; T cells can only recognize protein antigens. The B cell receptor interacts directly with the nominal antigen and recognizes its three-dimensional structure; in contrast, the T cell receptor does not interact with the nominal antigen directly. B cells can secrete their receptors and they can recognize soluble antigen, i.e., it is possible for the entire interaction of the B cell receptor to take place in solution. In sharp contrast,

0-8493-8967-4/95/$0.00+$.50

the T cell receptor is always membrane bound and it only recognizes antigen which is associated with MHC molecules and presented on the membrane of another cell, so-called antigen presenting cells (APC). Thus, T cell recognition occurs at the interface between two cells, a T cell and an APC. In molecular terms, it involves the creation of a trimolecular complex consisting of T cell receptor, antigen, and MHC (Fig. 1). T cells do not recognize large intact protein antigens, but rather short peptides obtained through proteolytic fragmentation ("antigen processing") of the protein antigen. In other words, the ligand of the T cell receptor consists of two components: an MHC molecule and an antigen derived peptide. It follows that an APC to generate the ligand of the T cell receptor must express MHC molecules and it must possess an active metabolic apparatus capable of performing "antigen processing" (reviewed in Unanue, 1972, 1981, 1984).

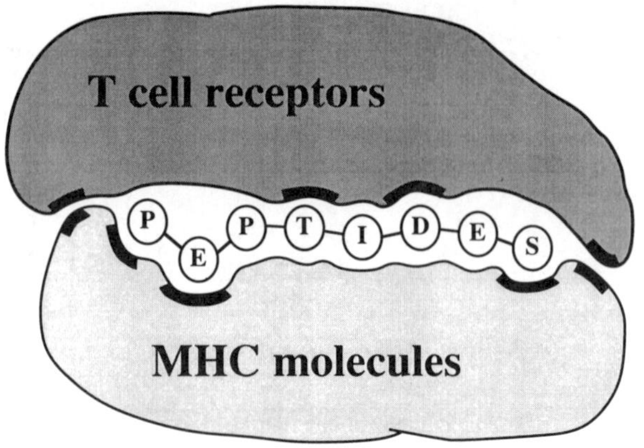

Fig. 1. An artists outline of the three molecular complex.

I.2. MHC restriction

A very large number of MHC alleles exists in the population; however, each individual only possesses one or two of these at each MHC locus (i.e., corresponding to being homozygous or heterozygous at this locus). How are MHC molecules involved in T cell recognition, and what is the relationship between antigen and MHC? The phenomenon of "MHC restriction" was established operationally in the early 1970s with the discovery that MHC molecules controlled T helper cell collaboration with B cells (Kindred and Shreffler, 1972), T helper cell collaboration with macrophages (Rosenthal and Shevach, 1973), and that they also controlled T cytotoxic cell recognition of target cells (Zinkernagel and Doherty, 1974). It was observed that T cells would only recognize antigen if it was presented in the context of one of the self-MHCs; no other MHC would do, and T cells were consequently said to be "MHC restricted". It was also observed that MHC molecules determined whether a particular individual could respond to a given antigen or not. To explain MHC restriction and T cell responder status, Rosenthal and Benacerraf in 1978 independently suggested that the function of MHC molecules is to specifically select some antigens and present them to T cells (Rosenthal, 1978; Benacerraf, 1978). According to this hypothesis, called "determinant selection", antigen-MHC interaction is specific, and antigens which are not selected by self-MHC cannot be seen by T cells. This hypothesis was provocative because it implied that the entire T cell defence of a given individual rested on the specificity of a few MHC molecules. This appeared paradoxical since any specificity of the MHC molecule would have to be broad, bordering nonspecificity, not to cripple T cell responses. An alternative hypothesis, called "holes in the T cell repertoire", proposed by Jan Klein and coworkers suggested that peptide-MHC specificity is only apparent and rather the result of T cells being educated during ontogeny to recognize antigen/self-MHC combinations and tolerized against self-antigen/self-MHC combinations. According to the latter hypothesis, the specificity lies at the level of the T cell, and the interaction between peptide and MHC is in itself not specific (Ishii et al., 1982). We now know that determinant selection and holes in the T cell repertoire act in concert to shape the T cell repertoire. The interaction between antigen and MHC is an absolute prerequisite for T cell recognition. Although the interaction is specific (i.e., determinant selection exists), the specificity is sufficiently broad to generate enough antigen presentation to support ample T cell immunity. On the other hand, the broad specificity of MHC does not allow it to distinguish between self and nonself. Thus, holes in the T cell repertoire are needed to avoid inappropriate responses directed against self-antigens presented by self-MHC.

II. THE LIGAND OF THE T CELL RECEPTOR

It took about 15 years of intense investigations to prove the existence of determinant selection, modify holes in the T cell repertoire and synthesize a unified hypothesis for the function of MHC molecules in antigen presentation. In the early 1970s, neither the T cell receptor nor its ligand were known, and the latter somehow included two components: antigen and MHC. The identity of the antigen was uncertain, and its relationship with the MHC was not understood. It is easy to understand why it took so long and why so much is still missing. The ligand of the T cell receptor cannot be understood and examined without a basic knowledge of how the antigen and MHC components are generated and meet within the APC. We shall therefore deal briefly with MHC structure/function and with antigen processing.

II.1. The major histocompatibility gene complex (MHC) molecules

The genes of the MHC are distributed over approximately 4 mega bases on chromosome number 6 in the human (Trowsdale et al., 1991). The MHC proper is subdivided into two classes: MHC class I and MHC class II, of which at least three different loci exist for each class. Several alleles exist for most of the MHC loci, making the MHC gene complex the most polymorphic gene region known [(Klein, 1986 (p 619); Parham et al., 1988)].

A schematic model of MHC class I and class II molecules is shown in Fig. 2 (reviewed in Strominger et al., 1981). It illustrates the heterodimeric nature of both MHC molecules and indicates their domain structure. Although class I and class II differ with respect to domain distribution, the two molecules have similar structures. MHC class I consists of a polymorphic 43,000 molecular weight (MW) transmembrane glycoprotein, the α-chain, which is noncovalently associated with an almost monomorphic 12,000 MW nonglycosylated protein, the β_2-microglobulin. The extracellular part of the class I heavy chain consists of three domains, α_1, α_2 and α_3, whereas the β_2-microglobulin consists of one domain. In contrast, MHC class II consists of two non-covalently associated polymorphic transmembrane glycoproteins, an α-chain of approximately 34,000 MW and a β-chain of approximately 29,000 MW. The extracellular part of each class II chain consists of two domains. Although noncovalently associated, class II molecules are extremely stable and must be heated to dissociate. In this respect, class I molecules are far more unstable.

The detailed structure of MHC class I molecules was solved at the X-ray crystallography level in 1987 (Björkman et al., 1987a) and the structure for class II was solved in 1993 (Brown et al., 1993). Despite differences in their domain organization, class I and class II have very similar structures: the distal domains of the molecule form a unique peptide binding structure which rests on top of two proximal immunoglobulin-like domains. The most striking feature of MHC molecules is the peptide binding structure which consists of two α helixes and eigth β-pleated strands. The two α-helixes form the walls and the eight β-pleated strands form the floor of the peptide binding groove (Fig. 3). The majority of the polymorphic residues points toward this peptide binding groove (Björkman et al., 1987b). The peptide binding groove can be subdivided into various pockets (A through F) of different size, shape and functionality

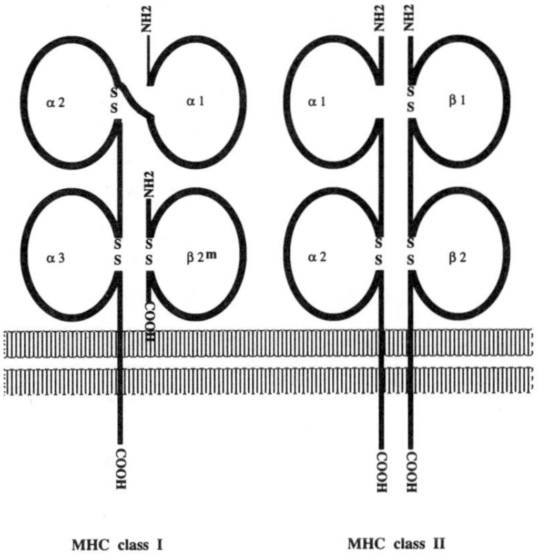

Fig. 2. A schematic representation of MHC class I and class II molecules.

(Garrett et al., 1989). The polymorphic residues lining the peptide binding groove determine the characteristics of the pockets and thus the specificity of the MHC-peptide interaction. Surprisingly, electrodense material was found occupying the groove in the initial X-ray crystallography structures, and it was suggested that this material consisted of peptides (Björkman et al., 1987a). This was also suggested by the finding that peptide material could be eluted off purified MHC, which could bind back to the appropriate MHC in an allele-specific manner (Buus et al., 1988). Subsequent X-ray crystallography studies at improved resolution demonstrated the presence of peptides in the groove and confirmed that the peptide was bound in an extended conformation (Madden et al., 1991; Fremont et al., 1992).

II.2. Antigen processing

T cells do not recognize the nominal antigen, but rather a modified version generated and presented by the APC (Seeger and Oppenheim, 1970; Waldron et al., 1973). Early on it was noted that the APCs needed to be alive and actively metabolizing (Waldron et al., 1974), and it appeared that the antigen was either internalized and/or modified in a time and temperature dependent process in such a way that it became enzyme resistant (Schmidtke and Unanue, 1971; Waldron et al., 1974; Ellner and Rosenthal, 1975; Ellner et al., 1977). Furthermore, it was known that T cells were unable to distinguish between native and denatured proteins (Gell and Benacerraf, 1959; Chesnut et al., 1980). This led to the idea that antigenic proteins were modified to some common denominator prior to T cell recognition. The handling of an antigen from the time it is offered to an APC until it is in a form that it, in association with MHC, can be recognized by T cells is collectively called antigen processing. This is an operational definition, and even to this day much of our information about antigen processing remains circumstantial. Antigen processing is thought to occur inside the APC under conditions which are ill defined, difficult to manipulate and virtually impossible to control.

In 1979 it was discovered that mildly fixed APCs could not present antigen if it was offered after the fixation, whereas the APCs were perfectly capable of presenting antigen processed prior to the fixation (Ziegler and Unanue, 1981; Shimonkevitz et al., 1983). This marked an important methodological advance since it allowed antigen processing and antigen display to be addressed as two separate events. In general, fixation assays are much better controlled, frequently also more robust and in some cases they even yield better signal-to-noise ratios. Fixation assays should be seriously considered whenever one needs to perform cellular assays for antigen processing and presentation. Fixation assays were first used to address presentation of exogenous protein antigens to class II restricted T cells. Antigen processing turned out to be temperature dependent, involving a lag time of approximately 30 min, and maximum presentation was obtained after 2–4 hours at 37°C (Ziegler and Unanue, 1981). This is compatible with the exogenous antigen being internalized, processed inside the cell and recycled to the cell surface in a processed version. Careful analysis of the nature of antigen processing also became possible with the fixation assay. Inhibitors of intracellular protein degradation turned out to be powerful inhibitors of antigen presentation (Ziegler and Unanue, 1982; Buus and Werdelin, 1986a). It was demonstrated that protein antigens digested in the test tube using enzymes like trypsin, chymotrypsin, etc. could stimulate T cell responses without any further processing requirements (Shimonkevitz et al., 1983). Peptides capable of replacing the intact protein antigen could be isolated from such a digest, identified by sequencing and made synthetically (Shimonkevitz et al., 1984). According to current consensus, antigen processing involves proteolysis and leads to peptide generation. Fragmentation is a more precise term than proteolysis because total degradation of the protein into its constituent amino acids will leave nothing specific for the T cells to recognize. The need for fragmentation implies that some mechanism(s) must exist to protect an antigen against complete degradation. The MHC molecule itself has been suggested to perform such a chaperone function, and it has been shown that a peptide is orders of magnitude more resistant to proteolysis once it is bound to MHC than when it is free in solution (Donermeyer and Allen, 1989; Mouritsen et al., 1992).

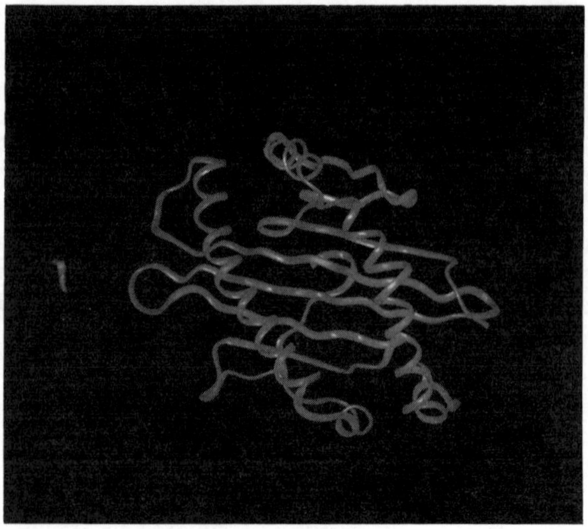

Fig. 3. The X-ray crystallographic structure of class I (in this case HLA-A2.1). (Adapted from Björkman et al., 1987a.)

There is a striking dichotomy in the T cell immune system. In general, cytotoxic T cells express the accessory molecule, CD8, and they tend to recognize endogenously derived antigens presented in association with MHC class I on target cells. In contrast, helper T cells predominantly express the accessory molecule, CD4, and they tend to recognize exogenously derived antigens presented in association with MHC class II on antigen presenting cells. Thus, a virus infected cell would present virally derived antigens in association with class I, which would target the infected cell for killing. In contrast, a cell which picks up antigen from the extracellular environment would present this antigen in association with class II, leading to recruitement of T helper cells. As mentioned above, MHC specificity is very broad and MHC molecules would in themselves not be able to see the difference between endogenously and exogenously derived antigen. Other factors relating to MHC synthesis, assembly, associated molecules or export pathways must explain the preference of class I and class II for endogenous and exogenous sources of antigen, respectively.

Although there is now compelling evidence for peptides being the end result of antigen processing, the exact enzymes, compartments, pathways, etc. involved in antigen processing are still not known. As indicated above, two separate pathways of antigen processing seem to exist: one for endogenously derived antigen leading to class I mediated presentation and one for exogenous antigen leading to class II mediated presen-

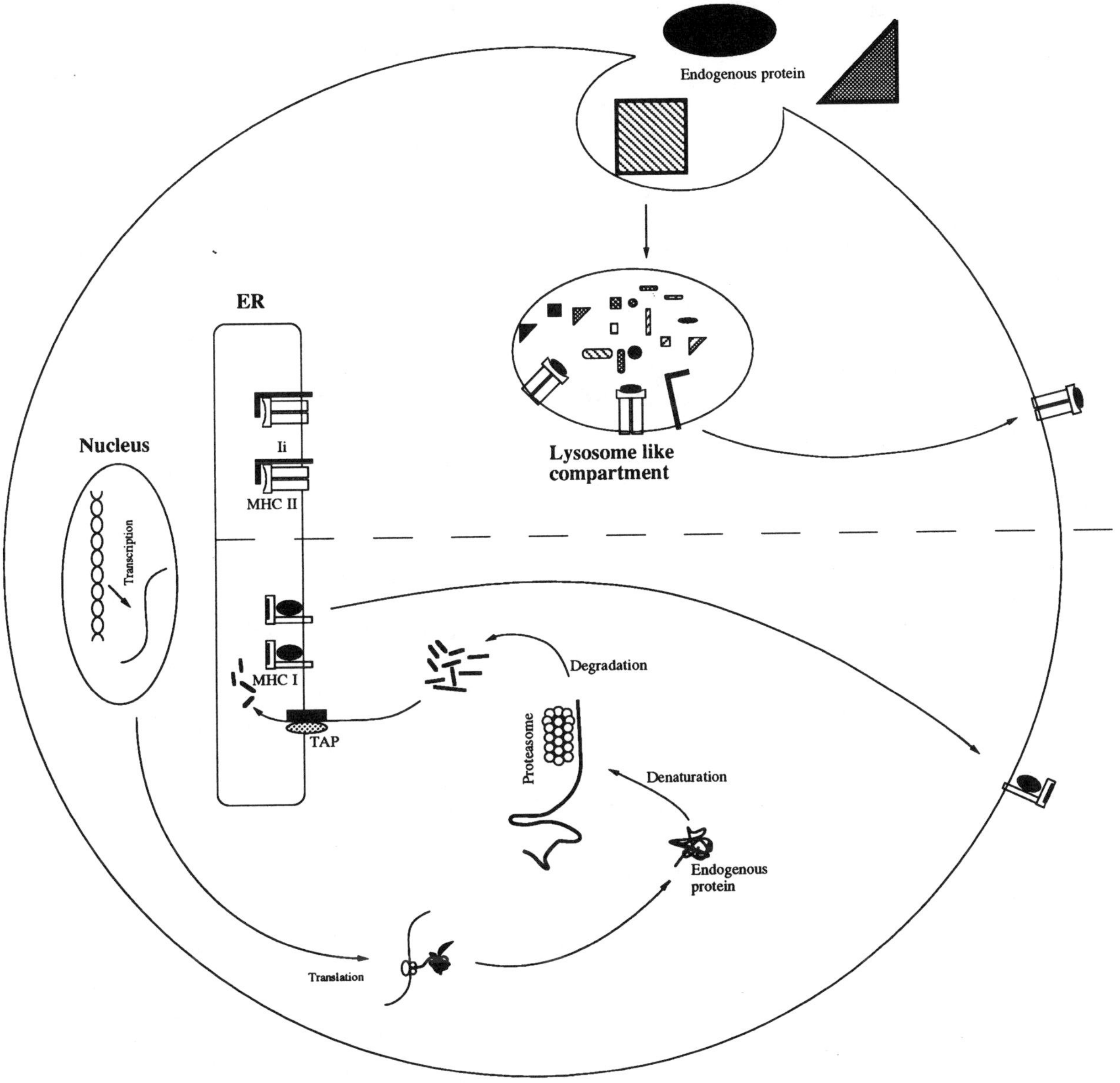

Fig. 4. The different pathways of antigen processing. The upper half illustrates how exogenous antigen is internalized and degraded in lysosome-like compartments. Some of the resulting fragments are bound to MHC class II, transported to the cell surface and presented to T cells. The lower half illustrates how endogenous antigen (e.g., viral antigen) is synthesized in the cytosol, whereafter part of it is degraded (possibly by proteasomes) and transported into the ER. Here, some of the resulting fragments are bound to MHC class I, transported to the cell surface and presented to T cells.

tation. A brief outline of the two pathways is given in Fig. 4. Endogenous protein is synthesized in the cytosol by the presenting cells own protein synthesis machinery. A fraction of all newly synthesized protein is fragmented, and the resulting peptides are transported across the ER membrane by ATP dependent transporters (called TAP for "transporters associated with antigen processing"). These peptides are picked up in a specific manner by *de novo* synthesized MHC class I molecules. Class I molecules are very unstable when empty and therefore are forced to acquire peptide already in the ER — and the available peptides will be of endogenous origin. Under physiological conditions, the majority of class I bound peptides will be about 8 to 10 amino acids long (Van Bleek and Nathenson, 1990; Rötzschke et al., 1990; Jardetzky et al., 1991; Hunt et al., 1992a). The reason for this preference is not yet clear; X-ray crystallography studies would indicate that the class I molecule itself prefers peptides of this narrow size distribution, but it is also possible that the processing and transporting machinery play a significant role in the selection of 8- to 10-mers. Class II mediated antigen processing is quite different. Class II molecules are also synthesized in the ER, but their peptide binding capacity is thought to be inhibited by the so-called invariant chain (Teyton et al., 1990). The invariant chain guides the virginal class II molecule to an endocytic compartment (a lysosomal compartment, or alternatively a specialized lysosome-like antigen processing compartment), where exogenous antigen is being processed. This compartment is acidic and contains proteolytic enzymes leading to cleavage and dissociation of the invariant chain liberating a fully functional class II molecule, which also happens to exhibit optimal peptide binding at acidic pH. The exogenous antigen is thought to be endocytosed and transported to the same endocytic compartment where the intact protein antigen is denatured and fragmented. Thus, peptide and class II are likely to meet one another in an endocytic compartment. Some of the resulting antigen fragments are specifically bound to the MHC class II and subsequently protected by class II against complete degradation. Under physiological conditions, the majority of class II bound peptides will be about 16 amino acids long (Rudensky et al., 1991; Chicz et al., 1992; Hunt et al., 1992b), and it is possible that these longer peptides can be further proteolytically trimmed while being bound to class II (Werdelin and Buus, 1988).

II.3. The specificity of MHC molecules ("motifs")

We can now bring peptide and MHC together. In the mid 80s, it became possible to measure class II binding biochemically (Babbitt et al., 1985; Buus et al., 1986c), and it is now also possible to measure class I binding biochemically (Olsen et al., 1994). It has now been demonstrated directly that peptide and MHC interact specifically and with a certain affinity (Babbitt et al., 1985; Buus et al., 1986c). T cells recognize peptide--MHC complexes, and no T cell response is possible without such specific complex formation (Buus et al., 1986c; Schaeffer et al., 1989). The specificity of MHC is rather broad, allowing about 20–30% of all randomly selected peptides to associate with any given MHC haplotype (Schaeffer et al., 1989; Sette et al., 1989b). Still MHC shows some degree of specificity, and the explanation for this apparent paradox is found by comparing unrelated peptides which all bind to the same MHC molecule. These peptides share a simple common motif which is absent in nonbinding peptides (Buus et al., 1987a; Sette et al., 1987). Presently, the definition of MHC binding peptide motifs is rather crude (see later). A few positions within a motif positively require the presence of one or a few dominating residues also known as "anchor residues" (Falk et al., 1991), while the remaining positions contain residues which are of secondary importance or simply act as spacers. Finally, some positions may contain residues which have a negative influence on binding (Boehncke et al., 1993). It is obvious that much still needs to be learned about the contribution of positive vs. negative interactions, of anchor vs. secondary, anchor vs. spacer residues, etc. X-ray crystallography studies of MHC class I have explained in stuctural terms how MHC binds broadly defined motifs with a rather high affinity (K_D in the nanomolar range) (Fremont et al., 1992; Matsumura et al., 1992). The peptide is bound in the groove in an extended conformation. The majority of the peptide is buried; only 20% of the total peptide surface, or about 300 $Å^2$, is potentially accessible to the T cell receptor. About 20 MHC residues, mostly of helical origin, interact with the peptide through multiple hydrogen bonds and one or more salt bridges. The large number of interactions explains the high affinity of peptide-MHC class I interactions. Characteristically, the MHC molecule participates with its amino acid side chains, whereas the peptide predominantly participates with its backbone carbonyl oxygens and amide nitrogens. This explains how one MHC class I molecule can bind so many different peptides: all peptides have a backbone. The ends of the groove appear spatially restricted, in particular the A pocket corresponding to the amino terminal of the peptide. Both the A and F pockets participate in an extended network of hydrogen bonds, and the F pocket, which corresponds to the C-terminal, participates in at least one salt bridge. Optimal binding requires that the charged amino and carboxylate termini of the peptide are available for binding to the corresponding ends of the MHC class I groove. This explains why MHC class I molecules prefer peptides around 9 amino acids long; this is energetically favored. In other words, the peptide appears to be strapped across the groove from the A to the F pocket. Shorter peptides can in some cases bind too if pockets in the floor of the groove can allow the peptide to make a short-cut, and in other cases longer peptides can bind because the helixes can be stretched and/or the middle of the peptide can bulge out of the groove. Finally, the structure partially explains how allele-specific binding is obtained since the side chains of the polymorphic amino acid residues form pockets of various depths and functionality which accommodate the "anchor" residues of the peptide. Binding of peptides to MHC class II is quite similar to the above-described binding of peptides to MHC class I (Brown et al., 1993). However, one major difference between

MHC class I and class II bound peptides pertaining to their size exist; class I prefers 8- to 10-mers, whereas class II prefers peptides which are about twice that size. This can be explained if the hydrogen networks found in class I and believed to bind the peptide termini are missing in class II. Indeed many of the conserved residues involved in creating the hydrogen networks of class I are absent in class II.

III. STRATEGIES TO EXAMINE AND CHARACTERIZE PEPTIDE-MHC INTERACTION

Peptide binding to MHC is an important intermediary step in the generation of a specific T cell response and thus to immune regulation in general. The ability to measure peptide binding to MHC molecules is useful when determining binding conditions and specificity. Knowledge about MHC specificity can be used to predict MHC binding peptides, thereby identifying potential T cell epitopes, and a binding assay can be used to verify such predictions. Examining peptide-MHC interactions is of great interest not only scientifically, but also practically and clinically (i.e., modulate the immune system for purposes such as vaccine development, enhance or block an immune response, adoptive T cell transfer, etc.).

Physiologically, peptide-MHC interaction takes place intracellularly under poorly defined and poorly controlled conditions. In particular, the identity of the naturally processsed peptide is difficult to ascertain. When designing an experiment on processing of intact antigen it is important to understand that proteolysis is a double-edged sword; too little degradation will not liberate the T cell epitope, too much degradation will destroy it. Blocking all proteolysis (e.g., using chloroquine) will completely block antigen processing and presentation. Blocking one particular protease with a highly specific inhibitor (e.g., using ZPADK) will block presentation if that enzyme is needed to produce the T cell epitope, but it will enhance presentation if the blocked enzyme normally would destroy the epitope (Buus and Werdelin, 1986a). When designing an experiment using preprocessed antigen such as a synthetic peptide, it is important to consider, and in many cases to control, proteolysis. Peptides can be extremely susceptible to proteases. Live cells, prefixed cells and even preparation of purified MHC molecules may be associated with sufficient protease activity to affect the outcome of an experiment both qualitatively and quantitatively. All too frequently is proteolysis left uncontrolled.

III.1. Functional approaches

Functional readout systems are indirect in the sense that they employ antigen-specific, MHC restricted T cells to detect the presence of functional peptide-MHC complexes. Presently these assays always involve cell culture condition with live T cells and frequently also live APC. They are difficult to control and to quantitate. Live APCs must be used when the purpose is *to address antigen processing*, however, preferably only in a pulse mode involving fixation at the end of the antigen pulse. This way antigen can be offered to isolated live APC, processing can be allowed to proceed under controlled conditions and the experiment can be terminated by fixation before live T cells are added. The APC can be manipulated with respect to temperature, time, metabolic inhibitors, etc. with a minimum effect on, and interference from, the T cell readout.

Inhibition of presentation can also be used *to address peptide-MHC class I or class II interactions*. In such an assay the APCs are used solely as antigen presenting particles. The APC are pulsed with increasing concentrations of the unknown peptide (the inhibitor) and subsequently a suboptimal concentration of the stimulating peptide. At the end of the experiment the APCs are washed and specific T cells are added. The T cell response is measured and the concentration of unkown peptide needed to inhibit the response significantly (say 50–75%) is determined. The less inhibitor needed to inhibit the response, the better the binding of the inhibitor to the restricting MHC molecule. This assay is semi-quantitative, but at least it depends on the exquisitive specificity of the T cell (other assays like direct binding of labelled peptide to live cells are of questionable specificity and should be avoided). It is important that the inhibitory and the stimulatory peptides are unrelated, otherwise the T cells may inadvertently be stimulated or antagonized (even though both events depend on MHC binding, the assay would be difficult to read and no longer quantitative). A more stringent version of the inhibition of presentation assay uses prefixed APCs as "inert" antigen presenting particles. Fixation prevents endocytosis and thereby most of the naturally occurring antigen processing. Finally, a cocktail of proteolytic inhibitors can be included during peptide pulsing of the prefixed APCs in order to further reduce the risk of inadvertent proteolytic modification of the inhibitory and/or stimulatory peptides (Buus and Werdelin, 1986b; Lamont et al., 1990). If one wishes to address peptide-MHC interaction using inhibition of antigen presentation, one should attempt to use prefixed APCs pulsed with peptide in the presence of a cocktail of protease inhibitors. If one wishes to address the potential effect of an inhibitory peptide say in a physiological setting, then one should include experiments omitting fixation, pulse mode and protease inhibitor cocktail. Peptides have different proteolytic susceptibility, and omitting these procedures might reveal peptides that despite efficient binding to MHC might be poor inhibitors of T cell stimulation under physiological conditions.

A currently popular functional assay to address peptide-MHC class I interaction uses mutant cell lines like the RMA-S and T2 which lack the TAP peptide transporter system, leaving their class I molecules empty and unstable (Townsend et al., 1989). Incubating these cells with a specific peptide for an extended period of time leads to a slow accumulation of peptide stabilized MHC class I molecules on the cell surface which can be detected by FACS analysis. Alternatively, peptides can be added to lysates of endogenously labelled mutant cell lines. This will also lead to class I stabilization which can be detected by immuno-precipitation followed by SDS-PAGE and autoradiography (Townsend et al., 1990). In both cases, the bindings detected are semi-quantitative and the binding conditions are difficult to control. To examine new MHC molecules one would have to generate mutant (RMA-S or the like) cell lines transfected with the appropriate MHC class I. Whether similar systems can be generated for class II is not known.

III.2. Biochemical approaches

Biochemical assays for peptide-MHC class I and class II binding are direct in the sense that one of the reactants, usually the peptide, is labelled and offered to the other, usually the MHC. Both reactants are soluble and purified. The interaction takes place in solution and it is determined in solution. This allows the assay to be conducted with reactants which are homogenous, specific and quantifiable. The assay conditions can be accurately defined and controlled. A biochemical assay should be used when the purpose is to address peptide-MHC class I or class II interactions, in particular when high stringency is required. Control of proteolysis is particularly important since contaminating proteolytic enzymes may change both the quantity and quality of the added peptide (Buus and Werdelin, 1986b; Lamont et al., 1990). Obviously, the peptide added to the reaction should also be the peptide resposible for the effect seen; however, many proposed assays using whole cells or crude cell lysates fall short of this requirement. Properly conducted, the biochemical binding assay yields more specific and sensitive results than the functional binding assay. Briefly, a direct biochemical assay is conducted as follows. Affinity purified MHC in detergent solution and labelled synthetic peptide are mixed in the presence of a protease inhibitor cocktail and incubated to allow equilibrium to be reached. Once associated with MHC, the complexes are sufficiently stable to allow quantitative separation of bound peptide from free peptide by gel filtration (Buus et al., 1986). A direct binding assay can, once a working peptide-MHC pair has been obtained, conveniently be converted into an inhibition assay; the less inhibitor needed to block the reaction, the better the binding. Such an inhibition assay is preferable since, once established, it is versatile and easy to use. For each MHC molecule to be examined, only one peptide needs to be labelled, and any unknown preparation can then be tested for inhibition. The inhibition assay is also very sensitive (the presence of 100 fmol of a good binder can be detected).

III.3. Examining what the MHC has already bound: an alternative approach

The first biochemical binding studies revealed that the majority of purified MHC molecules did not participate in peptide binding (Buus et al., 1987b). Of course, the purification procedure might have denatured most of the purified MHC molecules, but a more satisfying explanation would be that MHC molecules being of rather broad specificity had already bound a mixture of self-peptides during its biosynthesis and therefore were not fully available for binding. Indeed, material could be acid eluted off the MHC and identified as peptide capable of binding back to MHC in an allele-specific manner (Buus et al., 1988). In agreement, X-ray crystallography studies of MHC class I showed electro-dense material present in the peptide binding groove and later studies at higher resolution identified this material as peptide (Björkman et al., 1987a; Madden et al., 1991). The observation that MHC class I bound short peptides with the amino terminal in a fixed position relative to the MHC led to the revolutionary idea that the entire mixture of MHC class I bound peptides might be in register/synchronized in such a way that they could be eluted off the MHC and dominating amino acid residues may then be identified by N-terminal sequencing (Falk et al., 1991). Indeed, positions which contained only one or two amino acids, so-called anchor residues, were obtained. This method has turned out to be extremely efficient compared to the quite cumbersome way that MHC binding motifs were originally derived at in the class II system. The anchor residue motif, however, is only a crude MHC class I binding peptide motif, and the predictive power of this motif is not the best probably because the method tends to ignore "secondary anchors" and all together ignores negative interactions. Also, it is known that class I binds peptides from 8 to 10 amino acids in length which complicates interpretation because peptides of different length will not necessarily be in register and because each peptide size may be associated with subtle motif differences. However, the method is extremely important because it gives a good first approximation of the motif and it is very valuable in designing a peptide capable of binding in the direct biochemical assay.

The elution method can be extended to the identification of individual peptides bound to MHC class I and class II (Rötzschke et al., 1990; Jardetzky et al., 1991). After acid elution the peptide mixture is further purified, i.e., by HPLC, and individual peptides are identified by N-terminal sequencing or tandem mass spectrometry. This identifies singular peptides, mostly self-peptides, associated with the MHC in question and might be useful in the search for auto-antigens involved in autoimmunity. In the latter case, the sensitivity becomes an important issue, and a highly sensitive (30 fmol) tandem mass spectrometry method to purify and sequence peptides has recently been used successfully to identify peptides associated with both class I and class II (Hunt et al., 1992a, 1992b).

The sequencing of individual peptides is essential for work on MHC class II. Peptides associated with class II are not N-terminally synchronized with respect to MHC; their ends are said to be ragged. Thus, they cannot be subjected to N-terminal sequencing in bulk. However, individual class II associated peptides have been purified and sequenced (Rudensky et al., 1991). As expected, the peptides identified this way have contained the previously derived motif and the class II peptide ends have been ragged (Chicz et al., 1992; Hunt et al., 1992b).

IV. MHC MOTIFS

An MHC binding peptide motif is a short peptide sequence which confers MHC binding. The word motif is used to stress that MHC does not require identity, but merely a partial similarity between different binding epitopes. Motifs were discovered examining large sets of peptides all known to be strong binders to particular MHC molecules. It was noted that most good binders share a simple peptide sequence (the motif) which is absent in nonbinding peptides (Sette et al., 1987, 1988, 1989b). Aligning the motif bearing peptide sequences from several MHC binding epitopes reveals positions where only a few amino acids are found ("anchors residues") and other positions where allmost all amino acids are found (which we can call "spacer residues"). Different MHCs have different motifs, both with respect to length and composition, and this allotypic difference in motif recognition forms the molecular basis of determinant selection. The interest in motifs stems from the observation that accurate motif definition can be used to predict MHC binding. The I-A^d binding motif, "hydrophobic - x_4-hydrophobic", was present in about 75% of the known I-A^d binders and conversely absent from about 90% of the known I-A^d nonbinders (Sette et al., 1989b). A motif for I-E^d has subsequently been found, "basic - x_n - basic - x - basic" (Sette et al., 1989a). This motif is present in about 80% of known I-E^d binders and conversely absent from about 95% of the known I-E^d nonbinders. This is a considerably better success rate than that of more general predictions proposed. It is noteworthy that the two I-A^d and I-E^d motifs are completely dissimilar. This illustrates that each MHC has a unique specificity, which cannot be fitted into a common algorithm as otherwise proposed by some (DeLisi and Berzofsky, 1985; Rothbard and Taylor, 1988). In our view, each MHC specificity *should* be identified experimentally and only such data should be used to predict MHC restricted epitopes.

Motifs have now been defined for several MHC molecules, but the motifs are still rather crude and far from perfect. Presently, there is no consensus for how motifs should be defined. Many different approaches have been used; some motifs like the "basic-blank-basic" of the I-E^d can be scanned by eyeballing, whereas other approaches involve computer assisted alignments based on structural homology of amino acids (Sette et al., 1987, 1989b; Hammer et al., 1992). We believe that detailed information should be obtained for each MHC using several of the different methods. One future strategy to identify motifs may look like the following. The elution method described above will be the method of choice to obtain first-hand crude information about the motif and at the same time it will be extremely helpful in designing the peptide needed to generate a direct binding assay. After the direct binding assay has been established, the initial motif can be reevaluated and extended. The direct binding assay is used to determine (1) the core region of one or several strong binders (e.g., by truncation analysis), (2) the importance of residues within the core region (e.g., by analog substitution analysis) and (3) the binding of a panel of unrelated peptides. In the next steps, the core region is compared to the panel of unrelated peptides sorted into binders and nonbinders. In the initial phases, a simple matrix is used, such as Grantham tables (Grantham, 1974), assigning a "structural dissimilarity" value for the substitution of any two amino acids[See note 1]. All possible alignments are examined, and for each alignment a structural homology between the core region and the unrelated peptide is calculated (Padlan, 1977; Sette et al., 1987, 1989b). The best alignments of the core region with the good binders may reveal positions which are dominated by one or a few structurally related amino acids. In contrast, these anchor residues are missing in the best alignments of the non-binders[See note 2]. Thus, the initial motif generated by the elution procedure can be reevaluated and extended by systematic analysis using direct binding. We believe that a combination of the two approaches will be necessary. The motif generated by the elution method is, as described above, potentially misleading since it only identifies interactions which are absolutely needed for binding. In contrast, the direct binding method can evaluate the existence of secondary anchors as well as residues which are incompatible with binding. The relative distribution of positive and negative interactions in defining the total MHC specificity is not known, but they are probably both important. Therefore, pairing the two approaches, direct binding and sequencing of eluted peptides, will give us the total epitope definition of the future.

The work involved in the motif derivation as outlined previously is quite substantial. Recently, a couple of alternative and much less labor intensive techniques to determine the peptide binding motifs of individual MHC molecules have been developed. In one approach, a peptide library expressed as a fusion protein on a philamentous phage is selected by the MHC in question (Hammer et al., 1992). Phages of interest are amplified by a few consecutive rounds of selection, and finally a panel of phages are selected for sequencing of the peptide present in the fusion protein. The CGC program, PileUp, was used to establish a relationship (i.e., the motif) between the different peptides[See note 3]. This method holds out great promise for MHC class II motif definition, but it presently cannot be used for class I, which will not accept long peptides such as those fused to a phage protein.

V. PROTOCOLS

Given below are a few selected methods for detection of peptide-MHC interaction.

V.1. Peptide-MHC binding: Inhibition of antigen presentation

Functional assays rest on the availability of T cells restricted to the MHC in question. For the purpose of generating a functional assays for peptide-MHC interaction the T cells should preferably be unrelated to the peptides being examined to avoid cross-reactions and/or T cell receptor antagonism. Attempts should be made to generate T cell responses which will accept fixation of the APC. If possible, the APCs should be prefixed to prevent internalization, and the prefixed APCs should be pulsed with antigen in the presence of a cocktail of protease inhibitors to prevent uncontrolled peptide degradation. These measures on the other hand require that synthetic peptides (or at least proteolytic digests) derived from the stimulating protein antigen be available.

V.1.a. Inhibition of antigen presentation

1.) The APCs are taken from a culture growing in log phase. Wash the cells 3x in PBS and adjust the cells to 10^7 cells/ml.
2.) Make fresh paraformaldehyde as a 1% solution in PBS and dilute the cells 1:1 to a final cell concentration of 5 x 10^6/ml and a final paraformaldehyde concentration of 0.5%. Incubate at room temperature for 20 min. (Alternatively, adjust the cells to 5 x 10^6/ml, and fix them using glutaraldehyde at 0.05% final concentration for 30 s at RT.)
3.) Wash the cells 3x in 1:1 (PBS:0.15 M lysine) to quench further fixation.
4.) Adjust to 6 x 10^7/ml in PBS (e.g., 3x10^6 per 50 μl) with or without a protease inhibitor cocktail (see Section V.1.c).
5.) A dose range of inhibitor is added and incubated for 15 min at 37°C.
6.) Antigen in a suboptimal dose is added and the incubation continued for another 2 h at 37°C.
7.) The cells are washed 5x in PBS and adjusted to 5 x 10^6 cells/ml in tissue culture media.
8.) The prefixed and pulsed APC are co-cultured with T cells.
9.) The T cell response is read and the inhibitor concentration needed to inhibit the response 50% (for ^{3}H-thymidine incorporation) to 75% (for IL-2 secretion) is determined.

IC_{75}s are typically found to be around 5–50 μM.

V.1.b. Stock solutions of the protease inhibitor cocktail (cellular assays)

This inhibitor cocktail is a mixture of inhibitors against all four classes of proteases. They are toxic and should be treated with due care. The stock solutions are:

a.) Serine protease inhibitors:
PMSF (e.g., Sigma P-7626) 40.0 mg/ml in isopropanol (extremely toxic)
b.) Cysteine protease inhibitor:
NEM (e.g., Sigma E-3876) 12.5 mg/ml in isopropanol
c.) Metallo protease inhibitors:
EDTA (e.g., Sigma E-4884) 100.0 mg/ml in PBS (pH 7.2)
1.10 phenanthroline (e.g., Sigma P-9375) 26.0 mg/ml in isopropanol
d.) Aspartic acid protease inhibitor:
Pepstatin A (e.g., Sigma P-4265) 5.0 mg/ml in isopropanol
e.) Serine and cysteine protease inhibitors:
TLCK (e.g., Sigma T-7254) 20.0 mg/ml in water (smelly, use in hood)
TPCK (e.g., Sigma T-4376) 20.0 mg/ml in isopropanol

V.1.c. Intermediary dilution (15x) of the protease inhibitor cocktail (PI-mix, cellular assays)

The protease inhibitor cocktail is made up just before use as some of the components (in particular the PMSF) are unstable once the mixture is made. The intermediary dilution is diluted 15 times in the final

TABLE 1
PI-mix (15x)

For final set-up volume	1000 μl	1200 μl	1500 μl	2000 μl
EDTA (100 mg/ml)	30.0 μl	36.0 μl	45.0 μl	60.0 μl
NEM (12.5 mg/ml)	10.0 μl	12.0 μl	15.0 μl	20.0 μl
Pepstatin A (5 mg/ml)	10.0 μl	12.0 μl	15.0 μl	20.0 μl
1.10 phenanthroline (26 mg/ml)	10.0 μl	12.0 μl	15.0 μl	20.0 μl
TLCK (20 mg/ml)	2.5 μl	3.0 μl	3.8 μl	5.0 μl
TPCK (20 mg/ml)	2.5 μl	3.0 μl	3.8 μl	5.0 μl
PMSF (40 mg/ml)	4.3 μl	5.2 μl	6.5 μl	8.7 μl
Yield of PI-mix (15x)	69.3 μl	83.2 μl	104.1 μl	138.7 μl

set-up (e.g., 50 μl cell suspension + 3.6 μl PI-mix). Table 1 is given in microliters and yields a little bit more than what is needed for experiments of the indicated final volumes.

V.2. Peptide-MHC binding: A biochemical assay

To get started one would have to generate active preparations of the two reactants: MHC and peptide.

V.2.a. MHC

To obtain pure preparations of MHC molecules one would have to select a MHC producer cell line and an anti-MHC monoclonal antibody (the loosely associated light chain of MHC class I, β_2-microglobulin, can be obtained commercially). It is worthwhile to spend some effort on optimizing yields (search for workshop cell lines or search in the literature for homozygous cell lines, screening and/or FACS sorting for high expression, selection and test suitable antibodies etc.) since pure homogeneous preparations become much easier to produce with a good starting material. In general, you should aim at obtaining a homogeneous preparation of MHC. Cells that express several different MHC molecules are acceptable as long as the antibody used for purification can separate the different MHCs[See note 4, See note 5]. The MHC molecule is classically produced by a large-scale cell culture. This entails selecting a cell line which expressess large amounts of the desired MHC and which will grow resonably in *in vitro* cell culture (cell factories, roller bottles, spinner culture, fermentors), preferably in cheap serum. The cells are harvested, lysed in detergent and stored at -70°C until use.

V.2.a.1. MHC purification

1.) Frozen lysate (2–10 x 10^{10} cell equivalents) is thawed slowly in the refrigerator, cleared by high speed centrifugation and followed by 0.45 μ filtration. All steps are conducted in the cold.
2.) The lysate is applied slowly to an affinity column (5–20 ml) coupled with the relevant antibody (2–10 mg antibody/ml).
3.) The column is washed extensively (at least 15 column volumes PBS containing 0.5% NP-40 and 0.1% SDS followed by 15 column volumes PBS and finally equilibrated with 3 column volumes PBS containing 1% β-octyl glucoside).
4.) The MHC molecules are eluted at basic conditions (0.15 M NaCl containing 50 mM DEA and 1% β-octyl glucoside, pH 10.5 - 11.0) and neutralized in 1/20 volume of a 2 M Tris, pH 6.3. The eluate is collected in fractions.
5.) The fractions obtained are analyzed by SDS-PAGE
6.) The MHC containing fractions are pooled and concentrated by ultrafiltration to 250–500 ml (Schleicher and Schuell or Sartorius).
7.) The concentrated preparation (typically 0.5 -5 mg/ml) is quantitated and stored in azide at 4°C.

A typical yield from a mouse lymphoma cell is 50 μg/ 10^9 cells, whereas the yield from human EBV transformed cells lines is about 10 times higher. The activity of the MHC preparation varies somewhat from preparation to preparation, but 50% maximal peptide binding can be expected at around 100–1000 nM. Binding analyses are usually compatible with a single equilibrium reaction with a K_D from 2 nM and up. Only a minor fraction, usually between 2 and 5%, of the MHC is available for binding (see above).

V.2.b. Peptide

The next objective is to define an indicator peptide

which can be labelled, can bind to the MHC in question and is compatible with the separation method selected. The search for prospective binding peptides is based on known immunogenic epitopes reported in the literature. These epitopes may have been defined by random scanning of proteolytically (Shimonkevitz et al., 1983) or genetically (Townsend et al., 1986) derived fragments of native protein antigens. Alternatively, they may be based on sequencing of peptides eluted off the MHC (Falk et al., 1991). It should be noted that class I epitopes are very restricted in size, and that critical anchor residues must be accurately positioned within a given peptide epitope to match the class I in question. In comparison, the same size restriction does not apply to MHC class II binding peptides and the location of the MHC class II binding core region within a given peptide is very flexible. Thus, the digest approach is useful in the attempt to identify MHC class II, but not class I, binding peptides. In contrast, sequencing of peptide pools bound to class I is useful in the attempt to identify MHC class I binding peptides. This latter method can only be used for class II if singular peptides can be identified.

The best label is currently ^{125}I, and since it can label peptides to a high specific activity, the detection is sensitive and easy to quantitate[See note 6]. These prospective binding peptides are synthesized preferably with a tyrosine for labelling purposes. For class II binding peptides this tyrosine can be added either at the amino or carboxy terminal end. The class I size requirement dictates that the tyrosine should be somewhere in the motif. If it is not known where the tyrosine will be accepted, one may have to make several analogs scanning the tyrosine through the peptide.

Not all peptides are compatible with the Sephadex G50 size separation assay described below. Most peptides of interest are between 1000 and 4000 in molecular weight and should theoretically appear in the included volume of a Sephadex G50 column, whereas nothing (usually less than 1%) should appear in the void volume. Nonetheless, some peptides behave as though they are considerably larger (the worst we have seen was a peptide which appeared 70% in the void volume). We do not know the reason for this anomality, but we speculate that some peptides aggregate into larger complexes. This "background" problem renders some peptides useless as indicator peptides for MHC binding. Do not give up, be prepared to test a number of different peptides and aim for a peptide which binds with a binding of at least 10% of the offered label and has a background of less than 1%. Some of the best peptides that we have worked with have been bindable up to 100% with less than 1% background. It should be noted that some peptides appear less than 100% bindable, i.e., binding levels out and reach a plateau before 100% although you add more and more MHC. An MHC dose response should be made for each peptide preparation to determine the maximal bindable fraction of labelled peptide, and all results with this preparation should be corrected accordingly (i.e., if a peptide preparation is only 80% bindable and you in an experiment obtain 20% binding the latter figure is corrected to 20%/0.8 = 25%). The reason for this less than 100% bindability is probably modifications of the peptide caused by the labelling procedure. The loss of bindability is particularly noteworthy for β_2m which may be no more than 25–50% bindable when labelled by the procedure in Section V.2.b.1. We have observed that better β_2m labellings are obtained if the exposure to chloramine T is limited to 20 s and the pH is raised to 8 during the labelling (Lars Ø Pedersen, unpublished observation).

V.2.b.1. ^{125}I labelling of peptides (or β-2-microglobulin)

1.) The peptide is HPLC purified to homogeneity.
2.) About 2 nmol (2–4 μg) is taken up in 25 μl 0.3 M phosphate, pH 7.5 (for β_2m labelling the protein is taken up in 25 μl 50 mM Tris, pH 8).
3.) Add 10 μl ^{125}I (equalling 1 mCi).
4.) Add 5 μl 1 mg/ml chloramine T, and incubate for 45 s at RT.
5.) The reaction is quenced by adding 5 μl 1 mg/ml metabisulfite. Incubate for 45 s at RT.
6.) The labelled peptide is gel filtered over a 2 ml disposable Sephadex G10 column equilibrated in PBS containing 0.1% azide and 2% ethanol. To reduce loss due to absorption, prime the column with 100 μl 0.05% Tween 20 just before use.
7.) Collect 0.2 ml fractions, and pool the void volume containing the labelled peptide.
8.) Determine specific activity (usually 250 μCi/nmol peptide and 0.5–1 μCi/μl).
9.) Verify purity by HPLC followed γ-spectrometry. In our experience, there is less than 1% free radioactivity after one G10 separation. The peptide appears in one to two peaks (two peaks probably reflects a mixture of mono- and di-iodo--labelled tyrosin).
10.) Store at 4°C in a lead container.

The azide maintains sterility and ethanol acts as a scavenger of radiolysis. Note that a labelled peptide preparation has a rather short shelf life (about 3–6 weeks). It is time to relabel whenever the MHC binding drops or when free radioactivity/degradation products start to appear by HPLC analysis.

V.2.c. The biochemical binding assay

The interaction between peptide and MHC can be detected by equilibrium dialysis, but this classical technique is rather cumbersome and it demands quite a lot of material because of the volumes involved and the rather large degree of binding required to obtain accurate measurements. Easier, more efficient and economical methods are available. Peptide-MHC complexes are so stable that bound and free peptide can be separated even by nonequilibrium methods like simple size separation. The loss of complexes due to dissociation during the separation is insignificant, and size separation, once equilibrium has been reached, can be used to accurately determine the amount of bound and free peptide at equilibrium. Sephadex G50 size filtration is a fast, reliable and well established separation method for determining peptide-MHC interaction (Buus et al., 1986c). Similarly, Sephadex G75 can be used for determing β_2m binding to class I heavy chain (Hyafil and Strominger, 1979). Most peptides will be between 1000 and 4000 in molecular weight, whereas the peptide once bound to an MHC molecule (typically in a NP-40 detergent micelle of about 90,000 in molecular weigth) will be up to 150,000. Peptide associated with MHC will thus appear in the void volume and free peptide will appear in the included volume of a Sephadex G50 column (Fig. 5). Very little material is required because small volumes can be used and even low bindings can be determined accurately. All in all, the assay is sensitive, economical, simple and robust.

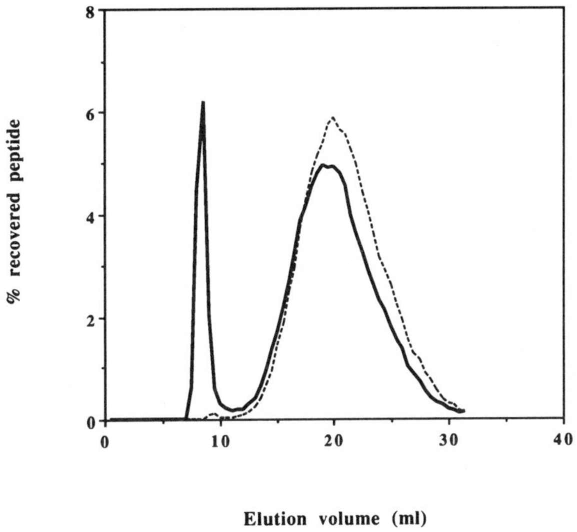

Fig. 5. Assay for peptide-MHC interaction. Radiolabelled peptide is offered to purified MHC molecules in detergent solution. Peptide-MHC complexes once formed are stable and can be assayed by Sephadex G50 gel filtration. Complexes appear in the void volume, whereas free peptide appears in the total volume. The fraction of peptide bound can be calculated as described in the text. The full line represents a peptide-MHC combination which led to binding, whereas the stippled line represents a peptide-MHC combination which did not lead to binding.

A dose response is made for each MHC preparation to determine the concentration of receptor to use in the subsequent experiments. Fig. 6 illustrates the theoretical outcome of simple equilibrium experiments. The left-hand panel shows the fraction of ligand bound (also defined as below) as a function of the concentration of ligand offered. Five different receptor concentrations have been used, A = 1/100, B = 1/10, C = 1, D = 10 and E = 100 times the K_D of the interaction. The figure clearly demonstrates that the maximum obtainable binding directly depends on the concentration of receptor used and its K_D. For this assay, a receptor concentration of at least 1/10 of the K_D must be used to obtain sufficient binding[See note 7]. The frequently used inhibition assay depends on accurate determination of the IC_{50} point, which again depends on a decent downward slope of the curve. Using less receptor than 1/10 of the K_D would lead to poor binding and inaccurate IC_{50} determinations. Heading for a binding in excess of 30% would also decrease the accuracy (Hulme and Birdsall, 1992) and it would be a waste of precious receptor. In particular, the IC_{50} determination will overestimate the K_D (underestimate the affinity) when excessive amounts of receptor are used. Thus, the concentration of receptor used should lead to a binding of between 15 and 30% of the ligand. In your preexperiment, you should also include a dose range of inhibitory peptide to verify that the MHC preparation is inhibitable. This will also assure that the IC_{50} is reproducible from preparation to preparation. The right-hand panel shows the fraction of receptor occupied with ligand as a function of the concentration of ligand offered. Five different receptor concentrations have been used, A = 1/100, B = 1/10, C = 1, D = 10 and E = 100 times the K_D of the interaction. Obviously, as more and more receptor is used, more and more ligand will have to be offered to saturate the receptor (that is why the IC_{50} overestimates the K_D under these conditions). Also shown is the characteristic merging of the occupancy curves when receptor concentrations below the K_D are used.

The optimal binding conditions vary depending on the MHC, and must be determined empirically. In general, the fastest association and dissociations occur at 37°C, but the MHC tends not to be stable at that temperature. Class I in particular is unstable at 37°C, leading to a biphasic association curve where binding peaks rapidly followed by a slow decline. This makes the assay difficult to handle and interpret. In general, we perform binding assays at 18°C, (alternatively, it can be done at RT) where MHC and peptide-MHC complexes are much more stable. A steady state is obtained at 18°C, and we can assume this reflects equilibrium. The time needed to obtain equilibrium ("steady state") must be examined and determined for the MHC in question. In general, we incubate for

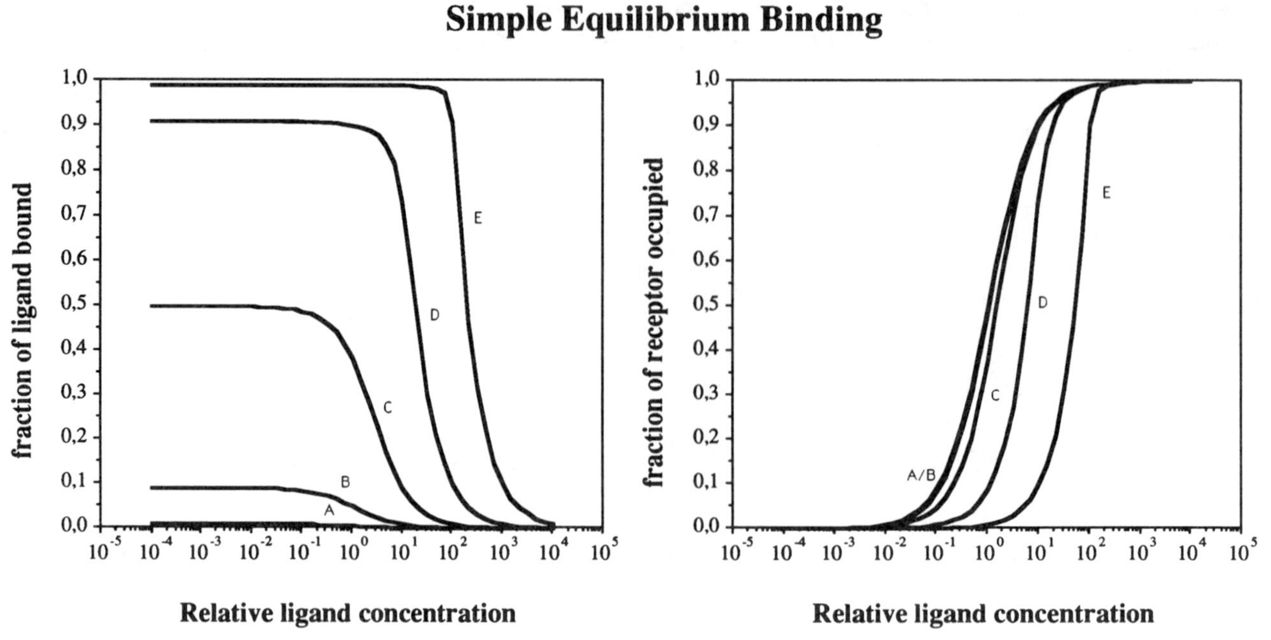

Fig. 6. The theoretical outcome of a simple equilibrium experiment. The concentration of complexes resulting from simple equilibrium is calculated as $C = 0.5 \times ((R_0 + L_0 + K_D) - ((R_0 + L_0 + K_D)^2 - 4 \times R_0 \times L_0)^{0.5})$ (Fazekas de St. Groth, 1979), where R_0 is the input receptor concentration, L_0 is the input ligand concentration and K_D is the dissociation equilibrium constant. The relative ligand concentration is L_0/K_D, the fraction of ligand bound is C/L_0 and the fraction of receptor occupied is C/R_0.

24–48 hours[See note 8]. Salt concentrations may affect binding and must be examined empirically. We have observed bindings which were inhibitable by isotonic salt concentration, and other bindings which were not affected by up to 0.5 M NaCl. In general, MHC class I binds optimally at pH 6 to 7, whereas MHC class II binds optimally at pH 4.5 to 5.5. We achieve the low pH by adding a citrate-phosphate buffer to the reaction mixture. It is also quite important to determine the optimal detergent and detergent concentration. We have found that 0.05% NP-40 in general is compatible with binding, but this does not preclude that other detergents might be better for a particular MHC [(e.g., high concentration of NP-40 and β-octyl glucoside has in some cases been incompatible with binding, whereas high concentrations of digitonin in one case (I-A^d) improved binding].

V.2.c.1. Set-up peptide-MHC class I interaction

1.) On ice, mix the following:
   x μl MHC (enough to lead to 10–30% binding)
   1 μl β_2m (1 mg/ml in PBS)
   y μl available (e.g., for inhibitor peptide) in PBS to make the volume 12 μl
   2 μl protease inhibitor — detergent cocktail (PI-De-mix; see Table 2)
   1 μl labelled indicator peptide (a total of 25,000 cpm)

The total volume should be 15 μl. The reaction is conveniently conducted in V-bottom 96-well microtiter plates. Seal the plate to prevent evaporation.

V.2.c.2. Set-up β2m-MHC class I interaction

1.) On ice, mix the following:
   μ ml MHC (enough to lead to 10–30% binding)
   y μl available in PBS to make the volume 12 μl
   2 μl protease inhibitor — detergent cocktail (PI-De-mix; see Table 2)
   1 μl labelled β_2m (a total of 25,000 cpm)

The total volume should be 15 μl. The reaction is conveniently conducted in V-bottom 96-well microtiter plates. Seal the plate to prevent evaporation.

V.2.c.3. Set-up peptide-MHC class II interaction

1.) On ice, mix the following:
   x μl MHC (enough to lead to 10–30% binding)
   5 μl of a citrate-phosphate buffer[See note 9] to adjust the pH to approximately 5
   y μl available (e.g., for inhibitor peptide) in PBS to make the volume 12 μl
   2 μl protease inhibitor — detergent cocktail (PI-De-mix; see Table 2)
   1 μl labelled indicator peptide (a total of 25,000 cpm)

The total volume should be 15 μl. The reaction is conveniently conducted in V-bottom 96-well microtiter plates. Seal the plate to prevent evaporation.

TABLE 2
PI-De-mix (7.5x)

For final set-up volume	1000 μl	1200 μl	1500 μl	2000 μl
EDTA (100 mg/ml)	30.0 μl	36.0 μl	45.0 μl	60.0 μl
NEM (12.5 mg/ml)	10.0 μl	12.0 μl	15.0 μl	20.0 μl
Pepstatin A (5 mg/ml)	10.0 μl	12.0 μl	15.0 μl	20.0 μl
1.10 phenanthroline (26 mg/ml)	10.0 μl	12.0 μl	15.0 μl	20.0 μl
TLCK (20 mg/ml)	2.5 μl	3.0 μl	3.8 μl	5.0 μl
TPCK (20 mg/ml)	2.5 μl	3.0 μl	3.8 μl	5.0 μl
PMSF (40 mg/ml)	4.3 μl	5.2 μl	6.5 μl	8.7 μl
NP-40 (0.75% in PBS)	66.7 μl	80.0 μl	100.0 μl	133.0 μl
PI-De-mix (7.5x)	136.0 μl	163.2 μl	204.1 μl	271.7 μl

V.2.c.4. Determining interaction by Sephadex size separation

2.) Incubate the reaction at 18°C for the time needed to obtain steady state.
3.) Prepare a 1 x 23 cm Sephadex G50 column and equilibrate it in 0.5% NP-40, 0.1% Azide, PBS (use Sephadex G75 for the β_2m assay!).
4.) Apply the 15 μl reaction mixture to the column and elute it in 0.5% NP-40, 0.1% Azide, PBS.
5.) Collect 1 ml effluent fractions and count them by γ spectrometry. The ratio of bound peptide to total peptide is calculated as: a (in %) = 100 x (total counts recovered in the void volume peak) divided with (total counts recovered in the void volume peak + total counts recovered in the included volume).
6.) Plot the concentration of inhibitor peptide vs. binding and interpolate to obtain the concentration of inhibitor needed to inhibit binding 50% (inhibitory concentration 50, IC_{50}). This value approximates the K_D of the inhibitory peptide as long as the receptor concentration is kept below the K_D.

V.2.c.5. Intermediary dilution (7.5x) of the protease inhibitor-detergent cocktail (PI-De-mix, biochemical assays)

The protease inhibitor cocktail is made up just before use, as some of the components (in particular the PMSF) are unstable once the mixture is made. The intermediary dilution is diluted 7.5 times in the final set-up (e.g., 10 μl PI-De-mix to a final set-up of 75 μl). Table 2 is given in microliters and yields a little bit more than what is needed for experiments of the indicated final volumes.

ABBREVIATIONS

MHC Major histocompatibility complex
APC Antigen presenting cell

VI. REFERENCES

Babbitt BP, Allen PM, Matsueda G et al. Binding of immunogenic peptides to Ia histocompatibility molecules. Nature 317:359, 1985

Benacerraf B. Opinion: A hypothesis to relate the specificity of T lymphocytes and the activity of I region-specific Ir genes in macrophages and B lymphocytes. J Immunol 120:1809, 1978

Björkman PJ, Saper MA, Samaoui B et al. Structure of the human class I histocompatibility antigen, HLA-A2. Nature 329:506, 1987a

Björkman PJ, Saper MA, Samraoui B et al. The foreign antigen binding site and T cell recognition regions of class I histocompatibility antigens. Nature 329:512, 1987b

Boehncke W-H, Takeshita T, Pendleton CD et al. The importance of dominant negative effects of amino acid side chain substitution in peptide-MHC molecule interactions and T cell recognition. J Immunol 150:331, 1993

Brown JH, Jardetzky TS, Gorga JC et al. Three-dimensional structure of the human class II histocompatibility antigen HLA-DR1. Nature 364, 1993

Buus S and Werdelin O. A group specific inhibitor of lysosomal cysteine proteinases selectively inhibits both proteolytic degradation and presentation of the antigen dinitrophenyl-poly-L-lysine by guinea pig accessory cells to T cells. J Immunol 136:452, 1986a

Buus S and Werdelin O. Oligopeptide antigens of the angiotensin lineage compete for presentation by paraformaldehyde-treated accessory cells to T cells. J Immunol 136:459, 1986b

Buus S, Sette A, Colon SM et al. Isolation and characterization of antigen-Ia complexes involved in T cell recognition. Cell 47:1071, 1986c

Buus S, Sette A, Colon SM et al. The relation between major histocompatibility complex restriction and the capacity of Ia to bind immunogenic peptides. Science 235:1353, 1987a

Buus S, Sette A and Grey HM. The interaction between protein-derived immunogenic peptides and Ia. Immunol Rev 98:115, 1987b

Buus S, Sette A, Colon SM et al. Autologous peptides constitutively occupy the antigen binding site on Ia. Science 242:1045, 1988

Chesnut RW, Endres RO and Grey HM. Antigen recognition by T cells and B cells: Recognition of cross-reactivity between native and denatured forms of globular antigens. Clin Immunol Immunopathol 15:397, 1980

Chicz RM, Urban RG, Lane WS et al. Predominant naturally processed peptides bound to HLA-DR1 are derived from MHC-related molecules and are heterogeneous in size. Nature 358:764, 1992

DeLisi C and Berzofsky JA. T cell antigenic sites tend to be amphipathic structures. Proc Natl Acad Sci USA 82:7048, 1985

Donermeyer DL and Allen PM. Binding to Ia protects an immunogenic peptide from proteolytic degradation. J Immunol 142:1063, 1989

Ellner JJ and Rosenthal AS. Quantitative and immunologic aspects of the handling of 2,4 dinitrophenyl guinea pig albumin by macrophages. J Immunol 114:1563, 1975

Ellner JJ, Lipsky PE and Rosenthal AS. Antigen handling by guinea pig macrophages: Further evidence for the sequestration of antigen relevant for activation of primed T lymphocytes. J Immunol 118:2053, 1977

Falk K, Rötzschke O, Stevanovic S et al. Allele-specific motifs revealed by sequencing of self-peptides eluted from MHC molecules. Nature 351:290, 1991

Fazekas de St Groth S. The quality of antibodies and cellular receptors. In: Lefkowitz Ed.: Immunological Methods, Academic Press, New York, 1979

Fremont DH, Matsumura M, Stura EA et al. Crystal structure of two viral peptides in complex with murine MHC class I H-2Kb. Science 257:919, 1992

Garrett TPJ, Saper MA, Björkman PJ et al. Specificity pockets for the side chains of peptide antigens in HLA-Aw68. Nature 342:692, 1989

Gell PGH and Benacerraf B. Studies on hypersensitivity. II. Delayed hypersensitivity to denatured proteins in guinea pigs. Immunology 2:64, 1959

Grantham R. Amino acid difference formula to help explain protein evolution. Science 185:862, 1974

Hammer J, Takacs B and Sinigaglia F. Identification of a motif for HLA.DR1 binding peptides using M13 display libraries. J Exp Med 176:1007, 1992

Hulme EC and Birdsall NJM. Receptor-Ligand interactions. Hulme EC, Ed: IRL Press, Oxford, 1992

Hunt DF, Henderson RA, Shabanowitz J et al. Characterization of peptides bound to the class I MHC molecule HLA-A2.1 by mass spectrometry. Science 255:1261, 1992a

Hunt DF, Michel HP, Dickinson TA et al. Peptides presented to the immune system by the murine class II major histocompatibility complex molecule, I-Ad. Science 256:1817, 1992b

Hyafil F and Strominger JL. Dissociation and exchange of the β2-microglobulin subunit of HLA-A and HLA-B antigens. Proc Natl Acad Sci USA 76:5834, 1979

Ishii N, Nagy ZA and Klein J. Restriction molecules involved in the interaction of T cells with allogeneic antigen-presenting cells. J Exp Med 156:622, 1982

Jardetzky TS, Lane WS, Robinson RA et al. Identification of self peptides bound to purified HLA-B27. Nature 353:326, 1991

Kindred B and Shreffler DC. H-2 dependence of co-operation between T and B cells *in vivo*. J Immunol 109:940, 1972

Klein J. Natural history of the major histocompatibility complex. J Wiley, New York, NY, 1986

Lamont AG, Powell MF, Colon SM et al. The use of peptide analogs with improved stability and MHC binding capacity to inhibit antigen presentation *in vitro* and *in vivo*. J Immunol 144:2493, 1990

Madden DR, Gorga JC, Strominger JL et al. The structure of HLA-B27 reveals nonamer self-peptides bound in an extended conformation. Nature 353:321, 1991

Matsumura M, Fremont DH, Peterson PA et al. Emerging principles for the recognition of peptide antigens by MHC class I molecules. Science 257, 1992

Mouritsen S, Meldal M, Werdelin O et al. MHC class II molecules protect T cell epitopes against proteolytic destruction. J Immunol 149:1987, 1992

Olsen AC, Pederson LØ, Stryhn A et al. A quantitative assay to measure the interaction between immunogenic peptides and MHC class I molecules. Eur J Immunol 24:385, 1994

Padlan EA. Structural implications of sequence variability in immunoglobulins. Proc Natl Acad Sci USA 74:2551, 1977

Parham P, Lomen CE, Lawlor DA et al. Nature of polymorphism in HLA-A, -B, and -C molecules. Proc Natl Acad Sci USA 85:4005, 1988

Rötzschke O, Falk K, Deres K et al. Isolation and analysis of naturally processed viral peptides as recognized by cytotoxic T cells. Nature 348:252, 1990

Rosenthal AS and Shevach EM. Function of macrophages in antigen recognition by guinea pig T lymphocytes. I. Requirement for histocompatible macrophages and lymphocytes. J Exp Med 138:1194, 1973

Rosenthal AS. Determinant selection and macrophage function in genetic control of the immune response. Immunol Rev 40:136, 1978

Rothbard JB and Taylor WR. A sequence pattern common to T cell epitopes. EMBO J 7:93, 1988

Rudensky AY, Preston-Hurlburt P, Hong S-C et al. Sequence analysis of peptides bound to MHC class II molecules. Nature 353:622, 1991

Schaeffer EB, Sette A, Johnson DL et al. Relative contribution of "determinant selection" and "holes in the T cell repertoire" to T-cell-responses. Proc Natl Acad Sci USA 86:4649, 1989

Schmidtke JR and Unanue ER. Macrophage-antigen interaction: Uptake, matabolism and immunogenicity of foreign albumin. J Immunol 107:331, 1971

Seeger RC and Oppenheim JJ. Synergistic interaction of macrophages and lymphocytes in antigen-induced transformation of lymphocytes. J Exp Med 132:44, 1970

Sette A, Buus S, Colon SM et al. Structural characteristics of an antigen required for its interaction with Ia and recognition by T cells. Nature 328:395, 1987

Sette A, Buus S, Colon S et al. I-Ad-binding peptides derived from unrelated protein antigens share a common structural motif. J Immunol 141:45, 1988

Sette A, Adorini L, Apella E et al. Structural requirements for the interaction between peptide antigens and I-Ed molecules. J Immunol 143:3289, 1989a

Sette A, Buus S, Apella E et al. Prediction of major histocompatibility complex binding regions of protein antigens by sequence pattern analysis. Proc Natl Acad Sci USA 86:3296, 1989b

Shimonkevitz R, Kappler JW, Marrack P et al. Antigen recognition by H-2-restricted T cells. I. Cell-free antigen processing. J Exp Med 158:303, 1983

Shimonkevitz R, Colon S, Kappler JW et al. Antigen recognition by H-2-restricted T cells. II. A tryptic ovalbumin peptide that substitutes for processed antigen. J Immunol 133:2067, 1984

Strominger JL, Engelhard VH, Fuks A et al. Biochemical Analysis of the Products of the MHC. Garland Press, New York, 1981

Teyton L, O'Sullivan D, Dickson PW et al. Invariant chain distinguishes between the exogenous and endogenous antigen presentation pathways. Nature 348:39, 1990

Townsend ARM, Rothbard J, and Gotch FM. The epitope of influenza nucleoprotein recognized by cytotoxic T lymphocytes can be defined with short synthetic peptides. Cell 44:959, 1986

Townsend A, Ohlen C, Bastin J et al. Association of class I major histocompatibility heavy and light chains induced by viral peptides. Nature 340:443, 1989

Townsend A, Elliott T, Cerundolo V et al. Assembly of MHC class I molecules analysed *in vitro*. Cell 62:285, 1990

Trowsdale J, Ragoussis J and Campbell RD. Map of the human MHC. Immunol Today 12:443, 1991

Unanue ER. The regulatory role of macrophages in antigen stimulation. Adv Immunol 15:95, 1972

Unanue ER. The regulatory role of macrophages in antigen stimulation. Part two: Symbiotic relationship between lymphocytes and macrophages. Adv Immunol 31:1, 1981

Unanue ER. Antigen-presenting function of the macrophage. Ann Rev Immunol 2:395, 1984

Van Bleek GM and Nathenson SG. Isolation of an endogenously processed immunodominant viral peptide from the class I H-2Kb molecule. Nature 348:213, 1990

Waldron JA, Horn RG and Rosenthal AS. Antigen-induced proliferation of guinea pig lymphocytes *in vitro*: Obligatory role of macrophages in the recognition of antigen by immune T lymphocytes. J Immunol 111:58, 1973

Waldron JA, Horn RG and Rosenthal AS. Antigen-induced proliferation of guinea pig lymphocytes *in vitro*: Functional aspects of antigen handling by macrophages. J Immunol 112:746, 1974

Werdelin O and Buus S. Evidence for two compartments in antigen presenting cells in which immunologically relevant antigen degradation may take place. In: Schook LB and Tew JG, Eds.: Antigen Presenting Cells: Diversity, Differentiation, and Regulation, Liss, Alan R, New York, NY, p 93, 1988

Ziegler HK and Unanue ER. Identification of a macrophage antigen-processing event required for I-region-restricted antigen presentation to T lymphocytes. J Immunol 127:1869, 1981

Ziegler HK and Unanue ER. Decrease in macrophage antigen catabolism cause by ammonia and chloroquine is associated with inhibition of antigen presentation to T cells. Proc Natl Acad Sci USA 79:175, 1982

Zinkernagel RM and Doherty PC. Restriction of *in vitro* T cell mediated cytotoxicity in lymphocytc choriomenigitis within a syngenic or semiallogenic system. Nature 248:701, 1974

Notes

1. No simple, nor truly satisfying, methods exist to establish a proper matrix for a given MHC. An unbiased matrix (e.g., Grantham tables) does not take into account the difference between the different positions within an MHC motif. A refined matrix can only be established experimentally, and it will differ from position to position and from MHC to MHC.

2. Ideally, a reiterative procedure should be used to optimize the ability of the motif to discriminate between binders and nonbinders. The first motif derived should be used to repeat the alignment and redefine the motif etc.

3. It should be noted that the PileUp program for its comparisons uses a matrix, which ideally should be altered to more accurately reflect the characteristics of the MHC in question.

4. So far, the use of preparations containing mixed MHC molecules has been successful in only one case where the specificity of the assay was resurrected using a labelled peptide which only bound to one of the MHC molecules present.

5. Recently, insect cell lines transfected with MHC genes have been reported to produce large amounts of empty MHC molecules.

6. Other labels like biotin followed by chemiluminescence or Europium etc. are in use, but usually the amount of free ligand is not measured, which in our hands lead to a significant loss in precision.

7. It also shows that the frequent use of "ratio of inhibitor to stimulator", or "excess of inhibitor", when reporting this kind of experiment is nonsense. In a properly conducted experiment the needed concentration of indicator or stimulator merely reflects the specific activity of your labelled ligand and the sensitivy of your counter, whereas the concentration of inhibitor needed reflects the K_D (or the receptor concentration) of the receptor-inhibitor interaction. The most accurate determination of the K_D is derived by a strict binding analysis. The second best choice is the IC_{50}, but never "excess of inhibitor", which at best is irrelevant.

8. Some MHC, in particular class I, only requires 18 hours. In one case more than 35 days seemed necessary.

9. This citrate-phosphate buffer contains 83 mM citric acid and 133 mM disodium hydrogen phosphate.

Chapter 6

PROMISCUITY AND SPECIFICITY OF T CELL EPITOPES

F. Sinigaglia, P. Panina-Bordignon, P. Valsasnini and J. Hammer

TABLE OF CONTENTS

I. INTRODUCTION

T lymphocytes recognize peptides derived from intracellular processing of antigens that have bound to major histocompatibility complex (MHC) molecules. Thus, the immunogenicity of a given epitope has three essential requirements: (i) the appropriate fragment must be generated by the processing machinery; (ii) an MHC molecule must be present and capable of binding this fragment; and (iii) specific T cells must be available in the repertoire. Data from experimental animal systems indicate that the capacity to respond to a given antigen is controlled by MHC gene products (Benacerraf, 1978) and that the lack of an MHC molecule capable of binding antigen is the most frequent cause of unresponsiveness. When an antigenic determinant is unable to interact with the MHC molecules possessed by a given individual, T cell clones specific for this particular combination of determinant and an MHC molecule cannot be activated and therefore not all individuals will respond to any one antigenic determinant.

MHC molecules are highly polymorphic, and it has been shown that peptides can bind to only one or a few murine class II alleles, but never to all (Babbitt et al., 1985; Guillet et al., 1986; Buus et al., 1986). Binding of peptides to human class II proteins, however, appears to be significantly more degenerate than binding to murine molecules. Promiscuous peptides binding to multiple allelic forms of HLA-DR molecules have indeed been described (Sinigaglia et al., 1988a; Panina-Bordignon et al., 1989; Ho et al., 1990).

0-8493-8967-4/95/$0.00+$.50

The purpose of this chapter is to review our own studies and those of others on promiscuous T cell determinants and to discuss new approaches to identify and characterize such epitopes.

II. PROMISCUOUS T CELL EPITOPES

II.1. A peptide to which most individuals can respond

The first evidence of extensive degenerate binding was an analysis of human T cell responses to peptides from the circumsporozoite protein of malaria (Sinigaglia et al., 1988a). The *Plasmodium falciparum* malaria circumsporozoite (CS) protein is the major sporozoite surface antigen and the best studied malaria vaccine candidate (Nussenzweig and Nussenzweig, 1989). We searched for T cell epitopes in the non repetitive regions of the CSprotein by testing the ability of different CS derived synthetic peptides to stimulate T cell proliferation in donors without a history of malaria infection. Using a restimulation assay, we could show specific proliferation of cells from 18 out of 33 donors in the presence of a synthetic peptide, CS.T3, corresponding to CS protein residues 378–398 (but with cysteines 384 and 389 replaced by alanine) (Sinigaglia et al., 1988b). These results suggested that CS.T3 can interact with a wide range of different MHC alleles, a finding confirmed by extensive analysis of cloned T cells. Analysis of the restriction specificity of the clones revealed that the peptide could be recognized in combination with at least eight different DR molecules including DRB1*0101, DRB1*1501, DRB1*0401, DRB1*1101, DRB1*1401, DRB1*0701, DRB1*0801 and DRB1*0901 (Sinigaglia et al., 1988a). We confirmed that peptide CS.T3 could bind to different DR antigens by testing it and several other peptides (comprising together about 50% of the CS protein) in a peptide competition assay (Kilgus et al., 1989). We found that CS.T3 was the only peptide able to compete with the binding of stimulator peptides to DRB1*1101 and DRB1*1401. CS.T3 was also the only peptide out of the five tested to bind to DRB1*0101 using a direct binding assay on soluble DRB1*0101 molecules (Jardetzky et al., 1990). The identification of a T cell determinant in the CS protein that can be presented to T cells by a broad spectrum of MHC proteins induced the hope to develop a subunit malaria vaccine applicable to the entire population.

II.2. *In vivo* studies

We examined whether the CS.T3 peptide could induce helper T cell function *in vivo*. We coupled it to the B cell determinant $(NANP)_3$. Mice from several strains, which were normally unable to respond to the NANP repeats, produced antibodies to the repeats as well as to sporozoites when injected with the CS.T3-$(NANP)_3$ conjugate (Sinigaglia et al., 1988a). This showed that the CS.T3 sequence was also recognized in association with many different mouse Ia molecules and that it can indeed function as carrier determinant for responses to other parts of the CS protein.

In a subsequent study the CS.T3 sequence was fused, by genetic engineering, to the 190L antigen from the blood stage of *Plasmodium falciparum* and tested for its ability to protect monkeys from parasite challenge. Immunization of Aotus monkeys with 190L alone resulted in poor protection against *Plasmodium falciparum* challenge. The addition of the CS.T3 epitope to 190L improved immunity, and as a result three of four monkeys were protected following challenge infection with blood-stage parasites (Herrera et al., 1992). Interestingly, in contrast to sera from nonprotected monkeys, sera from protected animals contained elevated levels of γ-interferon (IFN-γ), thus suggesting that the CS.T3 epitope, strongly bound to several different HLA-DR molecules, was involved *in vivo* in the activation of T lymphocytes to produce IFN-γ.

II.3. Other examples of promiscuous T cell determinants

Panina-Bordignon et al. identified two tetanus toxin-derived peptides (p2 and p30) that were recognized by all human donors who had been primed with tetanus toxin regardless of their MHC haplotypes (Panina-Bordignon et al., 1989). In the case of p30, corresponding to tetanus toxin residues 947–967, at least four different DR (DRB1*1101, DRB1*1102, DRB1*0701 and DRB1*0901) and two different DP alleles (DPB1*0201 and DPB1*0401) were used as restriction molecules. The T cell recognition of p2, corresponding to tetanus toxin residues 830–843, was even more degenerate. A large number of DR molecules were used as restriction elements: DRB1*0101, DRB1*1501, DRB1*0301, DRB1*0401, DRB1*1101, DRB1*1102, DRB1*1301, DRB1*0801, DRB1*0901, DRB3*0101, DRB3*0201. All these DR alleles were comparably efficient in peptide presentation. The analysis of deletion mutants revealed that p2 contained only one epitope (831–842) that was recognized in the context of each DR molecule tested. This was further supported by the analysis of several nonconservative substitutions that similarly affected T cell recognition in all DR combinations.

Ho et al. found two other regions in the tetanus toxin protein which also could be presented to their respective T cell clones by antigen presenting cells of many HLA class II specificities (Ho et al., 1990).

Using a cell surface binding assay, Busch et al. demonstrated that a biotinylated analoge of the influenza hemagglutinin (HA) peptide 307–319 could bind class II proteins on each of 22 homozygous B cell lines (Busch et al., 1990). Though not identical for each of the cell lines, the signal was shown to be specific for HLA-DR in each case by using cells transfected with DR molecules and by inhibition with anti-DR antibodies. Partial confirmation of this degeneracy in binding to HLA-DR proteins has been the demonstration that the HA peptide can bind purified DRB1*0101, DRB1*1101 and DRB1*0801 (Roche and Cresswell, 1990).

II.4. Is binding to human class II molecules more degenerate than binding to murine molecules?

The initial binding experiments using purified I-A and I-E molecules demonstrated great specificity in the binding. For example, the lysozyme peptide corresponding to residues 46–61 bound I-A^k, but not I-A^d, thus demonstrating a direct correlation between the capacity of a peptide to bind to MHC class II molecules and the capacity to induce an immune response (Babbitt et al., 1985). Similarly, Buus et al. showed that an immunogenic peptide of ovalbumin bound I-A^d, but not I-E^d, I-A^k or I-E^k (Buus et al., 1986). A more extensive study analyzing the capacity of 11 known determinants to bind I-A^d, I-A^k, I-E^d or I-E^k provided further evidence that the binding correlated with the MHC restriction pattern associated with the immune response to these peptides (Buus et al., 1987). The identification of different DR-restricted peptides that exhibited degeneracy at the binding level seemed to contradict the above-discussed binding results obtained with purified mouse MHC molecules. The discrepancy, however, is only apparent. Sette et al., using a large panel of synthetic peptides, found a significant correlation between the capacity of peptides to interact with different alleles of the same class II molecule (e.g., I-A^k, I-A^d) but no correlation with the capacity of peptides to interact with different isotypes within the same haplotype (e.g., I-A^d, I-E^d) (Sette et al., 1989). This suggests that different alleles of the same MHC molecule may actually recognize closely related structures whereas different isotypes may recognize unrelated, albeit not mutually exclusive, structures on the peptide molecule. DR proteins like I-E molecules have a conserved α-chain. It was thus tempting to speculate that promiscuous recognition might be dependent upon the presence of a monomorphic chain (Sinigaglia et al., 1988a).

III. THE MOLECULAR BASIS FOR SPECIFICITY AND PROMISCUITY OF PEPTIDE BINDING TO CLASS II MOLECULES

III.1. The principles for the recognition of peptides by MHC class I molecules

Studies of naturally processed peptides presented on MHC class I molecules have revealed sequence characteristics for MHC-bound peptides. Peptides binding to class I proteins are 8–10 amino acids long and in a certain position, relative to their N and C termini, one or few amino acid residues recur with considerably increased frequencies (Falk et al., 1991; Jardetzky et al., 1991; Hunt et al., 1992). Such residues, called "anchor residues", are characteristic for each class I allotype examined, and it has been shown that the side chains of these residues interact with allotype-specific residues in the peptide binding groove (Fremont et al., 1992; Madden et al., 1992). In contrast to class I, naturally processed peptides bound to self-class II molecules are longer (13–25 amino acids) and more heterogeneous than a single well-defined peptide bound to MHC class I molecules (Rudensky et al., 1991; Hunt et al., 1992; Chicz et al., 1992). Since the critical class II contact residues are variably positioned within different MHC binding peptides and because of their irregular length, it has been difficult to determine the requirements for the promiscuous or the specific recognition of peptide antigens by MHC class II molecules.

III.2. The phage library approach

To address the issue of how a single peptide can bind a broad range of different MHC class II alleles, we have chosen a new approach by screening a large, highly diverse peptide library expressed on the surface of filamentous phage (Hammer et al., 1992).

Epitope libraries comprise tens to hundreds of millions of short, variable amino acid sequences that are displayed on the surface of a bacteriophage virion (Scott and Smith, 1990; Devlin et al., 1990).

To construct the peptide display library we placed oligonucleotides encoding for random peptides of nine amino acid residues in length on the NH_2 terminus of the M13 phage protein III. Protein III is a minor coat protein of the M13 phage, and the region containing the peptide inserts is known to be exposed on the phage surface (Parmeley and Smith, 1988). The peptide sequences were flanked by two glycine spacers to maximize the flexibility of the peptide insert. DRB1*0101, DRB1*0401 and DRB1*1101 were affinity purified from human HLA-DR homozygous EBV-B cell lines, biotinylated and used to select phages from an M13 peptide display library (Fig. 1). The library, consisting of about 20 million independent phage recombinants, was screened in three cycles of panning, elution and amplification. The bound phages were then directly isolated after acid elution, and the sequences encoding the displayed peptides were determined. Table 1 lists the aligned sequences of the 9-mer peptide encoding region of 10 independent phages selected by each of the three DR alleles.

III.3. Promiscuous and allele-specific anchors in HLA-DR-bound peptides

Sequence analysis of the peptide encoding region of DR-bound phages led to the identification of position-specific anchor residues, defining motifs for peptide binding to DR molecules. The three DR motifs shared two anchor residues at relative positions 1 and 4, while allele-specific anchor residues were identified at position 6 (Hammer et al., 1993) (Table 2). Each of the aligned sequence pools showed a striking enrichment (>80%) of aromatic residues near the NH_2-terminus of the peptide inserts. A second general anchor residue was found at relative position 4, where the aligned peptide pools showed an enrichment of hydrophobic residues. Further constraints were found at position 6. In the DRB1*0401 selected peptides, we found mostly amino acids having hydroxyl groups (Thr and Ser). The positively charged amino acids Arg and Lys were enriched in the DRB1*1101 selected pool, and small amino acids (Ala and Gly) were found for the DRB1*0101 binding peptides. Such results appear to explain the basis for the promiscuity and the specificity of peptide recognition by HLA-DR molecules.

III.4. Implications for promiscuous recognition

The recent structural determination of the peptide binding groove of murine H-2K^b (Fremont et al., 1992) and HLA-B27 (Madden et al., 1992) class I molecules has revealed the presence of pockets of varying shapes that interact with anchor residues of binding peptides. The identification of anchor residues in class II binding peptides suggests that specific pockets also exist in MHC class II molecules. Moreover, computer-aided modelling predicts that the α- and the ß-chain of class II molecules each form approximately half of the peptide binding cleft (Brown et al., 1988). As previously discussed, unlike the ß-chain, the α-chain is constant in all DR molecules. Since the promiscuous anchor residues are situated in one-half of the peptides (positions 1 and 4), while the allele-specific residues are in the second half (position 6), it is possible that the promiscuous anchor residues bind primarily to the nonpolymorphic DRα-chain, while the variable ß-chain should contribute to polymorphic pockets for the variable anchor residues identified at position 6 (Fig. 2). Indeed, an aromatic binding pocket has been identified in the recently solved crystal structure of DRB1*0101 (Stern et al., 1994).

The identification of both conserved and allele-specific anchors in HLA-DR binding peptides suggests a simple mechanism to explain the promiscuity and the specificity of peptide binding to MHC class II molecules. The usage of some of these pockets in multiple combinations should allow HLA-DR molecules to bind a broad range of peptides in an allele-specific or in a promiscuous manner.

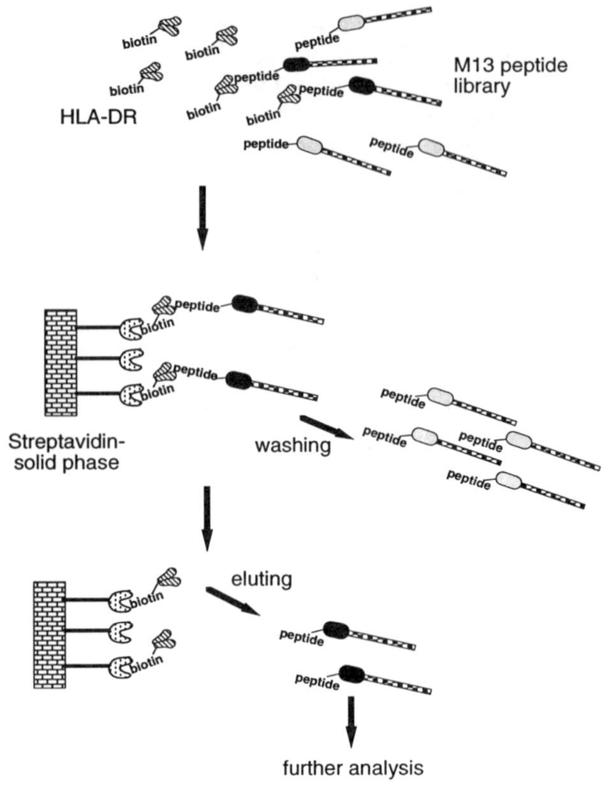

Fig. 1. Screening method for the isolation of HLA-DR binding phages. Phages from the library are reacted with biotynilated DR molecules, then immobilized on streptoavidin-agarose beads and extensively washed to remove unbound or weakly binding phages.

IV. FUTURE STRATEGIES FOR THE IDENTIFICATION OF PROMISCUOUS EPITOPES

Methods to identify regions of protein sequences capable of binding MHC proteins would be very valuable for many immunological applications. So far, it has been impossible to predict with reasonable accuracy whether a particular peptide sequence will be recognized by mouse or human T cells in association with a particular MHC molecule. The algorithms published by DeLisi and Berzofsky and Rothbard and Taylor (DeLisi and Berzofsky, 1985; Rothbard and Taylor, 1988), based on the detection of amphipathic α-helical segments and on sequence pattern recognition, respectively, have proven to be of limited use. Indeed, the X-ray structures of MHC peptide complexes have shown that the peptide is not bound to MHC molecules in a helical conformation. Madden et

TABLE 1
Aligned amino acid sequences of the random peptide region of DR selected phages

DRB1* 0101	**DRB1* 0401**	**DRB1* 1101**
Y R Q **M** S **A** P T L	G **W** G M **M** R **T** L R	M **W** Q E **L** V **R** I Q
Y K P **L** W **G** Q M T	K A **W** S R **M** G **S** M	T **W** G G **L** G **R** L L
Y L R **M** L Q I L K	N **W** R G **V** L **S** Q M	L **W** A Q **V** S **R** A I
M **Y** G L **M** L **A** Q R	**W** N Y I S **S** V Q S	**W** R W **M** S **R** V S S
Y K L **L** R **A** T Q M	**W** R I **A** Q **T** N H A	R **W** K L **M** Q **K** Q L
Y R G **M** S **A** F R A	**W** S T **A** R **T** L W Q	T **W** R T **L** W **R** Q V
Y K P **M** L **A** S V G	**Y** H A H R **T** L L Q	**Y** T Q **M** M **R** S F H
N **W** R A **L** R **G** A L	**Y** R H **A** V G Q L G	G **W** Q L T A **K** R T
Y Q A **L** R **A** Y W Q	A **W** A H **M** T **T** L R	**W** V S **L** G **R** L A A
R V **F** Q N **L** R **A** S	**W** Q N **M** V **T** T L Q	M **W** E M **L** W **R** P R

The phages were selected from the M13 peptide display library by DRB1*0101, DRB1*0401 and DRB1*1101 molecules and isolated after three rounds of screening (Hammer et al., 1993). Ten representative sequences for each DR allele are shown. Anchor residues are indicated in bold type.

al. have reported a crystallographic study of the human class I molecules HLA-B27 in which the peptides are bound in a largely extended conformation (Madden et al., 1992). In addition, Freemont et al. have determined the crystal structure of two peptides H-2K^b complexes. Both peptides bind in an extended conformation with most of their surface buried in the H-2K^b binding groove (Fremont et al., 1992).

The identification of promiscuous and allele-specific anchors in MHC binding peptides has clear implications for the prediction of T cell determinants on a given protein. Promiscuous binding peptides would require the presence of anchors interacting with corresponding pockets of several MHC alleles; it should, however, be taken into account the fact that some of these pockets may exhibit preferences, but not an absolute requirement, for a specific residue. Further complexity comes from the fact that consensus motifs are merely a minimal requirement for a peptide to be selected by an MHC molecule. Other constraints are likely to be essential for the binding. For example, by analyzing sequence characteristics of phage inserts binding to DRB1*0101, we observed that certain amino acid residues were significantly reduced in DRB1*0101 binders. In particular, we found that the negatively charged amino acids Glu and Asp were virtually excluded from the DRB1*0101 binding sequences (Fig. 3), thus indicating that, in addition to specific anchor residues, the overall amino acid composition of a given peptide is important for MHC binding.

It should be possible in the near future to develop a new generation of algorithms that take in to consideration all the factors described here, the sum of which will reflect the peptide's relative ability to bind MHC proteins.

V. PROTOCOLS FOR PEPTIDE BINDING ASSAYS

The method described here is based on previous tests for binding of peptides to MHC class II antigens (Sette et al., 1989; Sinigaglia et al., 1992), but it is adapted for screening of large numbers of peptides for binding to a given affinity-purified HLA-DR molecule.

V.1. Reagents

Stock solutions for affinity purification of HLA-DR antigens and peptide binding assay.

- *Lysis buffer*: 1% (v/v) NP-40; 25 mM iodoacetamide; 1 mM phenylmethylsulfonylfluoride (PMSF); 5 mM ϵ-amino-n-caproic acid; 10 μg/ml of each of soybean trypsin inhibitor, antipain, pepstatin, leupeptin and chymostatin; in 0.05 M sodium phosphate buffer, pH 7.5 containing 0.15 M NaCl. 5% sodium deoxycholate (DOC).
- *Wash buffer 1*: 50 mM Tris-Cl, pH 8; 0.15 M NaCl; 0.5% NP-40; 0.5% DOC; 10% glycerol; 0.03% NaN_3.
- *Wash buffer 2*: 50 mM Tris-Cl, pH 9; 0.5 M NaCl; 0.5% NP-40; 0.5% DOC; 10% glycerol; 0.03% NaN_3.
- *Wash buffer 3*: 2 mM Tris-Cl, pH 8; 1% octyl-β-D-glucopyranoside; 10% glycerol; 0.03% NaN_3.

TABLE 2
Allele-specific motifs of DR-bound peptides

DRB1*0101	**Y**	X	X	**M**	X	**A**	X	X	**L**
	F			**L**		**G**			
	W								
DRB1*0401	**W**	X	X	**M**	X	**T**	X	X	X
	Y			A		**S**			
				V					
				L					
DRB1*1101	**W**	X	X	**M**	X	**R**	X	X	X
				L		K			
				V					

Sequences of peptides from phages binding to DR molecules were aligned using the PileUp program of the GCG program package (Hammer et al., 1992). Residues recurring with increased frequency, relative to the nonselected library, at a given position of the random peptide inserts are shown. Residues whose frequency was higher than 20% are indicated in bold type.

o *Elution buffer*: 50 mM diethylamine-Cl, pH 11.5; 1% octyl-β-D-glucopyranoside; 0.15 M NaCl; 1 mM EDTA; 1 mM EGTA; 10% glycerol; 0.03% NaN_3; 10mM freshly added iodoacetamide.
o *Neutralizing solution*: 1 M Tris-Cl, pH 6.8; 0.15 M NaCl; 1% octyl-β-D-glucopyranoside.
o *Binding buffer*: 2% NP-40; 12 mM N-ethylmaleimide; 2 mM PMSF; 3.2 mM EDTA; 20 mg/ml of each of the protease inhibitors listed above for lysis buffer; in 0.05 M sodium phosphate buffer, pH 7.5 containing 0.15 M NaCl.
All of these solutions should be stored at 4°C, except for the DOC which can be kept at 20°C.

V.2. Immunoaffinity purification of DR molecules

V.2.a. Preparation of the antibody column
The mouse monoclonal anti-human HLA-DR (E.31) antibody, purified from mouse ascites by ammonium sulfate fractionation and DEAE anion exchange column chromatography as described previously (Sinigaglia et al., 1992), is first coupled to CNBr-activated Sepharose 4B (Pharmacia; Uppsala, Sweden) according to the manufacturers' instructions to give approximately 8 mg of antibody per milliliter of settled gel.

V.2.b. Isolation of DR molecules
o *Procedure*: About 10^{10} human HLA-DR homozygous Epstein Barr virus-transformed B cell lines are harvested by centrifugation and washed twice. Pelleted cells are resuspended to a cell density of 10^8/ml in lysis buffer and incubated on ice for 30 min. The lysate is cleared of nuclei and debris by centrifugation at 27,000 x g for 30 min. If not used immediately, cell lysates can be stored at -70°C. Add 0.2 volume of 5% sodium deoxycholate (DOC) to the post-nuclear supernatant. After 10 min of mixing the extract is centrifuged at 100,000 x g for 2 hours and the supernatant is filtered through a 0.45 μm poresize membrane. The lysate is mixed with 10 ml (settled volume) of Sepharose 4B-coupled anti-human HLA-DR antibody and rotated at 4°C for 3 hours. The gel mixture is transferred into a column and washed at 0.6 ml/min at 4°C with at least 20 column volumes of wash buffer 1, five column volumes of wash buffer 2 and five column volumes of wash buffer 3. DR antigens are then eluted at 0.4 ml/min with two column volumes of elution buffer. Two ml fractions are collected into tubes that contain 250 μl of neutralizing solution. Aliquots from each fraction are analyzed by SDS-PAGE. Fractions containing most of the DR antigens are pooled, aliquoted and kept frozen at -70°C until use.

V.3. Spin-column binding assay

o *Procedure:* Mix 25 μl (1 to 5 μg) of immunoaffinity-purified DR antigens, 25 μl of 2x binding buffer, 5 μl (5 ng) of ^{125}I-labeled immunogenic peptide, and 5 μl of PBS containing 0 or 10 ng of test peptide. The reaction mixtures can be incubated either at 20°C for 48 hours or at 37°C overnight. At the end of the incubation period a part or all of the incubation mixture is applied on top of Bio-Spin columns (Bio-Rad, Richmond, CA) and centrifuged at 1100 x g for 4 min. Excluded material that is collected into the collection tubes is counted directly in a γ-scintillation counter. The fraction of iodinated peptide bound to DR antigen is calculated by dividing the amount of radioactivity found in the void volume of the column by the total radioactivity applied.

VI. PROTOCOLS FOR THE IDENTIFICATION OF BINDING MOTIFS IN HLA-DR-BOUND PEPTIDES

The method described here is based on the recently developed phage display technology (Scott and Smith, 1990; Devlin et al., 1990), adapted for the isolation of large numbers of bacteriophage which display peptides able to bind to purified MHC class II antigens. The protocols are optimized for the screening of HLA-DRB1*0101 binding phages (Hammer et al., 1992). Small adjustments of binding and screening conditions might be necessary for other MHC class II molecules (Hammer et al., 1993).

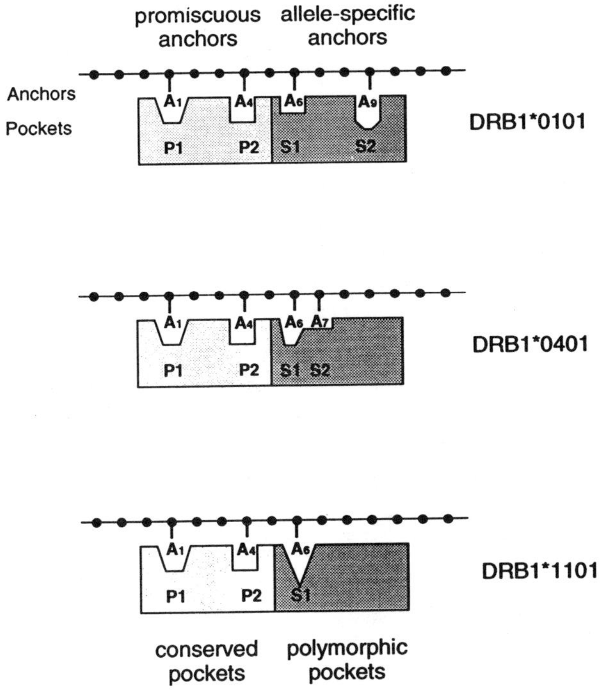

Fig. 2. Putative pockets in different HLA-DR alleles for binding of peptides. The model is based on peptide binding motifs identified through phage library screening (Hammer et al., 1993).

VI.1. Materials

VI.1.a. Reagents

o *Buffers and media*:

Binding buffer: 50 mM TRIS-HCl, pH 7.5; 150 mM NaCl; 2 mM EDTA and 0.2% Nonidet P40.

Biotinylation buffer: 0.25 M $NaHCO_3$, pH non-adjusted; 0.2% Nonidet P40.

Elution buffer: 0.1 N glycine-HCl, pH 2.2; 1 mg/ml BSA.

PSB (phage storage buffer: 50 mM TRIS-HCL, pH 7.5; 150 mM NaCl.

LB (Luria-Bertani) Medium: per liter: 10 g Bacto-tryptone; 5 g Bacto-yeast extract; 10 g NaCl; pH adjusted to 7.5 with NaOH.

o *Affinity purified HLA-DR antigen*: see above.

o *Amplified M13 peptide display library*: our library consisted of 2 x 10^7 independent clones. The random peptide inserts were nine amino acids in length and flanked by four glycine on each side.

o *Other reagents*: Biotin-XX-NHS (Calbiochem, La Jolla, CA): 25 mM in dimethyl-formamide; BSA: bovine serum albumin fraction V; *Echerichia coli* XL1Blue plating bacteria (OD = 0.5): XL1Blue can be obtained from Stratagene (LaJolla, CA); plating bacteria (OD = 0.5) are prepared as described (Sambrook et al., 1989); Microsep Microconcentrator, 30 k cutoff, (Scan AG, Basel, Switzerland); PD-10 columns: Sephadex G-25M (Pharmacia, Uppsala, Sweden); PEG solution: 20% polyethylene glycol (PEG 8000) in 2.5 M NaCl; streptavidin on 4% beaded agarose (Sigma, St.Louis, MO).

VI.2 Procedure

VI.2.a. Biotinylation of HLA-DR proteins

The following procedure results in approximately 80% biotinylation of HLA-DR antigens, which means that 80% of the DR molecules, binding a peptide *in vitro*, can be precipitated with streptavidin-agarose.

o *Procedure*: The buffer of 1 mg affinity-purified HLA-DR antigen solution (see previously) is exchanged into biotinylation buffer by gel filtration using two PD-10 columns sequentially, according to the manufacturer's instructions. The volume of the HLA-DR solution is reduced to 1 ml by ultradialyzation with Microsep Microconcentrators. To the concentrate 8 μl of a 25 mM solution of Biotin-XX-NHS is added. The mixture is rotated for 1 hour at room temperature (RT). The excess Biotin-XX-NHS is removed by gel-filtration using a PD-10 column equilibrated with binding buffer. The biotinylated DR antigens are concentrated by ultradialyzation (see previously), aliquoted and kept frozen at -70° C until used.

VI.2.b. Library screening

Different MHC class II alleles show distinct optimal peptide/phage binding characteristics. Therefore, optimization experiments are highly recommended for each MHC class II allele prior to screening. Optimization experiments may be performed using the spin-column binding assay (described previously) and/or M13 bacteriophage displaying peptides with known binding characteristics (Hammer et al., 1992).

o *Procedure*: Approximately 1 x 10^{10} bacteriophages of an amplified M13 peptide display library are incubated with 5 μg biotinylated HLA-DR molecules in binding buffer and 1 mM PMSF at RT for 48 hours (total volume: 60 μl). To reduce nonspecific binding, streptavidin on 4% beaded agarose is washed twice with 10 volumes binding buffer (1 volume = volume of settled streptavidin-agarose), incubated for 1 hour with 10 volumes binding buffer containing 5 mg/ml BSA and washed subsequently three times with 10 volumes binding buffer. 500 μl BSA blocked streptavidin on 4% beaded agarose in binding buffer (total 50 μl settled volume) is added to the library/HLA-DR incubution mix and rotated for 10 min at RT. The M13 phage/HLA-DR complexes are purified by washing the streptavidin solid phase eight times with binding buffer, once with 70 mM citrate-phosphate buffer, pH 3 and once more with binding buffer (volume: 1 ml per wash step). The bacteriophages are eluted with 1 ml elution buffer for 10 min and neutralized with 60 μl 2 M TRIS-base. For amplification, 300 μl of eluted bacteriophages are used to infect 1 ml of *E.coli* XL1-Blue plating bacteria (OD = 0.5). The infected cells are transferred to 7 ml LB medium in a 25 cm^2 tissue culture flask. After 1 hour incubation (200 rpm/37° C), ampicillin is added to a final concentration of 20 μg/ml. The flask is incubated for at least 20 hours at 37° C and 200 rpm. For harvesting and purification of the bacteriophages, the cell/phage mixture is transferred to 15 ml tubes (Falcon, polypropylene) and centrifuged at 3500 rpm for 25 min at RT. Five ml of supernatant is transferred into a new tube, and 1 ml PEG solution is added and mixed by inverting the tube several times. After 10 min incubation at RT, the precipitated bacteriophage particles are recovered by centrifugation using the same conditions as above. The supernatant is aspirated, the tubes are recentrifuged for 30 seconds and any residual supernatant is carefully removed. The phage pellet is resolved in 1.2 ml PSB and transferred into a 1.5 ml tube (Eppendorf). A 200 μl PEG solution is added, and the phage particles are recovered by centrifugation (15000 rpm, 10 min, RT). The phage pellet is resolved in 500 μl binding buffer and stored at 4°C until use. An aliquot of this phage solution (approximately 1 x 10^{10} phage particles) is used for the next round of screening and amplification. The eluate of the fourth round of screening is used to isolate and sequence individual M13 clones (Sambrook et al., 1989).

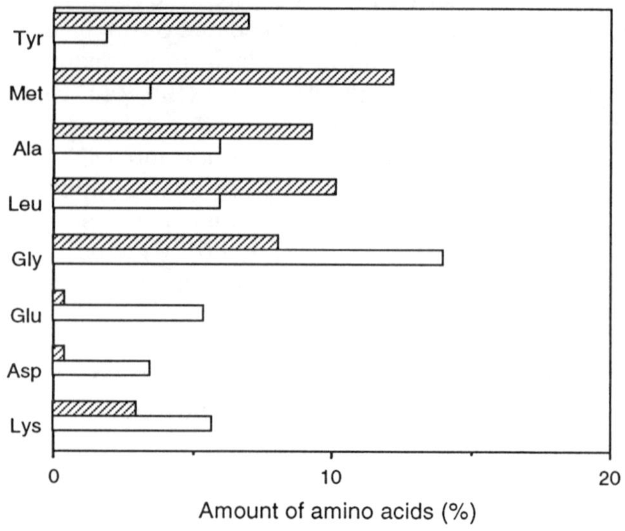

Fig. 3. Amino acid content of the peptide inserts before and after DRB1*0101 selection. The overall amino acid composition of the sequences of phage inserts selected by DRB1*0101 (dashed bars) was compared with sequences of control phages isolated from the original library (open bars). Only the amino acids significantly (>2%) increased or reduced are shown.

ABBREVIATIONS

BSA	Bovine serum albumin
CS	Circumsporozoite protein
DOC	Sodium deoxycholate
EDTA	(Ethylenedinitro)tetra-acetic acid
EGTA	[(Ethylene-bis(oxyethylenenitro)]tetra-acetic acid
HA	Hemagglutinin
IFN	Interferon
LB	Luria-Bertani
MHC	Major histocompatibility complex
PEG	Polyethylene glycol
PMSF	Phenylmethylsulfonylfluoride
PSB	Phage storage buffer

VII. REFERENCES

Babbitt B, Allen PM, Matsueda G et al. The binding of immunogenic peptides to Ia histocompatibility molecules. Nature 317:359, 1985

Benacerraf B. A hypothesis to relate the specificity of T lymphocytes and the activity of I region-specific Ir genes in macrophages and B lymphocytes. J Immunol 120:1809, 1978

Brown JH, Jardetzky T, Saper B et al. A hypothetical model of the foreign antigen binding site of class II histocompatibility molecules. Nature 332:845, 1988

Busch R, Strang G, Howland K et al. Degenerate binding of immunogenic peptides to HLA DR proteins on B cell surfaces. Int Immunol 2:443, 1990

Buus S, Sette A, Colon S et al. Isolation and characterization of antigen-Ia complexes in T cell recognition. Cell 47:1071, 1986

Buus S, Sette A, Colon S et al. The relation between major histocompatibility complex (MHC) restriction and the capacity of Ia to bind immunogenic peptides. Science 235:1353, 1987

Chicz RM, Urban RG, Lane WS et al. Predominant naturally processed peptides bound to HLA-DR1 are derived from MHC-related molecules and are heterogeneous in size. Nature 358:764, 1992

DeLisi C and Berzofsky J. T cell antigenic sites tend to be amphipathic structures. Proc Natl Acad Sci USA 82:7048, 1985

Devlin JJ, Panganiban LC, Devlin PE et al. Random peptide libraries: A source of specific protein binding molecules. Science 249:404, 1990

Falk K, Roetzschke O, Stefanovic S et al. Allele-specific motifs revealed by sequencing of self-peptides eluted from MHC molecules. Nature 351:290, 1991

Fremont DH, Matsumura M, Stura EA et al. Crystal structures of two viral peptides in complex with murine MHC class I H-$2K^b$. Science 257:919, 1992

Guillet JG, Lai MZ, Briner TJ et al. Interaction of peptide antigens and class II major histocompatibility antigens. Nature 324:260, 1986

Hammer J, Takacs B, Sinigaglia F et al. Identification of a motif for HLA-DR1 binding peptides using M13 display libraries. J Exp Med 176:1007, 1992

Hammer J, Valsasnini P, Tolba K et al. Promiscuous and allele-specific anchors in HLA-DR binding peptides. Cell 74:197, 1993

Herrera MA, Rosero F, Herrera S et al. Protection against malaria in Aotus monkeys immunized with a recombinant blood-stage antigen fused to a universal T-cell epitope: Correlation of serum gamma interferon levels with protection. Infection Immunity 60:154, 1992

Ho PC, Mutch DA, Winkel KD et al. Identification of two promiscuous T cell epitopes from tetanus toxin. Eur J Immunol 20:477, 1990

Hunt DF, Michel H, Dickinson TA et al. Peptides presented to the immune system by the murine class II major histocompatibility complex molecule IA^d. Science 256:1817, 1992

Jardetzky TS, Gorga JC, Busch R et al. Peptide binding to HLA-DR1: A peptide with most residues substituted to alanine retains MHC binding. EMBO J 9:1797, 1990

Jardetzky TS, Lane WS, Robinson RA et al. Identification of self peptides bound to purified HLA-B27. Nature 353:326, 1991

Kilgus J, Romagnoli P, Guttinger M et al. Vaccine T-cell epitope selection by a peptide competition assay. Proc Natl Acad Sci USA 86:1629, 1989

Madden DR, Gorga JC, Strominger JL et al. The three-dimensional structure of HLA-B27 at 2.1 Å resolution suggests a general mechanism for tight peptide binding to MHC. Cell 70:1035, 1992

Nussenzweig V and Nussenzweig RS. Rationale for the development of an engineered sporozoite malaria vaccine. Adv Immunol 45:283, 1989

Panina-Bordignon P, Tan A, Termijtelen A et al. Universally immunogenic T cell epitopes: Promiscuous binding to human MHC class II and promiscuous recognition by T cells. Eur J Immunol 19:2237, 1989

Parmeley SF and Smith GP. Antibody-selectable filamentous fd phage vectors: Affinity purification of target genes. Gene 73:305, 1988

Roche P and Cresswell P. High affinity binding of an influenza hemagglutinin derived peptide to purified HLA-DR. J Immunol 144:1849, 1990

Rothbard J and Taylor WR. A sequence pattern common to T cell epitopes. EMBO J 7:93, 1988

Rudensky AY, Preston-Hurlburt P, Hong S et al. Sequence analysis of peptides bound to MHC class II molecules. Nature 353:622, 1991

Sambrook J, Fritsch EF and Maniatis T. Molecular cloning: A laboratory manual, Second edition. Cold Spring Harbor Laboratory Press, Cold Spring Harbor, NY, 1989

Scott JK and Smith GP. Searching for peptide ligands with an epitope library. Science 249:386, 1990

Sette A, Buus S, Colon S et al. Structural analysis of peptides capable of binding to more than one Ia antigen. J Immunol 142:35, 1989

Sinigaglia F, Guttinger M, Kilgus J et al. A malaria T-cell epitope recognized in association with most mouse and human MHC class II molecules. Nature 336:778, 1988a

Sinigaglia F, Guttinger M, Gillessen D et al. Epitopes recognized by human T lymphocytes on malaria circumsporozoite protein. Eur J Immunol 18:633, 1988b

Sinigaglia F, Romagnoli P, Guttinger M et al. Selection of T-cell epitopes and vaccine engineering. Methods Enzymology 203:370, 1992

Stern LJ, Brown JH, Jardetzky TS et al. Chrystal structure of the human class II MHC protein HLA-DR1 complexed with an influenza virus peptide. Nature 368:215, 1994

Chapter 7

PROCESSING OF PROTEINS IN THE MHC CLASS II PATHWAY

J.M. van Noort

TABLE OF CONTENTS

I. PROCESSING REQUIREMENTS FOR PROTEIN RECOGNITION BY CLASS II-RESTRICTED T CELLS

In the early 1980s it became clear that the activation of antigen-specific helper T cells required more than the mere presence of whole antigen in conjunction with accessory or antigen-presenting cells (APC) of the corresponding MHC haplotype. Several observations indicated that uptake and intracellular handling ("processing") of a protein antigen by APC was a necessary first step before T cells could respond to it. Experiments performed by Emil Unanue and co-workers illustrated that mild fixation of live APC by paraformaldehyde, which prevents endocytic processes, blocked antigen recognition by helper T cells (Ziegler and Unanue, 1981, 1982; reviewed by Unanue, 1984 and Allen et al., 1987). Blocking, however, was only observed if fixation was performed prior to the addition of antigen. Once antigen and APC had been mixed, fixation no longer affected T cell activation. Similar effects were observed when APCs were treated with reagents that impair acidification and/or fusion of endocytic vesicles, such as chloroquin, ammoniumchloride or the ionophore monensin (reviewed by Unanue, 1984). Again, treatment of APC before adding protein antigen blocked T cell activation, whereas a similar treatment given after exposure to antigen had no effects. Apparently, protein antigens had to be taken up inside the endocytic pathway of APC before any immunogenic material could be formed. Later experiments confirmed that maintenance of low pH in the endocytic pathway was crucial to the ability of the APC to handle the antigen effectively (McCoy et al., 1989; Stossel et al., 1990). Studies of the intracellular pH within the endocytic route indicate that a pH gradient exists that ranges from about 5–6 within early endocytic vesicles to a little over 4.5 in late endosomes and lysosomes (Galloway et al., 1983; Helenius et al., 1983; Mellman et al., 1986).

Careful assessment of the time between APC encounter of antigen and exposure of immunogenic material confirmed the existence of a distinct lag-phase, consistent with a requirement for uptake and intracellular handling of antigen by the APC (Chesnut et al., 1982; Grey et al., 1982; Roosnek et al., 1988). This lag-phase ranges from 15 to about 60 min, depending on the type and complexity of the antigen used. The sequence of events between encounter of antigen by the APC and exposure of immunogenic material at the APC surface was termed "antigen processing".

II. EVIDENCE FOR PROTEOLYSIS DURING PROCESSING

In 1984, it was reported that pre-treatment of the protein antigen ovalbumin with trypsin prior to

0-8493-8967-4/95/$0.00+$.50

addition to APC circumvented the requirement for antigen processing (Shimonkevitz et al., 1983). Once ovalbumin had been treated with trypsin, neither fixation of APC nor the addition of chloroquin could block the activation of ovalbumin-specific helper T cells. After fractionation of trypsin-digested ovalbumin by reversed-phase HPLC, a single peptide fragment turned out to be sufficient for the activation of cloned helper T cells (Shimonkevitz et al., 1984). An identical finding was reported for a tryptic fragment of lysozyme (Allen et al., 1984). These observations provided the first direct evidence that a short protein fragment was all that was required for antigen recognition by a T cell. Since then, numerous findings have confirmed and extended this early observation in demonstrating that, in fact, peptide binding by MHC molecules was the molecular basis for the Ir gene control over T cell responses as had been suspected for over a decade (Rosenthal and Shevach, 1973; Babbitt et al., 1985; Buus et al., 1986,1987; Guillet et al., 1986).

The idea that processing of antigen by APC involved proteolysis was rapidly confirmed by the observation that treatment of APC with inhibitors of proteolysis, such as pepstatin or leupeptin, had similar effects as mild fixation or chloroquin treatment (Buus and Werdelin, 1986; Puri and Factorovich, 1988; Takahashi et al., 1988; Vidard et al., 1991, 1992b). However, despite the apparently straightforward explanation that was first given to these observations, subsequent experiments have revealed that much of the effects of inhibitors such as leupeptin can now be attributed to inhibitory effects upon the release of the invariant chain from immature class II MHC molecules rather than on blocking the degradation of antigen itself. Like a number of other intracellular events involving maturation and transport, removal of the invariant chain from class II MHC molecules is mediated by stepwise proteolytic degradation which is sensitive to leupeptin (Blum and Cresswell, 1988; Bakke and Dobberstein, 1990; Nguyen et al., 1989a, 1989b, 1993; Roche and Cresswell, 1991; Neefjes and Ploegh, 1992) and possibly mediated by cathepsin B (Reyes et al., 1991). Obviously, inhibition of this event may easily reduce the intracellular availability of empty, newly synthesized class II MHC molecules. Since protein fragments generated intracellularly largely depend on this newly synthesized pool of MHC molecules for presentation to T cells (Davidson et al., 1991; Lanzavecchia et al., 1992), leupeptin or more aspecific inhibitors of protein synthesis such as cycloheximide inhibit antigen presentation most likely by reducing the number of class II MHC molecules that are available inside the APC for presentation of protein fragments (Jensen, 1988; StPierre and Watts, 1990).

Nevertheless, the basic conclusion drawn from early experiments with protease inhibitors, *viz.* that proteolysis is an essential step in antigen processing, has remained unchallenged. Most likely, the main function of proteolysis is to achieve exposure of otherwise hidden functional groups in a polypeptide chain that are required for interacting with the antigen-binding groove of MHC molecules.

III. DISULFIDE BOND REDUCTION

Apart from proteolysis, another protein unfolding event is associated with processing. In many cases, peptides can be released from proteins only if disulfide bonds are broken (Feener et al., 1990; Collins et al., 1991; Hampl et al., 1992). Not only will disulfide bonds hamper proteolysis by holding together a protein molecule in a compact form, but they will also anchor amino acid sequences to the protein core. Thus, many protein fragments can only be presented if the protein antigen undergoes reduction first.

The lysozyme has been employed by several research groups as a model example of a protein antigen that contains disulfide bonds. Examination of the processing route for this model protein has led to the hypothesis that intracellular cysteine may act as a naturally occurring reducing agent within the processing compartment (Collins et al., 1991). Since this reducing capacity has been found in lysosomes but not in early endocytic vesicles, a requirement for disulfide bond reduction may explain why the lysozyme is processed mainly in lysosome-like vesicles and not in early compartments of the endocytic pathway (Harding et al., 1991, 1992). Our own observations have confirmed complete resistance of unreduced lysozyme to proteolytic cleavage *in vitro* (Van Noort and van der Drift, 1989), while reduced and unfolded lysozyme is readily cleaved by cathepsin D (Van Noort and Jacobs, 1994).

In summary, processing of protein antigens for presentation by class II MHC molecules involves both proteolysis and reduction of disulfide bonds at relatively low pH within the APC. Below, data are discussed on the exact intracellular location of processing and the type of proteases that could be involved in the process. These data only pertain to the processing route of antigen for presentation in the context of class II MHC molecules. With regard to the processing route for presentation in the class I MHC pathway, the reader is referred to reviews by others that have recently appeared on this rapidly evolving subject (Driscoll and Finley, 1992; Goldberg and Rock, 1992).

IV. SUBCELLULAR LOCALIZATION OF THE CLASS II PROCESSING MACHINERY

The subject of the exact intracellular site for processing in the class II pathway is a difficult one. Most likely, no distinct type of compartment exists to which antigen processing is exclusively limited and which is common to all types of cells that are capable of processing and presentation. Immunoelectron microscopy, together with biochemical studies, has clearly revealed the involvement of an endocytic compartment in which endocytosed antigen meets newly synthesized class II MHC molecules travelling to the cell surface (Cresswell, 1985; Harding and Unanue, 1989; Harding et al., 1985, 1990, 1991; Neefjes et al., 1990; Harding and Geuze, 1992)). However, two papers appearing shortly following each other (Gualiardi et al., 1990 and Peters et al., 1991, respectively) have highlighted a fundamental problem in the analysis which is the fact that research groups that have examined different types of antigen-presenting cells come up with quite different results. Apparently, different types of antigen- or MHC-carrying vesicles may be observed from one cell to another. In fact, many of the intracellular vesicles and bodies that make up the endocytic pathway are heterogeneous in appearance even within single cells. Among others, they include small clathrin-coated vesicles, large multi-vesicular bodies, dense phagosomes, highly reticulated tubules, large and apparently empty "sacs" and regular small bodies of low electron density (Marsh et al., 1986; Helenius et al., 1983; Harding and Geuze, 1992; Schmid et al., 1988; Schwartz, 1990). Presently, this variety in appearance renders it difficult to morphologically identify a distinct "processing compartment". Perhaps such a definition will turn out to be impossible. Rather, we may consider the endocytic compartment as a dynamic gradient of vesicles through which endocytosed material gradually traverses while being continuously redistributed (McVey Ward et al., 1989). Textbook routes involving discrete one-way traffic from one type of vesicle to another may well be only part of the story.

Neither morphological nor biochemical criteria have yet succeeded in identifying a single distinct processing compartment. Only the distinction between endosomal vesicles and classical lysosomal bodies can be made sufficiently clear when aided by the availability of classical lysosomal marker enzymes such as ß-hexosaminidase or specific receptors such as the mannose-6-phosphate receptor which are absent from early endosomes (Dahms et al., 1989). By using density gradient centrifugation, lysosomes of relatively high density can be separated from all other low density vesicles, including many of the heterogeneous endosomal vesicles (Van Renswoude et al., 1982; Beaumelle and Hopkins, 1989). Thus, the contribution to processing by both types of vesicles can also be examined by biochemical methods. Studies along these lines confirm morphological data; antigen as well as other components of the processing machinery such as acid-dependent proteolytic enzymes can be found in both endosomal and lysosomal vesicles (Van Renswoude et al., 1982; Galloway et al., 1983; Diment and Stahl, 1985; Van Noort et al., 1991; Ludwig et al., 1991). Thus, class II-associated processing appears to occur widely dispersed within endocytic vesicles, from early endosomes to late lysosome-like vesicles and dense phagolysosomes. While in some cases antigens need only to pass through early endosomes for the generation of immunogenic material (McCoy et al., 1993), passage through later endosomal/lysosomal compartments may be required in others (Niebling and Pierce, 1993). As stated previously, a requirement for disulfide bond reduction may at least partially explain differential requirements for intracellular routing since reduction can only be effected in lysosomes. Also the complexity of an antigen, i.e., whether it is a soluble protein or a bacterial particle, will probably pose restrictions upon the intracellular location where peptides can be effectively released. These data imply that peptides from various types of compartments should, in principle, all be able to meet class II MHC molecules on their way to the cell surface. Together, these data suggest widespread communication between vesicles belonging to both endocytic and exocytic pathways (Balch, 1989; McVey Ward et al., 1989; Schwartz, 1990). Alternatively, it may be envisaged that newly synthesized class II MHC molecules may be loaded into early endosomes and travel along with endocytosed antigen (Romagnoli et al., 1993). In this way, removal of the invariant chain could occur concomitantly with release of antigen fragments. Many of the fusion and/or routing steps in the class II MHC pathway, however, remain to be established. The general features of the pathway as discussed above are summarized in Fig. 1.

V. PROTEASES IN THE CLASS II PATHWAY

Several experimental approaches are open to collect information on the nature of proteolytic enzymes that are involved in the degradation of proteins along the class II processing pathway. First, the effects of exogenous class-specific protease inhibitors or protease-specific antibodies upon processing under *in vitro* culturing conditions may reveal contributions by distinct classes of proteases or individual enzymes. Second, identification of the processing compartment would allow isolation and identification of relevant proteases contained within that compartment. Third, immunocytochemical techniques allow the examination of intracellular compartments for the presence of distinct proteolytic enzymes and their co-localization with endocytosed antigen and other components of the processing and presentation machinery. Fourth, molecular biological techniques can be employed to create

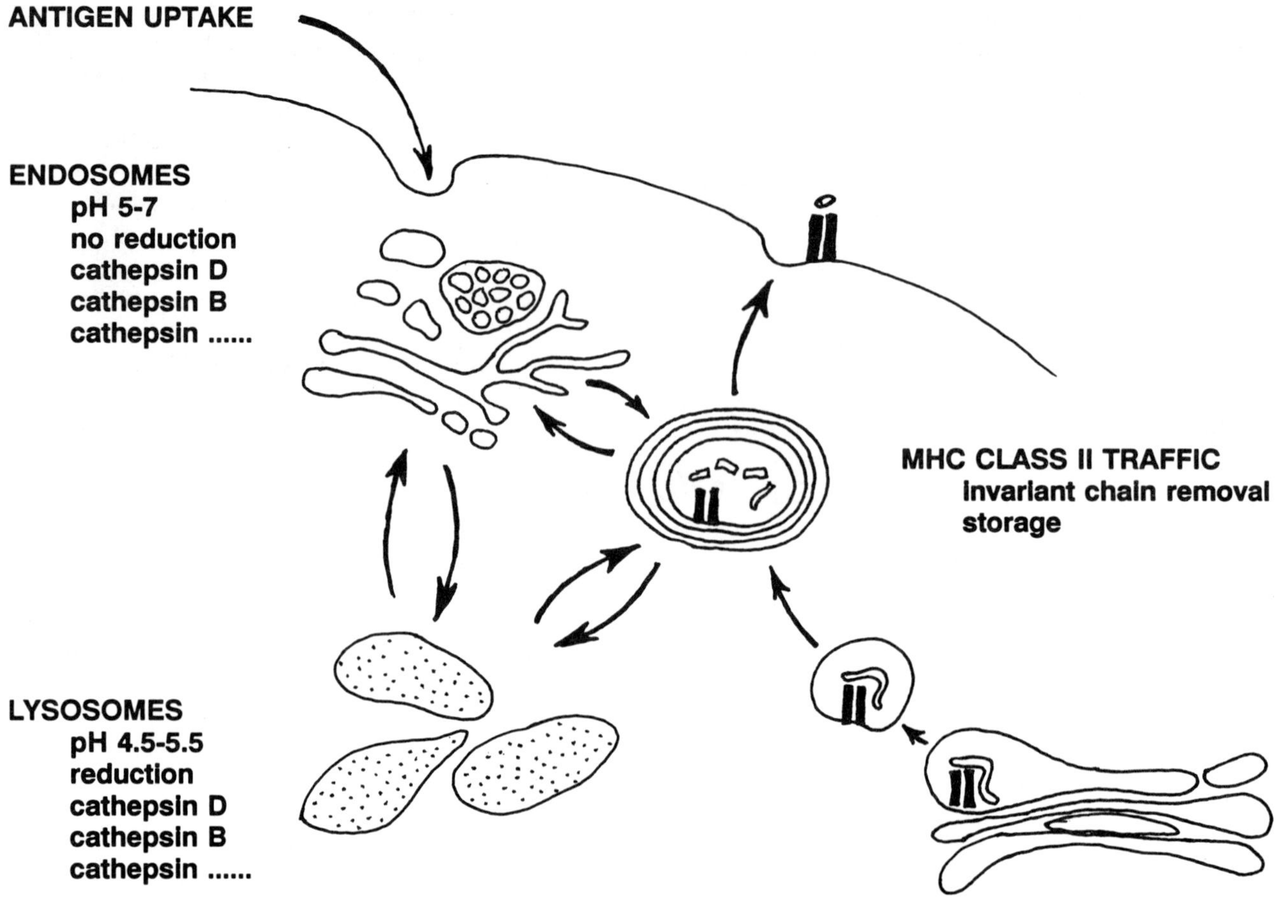

Fig. 1. General features of the class II processing pathway. Intracellular processing of protein antigen in the class II MHC pathway takes place within the endocytic pathway. This pathway comprises classical dense lysosomes together with a variety of heterogeneous endocytic vesicles and bodies generally referred to as endosomes. All or most endocytic vesicles contain acid-dependent proteases such as cathepsin D and cathepsin B, but only late lysosomes contain a significant reducing capacity, which may in several cases be required for release of peptides from antigen. The exocytic route of class II MHC molecules intersects the endocytic pathway. Separate MHC storage vesicles may communicate with the different types of endocytic vesicles. Alternatively, newly synthesized, immature class II MHC molecules may be delivered early into the endocytic pathway and travel in cohort with endocytosed antigen. Proteolytic enzymes involved in antigen cleavage may be similar to those involved in removal of the invariant chain.

processing mutants as cell lines or transgenic animals that lack one or more defined proteases, thus allowing evaluation of the selected enzyme's contribution. Although mutants have been described that are defective in the class II pathway but appear to have normal class II MHC molecules themselves (McCoy et al., 1989; Mellins et al., 1990; Ceman et al., 1992; Riberdy and Cresswell, 1992; Riberdy et al., 1992), none of these mutations have yet been attributed to proteases. Rather, they may affect proteins that are involved in transport, maturation and/or fusion steps. Finally, isolation of peptide species presented at the cell surface or formed within the endocytic compartment itself and identification of their sequences could reveal cleavage sites and provide information on the enzymes responsible. Later, the available evidence obtained by the various lines of research will be discussed.

VI. EFFECTS OF PROTEASE INHIBITORS

In several model systems, the effects of specific protease inhibitors have been examined. In these studies, inhibitory or stimulatory effects upon proliferative responses of antigen-specific helper T cell clones have been evaluated in order to obtain information on the intracellular antigen degradation step. Proteolytic enzymes are generally divided into four major classes, *viz*; serine proteases, metalloproteases, cysteine proteases and aspartyl proteases. Only the latter two classes are known to include several enzymes that are active at low pH (3–6); hence, most attention has been given to these cysteine and aspartyl proteases. In several studies, the inhibitor leupeptin has been used as a specific inhibitor of cysteine proteases. It should be kept in mind, however, that leupeptin is known to also inhibit some serine proteases (Bond and Butler, 1987). As a general inhibitor for aspartyl proteases, the inhibitor pepstatin can be used (McAdoo et al., 1973). Unfortunately, the strong hydrophobicity of pepstatin prevents it from proper solubilization into aqueous solution. Usually, pepstatin is dissolved in dimethylsulfoxide or some other organic solvent and diluted into the culturing medium. It is unclear, however, to what extent pepstatin diffuses into cellular membranes due to its hydrophobicity instead of remaining in solution for binding to proteases. In other words, no suitable control is possible over the effective dose of pepstatin that is supplied to APC under experimental conditions. Especially, if the concentration of proteases to be inhibited is substantial (intra-endosomal concentrations of cathepsin D, for example, may be as high as 30 mg/ml; Fruton, 1987), the lack of detectable inhibitory effects does not necessarily imply irrelevance of the corresponding target protease.

Apart from methodological problems in controlling effective doses of such inhibitors, an even greater problem lies in the possibility that protease inhibitors exert pleiotropic effects within live APC that could influence the presentation of processed material in other ways than by blocking antigen cleavage. The interpretation of experiments using leupeptin, a widely employed inhibitor, is complicated especially by the fact that leupeptin completely blocks release of the invariant chain from newly synthesized class II MHC molecules (Reyes et al., 1991). The serious implications of this effect are already discussed above.

Studies on the effects of protease inhibitors usually include control experiments to check for an intact presentation system. This is performed by demonstrating unaltered T cell responses to exogenous control antigens that do not require intracellular reduction and/or proteolysis, i.e., already fragmented proteins or peptides. However, while presentation of intracellularly generated protein fragments may be inhibited by a reduced availability of newly synthesized class II MHC molecules (Jensen, 1988; StPierre and Watts, 1990), presentation of exogenous peptides will be affected much less since the latter may bind directly to surface-exposed class II MHC molecules by displacement of weakly bound peptides. Consequently, inhibitory effects by leupeptin unrelated to the antigen cleavage step itself (e.g., by blocking the release of invariant chains) may not be adequately revealed by examining the presentation of exogenous peptides. Examination of the effects of leupeptin in the processing and presentation of ovalbumin has recently emphasized this pitfall (Diment, 1990; Rodriguez and Diment, 1992) which implies that any inhibitory effect by leupeptin is difficult to attribute unequivocally to the inhibition of antigen cleavage per se.

The effects of protease inhibitors upon the activation of class II-restricted T cells as reported by different groups differ from antigen to antigen and from T cell clone to T cell clone examined (Buus and Werdelin, 1986; Puri and Factorovich, 1988; Takahashi et al., 1988; Diment, 1990; Rodriguez and Diment, 1992). The most consistent inhibitory effects have been observed for leupeptin, an inhibitor of cysteine proteases. As stated above, however, it usually remains unclear to what extent this can be explained by inhibition of the antigen cleavage step itself. It is of interest to note that the possible role for cathepsin B in processing has been largely based on the above-described inhibition experiments that are difficult to interpret unequivocally.

Only few experiments have been performed to examine the effects of protease-specific antibodies upon processing of endocytosed antigen. One report describes full inhibition of the intraendosomal cleavage of bovine serum albumin by anti-cathepsin D antibodies (Dingle et al., 1973). The relevance of this finding, however, should be clarified by comparing this effect to that of antibodies directed against other proteases.

VII. INTRACELLULAR LOCALIZATION OF CLASS II PROCESSING

Immunocytochemical studies on the intracellular localization of proteases and their co-localization with other components of the class II MHC processing and presentation machinery have focused on cathepsin D and cathepsin B as the best-known and most abundant representatives of aspartyl and cysteine proteases, respectively. In most studies reported so far, both types of proteases have been found in a variety of endocytic vesicles, including electron-light endosomes and electron-dense lysosomal and autophagous vesicles, in line with biochemical data on the wide subcellular distribution of cathepsin D and cathepsin B (Diment and Stahl, 1985; Diment et al., 1988; Ludwig et al., 1991). Also, co-localization with class II MHC molecules has been reported for both proteases in several cases, consistent with a possible role in the generation of antigen-derived peptides for binding to class II molecules (Peters et al., 1991; Gualiardi et al., 1990; Harding and Geuze, 1992). Obviously, localization studies may only reveal an opportunity for involvement, but they cannot prove such an involvement. Possible co-localization of other proteolytic enzymes with components from the class II pathway awaits further studies with reagents specific for proteases other than cathepsin D and cathepsin B.

VIII. NATURALLY PROCESSED PEPTIDES BEFORE PRESENTATION: THE INTERMEDIATE PRODUCTS

Analysis of the peptides that are actually formed upon natural processing has proven difficult. Biochemical data on peptides formed after uptake by APC of model antigens such as cytochrome c (Marsh et al., 1986), myoglobin (McKean et al., 1983; Van Noort et al., 1991) or IgG molecules (Watts and Davidson, 1988) indicate that the extent of protein degradation within the time frame required for the formation of immunogenic material is surprisingly limited. T cell activation assays indicate that about 15 min is sufficient for the appearance of processed material on the APC surface that is derived from free soluble protein molecules. Yet, biochemical analyses reveal that within this period only a few percent of the ingested protein is usually degraded to any detectable extent (Marsh et al., 1986; Van Noort et al., 1991). Apparently, only mild degradation of an endocytosed pool of protein molecules is sufficient for the release of T cell stimulatory peptides.

It is not surprising, therefore, that only a few reports have appeared so far that describe a biochemical analysis of processed material. Several attempts to analyze naturally processed antigens prior to their binding by MHC molecules have been frustrated by the fact that there is so little material to analyze in the first place. Reversed-phase HPLC analysis of such material has only been reported for myoglobin (Van Noort et al., 1991). Somewhat unexpectedly, the profile revealed a discrete pattern of a few distinct peaks. This pattern could be mimicked accurately by incubating the same antigen with a preparation of permeabilized endosomal vesicles separated from lysosomal vesicles by density gradient centrifugation (Van Noort et al., 1991). It is important to note that mimicking the degradation event observed in live APC by this *in vitro* approach proved possible only if the proteases contained in the preparation were allowed to act on myoglobin for a short period of time. As shown in Fig. 2, longer incubations with the endosomal protease collection led to the appearance of a myriad of peptides inseparable by HPLC. These observations are consistent with the idea that limited rather than extensive degradation of protein antigen accompanies the release of T cell stimulatory peptides in the class II pathway.

In the above *in vitro* processing system of partially purified and permeabilized endosomal vesicles and by using actual appearance of myoglobin protein fragments as a read-out system, the effects of class-specific protease inhibitors like leupeptin and pepstatin were examined (Van Noort et al., 1991). The data obtained clearly pointed to aspartyl proteases as a crucial class of enzymes for the degradation of myoglobin; pepstatin completely blocked the appearance of any peptide from myoglobin, even upon extended incubation with the endosomal proteases. Leupeptin did have some inhibitory effects, but they appeared to be secondary, i.e., affected only subsequent degradation steps along the cleavage pathway but not initial cleavage of the protein. Consistent with these observations was the subsequent finding that treatment of myoglobin with purified cathepsin D, the most abundant representative of aspartyl proteases, led to cleavages at many of the same peptide bonds that were found to be cleaved by the complete collection of endosomal proteases at low pH. Treatment with cathepsin B, as the most abundant representative of cysteine proteases, did not affect intact myoglobin at all. Fragments of myoglobin, in contrast, were rapidly attacked by cathepsin B at their C-termini under the very same conditions. At least in this case, cathepsin B apparently fails to catalyze initial cleavage but may be quite active in trimming peptides once they are released by another protease such as cathepsin D. Subsequent studies performed in our laboratory have since confirmed the lack of any endoproteolytic activity of cathepsin B on several intact protein substrates under conditions where peptide substrates are rapidly trimmed down to small fragments. A display of exopeptidase-like activity rather than endoproteolytic activity by cathepsin B has been noted previously (McKay et al., 1983; Aronson and Barrett, 1978; Bond and Barrett, 1980). Endoproteolytic activity of cathepsin B does occur in some cases (McKay et al., 1983), but apparently not with every protein substrate. It should be noted that a preference for cleaving adjacent to basic amino acid residues by cathepsin B has been suggested on more than one occasion. Yet, no evidence for this has ever been reported, and the presumed preference for recognizing basic residues has been inferred only from the fact that cathepsin B recognizes a short synthetic tripeptide that contains such basic residues (Bond and Butler, 1987).

IX. NATURALLY PROCESSED AND CLASS II MHC-PRESENTED PEPTIDES

While the isolation and characterization of the intermediate products of processing (protein fragments released by endosomal proteases) have only been performed in a single case (Van Noort et al., 1991), the final products of processing and presentation (MHC-presented peptides) have recently been identified in several cases. By combining immunoaffinity chromatography to purify MHC-peptide complexes and

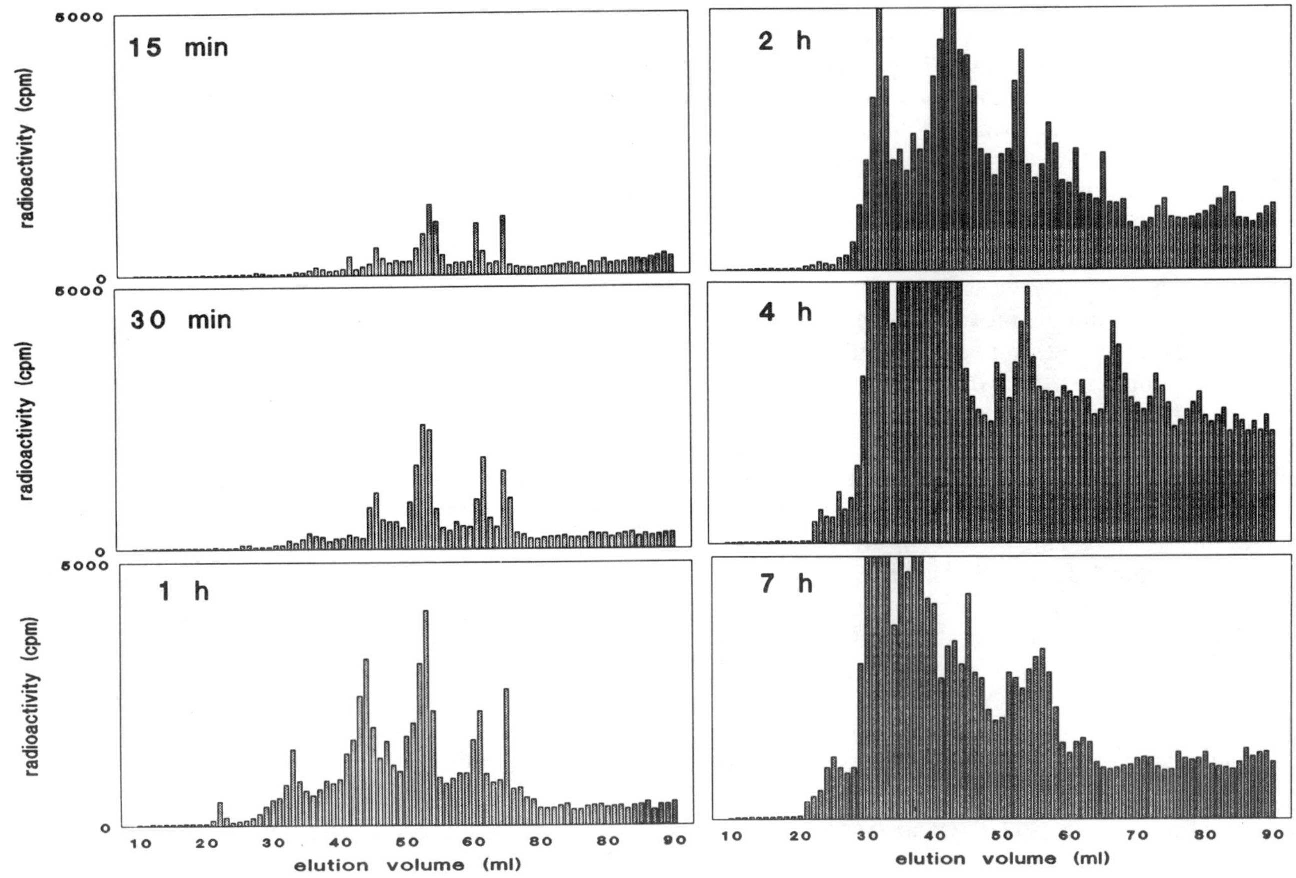

Fig. 2. *In vitro* degradation of myoglobin by collective endosomal proteases. A preparation of permeabilized macrophage endosomes was supplied with radiolabelled myoglobin and kept at pH 5.0 for several periods of time at 37°C. Reversed-phase HPLC analysis of the resulting myoglobin fragments shows the release of a distinct set of myoglobin fragments in the first stage of degradation. Only upon continued processing, a multitude of different fragments appear by HPLC, revealing the ultimately complete destruction of the antigen.

state-of-the-art sequencing equipment, a number of MHC-bound peptides were isolated and sequenced directly (Rudensky et al., 1991, 1992; Chicz et al., 1992; Nelson et al., 1992; Riberdy et al., 1992; Hunt et al., 1992; Sette et al., 1992), extending previous biochemical characterizations of naturally processed and presented antigen fragments (Demotz et al., 1989a; Srinivasan and Pierce, 1990). Data obtained have illustrated two major points. First, peptides eventually presented by class II MHC molecules are usually between 13 and 25 residues in length and quite heterogeneous in this respect when compared to the much stricter length distribution of class I MHC-bound peptides (between 8 and 11 residues).

Second, several partially overlapping peptide species have been found that differ from each other in both N- and C-termini boundaries. Although C-terminal ruffling of peptide termini seems a little more frequent (Rudensky et al., 1991), it has become clear that all peptides have been exposed to exopeptidase-like activities that attack both N- and C-termini of peptides. In view of this, it seems likely that the mere binding to MHC molecules is the reason for the ultimate survival of peptides and that exoproteolytic trimming continues until hampered by steric hindrance by the MHC molecule itself (Donermeyer and Allen, 1989; Mouritsen et al., 1992). It should be noted, however, that the phenomenon of extensive peptide trimming does not neccessarily imply such trimming to only occur intracellularly during processing. It is likely that also following exposure of MHC-peptide complexes at the cell surface, such trimming may continue, mediated by extracellular (serum) proteases which may have vivid activities upon relatively short peptides (Falo et al., 1992). Thus, the exact boundaries of peptides eluted from presenting MHC molecules do not neccessarily reflect endosomal processing alone but also subsequent extracellular activities.

When taken together, these features of naturally processed and presented peptides are consistent with the notion that, initially, antigen is degraded only mildly into relatively large protein fragments (about 20–30 residues). Next, these initial degradation products may bind to MHC molecules, which is likely to occur intracellularly (Harding et al., 1990; Neefjes et al., 1990; Lamb et al., 1991; Gualiardi et al., 1990; Peters et al., 1991). In this way, antigen fragments are rescued from continued degradation, including exoproteolytic trimming that is ultimately capable of destroying antigen up to the single amino acid level. This trimming may continue not only as long as the newly formed peptide-MHC complex travels along the exocytic pathway *en route* to the cell surface, but also following exposure at the cell surface. The sizes reported for naturally processed, MHC-presented peptides (13–25 residues) are significantly larger than the sizes of peptides that have routinely been found to fulfill the minimal requirements for MHC binding and T cell activation (7–8 residues). The minimally required sequence is usually taken to define a given T cell epitope. It is of interest, however, to note that peptides actually presented to T cells are usually twice as large as minimal determinants. The fact that the T cell receptor is forced to recognize a minimal determinant within the context of several flanking residues has been amply demonstrated to affect the fine specificity of many T cells.

X. HETEROGENEITY IN CLASS II PROCESSING

Occasionally, experimental data are reported which appear to suggest heterogeneity among different types of cells in class II-associated processing performance. Also, the existence of MHC-linked processing in the class II pathway has been suggested (De Koster et al; 1992; Demotz et al., 1989b; Michalek et al., 1989; Vidard et al., 1992a). It should be noted, however, that none of these studies have actually examined the degradation of the antigen itself, but rather they study (monoclonal) T cell responses to a selected antigen. While heterogeneity may exist in accessory factors that influence the activation of antigen-specific T cells by given sets of APC, no evidence has so far been collected to indicate functional heterogeneity for proteases in the class II processing pathway. In contrast, all evidence points to the involvement of regular lysosomal housekeeping enzymes such as cathepsin D and cathepsin B which renders it highly unlikely that significant heterogeneity at the functional level exists. Despite extensive studies of these enzymes over decades, no evidence has been reported so far that points to any functional heterogeneity or polymorphism. Detailed examination of the activities of cathepsin D from different tissues and species at the amino acid level has only revealed strongly conserved properties (Press et al., 1960; Ferguson et al., 1973; Offerman et al., 1983; Imoto et al., 1987; Fruton, 1987; Bond and Barrett, 1980; Lenarcic et al., 1988; Van Noort and van der Drift, 1989).

Also, from a functional point of view, the existence of extensive proteolytic heterogeneity in the class II processing pathway would make little sense. T cell help to IgG responses would be severely hampered if B cells were not able to generate the same peptide fragments from a given antigen as dendritic cells or macrophages, which are required to activate virgin T cells. Indeed, early studies have clearly shown similar processing performances by B cell tumors and macrophages (Chesnut et al., 1982).

One aspect of processing that may be prone to MHC-linked differences may be the extent to which certain MHC-entrapped peptides are trimmed down once they are bound. It may be envisaged that different modes of binding of a single peptide to different class II haplotypes may affect the range over which steric hindrance by the MHC molecule extends along the sequence. Thus, the activity of trimming exopeptidases may be limited to different residues in the context of different MHC molecules. This phenomenon may be detected by T cell clones sensitive to such minor variations in the structural context of a core epitope (Allen et al., 1987).

XI. STERIC FACTORS IN PROCESSING

The notion that MHC class II-associated antigen processing is a mild and selective event implies possible influences by steric factors such as complexation to antibody, binding to receptors or formation of quaternary structures. Much more than extensive degradation, mild cleavage will reflect accessibility of peptide bonds as well as the substrate selectivity of the enzymes involved. Any process or condition that affects peptide bond accessibility may be expected to alter the outcome of processing, presumably not by altering the locations of cleavages themselves but rather by changing the relative rates at which they occur.

Ample evidence points to the relevance of mere availability of processed peptide to presentation as one of the determining factors in epitope selection (Gammon et al., 1987; Li et al., 1992; Kim and Jang, 1992). Thus, by differentially affecting the relative rate at which different peptides are released from a given protein antigen, steric factors will be reflected in the set of peptides that are entrapped by MHC molecules and ultimately presented at the APC surface. Indeed, a significant number of reports have described such effects by antibody/ligand complexation or by quaternary interactions within an antigenic particle (Grey et al., 1982; Streicher et al., 1984; Manca et al., 1988, 1991;

Chu and Pizzo, 1993). Also, whether or not an antigen is administered as a free molecule or as part of a larger structure has been found to be of relevance to the selection of fragments presented to T cells (Boog et al., 1993).

Particular attention has been focused on IgG-mediated antigen uptake by B cells and subsequent processing. Several authors have observed differential effects upon epitope-specific responses caused by antigen complexation with antibody, covalently as well as non-covalently (Casten and Pierce, 1988; Davidson and Watts, 1989; Davidson et al., 1990, 1991). Especially the finding that the ratio of antibody/antigen in these complexes affects the outcome and that the binding of many antibodies to a single antigen inhibits presentation to T cells points to effects that steric factor may have on the outcome of the processing and presentation pathway (Manca et al., 1991). Also, under cell-free conditions, antibodies have been found to reduce the rate of proteolysis by relevant enzymes. Nevertheless, data obtained by us indicate that complexes between antigen and polyclonal antibodies eventually yield similar peptides, as do free antigen molecules albeit in much lower amounts (Van Noort et al., unpublished observations). Whereas polyclonal antibodies will have diffuse effects, monospecific antibodies may reduce the availability of distinct fragments and, thus, affect the nature of the epitopes presented by individual B cells. This effect has been taken to explain the phenomenon of so-called "directional help" (Manca et al., 1991).

XII. A PIVOTAL ROLE FOR CATHEPSIN D

When taken together, the above data allows some conclusions on the degradation pathway that are relevant to the release of protein fragments for presentation by class II MHC molecules. It is clear that more than just one or two different proteases are present in the class II processing pathway. These proteases are likely to be of lysosomal/endosomal origin, and their activity is acid dependent. Interestingly, both degradation of the antigen itself and the release of the invariant chain from newly synthesized class II MHC molecules share such requirements, suggesting that both processes may occur in close conjunction and perhaps even simultaneously in the same type of vesicle.

A considerable number of different proteases with the required characteristics are presently known, including cathepsin B, D, E, H, L, M, N, S and T (Bond and Butler, 1987). Among these lysosomal/endosomal proteases, cathepsin D and cathepsin B are by far the most abundantly present enzymes with intracellular concentrations, ranging up to about 30 mg/ml (Bond and Butler, 1987). Given the evidence that class II-associated processing is mild and, therefore, will reflect dominant enzyme activities, special attention for both cathepsin D and cathepsin B seems justified. Also, both proteases have been demonstrated by immunocytochemical methods to be localized in the same intracellular vesicles as endocytosed antigen and class II MHC molecules, consistent with their possible involvement in processing. Focused attention for cathepsin D and cathepsin B should, however, not rule out the possibility that isolated cases of processing may require contributions by other enzymes like cathepsin E (Bennett et al., 1992).

Together, endosomal/lysosomal proteases represent both endoproteolytic activities (a.o. cathepsin D, E and B) and exoproteolytic activities (a.o. cathepsin B, S, M), and they are able to fully degrade a protein if given the opportunity. Once again, however, in all cases examined by biochemical techniques, endocytosed protein only suffers limited degradation within the time frame relevant to the release of T cell activating peptides. In evaluating the available data for an involvement of any particular protease in the class II processing pathway, a distinction must be made between studies that actually analyze protein degradation and studies that employ indirect assays such as proliferation of antigen-specific T cells as a measure for antigen degradation. Data obtained in the latter type of study are difficult to interpret; many of the effects observed may result from other factors than changing the rate of intracellular antigen degradation. These considerations have been discussed previously.

Studies of actual antigen degradation imply cathepsin D as the single dominant endoproteolytic activity within endosomes. Anti-cathepsin D antibodies completely block protein degradation inside endosomal vesicles *in vivo* (Dingle et al., 1973), while the cathepsin D inhibitor pepstatin competely blocks endosomal protein degradation *in vitro* (Diment and Stahl, 1985; Van Lenten and Fogelman, 1990; Van Noort et al., 1991). Also, treatment of antigen with purified cathepsin D *in vitro* generates most of the cleavages in the model protein myoglobin that are found upon treatment with the complete collection of endosomal proteases isolated from macrophages (Van Noort et al., 1991). Comparison between the results of antigen treatment *in vitro* with either purified cathepsin D or cathepsin B has revealed that in most cases, including myoglobin, lysozyme, influenza virus nucleoprotein and foot-and-mouth disease virus, cathepsin B is simply unable to release any peptide from an intact protein while being quite active in trimming down peptides once they are released under the very same conditions

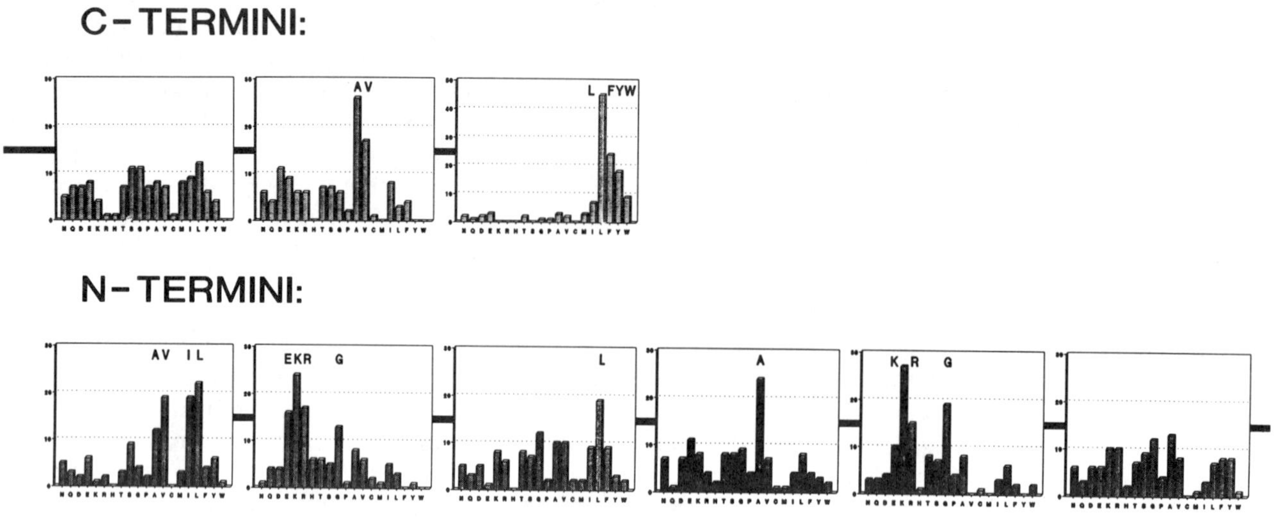

Fig. 3. Substrate specificity of cathepsin D. During the course of our studies, we have identified 113 peptide bonds in a large number of different proteins that are cleaved by purified bovine spleen cathepsin D under mild conditions, i.e., resulting in cleavage of less than 10% of the substrate protein. Fig. 3 shows a compilation of the sequences cleaved by giving the number of times any given amino acid was found at a particular position in the polypeptide chain relative to the cleavage site. Such a compilation reveals a distinct preference, albeit not absolute, for cleaving within a sequence motif that extends over eight amino acid residues.

(Van Noort et al., 1991 and partly unpublished observations). Thus, direct analyses of the activities of cathepsin D and B *in vivo* as well as *in vitro* reveal marked endoproteolytic activities by cathepsin D but not by cathepsin B. Interestingly, in all cases examined, the collective set of antigen fragments as released by cathepsin D did trigger responses by antigen-specific T cells.

Apart from quantitative evidence that points to a pivotal role for cathepsin D, qualitative evidence implies the enzyme in the release of T cell stimulatory peptides. Early evidence obtained with myoglobin as antigen (Van Noort et al., 1991) is currently being extended by other findings to indicate that several immunodominant T cell epitopes are located exactly adjacent to a cathepsin D cleavage site in such a way that it forms the N-terminus of a cathepsin D processed fragment (Van Noort et al., 1989, 1991; Boots et al., 1991; Van Noort and Jacobs, 1994). This qualitative phenomenon has been explained on the basis of the substrate specificity of the enzyme. Examination of the specificity of cathepsin D has revealed a distinct preference for cleaving within a particular amino acid sequence motif. In the course of our studies, we have identified 113 peptide bonds that are cleaved by purified cathepsin D under mild conditions. Fig. 3 shows a compilation of cleaved sequences and reveals the substrate specificity of cathepsin D, extending our previously reported work (Van Noort and van der Drift, 1989). The enzyme appears to recognize about eight amino acid residues within the sequence of its substrate. Such an extended substrate specificity is consistent with the relatively large substrate-binding groove of cathepsin D which it shares as a conserved property with all other aspartyl proteases (Fruton, 1987; Suguna et al., 1987; Navia et al., 1989; Metcalf and Fusek, 1993). Other types of proteases such as cysteine or serine protease do not possess such a large substrate-binding groove, but instead, they focus recognition of their substrate only at one or a few amino acid side chains at most. The strongest preference of cathepsin D is observed for the P1 position (the residue prior to the scissile bond) which is leucine in almost 50% of all sequences cleaved. Other positions are subject to a less strict selection which illustrates the enzyme's ability to accommodate a large number of different sequences. Yet, a clear preference exists for having a certain type of amino acid residue at a certain position relative to the cleavage site. At positions P2' and P5' (two and five residues, respectively, following the scissile bond), charged, polar or small residues like Gly are preferred, while in the other positions hydrophobic residues are over-represented. It should be stressed, however, that not all characteristics of the apparently multi-ligand interaction may be reflected in simple properties of amino acid side chains.

As pointed out previously by us (Van Noort and van der Drift, 1989), the structure of this motif is reminiscent of previously reported structural motifs in T cell stimulatory peptides (Spouge et al., 1987; Rothbard and Taylor, 1988) and MHC class II-binding peptides (Sette et al., 1988, 1989). Together, these observations have led us to the idea that cathepsin D and class II MHC molecules may share some ligand requirements. It seems an interesting thought that MHC molecules could have evolved to recognize some of the structural elements in a polypeptide chain that are generated by cathepsin D as a common property among processed peptides derived from an endless variety of proteins. Not only would cathepsin D-released peptides be endowed with a more that average propensity to bind to any given class II MHC molecule, but they would also be likely to display promiscuous behavior. In this context, it is of interest to note the highly conserved specificity of cathepsin D among different species (Press et al., 1960; Ferguson et al., 1973; Offerman et al., 1983; Imoto et al., 1987; Fruton, 1987; Bond and Barrett, 1980; Lenarcic et al., 1988; Van Noort and van der Drift, 1989).

The relationship we envisage between antigen recognition by cathepsin D and class II MHC molecules, respectively, is depicted in Fig. 4.

XIII. *IN VITRO* USE OF CATHEPSIN D AS EPITOPE GENERATOR

A pivotal role of cathepsin D in the class II processing pathway, as suggested by both quantitative and qualitative evidence, is consistent with its abundant presence and wide distribution over different types of cells and tissues as well as within different types of intracellular compartments (Bond and Butler, 1987; Fruton, 1987). Both early endosomes and late lysosome-like vesicles contain cathepsin D (Gualiardi et al., 1990; Harding and Geuze, 1992). Thus, without implying that all T cell epitopes will be linked to potential sites of cathepsin D cleavage, it seems justified to consider the enzyme as a suitable experimental tool for the generation of possibly immunogenic protein fragments *in vitro*.

Elsewhere in this volume (chapters 1–6), a more detailed description is presented of methods to identify T cell stimulatory sequences. The application of PEPSCAN technology requires knowledge of the primary structure of the protein antigen of interest. Elution and sequencing of MHC-bound antigen fragments are still technically difficult and require both specific reagents (monospecific T cell lines or clones) and sophisticated equipment. In comparison to these methods, *in vitro* degradation of protein antigen with cathepsin D may be considered as an alternative to the identification of T cell epitopes that requires no information of the primary structure of the antigen of interest nor the availability of expensive equipment. The sole requirement for application of this strategy is the availability of protein in milligram amounts. Then, a mild degradation followed by reversed-phase HPLC fractionation of the digest will yield a set of protein fragments that may be expected to be quite similar to the initial products of natural processing. Application of routine reversed-phase HPLC techniques allows the fractionation of peptides and screening on T cell stimulatory potential. Subsequently, more refined fractionation steps will allow the identification of the active peptide components. Data obtained by us in collaboration with several other research groups have illustrated a remarkable effectiveness of cathepsin D in releasing T cell stimulatory peptides from a variety of antigens from different sources and in different species. Cathepsin D-generated protein digests have been found to contain numerous fragments that trigger T cell responses, and even small amounts of such fragments have been found quite effective in mounting a proliferative response of antigen-specific T cells. By applying this strategy, we have succeeded in defining regions in the influenza virus nucleoprotein that are stimulatory to human T cells (Van Noort et al., unpublished), in the mycobacterial 65 kDa heat shock protein that are stimulatory to rat T cells (Boog et al., 1993), in coat proteins from foot-and-mouth disease virus that are stimulatory to bovine T cells (Van Lierop et al., 1994; Van Noort et al., 1994) and, finally, of lysozyme that are stimulatory to murine T cells (Van Noort and Jacobs, 1994). Clearly, useful results can be obtained in different species since the substrate specificity of cathepsin D is a strongly conserved property. A particularly useful application of cathepsin D as an *in vitro* "generator" of T cell stimulatory protein fragments can be found in the field of autoimmune T cell responses (Boog et al., 1993; Van Noort et al., 1994). Whether relevant to the development of multiple sclerosis, rheumatoid arthritis or insulin-dependent diabetes, several autoimmune T cell responses are organ directed, but, in many cases, the disease-associated target antigen (if there is one) has remained unidentified. Especially in such cases may T cells of interest be focused on organ-specific proteins with unidentified primary structures. Upon extraction of protein from the target tissue, a mild treatment with cathepsin D may generate an experimentally useful set of peptides. When recognition of a given peptide by relevant T cells is demonstrated, such a peptide can then be isolated and sequenced. In this way, T cell

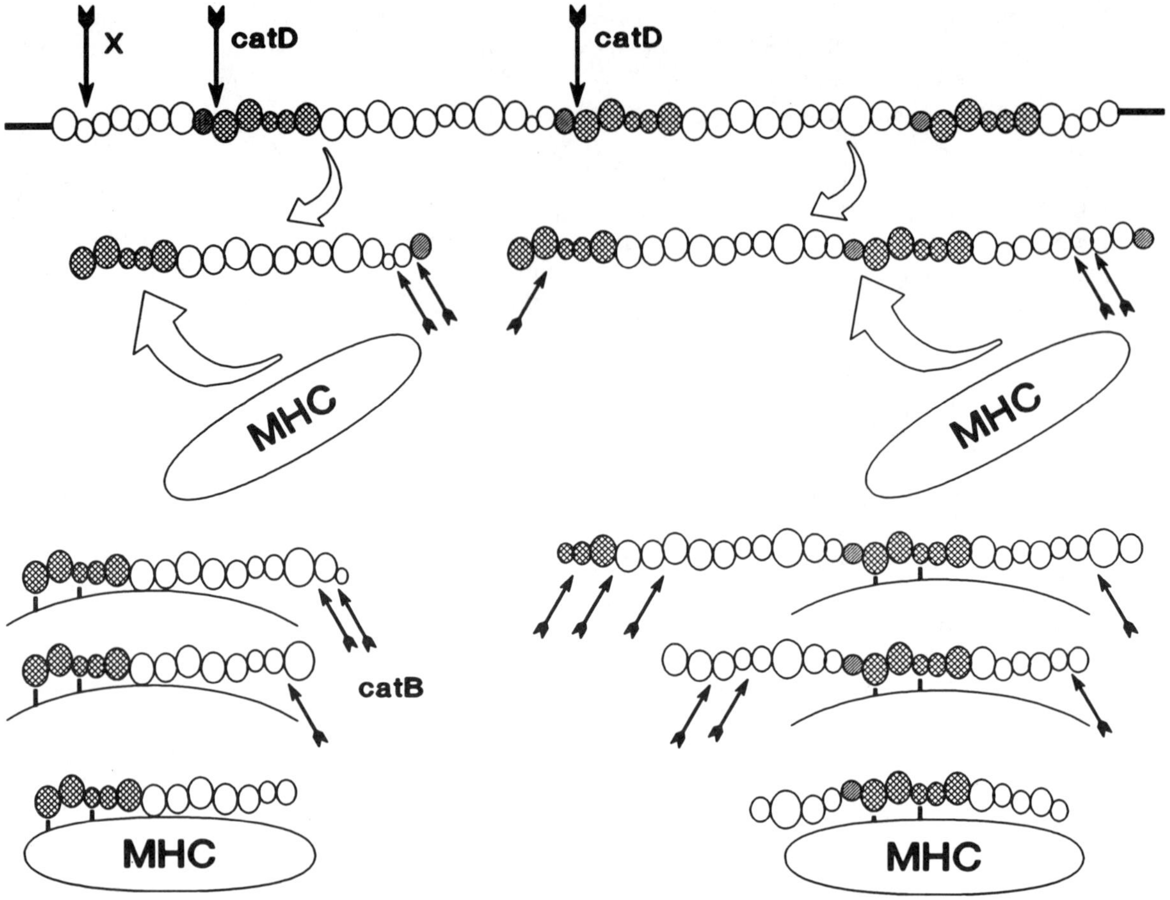

Fig. 4. Model for a structural link between antigen recognition by cathepsin D and by class II MHC molecules. Several lines of evidence point to cathepsin D as a protease that plays a dominant role in processing and may be responsible for many of the initial cleavages in antigen. The enzyme's selectivity generates common terminal structures in processed antigen, especially at N-termini where they extend over five residues. It is proposed that class II MHC molecules recognize part of these structures that may appear as N-termini of processed peptides (left part of Fig. 4) or as internal segments of partially processed fragments (right part of Fig. 4) as anchor residues in the peptides they bind. Other processing enzymes, such as cathepsin B, display dominant activities in exoproteolytic trimming which may occur at both N- and C-terminal boundaries (smaller arrows). Binding of initially released antigen fragments by MHC molecules will protect a core region of the sequence that is buried in the groove from continued trimming. According to the model, this core region will frequently overlap a site of preferential cathepsin D cleavage.

determinants can be identified without one having to purify and sequence the original protein antigen first. In our laboratory, we are currently following this approach in the study of autoreactive T cells to myelin, the insulating sheath surrounding axons and the target of destruction in multiple sclerosis. By applying cathepsin D to generate peptides from myelin or myelin proteins *in vitro*, comparison between healthy and multiple sclerosis-associated T cell responses at the level of individual determinants can be performed without protein purification and sequencing or peptide synthesis. Many other applications of this approach can be envisaged, and in this way information on the nature of processing enzymes, albeit incomplete, can be made useful to laboratory practice.

ABBREVIATIONS

APC Antigen-presenting cell
HPLC High preformance liquid chromatography
MHC Major histocompatibility complex

XIV. REFERENCES

Allen PM, Strydom DJ and Unanue EJ. Processing of lysozyme by macrophages: Identification of the determinant recognized by two T cell hybridomas. Proc Natl Acad Sci USA 81:2489, 1984

Allen PM, Babbitt BP and Unanue ER. T cell recognition of lysozyme: The biochemical basis for presentation. Immunol Rev 98:171, 1987

Aronson NN and Barrett AJ. The specificity of cathepsin B. Biochem J 171:759, 1978

Babbitt BP, Allen PM, Matsueda G et al. Binding of immunogenic peptides to Ia histocompatibility molecules. Nature 317:359, 1985

Bakke O and Dobberstein B. MHC class II-associated in variant chain contains a sortingsignal for endosomal compartments. Cell 63:707, 1990

Balch WE. Biochemistry of interorganelle transport. J Biol Chem 264:16965, 1989

Beaumelle BD and Hopkins CR. High-yield isolation of functionally competent endosomes from mouse lymphocytes. Biochem J 264:137, 1989

Bennett K, Levine T, Ellis JS et al. Antigen processing for presentation by class II major histocompatibility complex requires cleavage by cathepsin E. Eur J Immunol 22:1519, 1992

Blum J and Cresswell P. Role for intracellular proteases in the processing and transport of class II HLA antigens. Proc Natl Acad Sci USA 85:3975, 1988

Bond JS and Barrett AJ. Degradation of fructose-1, 6-bisphosphate aldolase by cathepsin B. A further example of peptidyl dipeptidase activity of this proteinase. Biochem J 189:17, 1980

Bond JS and Butler PE. Intracellular proteases. Ann Rev Biochem 56:333, 1987

Boog CJP, Wagenaar JPA, van Holten C et al. Differential rat T cell recognition of cathepsin D-released fragments of mycobacterial hsp65 fragments by arthritic versus healthy Lewis rats. Transpl Proc 25:2840, 1993

Boots, AMH, Kusters JM, van Noort JM et al. Identification of a T cell epitope of avian coronavirus. Immunology 74:8, 1991

Buus S and Werdelin O. A group-specific inhibitor of lysosomal cysteine proteases selectively inhibits both proteolytic degradation and presentation of the antigen dinitrophenyl-poly-L-lysine by guinea pig accessory cells to T cells. J Immunol 136:452, 1986

Buus S, Colon C, Smith C et al. Interaction between a "processed" ovalbumin peptide and Ia molecules. Proc Natl Acad Sci USA 83:3968, 1986

Buus S, Sette A, Colon S. et al. The relationship between major histocompatibility complex (MHC) restriction and the capacity of Ia to bind immunogenic peptides. Science 235:1353, 1987

Casten LA and Pierce SK. Receptor-mediated antigen processing: Increased antigenicity of a globular protein covalently coupled to antibodies specific for B-cell surface structures. J Immunol 140:404, 1988

Ceman S, Rudersdorf R, Long EA et al. MHC class II deletion mutant expresses normal levels of transgene encoded class II molecules that have abnormal conformation and impaired antigen presentation ability. J Immunol 149:754, 1992

Chesnut RW, Colon SM and Grey HM. Requirements for the processing of antigens by antigen-presenting B cells. I. Functional comparison of B cell tumors and macrophages. J Immunol 129:2382, 1982

Chicz RM, Urban RG, Lane WS et al. Predominant naturally processed peptides bound to HLA-DR1 are derived from MHC-related molecules and are heterogeneous in size. Nature 358:764, 1992

Chu CT and Pizzo SV. Receptor-mediated antigen delivery into macrophages. Complexing antigen to alpha2-macroglobulin enhances presentation to T cells. J Immunol 150:48, 1993

Collins DS, Unanue RE and Harding CV. Reduction of disulfide bonds is a key step in antigen processing. J Immunol 147:4054, 1991

Cresswell P. Intracellular class II HLA antigens are accessible to transferrin-neuraminidase conjugates internalized by receptor mediated endocytosis. Proc Natl Acad Sci USA 82:8188, 1985

Dahms NM, Lobel P and Kornfeld S. Mannose 6-phosphate receptors and lysosomal targeting. J Biol Chem 264:12115, 1989

Davidson HW, West MA and Watts C. Endocytosis, intracellular trafficking and processing of membrane IgG and monovalent antigen/membrane IgG complexes in B lymphocytes. J Immunol 144:4101, 1990

Davidson HW, Reid PA, Lanzavecchia A et al. Processed antigen binds to newly synthesized MHC class II molecules in antigen-specific B lymphocytes. Cell 67:105, 1991

Davidson HW and Watts C. Epitope-specific processing of specific antigen by B lymphocytes. J Cell Biol 109:85, 1989

De Koster HS, van Rood JJ and Termijtelen A. HLA-DR peptide-induced alloreactive T cell lines reveal an HLA-DR sequence that can be both "dominant" and "cryptic": Evidence for allele-specific processing. Eur J Immunol 22:1531, 1992

Demotz S, Grey HM, Appella E et al. Characterization of a naturally processed MHC class II-restricted T cell determinant of hen egg lysozyme. Nature 343:682, 1989a

Demotz S, Matricardi PM, Irle C et al. Processing of tetanus toxin by human antigen-presenting cells. Evidence for donor and epitope-specific processing pathways. J Immunol 143:3881, 1989b

Diment S and Stahl P. Macrophage endosomes contain proteases which degrade endocytosed protein ligands. J Biol Chem 260:15311, 1985

Diment S, Leech MS and Stahl PD. Cathepsin D is membrane-associated in macrophage endosomes. J Biol Chem 263:6901, 1988

Diment S. Differential roles for thiol and aspartyl proteases in antigen presentation of ovalbumin. J Immunol 145:417, 1990

Dingle JT, Poole AR, Lazarus GS et al. Immunoinhibition of intracellular protein digestion in macrophages. J Exp Med 137:1124, 1973

Donermeyer DL and Allen PM. Binding to Ia protects an immunogenic peptide from proteolytic degradation. J Immunol 142: 1063, 1989

Driscoll J and Finley D. A controlled breakdown: Antigen processing and the turnover of viral proteins. Cell 68:823, 1992

Eisenlohr LC, Gerhard W and Hackett CJ. Individual class II-restricted antigenic determinants of the same protein exhitbit distinct kinetics of appearance and persistence on antigen-presenting cells. J Immunol 141:2581, 1988

Falo LD, Colarusso LJ, Benacerraf B et al. Serum proteases alter the antigenicity of peptides presented by class I major histocompatibility complex molecules. Proc Natl Acad Sci USA 89:8347, 1992

Feener EP, Shen WC and Ryser HJP. Cleavage of disulfide bonds in endocytosed macromolecules. J Biol Chem 265:18780, 1990

Ferguson JB, Andrews JR, Voynick IM et al. The specificity of cathepsin D. J Biol Chem 248:6701, 1973

Fruton JS. Aspartyl proteinases. In: Neuberger A and Brocklehurst K Eds: Hydrolytic Enzymes. Elseviers, Amsterdam, p 1, 1987

Galloway CJ, Dean GE, Marsh M et al. Acidification of macrophage and fibroblast endocytic vesicles in vitro. Proc Natl Acad Sci USA 80:3334, 1983

Gammon G, Shastri N, Cogswell J et al. The choice of T cell epitopes utilized on a protein antigen depends on multiple factors distant from, as well as at the determinant site. Immunol Rev 98:53, 1987

Goldberg AL and Rock KL. Proteolysis, proteasomes and antigen presentation. Nature 357:375, 1992

Gradehandt G and Ruede E. The endo/lysosomal protease cathepsin B is able to process conalbumin fragments for presentation to T cells. Immunology 74:393, 1991

Grey HM, Colon S and Chesnut RW. Requirements for the processing of antigen by antigen-presenting B cells. II. Biochemical comparison of the fate of antigen in B cell tumours and macrophages. J Immunol 129:2389, 1982

Gualiardi LE, Koppelman B, Blum JS et al. Co-localization of molecules involved in antigen processing and presentation in an early endocytic compartment. Nature 343:133, 1990

Guillet JG, Lai MZ, Briner TJ et al. Interaction of peptides antigens and class II major histocompatibility complex antigens. Nature 324:260, 1986

Hampl J, Gradehandt G, Kalbacher H et al. In vitro processing of insulin for recognition by murine T cells results in the generation of A chains with free CysSH. J Immunol 148:2664, 1992

Harding C, Levy MA and Stahl P. Morphological analysis of ligand uptake and processing: The role of multivesicular endosomes and CURL in receptor-ligand processing. Eur J Cell Biol 36:230, 1985

Harding CV and Unanue ER. Antigen processing and intracellular Ia. J Immunol 142:12, 1989

Harding CV, Unanue ER, Slot JW et al. Functional and structural evidence for intracellular formation of major histocompatibility conplex class II-peptide complexes during antigen processing. Proc Natl Acad Sci USA 87:5553, 1990

Harding CV, Collins DS, Slot JW et al. Liposome-encapsulated antigens are processed in lysosomes, recycled and presented to T cells. Cell 64:393, 1991

Harding CV and Geuze HJ. Class II MHC molecules are present in macrophage lysosomes and phagolysosomes that function in the phagocytic processing of Listeria monocytogenes for presentation to T cells. J Cell Biol 119:531, 1992

Helenius A, Mellman I, Wall D et al. Endosomes. Trends Biochem Sci 8:245, 1983

Hunt DF, Michel H, Dickinson TA et al. Peptides presented to the immune system by the murine class II major histocompatibility complex molecule I-A^d. Science 256:1817, 1992

Imoto T, Okazaki K, Koga H et al. Specificity of rat liver cathepsin D. J Biochem 101:575,1987

Jensen P. Protein synthesis in antigen processing. J Immunol 141:2545, 1988.

Kim BS and Jang YS. Constraints in antigen processing result in unresponsiveness to a T cell epitope of hen egg lysozyme in C57BL/6 mice. Eur J Immunol 22:775, 1992

Lamb CA, Yewdell JW, Bennink JW et al. Invariant chain targets HLA class II molecules to acidic endosomes containing internalized influenza virus. Proc Natl Acad Sci USA 88:5998, 1991

Lanzavecchia A, Reid PA and Watts C. Irreversible association of peptides with class II MHC molecules in living cells. Nature 357:249, 1992

Lenarcic B, Dolenc I, Lucovnik P et al. Cathepsin D inactivates cystein proteinase inhibitors cystatins. Biochem Biophys Res Commun 154:765, 1988

Li WF, Fan MD, Pan CB et al. T cell epitope selection: Dominance may be determined by both affinity for major histocompatibility complex and stoichiometry of epitope. Eur J Immunol 22:943, 1992

Ludwig T, Griffiths G and Hoflack B. Distribution of newly synthesized lysosomal enzymes in the endocytic pathway of normal rat kidney cells. J Cell Biol 115:1561, 1991

Manca F, Fenoglio D, Kunkl A et al. Differential activation of T cell clones stimulated by macrophages exposed to antigen complexed with monoclonal antibodies. J Immunol 140:3893, 1988

Manca F, Fenoglio D, Li Pira G et al. Effect of antigen/antibody ratio on macrophage uptake, processing and presentation to T cells of antigen complexed with polyclonal antibodies. J Exp Med 173:37, 1991

Marsh EW, Dalke DP and Pierce SK. Biochemical evidence for the rapid assembly and disassembly of processed major-histocompatibility complex class II molecules in acidic vesicles of B cells. J Exp Med 175:425, 1992

Marsh M, Griffiths G, Dean GE et al. Three-dimensional structure of endosomes in BHK-21 cells. Proc Natl Acad Sci USA 83:2899, 1986

McAdoo MH, Dannenberg AM, Hayes CJ et al. Inhibitor of the cathepsin D-type proteinase of macrophages by pepstatin, a specific pepsin inhibitor and other substances. Infect Immun 7:655, 1973

McCoy KL, Miller J, Jenkins M et al. Diminished antigen processing by endosomal acidification mutant antigen-presenting cells. J Immunol 143:29, 1989

McCoy KL, Noone M, Inman JK et al. Exogenous antigens internalized through transferrin receptors activate CD4$^+$ T cells. J Immunol 150:1691, 1993

McKay MJ, Offerman MK, Barrett AJ et al. Action of human liver cathepsin B on the oxidized insulin B chain. Biochem J 213:467, 1983

McKean DJA, Nilson A, Infante AJ et al. Biochemical characterization of of B lymphoma cell antigen processing and presentation to antigen-reactive T cells. J Immunol 131:2726, 1983

McVey Ward D, Ajioka R and Kaplan J. Cohort movement of different ligands and receptors in the intracellular endocytic pathway of alveolar macrophages. J Biol Chem 264:8164, 1989

Mellins E, Smith L, Arp B et al. Defective processing and presentation of exogenous antigens in mutants with normal HLA class II genes. Nature 343:71, 1990

Mellman I, Fuchs R and Helenius A. Acidification of the endocytic and exocytic pathways. Ann Rev Biochem 55:663, 1986

Metcalf P and Fusek M. Two crystal structures for cathepsin D: The lysosomal targeting signal and active site. EMBO J 12:1293, 1993

Michalek MT, Benacerraf B and Rock KL. Two genetically identical antigen-presenting cell clones display heterogeneity in antigen processing. Proc Natl Acad Sci USA 86:3316, 1989

Mouritsen S, Meldal M, Werdelin O et al. MHC molecules protect T cell epitopes against proteolytic destruction. J Immunol 149: 1987, 1992

Navia MA, Fitzgerald PMD, McKeever BM et al. Three-dimensional structure of aspartyl protease from human immunodeficiency virus HIV-1. Nature 337:615, 1989

Neefjes JJ, Stollorz V, Peters PJ et al. The biosynthetic pathway of MHC class II but not class I molecules intersects the endocytic route. Cell 61:171, 1990

Neefjes JJ and Ploegh HL. Inhibition of endosomal endoproteolytic activity by leupeptin blocks surface expression of MHC class II molecules and their conversion to SDS resistance alpha-beta heterodimers in endosomes. EMBO J 11:411, 1992

Nelson CA, Roof, RW, McCourt DW et al. Identification of the naturally processed form of hen egg white lysozyme bound to the murine major histocompatibility complex class II molecule Ia^k. Proc Natl Acad Sci USA 89:7380, 1992

Nguyen QV, Knapp W and Humphreys RE. Inhibition by leupeptin and antipain of the intracellular proteolysis of Ii. Hum Immunol 24:153, 1989a

Nguyen QV and Humphreys RE. Time course of intracellular association, processing and cleavage of Ii forms and class II major histocompatibility complexes. J Biol Chem 264:1631, 1989b

Nguyen QV, Roskey A and Humphreys RE. Effects of brefeldin A on cleavage of invariant chain to p21 and p10 and the appearance of Ii-freed class II MHC dimers. Mol Immunol 30:137, 1993

Niebling WL and Pierce SK. Antigen entry into early endosomes is insufficient for class II processing. J Immunol 150:2687, 1993

Offerman MK, Chlebowski JF and Bond JS. Action of cathepsin D on fructose-1, 6-bisphosphate aldolase. Biochem J 211:529, 1983

Peters PJ, Neefjes JJ, Oorschot V et al. Segregation of MHC class II molecules from MHC class I molecules in the Golgi complex for transport to lysosomal compartments. Nature 349:669, 1991

Press EM, Proter RR and Cebra J. The activity of cathepsin D. Biochem J 74:501, 1960

Puri J and Factorovich Y. Selective inhibition of antigen presentation to cloned T cells by protease inhibitors. J Immunol 141:3313, 1988

Reyes VE, Lu S and Humphreys RE. Cathepsin B cleavage of Ii from class II MHC alpha- and beta-chains. J Immunol 146:3877, 1991

Riberdy JM and Cresswell. The antigen-processing mutant T2 suggests a role for MHC-linked genes in class II antigen presentation. J Immunol 148:2586, 1992

Riberdy JM, Newcomb JR, Surman MJ et al. HLA-DR molecules from an antigen-processing mutant cell line are associated with invariant chain peptides. Nature 360:474, 1992

Roche PA and Cresswell P. Proteolysis of the class II-associated invariant chain generates a peptide binding site in intracellular HLA-DR molecules. Proc Natl Acad Sci USA 88:3150, 1991

Rodriguez GM and Diment S. Role of cathepsin D in antigen presentation of ovalbumin. J Immunol 149:2894, 1992

Romagnoli P, Layet C, Yewdell J et al. Relationship between invariant chain expression and major histocompatibility complex class II transport into early and late endocytic compartments. J Exp Med 177:583, 1993

Roosnek E, Demotz S, Corradin G et al. Kinetics of MHC-antigen complex formation on antigen-presenting cells. J Immunol 140:4079, 1988

Rosenthal AS and Shevach ES. Function of macrophages in antigen recognition by guinea pig T lymphocytes. J Exp Med 138:1194, 1973

Rothbard JB and Taylor WR. A sequence pattern common to T cell epitopes. EMBO J 7:93, 1988

Rudensky AY, Preston-Hurlburt P, Hong SC et al. Sequence analysis of peptides bound to MHC class II molecules. Nature 353:622, 1991

Rudensky AY, Preston-Hurlburt P, Al-Ramadi BK et al. Truncation variants of peptides isolated from MHC class II molecules suggest sequence motifs. Nature 359:429, 1992

Schmid SL, Fuchs R, Male P et al. Two distinct subpopulations of endosomes involved in membrane recycling and transport to lysosomes. Cell 52:73, 1988

Schwartz AL. Cell biology of intracellular protein trafficking. Ann Rev Immunol 8:195, 1990

Sette A, Buus S, Colon S et al. I-A^d binding peptides derived from unrelated protein antigens share a common structural motif. J Immunol 141:45, 1988

Sette A, Adorini L, Appella E. et al., Structural requirements for the interaction between peptide antigens and I-E^d molecules. J Immunol 143:3289, 1989

Sette, A., Ceman, S, Kubo RT et al. Invariant chain peptides in most HLA-DR molecules of an antigen-processing mutant. Science 258:1801, 1992

Shimonkevitz R, Kappler JW, Marrack P et al. Antigen recognition by H-2 restricted T cells. I. Cell-free antigen processing. J Exp Med 158:303, 1983

Shimonkevitz R, Colon S, Kappler JW et al. Antigen recognition by H-2 restricted T cells. II. A tryptic ovalbumin peptide that substitutes for processed antigen. J Immunol 133:2067, 1984

Spouge JL, Guy RH, Cornette JL et al. Strong conformational propensities enhance T cell antigenicity. J Immunol 138:204, 1987

Srinivasan M and Pierce SK. Isolation of a functional antigen-Ia complex. Proc Natl Acad Sci USA 87:919, 1990

Stossel H, Koch F, Kampgen E et al. Disappearance of certain acidic organelles (endosomes and Langerhans cell granules) accompanies loss of antigen processing capacity upon culture of epidermal Langerhans cells. J Exp Med 172:1471, 1990

StPierre Y and Watts TH. MHC class II-restricted presentation of native protein antigens by B cells is inhibitable by cycloheximide and brefeldin A. J Immunol 145:812, 1990

Streicher HZ, Berkower IJ, Busch M et al. Antigen conformation determines processing requirements for T cell activation. Proc Natl Acad Sci USA 81:6831, 1984

Suguna K. Padlan EA, Smith CW et al. Binding of a reduced peptide inhibitor to the aspartic proteinase from rhizopus chinensin: Implications for a mechanism of action. Proc Natl Acad Sci USA 84:7009, 1987

Takahashi, H, Cease KB and Berzofsky JA. Identification of proteases that process distinct epitopes on the same protein. J Immunol 141:3313, 1988

Unanue ER. The antigen-presenting function of the macrophage. Ann Rev Immunol 2:395, 1984

Van Lenten BJ and Fogelman AM. Processing of lipoproteins in human monocyte-macrophages. J Lipid Res 31:1455, 1990

Van Lierop MJ, Van Noort JM, Wagenaar JPA et al. T cell stimulatory fragments of foot-and-mouth disease virus realeased by mild treatment with cathepsin D. J Gen Virol 75:2937, 1994

Van Noort JM and van der Drift ACM. The selectivity of cathepsin D suggests an involvement of the enzyme in the generation of T cell epitopes. J Biol Chem 264:14159, 1989

Van Noort JM, Boon J, van der Drift ACM et al. Antigen processing by endosomal proteases determines which sites of sperm whale myoglobin are eventually recognized by T cells. Eur J Immunol 21:1989, 1991

Van Noort JM, Anderton SM, Wagenaar JPA et al. Differential rat T cell recognition of cathepsin D-released fragments of mycobacterial 65 kDa heat shock protein after immunization with either the recombinant protein or whole mycobacteria. Int Immunol 6:603, 1994

Van Noort JM and Jacobs MJM. Cathepsin D, but not cathepsin B, releases T cell stimulatory lysozyme that are functional in the context of multiple murine class II MHC molecules. Eur J Immunol 24:2175, 1994

Van Renswoude J, Bridges KR, Harford JB et al. Receptor-mediated endocytosis of transferrin and the uptake of Fe in K562 cells: Identification of a nonlysosomal acidic compartment. Proc Natl Acad Sci USA 79:6186, 1982

Vidard L, Rock KL and Benacerraf B. The generation of immunogenic peptides can be selectively increased or decreased by proteolytic enzyme inhibitors. J Immunol 147:1786, 1991

Vidard L, Rock KL and Benacerraf B. Heterogeneity in antigen processing by different types of antigen-presenting cells. J Immunol 149:1905, 1992a

Vidard L, Rock KL and Benacerraf B. Diversity in MHC class II ovalbumin T cell epitopes generated by distinct proteases. J Immunol 149:498, 1992b

Watts C and Davidson HW. Endocytosis and recycling of specific antigen by human B cell lines. EMBO J 7:1937, 1988

Ziegler HK and Unanue ER. Identification of a macrophage antigen-processing event required for I-region restricted antigen presentation to T lymphocytes. J Immunol 127:1869, 1981

Ziegler HK and Unanue ER. Decrease in macrophage antigen catabolism by ammonia and chloroquine is associated with inhibition of antigen presentation to T cells. Proc Natl Acad Sci USA 79:175, 1982

Chapter 8

USE OF T CELL EPITOPES IN RAISING IMMUNE RESPONSES

Netty D. Zegers and Wim J.A. Boersma

TABLE OF CONTENTS

I. INTRODUCTION

Synthetic peptides are valuable tools to raise antibodies crossreactive with the native proteins from which the sequences were derived. Most randomly chosen short peptides consist of a B cell determinant which, by definition, does not induce an epitope-specific immune response by itself. Introduction of a T cell epitope is necessary to offer T cell help and memory for a humoral immune response. A T cell determinant can activate specific T cells to synthesize certain lymphokines necessary for T and B cell activation, interaction and proliferation. The T cell determinant needs to be presented to T cells bound in the groove of an MHC molecule at the cell surface of antigen presenting cells. The peptide-MHC complex is recognized by the T cell receptor (TCR) at the T cell. Therefore, a T cell determinant contains at least two domains: one which binds to the groove of the MHC molecule and one for interaction with the TCR. In peptide immunization, the T cell epitope is usually supplied by a carrier protein to which synthetic peptides are covalently coupled. Alternatively, a short synthetic determinant with a distinct sequence motif which can be presented by MHC class II to T cells can be linked directly to a B cell epitope.

Milich et al. (1986) were among the first to demonstrate that a synthetic peptide immunogen, encompassing one T cell epitope and one B cell epitope, could be used to raise anti-native protein antibodies. These epitopes were non-overlapping peptide bond linked determinants situated in the same region of the pre-S(2) protein of hepatitis B surface antigen. With constructs of a similar synthetic peptide containing a T cell determinant of the circumsporozoite protein of the *Plasmodium falciparum* and a major B cell determinant from the same protein covalently linked with MBS, it was possible to elicit high-titer antibodies against the B cell epitope (Good et al., 1987). Hybrid constructs of T and B cell determinants derived from different proteins made by linear synthesis (Francis et al., 1987a; Cox et al., 1988; Sinigaglia et al., 1988; Borras-Cuesta et al., 1987; Palker et al., 1989) or linked via glutaraldehyde (Leclerc et al., 1987; Jolivet et al., 1990) induced antibodies against the B cell determinant of the hybrid peptides which were crossreactive with the native protein. This proved that the anti-B cell determinant responses also may be elicited with foreign helper T cell determinants.

An immunogen consisting of only one B cell epitope and one T cell epitope may be advantageous over a large carrier-hapten complex immunogen. Since the number of antigenic sites in the peptide immunogen is restricted, implicating the lack of antigenic competition, such a construct may be useful to produce antibodies against weakly immunogenic regions of a large protein. Also immunodominance resulting from the carrier protein could be circumvented.

Here, we will describe some aspects concerning the use of T cell determinants for MHC class II presentation to T cells to raise high-titer native protein specific antibodies.

0-8493-8967-4/95/$0.00+$.50

II. SELECTION OF T CELL EPITOPES

Dependent on the specific application of a synthetic antigen to be used for immunization, the choice of a T cell epitope sequence for incorporation in a synthetic antigen is more or less fixed. For vaccines, T cell determinants have to be derived from the pathogen sequence, since a pathogen-specific memory response is needed at the time a pathogen enters the body. Identification of strong T cell epitope fragments in protein sequences is therefore of importance. These T cell determinants preferably should be promiscuous, i.e., suitable for stimulation of multiple MHC haplotypes. A vaccine constructed with such a promiscuous T cell determinant then is active in a major part of the population.

To raise protein-specific antibodies applicable for research or diagnostic purposes, it is important to select a T cell determinant that is dominant in the target animal and restricted for the target animal's MHC class II haplotype. In this case, the T cell epitope sequence may be derived from any other protein than the protein from which the B cell epitope is derived. A limited number of known T helper cell determinants for use in different strains of mice is listed in Table 1.

A number of methods have been proposed for prediction of T cell determinants and are described in detail elsewhere in this edition (Chapters 1–6). A brief summary is made in this paragraph. Prediction only is possible if the exact amino acid sequence of this protein is known. On the other hand, T cell epitopes can be assigned in a more pragmatic approach by using T cell stimulation assays with protein fragments (see below). It may be clear that only linear T cell epitopes can be predicted this way.

T cells recognize peptide fragments presented in a groove of the MHC class II molecules at the cell surface of antigen presenting cells. These peptide fragments which originate from processing of a protein by an antigen presenting cell are usually 8–12 residues in length. T cells recognize only the primary and secondary structure of the protein, but not the tertiary structure of the native protein. As a result, prediction of T cell recognition sites can be based on secondary structure and the amino acid sequence of the native protein. Therefore, most theoretical prediction methods are merely based on the feasibility of peptide binding in the groove of the MHC molecule, ignoring the interaction with the TCR.

By analyzing functionally recognized T helper determinants, the hypothesis was formulated that T cell determinants tend to be amphipatic helices, α-helices with hydrophobic residues on one side and hydrophilic residues on the other side along the helix (DeLisi and Berzofsky, 1985; Berkower et al., 1986; Spouge et al., 1987). Based on this hypothesis, an algorithm called AMPHI was developed for predicting such T cell determinants (Margalit et al., 1987). The computer program identifies regions that tend to form amphipatic helices by scoring both the strength of hydrophobic moments as a function of the periodicity, within seven amino acids, and the lengths of positive segments.

An alternative method was developed by the group of Humphreys (Elliott et al., 1987; Stille et al., 1987). They demonstrated that the strip-of-helix hydrophobicity index (SOHHI) (mean hydrophobicity of residues at positions n, $n+4$, $n+7$, $n+11$, etc.) identified T cell-presented epitopes, presumably on the basis of quantifying the cooperativity among hydrophobic residues in such longitudinal hydrophobic strips. Based on the SOHHI, they developed the strip-of-helix hydrophobicity algorithm (SOHHA). The degree of hydrophilicity of the non-hydrophobic side of the helix was not relevant to the selection of the T cell-presented epitope. The algorithm was proposed to select both class I and class II restricted epitopes (Reyes et al., 1988).

A third method for selection of T cell determinants was the method developed by Rothbard and Taylor (1988). They formulated an algorithm based on the motif (charged or Gly)-(hydrophobic)-(hydrophobic)-(hydrophobic or Pro)-(polar or Gly). The motif was postulated on the basis of analysis of functional T cell epitopes. Sette et al. postulated a similar motif for interaction of peptides with I-A^d or I-E^d (Sette et al., 1987, 1988, 1989a, 1989b, 1989c). For I-E^d, the motif was composed of three basic residues. The first basic residue was found either in positions 1, 2 or 3. The other basic residues were found in position 4 and 6, whereas position 5 was usually occupied by a non-charged or somewhat hydrophobic amino acid. Using this motif based method, they predicted three overlapping I-E^d binding sites in dynorphin, an opioid peptide. The peptide 1–13 of dynorphin was found to bind with more than a 10-fold higher affinity than other good I-E^d binding peptides (Sette et al., 1990).

The efficacy of three methods for predicting T cell epitopes (AMPHI, SOHHA and the motif of Rothbard and Taylor) was evaluated (Reyes et al., 1990). By comparing functional T cell determinants and predicted sequences, the SOHHA methods scored higher in terms of overlap (ratio predicted over experimental sequence residues ≥ 0.5), sensitivity (correct predictions/number of reported T cell-presented sequences) and efficiency (correct predictions/number of predictions). However, it is not clear which methods were used for tracing the known T cell determinants involved in this analysis (Reyes et al., 1990). The T cell epitopes used in this study for comparison with predicted sequences were from the

TABLE 1
Murine MHC restricted T cell epitopes

MHC	AA residues	Sequence	References
Hen egg lysozyme			
H-2[a]	46–61	NTDGSTDYGILQINSR	Gammon et al., 1991
H-2[d]	20–35	YRGYSLGNWVCAAKFE	"
H-2[d]	116–129	KGTDVQAWIRGCRL	"
H-2[b]	74–90	NLCNIPCSALLSSDITA	"
H-2[b]	30–53	CAAKFESNFNTQATNRNTDGSTDY	"
H-2[b]	81–96	CSALLSSDITASVNCAK	"
I-E[k]	1–18	KVFGRCELAAAMKRHGLD	Adorini et al., 1988b
I-A[k]	51–61	TDYGILQINSR	"
I-A[k]	112–129	RNRCKGTDVQAWIRGCRL	"
H-2[d]	108–120	WVAWRNRCKGTDV	Gammon et al., 1991
H-2[d]	105–129	MNAWVAWRNRCKGTDVQAWIRGCRL	Adorini et al., 1988a (2 apparent I-E[d] sites)
Ovalbumin			
H-2[d]	323–339	ISQAVHAAHAEINEAGR	Shimonkevitz et al., 1984
Sperm whale myoglobin			
H-2[k]	26–40	QDILIRLFKSHPETL	Kim et al., 1990
I-A[d]	102–118	KYLEFISEAIIHVLHSR	Cease et al., 1986
I-E[d]	136–146	ELFRKDIAAKY	Berkower et al., 1986
Pigeon cytochrome *c*			
I-E[k]	95–104	LIAYLKQATAK	Carbone et al., 1987
λ Repressor protein			
I-A[d]; I-E[k]	12–26	LEDARRLKAIYEKKK	Guillet et al., 1986 Lai et al., 1987
I-E[d]	12–26	LEDARRLKAIYEKKK	Sette et al., 1990
Influenza hemagglutinin protein (A/PR8/48)			
H-2[d]	111–119	FERFEIFPK	Hackett et al., 1985
I-A[k]	129–140	NGVTAACSHEGK	Sette et al., 1989a
Plasmodium falciparum circumsporozoite coat protein			
I-A[b]	365–380	NANPNANPNANPNANP	Good et al., 1987
Promiscuous	378–398	DIEKKIAKMEKASSVFNVVMS	Sinigaglia et al., 1988

literature communicated by multiple researchers which have used a variety of functional assays. The number and sequence of the experimentally determined known T cell epitopes is dependent on the assays and strains of mice used (Gammon et al., 1991). More extensive, precise, experimental localization of T cell epitopes may thus alter the outcome of the comparison of the three methods.

Prediction of T cell determinants does not allow for MHC polymorphisms. Many known T cell determinants are MHC restricted. Moreover, predicted sequences, capable of binding to MHC molecules, are not always immunogenic because of a deficient T cell repertoire. For these reasons, some investigators favor the empirical approach for localization of T cell determinants. Analysis of a protein is carried out by stimulation of protein primed T cells with overlapping peptides. The most comprehensive method uses the PEPSCAN strategy initially developed by Geysen et al. (1984) and since then many times improved and refined (van der Zee et al., 1989). The advantage of the PEPSCAN method using peptide synthesis onto pins is the possibility of synthesizing a high number of different peptides. Scanning a protein sequence in single amino acid steps with a series of sequential peptides allows exact localization of the T cell determinants present in the protein. Complete scanning of lysozyme using this method for three strains of mice with different haplotypes showed markedly different T cell determinants for the strains used (Gammon et al., 1991).

An alternative functional method for tracing MHC class II restricted T cell determinants was developed in our laboratory (van Noort and van der Drift, 1989). This method is of major importance when the amino acid sequence of the protein is not known, but it is also applicable for known sequences. The protein is mildly digested by cathepsin D, a lysosomal/endosomal enzyme involved with cellular antigen processing. Subsequently, the fragments are separated and used in a T cell stimulation assay. Using this method, several proteins were analyzed for the presence of T cell determinants (van Noort et al., 1991; Boots et al., 1991; Boog et al., 1993). Cathepsin D recognizes a certain motif of eight residues as substrate for digestion. Retrospectively, based on experimental data, the motif (hydrophobic)-(Leu)-(hydrophobic)-(charged, polar or small)-(hydrophobic)-(hydrophobic)-(charged, polar or small)-(hydrophobic) was postulated (Van Noort and Van der Drift, 1989; see also Chapter 7 in this edition). In 50% of all sequences cleaved, a Leu residue is found at position P1 (the residue prior to the scissile bond). The N-terminus of a processed T cell determinant is located adjacent to the cleavage site formed by cathepsin D. T cell determinants may be predicted in this way on the basis of the cathepsin D substrate motif.

Fundamental questions about the length of the T cell determinant are difficult to answer (see also the following paragraphs). For MHC class II with an open peptide binding groove, a variety of peptide lengths can be found, but the minimal length appeared to be 8–10 amino acids (Brown et al., 1993; Rudensky et al., 1991; Chicz et al., 1993).

In conclusion, it is clear that present prediction methods are not fully efficient due to the fact that they are based on peptide-MHC interaction. For a functional T cell determinant, the interaction of this complex with the TCR is highly important. This interaction is, however, hard to predict. Selection of T cell determinants in a functional assay is most efficient, with the advantage that the target animal of choice can be involved. Especially for application in vaccines, experimental determination of useful sequences appears to be the method of choice.

III. T CELL EPITOPE RESTRICTION

Prediction methods for T cell epitopes do not take MHC polymorphism into account. Apart from MHC restricted T cell determinants (see Table 1), numerous promiscuous T cell determinants, capable of binding to different MHC haplotypes, were described also. Milich et al. (1988) demonstrated that two different MHC class II restricted T cell determinants combined with a B cell determinant incorporated in one construct yielded an immunogen which was capable of eliciting B cell determinant-specific antibodies in mouse strains with different MHC haplotypes (Table 2).

High-titer NANP-specific antibodies could be raised in both H-2^b and H-2^k mice after immunization with hybrid two-component peptides consisting of the NANP repeat from the immunodominant surface antigen of circumsporozoite protein (CSP) from *Plasmodium falciparum* and five different epitopes (T cell determinants) of the merozoite surface antigen 2 (MSA2) of *P. falciparum* (Rzepczyk et al., 1990). The NANP repeat itself is poorly immunogenic in H-2^b mice only (DelGuidice et al., 1986; Good et al., 1987). Even in studies using the $(NANP)_n$ repeat coupled to a carrier protein, only a low level of anti-$(NANP)_1$ was induced. This study also confirms the finding that genetic restriction can be overcome by addition of the proper epitope. A more striking finding was that the simple addition of a cysteine to the N-terminus of the peptide $(NANP)_6$ markedly enhanced the immunogenicity in H-2^b mice and generated antibodies in the normally non-responder H-2^k mice. By introducing a Cys, dimers could possibly be formed, which could account for better T cell stimulation (see also paragraph on multiple copies), or a new T cell epitope could be formed in the junctional sequence of the dimer. There appeared to be some proof for the formation of dimers, since blocking the free -SH group renders the peptide non-immunogenic. Enhanced immunogenicity after the addition of a cysteine residue was reported for poliovirus peptides, hepatitis B surface Ag peptides and foot-and-mouth disease virus

TABLE 2
Antibody responses to peptide constructs with MHC restricted T cell epitopes

Construct	Anti-BD antibodies in mice with MHC haplotype	Remarks	References
TD^x-TD^y-BD	x; y		Milich et al., 1988
$TD^k(1)$-BD	k; b	BD=TD in H-2^b mice	Rzepczyk et al., 1990
$TD^k(2)$-BD	k; b		
$TD^k(3)$-BD	k; b		
$TD^k(4)$-BD	k; b		
$TD^k(5)$-BD	k; b		
TD^b-TD^k-BD	b; k	Also Abs to TD^b and TD^k	Londoño et al., 1990
TD^k-BD	k; d	New H-2^d TD possibly formed in conjunctional sequence	
$TD^{promisc}$-BD	b; d; f; k; q; r; s		Sinigaglia et al., 1988

x and y can be any haplotype.

peptides (Ferguson et al., 1985; Dreesman et al., 1982; Brown et al., 1984; Francis et al., 1987b).

The action of different MHC class II restricted T cell epitopes together in one hybrid peptide construct was explored in studies described by Londoño et al. (1990). They synthesized a tri-component peptide composed of (1) an H-2^b restricted tandemly repeated tetrapeptide sequence (NANP) from the immunodominant surface antigen of circumsporozoite protein (CSP) from *Plasmodium falciparum*, (2) an H-2^k restricted T helper cell epitope (referred to as Th2R) from the non-repetitive part of the CSP sequence, and (3) a pure B cell determinant from *P. falciparum* liver stage Ag, the LSA, a 17 residue repeat sequence. Special attention was paid to the connection of the three sequences to obtain a reciprocal reinforcement of the helical potentials of the Th2R and LSA helical components. With this tri-component peptide $(NPNA)_4$N-Th2R-LSA_{27}, high and long lasting levels of antibodies were raised to all three components after a single immunization without adjuvant in mice with H-2^b (B6 and B10 mice), H-2^k (B10BR and C3H/He mice) and also H-$2^{d/b}$ (F1(BALB/cxB6) mice) background, but not in H-2^d mice (Table 2). The experiments indicated that each sequence containing an active T helper site induced antibodies against itself and helped antibody production to neighboring sequences, overcoming the genetic restriction of the immune response to the latter. Surprisingly, a di-hybrid peptide Th2R-LSA_{27} induced antibodies recognizing the LSA sequence in H-2^d mice in addition to antibodies against Th2R and LSA in H-2^k mice. A newly formed H-2^d epitope may be formed in the overlapping portion of the hybrid peptide. This H-2^d epitope was, however, not active in the tri-component peptide. Some hypotheses were proposed for this phenomenon. First, the NPNA repeats could induce in H-2^d mice a specific T cell mediated suppression of the antibody response against the rest of the determinants in the tri-component hybrid. Possible suppression in H-2^d mice might be overcome by conjugating the tri-component peptide to carrier proteins. Other explanations for the differences in the priming capacity between the hybrids include differences at the level of the cognate interaction between T and B cells, differences in antigen processing and presentation or major alterations in the tri-hybrid peptide of the predicted α-helix secondary structure of Th2R-LSA_{27}. These experiments clearly demonstrate the necessity to carefully explore all possible actions of the constituting components of a hybrid construct.

Sinigaglia et al. (1988) localized a promiscuous T cell determinant in the CSP of the malaria parasite *P. falciparum* (residues 378–398 made with two Cys→Ala substitutions to prevent formation of a disulphide) which is recognized in association with most mouse and human MHC class II molecules. In mice with different genetic backgrounds (haplotypes b, d, f, k, q, r and s), antibody titers specific for $(NANP)_3$ were raised with a synthetic hybrid peptide composed of the

$(NANP)_3$ sequence, which is a B cell determinant, and the T cell determinant AA 378–398. The elicited antibodies were also crossreactive with sporozoites. T cell clones from eight non-immune human donors with different haplotypes (DR1, DR2, DR4, DR5, DRw6, DR7 and DR9) were derived upon *in vitro* stimulation with that T cell determinant. Such a promiscuous T cell determinant may be valuable for incorporation in a synthetic anti-malaria vaccine, since individuals with different MHC class II alleles would respond to such a vaccine.

In a study performed by Su and Caldwell (1992), a chimeric peptide composed of a T helper and a B cell determinant of the *Chlamydia trachomatis* major outer membrane protein induced antibodies in strains of mice with different H-2 haplotypes. The peptide sequence used as the T cell determinant was rather long (25 residues) and might therefore contain more than one T cell determinant. The antibodies all recognized the B cell determinant regardless the strain of mice used, but neutralizing antibodies were only found in some strains. The fine specificity of the antibodies as determined with the PEPSCAN method (Geysen et al., 1984, 1987) did not appear to be MHC related, since both neutralizing and non-neutralizing antibodies were detected in strains of the same MHC haplotype. The difference in fine specificity may be a B cell related phenomenon, such as a "hole" in the B cell repertoire, or differences in genes that control the production of antibody to the epitope.

In conclusion, MHC restriction to a certain synthetic construct may be overcome by incorporating a T cell epitope with the proper MHC restriction (Milich et al., 1988; Rzepczyk et al., 1990; Londoño et al., 1990) or a promiscuous T cell epitope (Sinigaglia et al., 1988; Su and Caldwell, 1992). Special attention should be paid to the conjunctional site of two components, since new T cell epitopes may be formed or existing T cell epitopes may be lost (see also the following sections).

IV. INFLUENCE OF CHAIN ELONGATION ON T CELL EPITOPES

Ample evidence for the positive influence of elongated T helper cell determinants concerning T cell and antibody responses has been collected in several studies.

Childerstone et al. (1990) determined the minimal peptide sequence from the N-terminal region of a 3800 MW streptococcal antigen (residues 6–15) that stimulated T cells. Elongation of this T cell epitope with six residues (to residues 1–16) enhanced the proliferative T cell response roughly with a factor 2. These results were confirmed in a study in which an extended peptide was made to increase the amphipathic score of a T cell determinant, residues 26–40 from sperm whale myoglobin (Kim et al., 1990). These latter elongated peptide (residues 24–42) induced proliferation of specific T cell clones at a 30-fold lower concentration than the short peptide.

Ertl et al. (1991) showed that even very short peptides of 3–5 amino acids from the immunodominant region of the viral nucleoprotein of rabies virus (AA 404–418) were able to stimulate a T cell clone generated with intact rabies virus. They tested a series of peptides varying in length (3–32 amino acids long). The ability of peptides to induce a response was inversely correlated with their lengths. Short peptides had to be used at 10^6 times higher concentrations as compared to the long peptides (15 or 32 amino acids long). The specificity of the T cell response, however, was directly correlated to the length of the peptides. The response to 15 amino acids long peptides showed a high degree of specificity and the response to 3 to 5 amino acid long peptides showed a high degree of degeneration (flexibility). The long as well as the short peptides had to be presented in association with the same MHC class II haplotype (I-E^k).

In contrast to the studies described above, truncation of the N-terminus with one or two amino acids (Ser or Ser-Asp) of a minimal class I cytotoxic T lymphocyte (CTL) epitope (SDYEGRLI) from the nucleoprotein of influenza A/PR resulted in a higher binding affinity for a class I molecule than the epitope (Cossins et al., 1993). The minimal CTL epitope was determined in a functional ^{51}Cr release assay for recognition of a nested series of peptides by a specific class I restricted CTL clone in the presence of target cells. The relative binding affinities were measured using an indirect functional competition assay. Briefly, a suboptimal epitope (IEGGWTGMIDGW) and varying concentrations of competing peptides were incubated with specific (for IEGGWTGMI) effector cells and labeled target cells and subsequently ^{51}Cr release was measured. The N-terminal truncated peptides showed a higher inhibition of lysis than the minimal CTL epitope. The truncated peptides could not be recognized, however, by a class I restricted CTL clone. This study emphasizes the importance of choosing the proper functional read-out system when assigning the consequences of an engineered peptide construct on immune responses.

Differential effects of chain elongation of T cell epitopes on the specificity of the antibody response in the mouse appeared to be strain related (Partidos and Steward, 1992c). A peptide sequence derived from the fusion protein of measles virus (residues 240–252, GDINKVLEKLGYS(C)) induced both T and B cell reactivity in several strains of mice (C57BL/6, CBA, A/J and TO), but it elicited no reactivity in BALB/c and SWR mice. Elongation of this peptide sequence with six amino acids (GGDLLG(C)) at the C-terminus showed both positive and negative effects. Extension was based on the presence of residues of moderate hydrophilicity, flexibility and protrusion index along with the presence of a β-turn. A cysteine residue at the carboxyl end of each peptide was introduced to increase immunogenicity (Francis et al., 1987b). The elongated peptide did induce an antibody response in BALB/c and SWR mice, but the B and T cell responsiveness disappeared in C57BL/6 mice. In addition, the antibody response was significantly higher after immunization with the elongated peptide than with the shorter one.

Using a major immunogenic T cell determinant of pigeon cytochrome *c* (residues 92–103), it was reported that N-terminal extension of the minimally required 12 to 23 residues (stepwise addition of one amino acid was carried out) maximally stimulated cytochrome *c*-specific T cell hybridomas (Srinivasan et al., 1993). The elongated peptide had the ability to bind directly to MHC class II molecules and did not require processing. Even the addition of a second T cell epitope of tetanus toxoid to the cytochrome *c* sequence, which resulted in a peptide of 51 residues, further enhanced stimulation of cytochrome *c*-specific T cells without processing. Thus, regions of the antigenic peptide which lie outside the proposed peptide binding groove of the class II molecule can influence presentation *in vitro* and subsequently the stimulation of specific T cells.

Elucidation of the structure of the MHC class I (Bjorkman et al., 1987) and class II molecules (Brown et al., 1993) together with binding studies revealed some striking differences between the two MHC classes apart from similarities. The class I peptide-binding groove is blocked at either end, implicating that only peptides with restricted size (8–10 residues) can bind. The class II binding groove is open at both ends and allows binding of a peptide which may stick out at both ends of the groove. Peptides varying in size (13–25 residues) could be extracted from class II molecules (Rudensky et al., 1991; Chicz et al., 1993).

In conclusion, interpretation of the results discussed with respect to determination of T cell epitope length is hampered, since in only a few experiments was the exact optimum T epitope determined. Elongation of T cell determinants may have a dual effect. On the one hand, increased stability of a helix as a result of a longer peptide chain may enhance peptide binding to class II binding. The composition of the sequences used for elongation is, of course, of importance in achieving a helix conformation. This is not the case for class I molecules, since the length of the bound peptide is fixed. On the other hand, elongation may more effectively result in an enhanced immune response in animals with a different MHC haplotype than with a shorter TD. Keeping this in mind, extension of a T cell determinant probably will be most effective for class II determinants. Elongation of T cell epitopes positively influences the strength of the immune response and may influence the restriction pattern to MHC binding.

V. POSITIONING OF T CELL EPITOPES IN CHIMERIC PROTEINS

Different combinations of a T cell determinant (residues 65–85) and a B cell determinant (residues 422–436), both derived from the mycobacterial 65-kD protein, were used to study whether it was possible to stimulate the antibody response against a poor immunogenic B cell determinant (BD) by coupling this determinant to a potent T cell determinant (TD) (Cox et al., 1988). The orientation of the two determinants proved to have a profound effect on the magnitude and specificity of the antibody response. Only the BD-TD construct and not the TD-BD construct elicited antibodies specific for the B cell determinant (Table 3). Strikingly, both tested hybrid peptides stimulated T cell proliferation equally well. The TD monomer as well as the hybrid peptides (TD-BD and BD-TD) induced TD specific antibodies. The extent of the TD-specific antibody response was substantially decreased after immunization with the TD-BD construct. The TD, in monomer and hybrid, has a well-defined α-helical structure. The proline residue at the N-terminus of the B cell determinant is likely to disturb the α-helical nature of the hybrid TD-BD construct, resulting in an abrogated BD-specific antibody response in case of the TD-BD. The role of the proline at the N-terminus of the BD-TD construct is not important for the secondary structure of the construct, and antibodies specific for the B cell determinant can be elicited.

In another study, described by Golvano et al. (1990), combinations of two T cell determinants and two B cell determinants in hybrid constructs were evaluated for their ability to raise antibody responses. It was demonstrated that TD-BD constructs elicited antibodies specific for the B cell determinant and no, or low levels of, antibodies specific for the T cell determinant. Contrary to this, the BD-TD constructs induced antibodies specifically directed against the T cell determinant and not, or only at a low level, against the B cell determinant. It was hypothesized that enzymatic processing of the constructs by antigen presenting cells probably might be responsible for the difference in antibody responses. The lysine or arginine at the C-terminus of the T cell determinant may provide a cleavage site for certain proteases. Such a

TABLE 3
Antibody responses elicited with hybrid TD-BD constructs

Immunogen	Anti-TD	Anti-BD	Remarks	References
TD-BD	+	-	Proline at N-	Cox et al., 1988
BD-TD	+	+	terminus BD	
TD-BD	-	+		Golvano et al., 1990
BD-TD	+	-		
TD-BD	-	+	Linear	Levely et al., 1990
BD-TD	+	-	Linker (BMH coupling)	
TD-BD		+		Partidos et al., 1992a
BD-TD	+		Appeared to be strain related	Partidos et al., 1992b
(151–174)-(133–143) =TD-BD	-	+		Milich et al., 1990
133–174 =BD-TD	+	-		

site is not present in the BD-TD constructs; here, the lysine or arginine is at the C-terminus of the complete construct.

Levely et al. (1990) compared a linearly synthesized construct yielding a TD-BD configuration with a construct in which the T cell and the B cell determinant were chemically coupled to each other via their terminal cysteines with the linker bismaleimidohexane yielding a BD-(linker)-TD construct. The TD-BD construct elicited BD-specific antibodies, but with the BD-(linker)-TD construct TD-specific antibodies were found only. Proliferative responses were similar for both constructs. The orientation of the TD appeared to determine the specificity of the antibody response. However, it is not quite clear whether this is due to the method used for coupling of the determinants or a mere consequence of the polarity of the hybrid peptide.

The importance of the mutual orientation of T and B cell determinants in one hybrid construct on the antibody response induced with such constructs was confirmed by Partidos et al. (1992a). Synthetic constructs were used with a T cell determinant (residues 288–302) and a B cell determinant (residues 404–414) from the F protein of measles virus. Chimeric peptides in which the T cell epitope was synthesized to the N-terminus of a B cell epitope induced primarily antibodies which recognized the B cell determinant. In contrast, localization of the T cell epitope at the C-terminus of a B cell epitope induced antibodies preferentially reactive with the T cell epitope. In a later study, it was found that positioning of the T cell epitope was strain related (Partidos et al., 1992b). In TO mice (H-2^s), the N-terminal orientation of the T cell epitope with respect to the B cell epitope was the most effective, whereas in CBA mice (H-2^k) the C-terminal orientation was the best in inducing high levels of antibodies specific for the B cell epitope. This may reflect differences in processing and presentation of the chimeras by the two mouse strains. The strain dependency was previously reported in a study using multiple antigen peptides with epitopes from the *P. falciparum* circumsporozoite (Munesinghe et al., 1991).

Both the orientation of the T and B cell determinants and the context of the T cell site within the larger composite peptide was found to influence both antibody fine specificity and T cell fine specificity (Milich et al., 1990). A hybrid peptide consisting of the dominant T cell determinant of the pre-S(2) region of the hepatitis B surface antigen (residues 151–174) positioned N-terminally to a dominant B cell determinant (residues 133–143) from the same protein yielded an effective pre-S(2) synthetic immunogen (151–174)-(133–143) = TD-BD. Antibody and T cell responses obtained with this hybrid construct were compared with the responses obtained with a linear peptide from the native sequence (residues 133–174) in several strains of mice with different MHC class II haplotypes.

TABLE 4
Antibody responses elicited with fusion proteins

Immunogen	Anti-BD1	Anti-BD2
BD1-spacer-TD-BD2	+++	-
BD1-spacer-$(TD)_2$-BD2	+++	-
BD1-spacer-$(TD)_4$-BD2	++++	+
BD1-TD-spacer-BD2	++	-
BD1-$(TD)_2$-spacer-BD2	+++	±
BD1-$(TD)_4$-spacer-BD2	++++	+++

From Löwenadler et al. (1992).

The 133–174 peptide induced anti-peptide antibodies in all strains, but these antibodies were not cross-reactive with the native protein. The level of antibodies specific for the 156–174 peptide (a minor antibody site after immunization with HBsAg) was significantly higher than the levels of antibodies specific for the 120–145 peptide, which contains two dominant group-specific pre-S(2) antibody sites 133–139 and 137–143. Strikingly, reversion of the T and B cell determinants induced a qualitatively different immune response. The (151–174)-(133–143) construct (=TD-BD) generated antibodies primarily specific for the 133–143 sequence. Above all, these antibodies recognized the native HBsAg protein. A low level of antibodies specific for the 156–174 region was found. Also T cell responses of lymph node cells from mice primed with one of the peptides qualitatively differed. Two overlapping T cell epitopes, 156–170 and 161–174, were found in the 151–174 region. The 133–174 primed T cells recognized the 156–170 site, but not the 161–174 site. Conversely, (151–174)-(133–143) primed T cells preferentially recognized the 161–174 site, and the 156–170 site was less well recognized. In addition, it was observed that differences in T cell specificity occurred after elongation or shortening of the T cell determinant (148–174, 156–174, 161–174).

Löwenadler et al. (1992) investigated how individual B and T cell epitopes are functionally linked to each other in a multi-determinant fusion protein. They constructed fusion proteins with two different B cell determinants derived from different proteins (Table 4). The two B cell determinants were separated by one, two or four copies of a dominant T cell epitope (ovalbumin residues 323–339) (Shimonkevitz et al., 1984; Buus et al., 1987) dominant in H-2^d mice (BALB/c). A peptide spacer (an IgG binding peptide) was positioned upstream or downstream of the T cell epitopes. The immunological relationship between T and B cell epitopes was examined by measuring specific T cell activation and antibody responses in animals immunized with chimeric peptides containing the inserts in varying numbers (1, 2 and 4 copies). The antibody responses against B cell determinants at the amino and the carboxy terminus, respectively, were differently influenced by the molecular localization of the inserted Th determinant (Table 4). All fusion proteins containing the Th epitopes induced antibody production against the B cell determinant at the amino terminal end irrespective of the positioning of the T cell determinants. Only a very weak antibody response was observed against the B cell determinant at the carboxy terminus after immunization with fusion proteins with single or multiple copies of the T cell determinant in the adjacent position. Similar results were obtained with constructs containing one or two copies of the T cell determinant in a distal position. This B cell determinant proved to be a minor epitope in these constructs. However, four copies of the T cell determinant in a distal position markedly enhanced the antibody response. The authors concluded that specific T cell help for antibody production was more effective when the T cell determinants were placed in a distal position. Furthermore, a fusion protein containing four copies of the ovalbumin (OVA) T cell determinant effectively elicited T cell help for high levels of antibody production against both B cell determinants examined, showing that activated Th cells recognizing a single epitope simultaneously provide help for distinct sets of B cells specific for widely spaced epitopes within a protein. Only the influence of inserting a spacer between the T cell determinant and the B cell determinant was investigated in these experiments. They did not investigate the influence of positioning the T cell determinant(s) N- or C-terminally of the same B cell determinant.

The examples above (Table 3) demonstrate that in most cases the antibodies are elicited against the C-terminal part of the hybrid constructs. The investigation of Cox et al. (1988) is an exception. With the TD-BD construct, TD-specific antibodies were found in their study. It was hypothesized that the conformation of the construct might play a role in directing an antibody response. The α-helical conformation in the TD-BD construct is disturbed by a proline situated at the N-terminus of the BD. Altered antigen processing and/or presentation may be the result. Differences in antigen processing and presentation were also assumed for the strain dependant antibody response, as was observed by Partidos et al. (1992b). Strikingly, the T cell proliferation was similar for both orientations of a hybrid construct (Cox et al., 1988; Levely et al., 1990).

In conclusion, a light preference for the N-terminal position of a TD seems to exist. No generals rules can be applied, however, for the best localization of a TD in a hybrid (TD/BD) construct with the aim of optimal BD-specific responses. Once again, to date, experimental testing of all possible combinations is the best approach in designing immunogenic constructs.

VI. IMMUNODOMINANCE

Despite the ability of each MHC molecule to bind many different peptides, the T cell response after immunization with a whole protein is directed to a few and often only to one of the processed peptide determinants (Roy et al., 1989). Several mechanisms might explain this observed immunodominance of certain T cell determinants: (a) the relative availability of the antigenic determinants after processing (Gammon et al., 1987; Adorini et al., 1988b); (b) the relative affinity of their binding to MHC molecules (Buus et al., 1987); (c) competition at the levels of antigen processing, binding and presentation among different determinants restricted to the same or different MHC molecules (Guillet et al., 1987; Adorini et al., 1988b); (d) the frequency in the individual's peripheral T cell repertoire of clones that can be stimulated by each determinant class II complex (Schaeffer et al., 1989; Adorini et al., 1988b; Gammon et al., 1990; Perkins et al., 1991b); (e) regulatory mechanisms such as suppression (Sercarz and Krzych, 1991).

Elongation of a selected T cell determinant with a sequence representing, e.g., a B cell determinant may strongly influence structural and physicochemical features of the T cell determinant with altered antigen presentation as a result. Immunodominance of a newly formed T cell determinant may emerge from such combined sequences (Bhardwaj et al., 1992). In fact, it is often seen that in chimeric peptides an overlapping new TD is formed (Perkins et al., 1991a; Wang et al., 1992a, 1992b). This newly formed junctional T cell determinant of a hybrid peptide was found to be immunodominant in several strains of mice with a different H-2 haplotype (Wang et al., 1992b). They also found that chimeric peptides did not require processing (Perkins et al., 1991a; Wang et al., 1992a; Bhardwaj et al., 1992), and hence, the enhanced immunogenicity of certain determinants may be due to competition for stimulation of T cells superimposed upon a hierarchy of competitive binding among different TD sequences of the complete peptides into the groove of the MHC molecule. When a peptide spanning the sequence of the overlapping new TD was covalently linked to a different previously defined TD, antigenicity of the immunodominant junctional region was silenced and a new epitope assumed immunodominance (Wang et al., 1992b). The immunodominant epitope is determined by complex interactions among the epitopes, which most likely depend on the structural conformation of the composite peptide.

For MHC class II, there seems to be a preference for binding of longer peptides at a central position rather than at either end. To avoid central binding, a hybrid peptide of 51 residues, composed of two known (α-helical peptides) T cell determinants joined by a β-turn to allow stable association of the α-helical peptides, was constructed (Srinivasan et al., 1993). This long peptide construct retained its ability to trigger a T cell response specific for one of the original T cell determinants. Therefore, conjunctions may be constructed keeping the two separated T cell determinants functional with the constraint that the structural units are taken into account.

Smolenski et al. (1990) found that several peptides from different proteins could compete *in vitro* for presented antigen binding sites on antigen presenting cells with processed cytochrome *c*. The cytochrome *c*-specific T cell response could be blocked in an MHC unrestricted manner. They speculated that peptides may be bound to APC structures other than MHC class II molecules prior to the association with MHC class II. In addition, an important role in the immunodominance phenomenon may be attributed to the MHC class II associated invariant chain Ii. Recently, it was demonstrated that particular epitopes derived from hen egg lysozyme (HEL) could be preferentially presented probably due to Ii modulation (Momburg et al., 1993). Maybe the Ii is the APC structure speculated about by Smolenski and co-workers.

These studies demonstrate the so far unpredictable effect of flanking sequences on peptide recognition. Induced changes in peptide conformation, which either promote or inhibit appropriate interaction with the class II molecule or the TCR, may be the cause of this effect. Newly formed immunodominant T cell epitopes in the junction of a TD and BD are highly undesirable for application in vaccines. As demonstrated by Srinivasan et al. (1993), this effect can be avoided by keeping the structural units (α-helical peptides) intact. Keeping these investigations in mind, it is highly recommended that elongation of a selected T cell determinant with other T cell determinants or a B cell determinant with the aim to construct a strong and/or promiscuous immunogen should be accompanied by a thorough investigation into the effect on T cell determinant usage in the system under study.

VII. INFLUENCE OF MULTIPLE COPIES OF T CELL EPITOPES

Many investigators have demonstrated the beneficial effect of inserting multiple copies of a T cell epitope in a synthetic immunogen for raising a proper antibody response (Leclerc et al., 1987; Borras-Cuesta et al., 1988; Cox et al., 1988; Löwenadler et al., 1990; Munesinghe et al., 1991; Löwenadler et al., 1992; Partidos et al., 1992b). The mechanisms responsible for this property could be multiple and were the subject of many studies.

Linear homopolymers of peptides (the monomers consisted of only a B cell determinant or the B cell determinant was extended N-terminally with a T cell determinant), made by using a carbodiimide coupling with side chain protected peptides, were used to study immunogenicity (Borras-Cuesta et al., 1988). Two different B cell determinants were used. Three of the four polymers showed enhanced antibody responses. The generation of a new T cell epitope in the junction of the monomers was suggested by the investigators. They used several prediction methods to validate their assumption. One method (DeLisi and Berzofsky, 1985) indeed predicted a new T cell determinant for two of the three polymer peptides that showed an enhanced immunogenicity. With another method (Margalit et al., 1987), a predicted reinforced amphipathic pattern was found for all three polymer peptides, but this method could not predict a new T cell determinant.

Also, fusion proteins containing multiple copies of the major antigenic determinant of foot-and-mouth disease virus were shown to have enhanced immunogenicity (Broekhuijsen et al., 1987).

Löwenadler et al (1992) constructed fusion proteins with two different B cell determinants derived from different proteins. The two B cell determinants were separated by one, two or four copies of a dominant T cell epitope. A peptide spacer was positioned upstream or downstream of the T cell epitopes (see Table 4). Multiple copies of the T cell determinant in any position led to increased levels of antibody production against the N-terminal B cell epitope, confirming earlier findings of a stoichiometric relationship between T and B cell epitopes in B cell activation (Löwenadler et al., 1990). Four copies of the T cell determinant in a distal position improved the poor antibody response to the C-terminal B cell determinant. Lymph node cells of BALB/c mice primed with the different chimeric fusion proteins proliferated, *in vitro*, to a similar extent in response to a synthetic peptide containing the T cell epitope and thus did not reveal any differences in priming efficiencies of the various constructs. T cell activation was also examined by measuring IL-2 production by a T cell hybridoma specific for the T helper determinant used in their constructs. The level of IL-2 production by this T cell line was correlated to the number of T cell epitopes in the fusion peptide. Furthermore, there was an absolute requirement for uptake and processing of the constructs since fixed antigen presenting cells failed to trigger IL-2 production. In contrast, presentation of the isolated synthetic T cell peptide was only partially inhibited in fixed antigen presenting cells. To exclude the possibility that any newly formed T cell epitope in the junction of two fused sequences contributed to T cell activation, the proliferative response to a junctional peptide spanning six amino acids on either side of the fusion point was examined. Lymph node cells of mice primed with any of the constructs did not recognize the junctional peptide.

These experiments demonstrate the stoichiometric relationship between the number of copies of T cell determinants and antibody production which proved to be based on an increased level of IL-2 production rather than on an enhanced T cell proliferation. The results, however, were obtained from *in vitro* proliferation data. It is not clear whether these conclusions are also valid under physiological conditions. The absolute numbers of specific T cells after immunization with the different constructs were not determined. On the other hand, the increase in antibody production also may be the result of increased T cell cytokine production rather than the expansion of specific Th cells by proliferation, since, *in vivo*, proteins with multiple copies of a T cell determinant may trigger increased production of other cytokines, e.g., IL-4, that are important in regulating B cell activity. Indeed, it was shown by Evavold and Allen (1991) that antigen recognition may result in IL-4 production by T cells even in the absence of proliferation.

Sette et al. (1990) incorporated the core region crucial for class II binding in multiple copies in the same peptide ("reiterative motif") and measured the binding capacity of this peptide for purified MHC class II molecules. They showed that the relative binding capacity increased a factor of 27.7 for the reiterative motif VHAAHAVHAAHAVHA derived from the I-A^d

restricted T cell determinant of ovalbumin (AA 323–339) ISQAVHAAHAEINEAGR (Sette et al., 1987). This binding remained I-A^d specific, because the relative binding capacity for I-E^d was less than 1% of the binding capacity for I-A^d. When the peptide spacer EIN (AA 333–335) was inserted between two copies of the core motif, the relative binding capacity for I-A^d decreased to 1.6. In contrast, positioning the peptide spacer EIN at the end of two copies of the core motif yielded a binding capacity of 24.8. It may be questioned if two core motifs subsequently repeated are sufficient for maximum binding capacity. Similar results were obtained with the I-E^d restricted T cell determinant from HEL (AA 105–120) and the I-E^d restricted T cell determinant from the λ-repressor protein (AA 12–26). These data showing enhanced binding capacity of the reiterative motif support the hypothesis that cooperative stabilization of peptide secondary structures is favorable for class II binding. Insertion of the tripeptide spacer prevented the stabilization, but placing the spacer C-terminally of the two copies did not influence the stabilization. Since it was known from earlier studies (Sette et al., 1989b) that a Val → Ala substitution at position 327 did not affect the binding of the peptide to I-A^d, the analog reiterative motif peptide composed of only two amino acids (AHAAH-AAHAAHAAHAA) was tested for I-A^d binding. This peptide even appeared to bind very strongly to I-A^d with an enhanced relative binding capacity of 34.

The OVA reiterative core peptide (VHAAHA-VHAAHAVHA) also inhibited antigen presentation to an OVA-specific T cell line of the complete natural T cell determinant by fixed antigen presenting cells. This suggests MHC binding at the same site which is involved in T cell activation. Reiterative core motifs do not directly stimulate T cells since a part of the complete natural T cell epitope is missing.

Naturally occurring clustered class II binding sites have been identified. This implies a possible role for reiterative-like structures in the generation of immunodominant regions. Three overlapping I-E^d motifs have been found in dynorphin AA 1–13 (Sette et al., 1989b). The 327–332 OVA core motif is immediately preceded by another I-A^d motif 321–326. Other examples are the influenza virus hemagglutinin 121–146 region (two apparent I-A^k sites (Sette et al., 1989a); the HEL 105–129 region (two apparent I-E^d sites) (Adorini et al., 1988a); the pigeon cytochrome *c* 81–104 region (two apparent I-E^k sites)(Carbone et al., 1987); the staphylococcal nuclease 1–40 region (at least two independent I-E^k sites)(Schaeffer et al., 1989); and in some regions of the VP1, VP2 and VP3 capsid proteins of human rhinovirus (Hastings et al., 1993).

We think that for an estimate of the practical implications, these findings should be further investigated *in vivo* by analyzing qualitatively and quantitatively the antibody response after immunization of mice with peptide constructs composed of either (a) the complete natural T cell determinant, or (b) the repeated core motif extended with the necessary flanking sequence to obtain a complete T cell determinant, and both (a) and (b) combined with a B cell determinant.

Elongation of a T cell epitope sequence may increase the amphipatic score in a predictive algorithm (AMPHI) or may enhance the T cell response as measured in a proliferation assay (Ertl at al., 1991; Kim et al., 1990). These results are further extended evidence for increased conformational stability of the elongated T cell epitope and therefore for increased efficiency.

Using a multiple antigen peptide containing four copies of tandemly arranged T and B cell epitopes, from the repeat region of the *P. falciparum* circumsporozoite protein, optimal antibody responses were obtained (Munesinghe et al., 1991).

Both the level of specific antibodies and the affinity for the B cell determinant of these antibodies were enhanced after immunization with hybrid peptide constructs containing a double copy of a T cell determinant located at the N- or C-terminus of a B cell determinant as compared to constructs with a single copy of the T cell determinant (Partidos et al., 1992b). In these constructs, the B and T cell determinants were separated by a gly-gly dipeptide spacer. In this study, position of the T cell determinant at the N-terminus of the B cell determinant appeared to be more advantageous in terms of affinity.

The examples above demonstrate that multiple copies of a T cell determinant may lead to an increase of the immune response, which could be both quantitatively higher and qualitatively better in terms of affinity. Several mechanisms can contribute to this phenomenon. An increased conformational stability of the T cell determinant as a result of cooperative action of multiple copies within one molecule may lead to a higher binding constant between the MHC molecule and the T cell determinant. In addition, new potential T cell epitopes may be generated in the junction of the monomers. A stoichiometric relationship between copy number of T cell epitopes and the number of MHC molecules which are occupied with a Th peptide exposed at the cell membrane of the antigen presenting cell may lead to a more efficient activation of T cells at higher T/B cell epitope ratios. It is also conceivable that the level of IL-2 production and/or of other cytokines increases upon antigen presentation of a construct with multiple copies of a T cell epitope with more efficient activation of antibody forming cells as a consequence.

VIII. INFLUENCE OF COUPLING MODES ON T CELL EPITOPES

Linear constructs consisting of a selected T cell determinant and a B cell determinant can be made by linear synthesis. A linearly synthesized hybrid peptide often consists of more than 30 residues. An alternative method may be linking the two determinants after synthesizing them separately. This latter method offers the advantage of the combination of a stock of a well documented T cell determinant with various B cell determinants. A large batch of the T cell determinant ensures a constant quality when comparing different B cell determinants in hybrid constructs. Another advantage is that it is then possible to synthesize shorter peptides, since efficiency of peptide synthesis gradually decreases with length.

Coupling of the two determinants via their N-terminal amino group, their C-terminal carboxyl group or via a functional group of one of their amino acids, however, may influence the epitope orientation and the secondary structure of the composed peptides or may alter the immunogenic epitope. Examples from the literature making use of different coupling methods resulting in both effective or useless constructs will be discussed.

One of the first experiments using peptide constructs, which encompass both a T cell and a B cell determinant, made by covalent coupling via a linker, was described by Good et al. (1987). They coupled a predicted and subsequently proven T cell determinant from the malaria circumsporozoite protein via its C-terminal cysteine with the aid of MBS to the N-terminal amino group of the $NP(NANP)_{15}NA$ peptide, a B cell determinant from the same protein. This construct elicited high-titer antibodies specific for $NP(NANP)_{15}NP$. However, in their experiments, they did not differentiate between the antibody response specifically directed to the T cell determinant or B cell determinant, nor did they analyze the functionality of the elicited antibodies in a neutralization assay. Nevertheless, this study is the first proof of the ability of a synthetic construct made by conjugation via a linker to elicit B cell determinant-specific antibodies.

Co-polymers of a B and a T cell determinant derived from different proteins, made by glutaraldehyde, elicited B cell determinant-specific antibodies (Leclerc et al., 1987). In a study performed by Jolivet et al. (1990), co-polymers of four different peptides of which at least one of the peptides represented a T cell determinant were made by glutaraldehyde. The polyvalent synthetic peptides induced antibodies directed against each of the four component peptides in outbred guinea pigs. Linkage via glutaraldehyde makes use of amino groups. Only one of the four peptides contained several lysines scattered through the whole sequence. But despite linkage via the lysine residues which may destroy the epitope, and despite random orientation of this peptide antibody, induction against this epitope was not prevented.

In a study from Levely et al. (1990), two methods of construction of hybrid peptides were compared. They linearly synthesized a peptide composed of a T cell determinant, at the N-terminal side of a B cell determinant both derived from proteins from the respiratory syncytial virus separated by a gly-gly spacer. These determinants were also chemically conjugated by means of a natural N-terminally located cysteine in the T cell determinant and an added N-terminal cysteine in the B cell determinant. The cysteines were coupled by the homo-bifunctional linker bismaleimidohexane (BMH). The chimeric peptides induced different specific antibody responses. Also, the free T cell determinant induced peptide-specific antibodies. The antibodies found after immunization with the linearly synthesized hybrid peptide were mainly directed to the B cell determinant, while the antibodies found after immunization with the chemically constructed peptides were principally directed against the T cell determinant. The consequences are discussed in Section V.

Similar results were found in experiments performed in our laboratory (Zegers et al., 1993). We constructed hybrid peptides by linear synthesis and by different chemical conjugation. The T and B cell determinants were coupled by the homo-bifunctional linker BMH or by the hetero-bifunctional linker MBS. Additional cysteines were incorporated in the sequences of the determinants for coupling purposes. The cysteine was placed C-terminally in the T cell determinant and C- or N-terminally in two different B cell determinants. First, with MBS conjugation, the thiol group of the B cell determinant was blocked with ethylmaleimide. After reaction of MBS with one of the amino groups of the B cell determinant, the T cell determinant was coupled. After immunization with these different constructs, only T cell determinant-specific antibodies were found with the chemical constructed peptides, while both T- and B cell determinant-specific antibodies were found with the linearly synthesized peptides.

Several explanations can be given why in this case we did not find B cell determinant-specific antibodies with chemically linked constructs. First, chemical modification of the B cell determinant, introduced during coupling procedures, might have altered or destroyed the immunogenic determinant. We therefore analyzed the antibody response in an enzyme-linked immunosorbent assay on isolated chemically modified B cell determinants. However, antibodies specific for a modified B cell epitope were not found. Previously, high-titer antibodies specific for the B cell determinant were elicited with conjugates of keyhole limpet hemocyanine (KLH) and the B cell determinants made by coupling with MBS. For these B cell determinants, the thiol group of an added cysteine to the B cell epitope was used. Thus, chemical crosslinking using thiol groups did not destroy immunogenicity by itself. However, antibodies specific for the B cell determinant could not be raised with a KLH conjugate made by coupling with glutaraldehyde. We then concluded that lysine residues in this specific B cell determinant are important for immunogenicity and therefore should not be used for coupling (Zegers et al., 1990). Second, antigen processing or presentation also may have impaired the antibody response against the B cell determinant after immunization with chemically linked constructs. It is conceivable that these TD-specific antibodies are elicited by the free TD molecules still present in the preparation of the hybrid construct. On the other hand, the chemically conjugated construct might not be processed properly which impairs binding of TD to MHC molecules. Antigen presentation is thus hampered, resulting in the absence of BD-specific antibodies. Several investigators have proposed that adjacent sequences may alter the conformation of an epitope, resulting in significant altered antigen processing or changes in T cell recognition (Perkins et al., 1991a; Wang et al., 1992a, 1992b; Kim and Jang, 1992).

As appears from the examples above, in appropriate designed experiments, chemically coupled TD-BD constructs may lead to the induction of BD-specific antibodies (Good et al., 1987; Leclerc et al., 1987; Jolivet et al., 1990). Glutaraldehyde or MBS are the linkers of choice. However, these coupling reagents did not appear to be useful if amino groups of residues present in the sequence of the BD were used for coupling (Zegers et al., 1993). Conjugation with BMH did not yield constructs which elicited BD-specific antibodies (Levely et al., 1990; Zegers et al., 1993).

IX. IS COVALENT COUPLING OF T AND B CELL EPITOPES NECESSARY?

In order to obtain an optimal humoral immune response, B and T cells must cooperate in close conjunction (Kupfer et al., 1986; Vitetta et al., 1987; Noelle and Snow, 1990; Clark and Ledbetter, 1994). The cognate interaction between helper T and B cells involves the T cell receptor which recognizes the antigen presented in association with MHC class II molecules at the cell surface of B cells. Also, the CD3 and the CD4 molecules are associated with this complex. Besides this cognate interaction, second costimulatory signals between T and B cells are necessary for activation of B cells to differentiate and proliferate. CD28 and B7 interaction plays a major role, and recently, the interaction between gp39 (activated T helper cells) and CD40 (B cells) was described (Noelle et al., 1992). Moreover, certain lymphokines produced by activated T cells are essential factors involved in antibody production. Lymphokines are active only in high local concentrations and are therefore optimally inductive only within short distances from their production site. Bartlett et al. (1989) showed *in vitro* that "bystander" B cells were less frequently induced into cell cycle than B cells in contact with T cells. Moreover, the B cells in physical contact with T cells had a higher RNA content and were larger than the bystander B cells present in the same cultures. Entry of B cells into the G_1 phase appeared to be antigen and T cell contact dependant, but not lymphokine dependant. Lymphokines were needed for cycle progression to S phase. Several other reports support the view that conjunction of B and T cells is required for B cell triggering (Kupfer et al., 1986; Vitetta et al., 1987; Noelle et al., 1983; Krusemeier and Snow, 1987). There are, however, also reports which describe lymphokine induced polyclonal B cell cycle entry in an MHC unrestricted manner and subsequent progression to S phase for 20% of the B cells (Leclercq et al., 1984, 1986; Defranco et al., 1984). A fraction of these cells differentiates and produces Ig, although at a lower rate than terminally differentiated plasma cells (Leclercq et al., 1986).

B cell activation via CD40 is required for specific antibody production and can also be induced *in vitro*, besides through T cell interaction, by a soluble recombinant gp39 or by an anti-CD40 antibody in the presence of cytokines (Hollenbaugh et al., 1992; Gascan et al., 1991; Nonoyama et al., 1993). Also, *in vivo*, the involvement of CD40-gp39 interaction was established to be essential for thymus dependent humoral immunity in our own laboratory (Van den Eertwegh et al., 1993).

Minimal requirements for antibody production to peptide antigens were defined in a study by Goodman-Snitkoff et al. (1990, 1991). They crosslinked a T cell as well as a B cell epitope to a hydrophobic anchor and complexed these epitopes into the phospholipid bilayer of liposomes resulting in a multivalent configuration. These immunogens were able to induce production of antibodies specific for the B cell epitope. Liposomes with a linear peptide comprising a T and a B cell determinant also induce B cell determinant-specific antibodies. These data suggest that Th epitopes do not have to be covalently linked to the B cell epitope to be active, but that intrastructural help is equal to intramolecular help.

Recently, some papers were published describing antibody production specific for B cell determinants after immunization with mixed uncoupled T and B cell determinants (Sarobe et al., 1991; Partidos et al., 1992d; Shaw et al., 1993). This antibody production can only be explained by assuming direct binding (or after internalization) of the T cell determinant to class II molecules of the B cell having recognized the B cell determinant. Direct binding to MHC class II molecules could take place by competition of the T cell determinant with a (self) peptide that is already present in the cleft of the MHC class II molecule. To accomplish this, the affinity of the T cell determinant for the class II molecule should be much higher than the affinity of the peptide bound in the MHC molecule. Stimulation via this pathway is probably very inefficient. A (locally) very high concentration of the Th peptides is thus required. Binding of the TD to the class II molecule after internalization by the B cell favors the hypothesis of intermolecular/intrastructural help (Lake and Mitchison, 1976; Milich et al., 1987; Goodman-Snitkoff et al., 1990,1991). In the studies which describe co-immunization of the TD and the BD in a mixture, the peptides were emulsified in (complete) Freund's adjuvant. Both the TD and the BD peptide should be entrapped in the same micro-droplet to give intrastructural help. Evidence for the assumption of simultaneous presence of both T and B cell determinants in one droplet came from the study of Partidos et al. (1992d). B cell-specific antibodies only were detected after immunization with both BD and TD peptides mixed in the Freund's adjuvant emulsion. Intraperitoneal immunization of the TD combined with subcutaneous immunization of the BD at the same moment did not result in BD-specific antibodies. The possibility of associated, non-covalently bound TD-BD complex formation cannot be excluded. Co-immunization of a poor immunogenic peptide from the malaria circumsporozoite protein together with keyhole limpet hemocyanin raised peptide-specific antibodies (Good et al., 1988). In contrast to this finding, we observed that "pseudo conjugates", formed by non-specific adherence of peptides to carrier molecules, were not able to elicit a peptide-specific antibody response (Deen et al., 1990). Similarly, immunization with a mixture of the nucleocapsid (strong immunogen) of the hepatitis B virus with envelope antigen (not immunogenic) of the virus did not result in anti-envelope antigen-specific antibodies, whereas a particle (virion) encompassing both proteins did (Milich et al., 1987). These effects can hardly be explained as the result of bystander help as suggested by Partidos et al. (1992d) since only a small proportion of the B cells can be activated via this pathway (Bartlett et al., 1989). Moreover, the level and affinity of the antibodies to the B cell epitopes following immunization with mixed, non-covalently bound TD and BD peptides was lower than that obtained following immunization with an analogous chimeric TD-BD construct (Shaw et al., 1993).

In our laboratory, we were not successful in applying co-immunization with TD and BD peptides (Zegers et al., 1993). Our experiments indicated that induction of BD-specific antibodies with a mixture of free TD and BD peptides is certainly not possible with all combinations of T and B cell determinants. In contrast, a covalent bond between T cell and B cell determinants in general is required to induce high affinity anti-BD antibodies crossreactive with the native protein.

From the investigations described above, it is clear that for efficient antigen presentation and subsequent antibody production, a cognate interaction between B and T cells is required. This will assure antigen-specific T cell help resulting in a specific memory response. To accomplish this, a covalently bound TD-BD construct, or a liposome, a virion or another particulate structure with both a TD and a BD built in to give intrastructural help, is the most applicated and relatively effective manner. Also, a water-in-oil emulsion, in which a TD and BD are mixed, will be appropriate, if the TD and BD are entrapped in the same micro-droplet to give intrastructural help. However, entrapping in the same droplet is a chance process and will be less efficient. This might have caused the affinity and titer of the antibodies raised with the mixed TD and BD to be lower than that obtained with a covalently bound TD-BD construct.

Taken together, to obtain a high level of BD-specific antibodies with a high affinity and cross-reactive with the native protein, a covalently bound BD-TD construct should be used for immunization.

X. CONCLUDING REMARKS

Coupling of B cell determinants to carrier proteins may serve several goals. The carrier protein provides T cell help to anti-peptide antibody forming cells. Processing of a carrier-peptide conjugate by the antigen presenting cells generates different T cell determinants which can stimulate different T cells. The immunogenicity of a peptide-carrier may thus be increased by stimulating a larger T cell pool. Also through the size of a carrier-peptide conjugate, the capture by phagocytes may be increased.

On the other hand, it may be advantageous to work with a simple and precisely defined immunogenic construct. Using a rather small peptide containing one or several (the same or different) copies of a well selected T cell determinant and a B cell determinant for immunization provides the ability to manipulate the specificity, level and affinity of the immune response. Moreover, immunodominance of TDs from a carrier protein can be excluded when only TD-BD constructs are used for immunization. Especially for vaccines, such an approach offers the possibility to immunize with strong acting T cell determinants with a broad MHC haplotype spectrum. Cryptic T cell determinants hidden in a pathogen-derived native protein are circumvented in this way. In addition, it is highly desirable to stimulate pathogen-specific T cells for vaccine applications.

Selection of T cell determinants on the basis of prediction methods is not fully efficient. The action of a selected T cell determinant should be defined in a functional assay with cells from the target animal. For vaccine applications, it is recommended that a promiscuous T cell epitope is chosen, certainly for outbred populations and human use.

A light preference for the N-terminal position of a TD in a construct exists, since the antibody response is directed to the C-terminal part of a construct. However, this is not a general rule, since processing of a construct and antigen presentation largely depends on the sequence of the overlapping junction between the two determinants. In animals, this is strain dependent. Special attention, therefore, should be paid to the conjunctional site of the two components. New T cell epitopes may emerge, which is not advantageous since such a newly formed TD is not pathogen or protein derived and alternatively existing T cell epitopes may be lost. This effect can be avoided by keeping the structural units (α-helices) intact, e.g., separated by a helix breaker.

Chain elongation of a TD may increase the stability of a helix and, as a result, may enhance peptide binding to MHC class II molecules. Higher titers of antibodies were reported with a broader MHC restriction spectrum. On the other hand, the response in an animal with a certain MHC haplotype may disappear. Also in this respect, special attention should be paid to the TD and BD junction.

Multiple copies of the TD may often lead to an enhancement of the immune response. An increased conformational stability of the TD favors binding to an MHC molecule. Processing of the constructs yields more molecules of the TD. A stoichiometric relationship between the copy number of TDs and the number of MHC molecules occupied with a TD and exposed at the cell membrane of the antigen presenting cell may lead to a more efficient activation of T cells. In addition, the level of Il-2 production may enhance upon better T cell activation with the more efficient activation of antibody cells as an effect.

Chemical coupling of a TD and a BD with linkers not always results in an efficient immunogen. Dependent on the availability of functional groups in the TD and BD, random orientation of the determinants may be generated. This may be disastrous for the immunogenicity. We are of the opinion that linear constructs with a natural peptide bond are most appropriate. This may be accomplished by linear synthesis, which becomes more difficult, however, with increasing length of the sequence. Condensation of fully protected fragments offers an alternative synthesis method.

In general, it is a prerequisite for an efficient immunogen to build the TD and BD in one construct or in a structural unit, such as a liposome or another particulate structure. Antibody production was reported upon immunization with just a mixture of free TD and BD, but the affinity and titer of the antibodies raised were lower than those obtained with a covalently bound TD-BD construct.

It is clear that careful analysis should be made of the structural influences of all components in designing immunogenic constructs. As shown above, many factors may play an important role. Just adding two sequences in one construct generally does not always lead to an optimally active immunogen. The construct which has been found most appropriate on the basis of theoretical choices still needs confirmation of the putative superiority in a functional assay where it is compared to alternative simple constructs.

ABBREVIATIONS

AA	Amino acid
BD	B cell determinant
BMH	Bismaleimidohexane
CSP	Circumsporozoite protein
CTL	Cytotoxic T lymphocyte
HEL	Hen egg lysozyme
KLH	Keyhole limpet hemocyanin
MBS	m-Maleimidobenzoyl-succinimide ester
MHC	Major histocompatibility complex
OVA	Ovalbumin
TCR	T cell receptor
TD	T cell determinant
Th	T helper

XI. REFERENCES

Adorini L, Sette A, Buus S et al. Interaction of an immunodominant epitope with Ia molecules in T cell activation. Proc Natl Acad Sci USA 85:5181, 1988a

Adorini L, Appella E, Doria G et al. Mechanisms influencing the immunodominance of T cell determinants. J Exp Med 168:2091, 1988b

Bartlett WC, Michael A, McCann J et al. Cognate interactions between helper T cells and B cells. II. Dissection of cognate help by using a class-II restricted, antigen-specific, IL-2-dependent helper T cell clone. J Immunol 143:1745, 1989

Berkower I, Buckenmeyer GK, Berzofsky JA. Molecular mapping of a histocompatibility-restricted immunodominant T cell epitope with synthetic and natural peptides: Implications for T cell antigenic structure. J Immunol 136:2498, 1986

Bhardwaj V, Kumar V, Geyson HM et al. Subjugation of dominant immunogenic determinants within a chimeric peptide. Eur J Immunol 22:2009, 1992

Bjorkman PJ, Saper MA, Samraoui B et al. Structure of the human class I histocompatibility antigen, HLA-A2. Nature 329:506, 1987

Boog CJP, van Holten C, Wagenaar JPA et al. Differential rat T cell recognition of cathepsin D-released fragments of mycobacterial hsp65 by arthritic versus healthy rats. Transpl Proc 25:2840, 1993

Boots AMH, Kusters JM, Van Noort JM et al. Identification of a T cell epitope of avian coronavirus. Immunology 74:8, 1991

Borras-Cuesta F, Petit-Camurdan A and Fedon Y. Engineering of immunogenic peptides by co-linear synthesis of determinants recognized by B and T cells. Eur J Immunol 17:1213, 1987

Borras-Cuesta F, Fedon Y and Petit-Camurdan A. Enhancement of peptide immunogenicity by linear polymerization. Eur J Immunol 18:199, 1988

Broekhuijsen MP, Van Rijn JMM, Blom AJM et al. Fusion proteins with multiple copies of the major antigenic determinant of foot and mouth disease virus protect both the natural host and laboratory animals. J Gen Virol 68:3137, 1987

Brown JH, Jardetzky TS, Gorga JC et al. Three-dimensional structure of the human class II histocompatibility antigen HLA-DR1. Nature 364:33, 1993

Brown SE, Howard CR, Zuckerman AJ et al. Affinity of antibody response in man to hepatitis B vaccine determined with synthetic peptides. Lancet 2:184, 1984

Buus S, Sette A, Colon SM et al. The relation between major histocompatibility complex (MHC) restriction and the capacity of Ia to bind immunogenic peptides. Science 235:1353, 1987

Carbone FR, Staerz UD, and Paterson Y. A new T helper cell specificity within the pigeon cytochrome *c* determinant 95-104. Eur J Immunol 17:897, 1987

Cease J, Berkower I, York-Jolley J et al. T cell clones specific for an amphipathic α-helical region of sperm whale myoglobin show differing fine specificities for synthetic peptides. J Exp Med 164:1779, 1986

Chicz RM, Urban RG, Gorga JC et al. Specificity and promiscuity among naturally processed peptides bound to HLA-DR alleles. J Exp Med 178:27, 1993

Childerstone A, Haron J, and Lehner T. The reactivity of naturally sensitized human CD4 cells and IgG antibodies to synthetic peptides derived from the amino terminal sequences of a 3800 MW *Streptococcus mutans* antigen. Immunology 69:177, 1990

Clark EA and Ledbetter JA. How B and T cells talk to each other. Nature 367:425, 1994

Cossins J, Gould K, and Brownlee GG. Peptides shorter than a minimal CTL epitope may have a higher binding affinity than the epitope for the class I K^k molecule. Virology 195:851, 1993

Cox JH, Ivanyi J, Young DB et al. Orientation of epitopes influences the immunogenicity of synthetic peptide dimers. Eur J Immunol 18:2015, 1988

Deen C, Claassen E, Gerritse K et al. An improved and modified carbodiimide coupling method for synthetic peptides: Enhanced antipeptide antibody responses. J Immunol Meth 129:119, 1990

DeFranco AL, Ashwell JD, Schwartz RH et al. Polyclonal stimulation of resting B lymphocytes by antigen-specific T lymphocytes. J Exp Med 159:861, 1984

DelGuidice G, Cooper JA, Merino J et al. The antibody response in mice to carrier-free synthetic polymers of *Plasmodium falciparum* circumsporozoite repetitive epitope is IA^b-restricted: Possible implications for malaria vaccines, J Immunol 137:2952, 1986

DeLisi C and Berzofsky JA. T cell antigenic sites tend to be amphipatic structures. Proc Natl Acad Sci USA 82:7048, 1985

Dreesman GR, Sanchez Y, Ionescu-Matiu I et al. Antibody response to hepatitis B surface antigen after a single inoculation of uncoupled synthetic HBsAg peptides. Nature 295:158, 1982

Van Den Eertwegh AJM, Noelle RJ, Roy M et al. *In vivo* CD40-gp39 interactions are essential for thymus-dependent humoral immunity. I. *In vivo* expression of CD40 ligand, cytokines and antibody production delineates sites of cognate T-B cell interactions. J Exp Med 178:1555, 1993

Elliott WL, Stille CJ, Thomas LJ et al. An hypothesis on the binding of an amphipatic, alpha helical sequence in I_i to the desetope of class II antigens. J Immunol 138:2949, 1987

Ertl HCJ, Dietzschold B, and Otvos L. T helper cell epitope of rabies virus nucleoprotein defined by tri- and tetrapeptides. Eur J Immunol 21:1, 1991

Evavold BD and Allen PM. Separation of IL-4 production from Th cell proliferation by an altered T cell receptor ligand. Science 252:1308, 1991

Ferguson M, Evans DMA, Magrath DI et al. Induction by synthetic peptides of broadly reactive, type-specific neutralizing antibody to poliovirus type 3. Virology 143:505, 1985

Francis MJ, Hastings GZ, Syred AD et al. Non-responsiveness to a foot-and-mouth disease virus peptide overcome by addition of foreign helper T cell determinants. Nature 300:168, 1987a

Francis MJ, Fry CM, Rowlands DJ et al. Immune response to uncoupled peptides of foot-and-mouth disease virus. Immunology 61:1, 1987b

Gammon G, Shastri N, Cogswell J et al. The choice of T cell epitopes utilized on a protein antigen depends on multiple factors distant from, as well as at the determinant site. Immun Rev 98:53, 1987

Gammon G, Klotz J, Ando D et al. The T cell repertoire to a multideterminant antigen. J Immunol 144:1571, 1990

Gammon G, Geysen HM, Apple RJ et al. T cell determinant structure: Cores and determinant envelopes in three mouse major histocompatibility complex haplotypes. J Exp Med 173:609, 1991

Gascan H, Gauchat JF, Aversa G et al. Anti-CD40 monoclonal antibodies or $CD4^+$ T cell clones and IL-4 induced IgG4 and IgE switching in purified human B cells via different signaling pathways. J Immunol 147:8, 1991

Geysen HM, Meloen RH, and Barteling SJ. Use of peptide synthesis to probe viral antigens for epitopes to a resolution of a single amino acid. Proc Natl Acad Sci USA 81:3998, 1984

Geysen HM, Rodda SJ, Mason TJ et al. Strategies for epitope analysis using peptide synthesis. J Immunol Meth 102:259, 1987

Golvano J, Lasarte JJ, Sarobe P et al. Polarity of immunogens: Implications for vaccine design. Eur J Immunol 20:2363, 1990

Good MF, Maloy WL, Lunde MN et al. Construction of synthetic immunogen: Use of new T-helper epitope on malaria circumsporozoite protein. Science 235:1059, 1987

Good MF, Pombo D, Lunde MN et al. Recombinant human IL-2 overcomes genetic nonresponsiveness to malaria sporozoite peptides. J Immunol 141:972, 1988

Goodman-Snitkoff G, Eisele LE, Heimer EP et al. Defining minimal requirements for antibody production to peptide antigens. Vaccine 8:257, 1990

Goodman-Snitkoff G, Good MF, Berzofsky JA et al. Role of intrastructural/intermolecular help in immunization with peptide-phospholipid complexes. J Immunol 147:410, 1991

Guillet JG, Lai MZ, Briner TJ et al. Interaction of peptide antigens and class II major histocompatibility complex antigens. Nature 324:260, 1986

Guillet JG, Lai MZ, Briner TJ et al. Immunological self, nonself discrimination. Science 235:865, 1987

Hackett CJ, Hurwitz JL and Dietzschold B et al. A synthetic decapeptide of influenza virus hemagglutinin elicits helper T cell with the same fine recognition specificities as occur in response to whole virus. J Immunol 135:1391, 1985

Hastings GZ, Francis MJ, Rowlands DJ et al. Epitope analysis of the T cell response to a complex antigen: Proliferative responses to human rhinovirus capsids. Eur J Immunol 23:2300, 1993

Hollenbaugh D, Grosmaire LS, Kullas CD et al. The human T cell antigen gp39, a member of the TNF gene family, is a ligand for the CD40 receptor: Expression of a soluble form of gp39 with B cell costimulatory activity. EMBO J 11:4313, 1992

Jolivet M, Lise L, Gras-Masse H, Tartar A et al. Polyvalent synthetic vaccines: Relationship between T epitopes and immunogenicity. Vaccine 8:35, 1990

Kim BS and Jang Y-S. Constraints in antigen processing result in unresponsiveness to a T cell epitope of hen egg lysozyme in C57BL/6 mice. Eur J Immunol 22:775, 1992

Kim JE, Kojima M, Houghten R et al. Characterization of helper T cell epitope recognized by mice of a low responder major histocompatibility type. Mol Immunol 27:941, 1990

Krusemeier M and Snow EC. Induction of lymphokine responsiveness of hapten-specific B lymphocytes promoted through an antigen-mediated T helper lymphocyte interaction. J Immunol 140:367, 1987

Kupfer A, Swain SL, Janeway Jr CA et al. The specific direct interaction of helper T cells and antigen-presenting B cells. Proc Natl Acad Sci USA 83:6080, 1986

Lai M-Z, Ross DT, Guillet J-G et al. T lymphocyte response to bacteriophage λ repressor cI protein. J Immunol 139:3973, 1987

Lake P and Mitchison NA. Regulatory mechanisms in the immune response to cell-surface antigens. Cold Spring Harbor Symp Quant Biol 41:589, 1976

Leclercq L, Bismuth G and Thèze J. Antigen-specific helper T cell clone supernatant is sufficient to induce both polyclonal proliferation and differentiation of small resting B lymphocytes. Proc Natl Acad Sci USA 81:6491, 1984

Leclercq L, Cambier JC, Mishal Z et al. Supernatant from a cloned helper T cell stimulates most small resting B cells to undergo increased I-A expression, blastogenesis, and progression through cell cycle. J Immunol 136:539, 1986

Leclerc C, Przewlocki G, Schutze M-P et al. A synthetic vaccine constructed by copolymerization of B and T cell determinants. Eur J Immunol 17:269, 1987

Levely ME, Mitchell MA and Nicholas JA. Synthetic immunogens constructed from T cell and B-cell stimulating peptides (T:B chimeras): Preferential stimulation of unique T- and B-cell specificities is influenced by immunogen configuration. Cell Immunol 125:65, 1990

Londoño JA, Gras-masse H, Dubeaux C et al. Secondary structure and immunogenicity of hybrid synthetic peptides derived from two *Plasmodium falciparum* pre-erythrocytic antigens. J Immunol 145:1557, 1990

Löwenadler B, Svennerholm AM, Gidlund M et al. Enhanced immunogenicity of recombinant peptide fusions containing multiple copies of a heterologous T helper epitope. Eur J Immunol 20:1541, 1990

Löwenadler B, Lycke N, Svanhol C et al. T and B cell responses to chimeric proteins containing heterologous T helper epitopes inserted at different positions. Mol Immunol 29:1185, 1992

Margalit H, Spouge JL, Cornette JL et al. Prediction of immunodominant helper T cell antigenic sites from the primary sequence. J Immunol 138:2213, 1987

Milich DR, McLachlan A, Chisari FV et al. Nonoverlapping T and B cell determinants on an hepatitis B surface antigen pre-S(2) region synthetic peptide. J Exp Med 164:532, 1986

Milich DR, McLachlan A, Thornton GB et al. Antibody production to the nucleocapsid and envelop of the hepatitis B virus primed by a single synthetic T cell site. Nature 329:547, 1987

Milich DR, Hughes JL, McLachlan A et al. Hepatitis B synthetic immunogen comprised of nucleocapsid T cell sites and an envelope B-cell epitope. Proc Natl Acad Sci USA 85:1610, 1988

Milich DR, Jones JE, McLachlan A et al. Importance of subtype in the immune response to the pre-S(2) region of the hepatitis B surface antigen. II. Synthetic Pre-S(2) immunogen. J Immunol 144:3544, 1990

Momburg F, Fuchs S, Drexler J et al. Epitope-specific enhancement of antigen presentation by invariant chain. J Exp Med 178:1453, 1993

Munesinghe DY, Clavijo P, Calle MC et al. Immunogenicity of multiple antigen peptides (MAP) containing T and B cell epitopes of the repeat region of *P. falciparum* circumsporozoite protein. Eur J Immunol 21:3015, 1991

Noelle RJ, Snow EC, Uhr JW et al. Activation of antigen-specific B cells: Role of T cells, cytokines, and antigen in induction of growth and differentiation. Proc Natl Acad Sci USA 80:6628, 1983

Noelle RJ and Snow EC. Cognate interactions between helper T cells and B cells. Immunol Today 11:361, 1990

Noelle RJ, Roy M, Shepard DM et al. A 39-kDa protein on activated helper T cells binds CD40 and transduces the signal for cognate activation of B cells. Proc Natl Acad Sci USA 89:6550, 1992

Nonoyama S, Hollenbaugh D, Aruffo A et al. B cell activation via CD40 is required for specific antibody production by antigen-stimulated human B cells. J Exp Med 178:1097, 1993

van Noort JM and van der Drift ACM. The selectivity of cathepsin D suggests an involvement of the enzyme in the generation of T cell epitopes. J Biol Chem 264:14159, 1989

van Noort JM, Boon J, van der Drift ACM et al. Antigen processing by endosomal proteases determines which site of sperm whale myoglobin are eventually recognized by T cells. Eur J Immunol 21:1989, 1991

Palker TJ, Matthews TJ, Langlois A et al. Polyvalent human immunodefiency virus synthetic immunogen comprised of envelope gp120 T helper cell sites and B cell neutralization epitopes. J Immunol 142:3612, 1989

Partidos CD, Stanley CM, and Steward MW. The effect of orientation on the immunogenicity of chimeric synthetic peptides representing measles virus protein sequences. Mol Immunol 29:651, 1992a

Partidos CD, Stanley CM and Steward MW. The influence of orientation and number of copies of T and B cell epitopes on the specificity and affinity of antibodies induced by chimeric peptides. Eur J Immunol 22:2675, 1992b

Partidos CD and Steward MW. The effects of a flanking sequence on the immune response to a B and a T cell epitope from the fusion protein of measles virus. J Gen Virol 73:1987, 1992c

Partidos CD, Obeid OE and Steward MW. Antibody responses to non-immunogenic synthetic peptides induced by co-immunization with immunogenic peptides. Immunology 77:262, 1992d

Perkins DL, Berritz G, Kamradt T et al. Immunodominance: Intramolecular competition between T cell epitopes. J Immunol 146:2137, 1991a

Perkins DL, Wang Y, Fruman D et al. Immunodominance is altered in T cell receptor (β-chain) transgenic mice without the generation of a hole in the repertoire. J Immunol 146:2960, 1991b

Reyes VE, Chin LT and Humphreys RE. Selection of class I MHC-restricted peptides with the strip-of-helix hydrophobicity algorithm. Mol Immunol 25:867, 1988

Reyes VE, Fowlie EJ, Lu S et al. Comparison of three related methods to select T cell-presented sequences of protein antigens. Mol Immunol 27:1021, 1990

Ria F, Chan BMC, Scherer MT et al. Immunological activity of linked T cell epitopes. Nature 343:381, 1990

Rothbard JB and Taylor WR. A sequence pattern common to T cell epitopes. EMBO J 7:93, 1988

Roy S, Scherer MT, Briner TJ et al. Murine MHC polymorphism and T cell specificities. Science 244:572, 1989

Rudensky AY, Preston-Hurlburt P, Hong SC et al. Sequence analysis of peptides bound to MHC class II molecules. Nature 353:622, 1991

Rzepczyk C, Csurhes PA, Lord R et al. Synthetic peptide immunogens eliciting antibodies to *Plasmodium falciparum* sporozoite and merozoite surface antigens in H-2^b and H-2^k mice. J Immunol 145:2691, 1990

Sarobe P, Lasarte J-J, Golvano J et al. Induction of antibodies against a peptide hapten does not require covalent linkage between the hapten and a class II presentable helper peptide. Eur J Immunol 21:1555, 1991

Schaeffer EB, Sette A, Johnson DL et al. Relative contribution of "determinant selection" and "holes in the T cell repertoire" to T cell responses. Proc Natl Acad Sci USA 86:4649, 1989

Sercarz E and Krzych U. The distinctive specificity of antigen-specific suppressor cells. Immunol Today 12:111, 1991

Sette A, Buus S, Colón S et al. Structural characteristics of an antigen required for its interaction with Ia and recognition by T cells. Nature 328:395, 1987

Sette A, Buus S, Colón S et al. I-A^d-binding peptides derived from unrelated protein antigens share a common structural motif. J Immunol 141:45, 1988

Sette A, Buus S, Colón S et al. Structural analysis of peptides capable of binding to more than one Ia antigen. J Immunol 142:35, 1989a

Sette A, Buus S, Appella E et al. Prediction of MHC-binding regions of protein antigens by sequence pattern analysis. Proc Natl Acad Sci USA 86:4649, 1989b

Sette A, Adorini L, Appella E et al. Structural requirements for the interaction between peptide antigens and I-E^d molecules. J Immunol 143:3289, 1989c

Sette A, Sidney J, Albertson M et al. A novel approach to the generation of high affinity class II-binding peptides. J Immunol 145:1809, 1990

Shaw DM, Stanley CM, Partidos CD et al. Influence of the T-helper epitope on the titre and affinity of antibodies to B cell epitopes after co-immunization. Mol Immunol 30:961, 1993

Shimonkevitz R, Colón SM, Kappler JW et al. Antigen recognition by H-2-restricted T cells. II. A tryptic ovalbumin peptide that substitutes for processed antigen. J Immunol 133:2067, 1984

Sinigaglia F, Guttinger M, Kilgus J et al. A malaria T cell epitope recognized in association with most mouse and human MHC class II molecules. Nature 336:778, 1988

Smolenski LA, Kaumaya P, Atassi MZ et al. Characteristics of peptides which compete for presented antigen-binding sites on antigen-presenting cells. Eur J Immunol 20:953, 1990

Spouge JL, Guy HR, Cornette JL et al. Strong conformational propensities enhance T cell antigenicity. J Immunol 138:204, 1987

Srinivasan M, Domanico SZ, Kaumaya PTP et al. Peptides of 23 residues or greater are required to stimulate a high affinity class II-restricted T cell response. Eur J Immunol 23:1011, 1993

Stille CJ, Thomas LJ, Reyes VE et al. Hydrophobic strip-of-helix algorithm for selection of T cell-presented peptides. Mol Immunol 24:1021, 1987

Su H and Caldwell HD. Immunogenicity of a chimeric peptide corresponding to T helper and B cell epitopes of the *Chlamydia trachomatis* major outer membrane protein. J Exp Med 175:227, 1992

Vitetta ES, Bossie A, Fernandez-Botran R et al. Interaction and activation of antigen-specific T and B cells. Immunol Rev 99:193, 1987

Wang Y, Smith JA, Kamradt T et al. Silencing of immunodominant epitopes by contiguous sequences in complex synthetic peptides. Cell Immunol 143:284, 1992a

Wang Y, Smith JA, Gefter ML et al. Immunodominance: Intermolecular competition between MHC class II molecules by covalently linked T cell epitopes. J Immunol 148:3034, 1992b

Van der Zee R , Van Eden W , Meloen RH et al. Efficient mapping and characterization of a T cell epitope by the simultaneous synthesis of multiple peptides. Eur J Immunol 19:43, 1989

Zegers N, Gerritse K, Deen C et al. An improved conjugation method for controlled covalent coupling of synthetic peptides to proteins using glutaraldehyde in a dialysis method. J Immunol Meth 130:195, 1990

Zegers ND, van Holten C, Boersma WJA et al. Peptide induced memory (IgG) response, crossreactive with native proteins, requires covalent linkage of a specific B cell epitope with a T cell epitope. Eur J Immunol 23:630, 1993

Chapter 9

PROGRESS AND FRONTIERS IN PREDICTION OF B AND T CELL EPITOPES

Susanne Modrow and Hans Wolf

TABLE OF CONTENTS

I. INTRODUCTION

During the past few years, much knowledge has accumulated concerning the sequences and also the structures of proteins. Automated methods for DNA and protein sequencing made the primary information of biological molecules easily accessible for research in all fields of molecular and microbiology. For many purposes, particularly, but not exclusively, in studying the biology of infections agents, development of new types of vaccines or of reagents for serological testing, the prediction of the relevant epitopes or antigenic sites in the amino acid sequence of a protein component is of major importance. This is extremely important and unavoidable for the production of applicable biological substances and successful progress in research. Of interest are two major subgroups, B cell epitopes and T cell epitopes. B cell epitopes refer to the sites in a protein which interact with the binding sites of an antibody molecule. During the normal process of induction of the synthesis of specifically binding antibodies in an organism in response to a foreign protein, the variable domains of immunoglobulins recognize defined segments of the amino acid sequence in an intact protein molecule. In contrast to the recognition of B cell epitopes as parts of complex proteins, T cell epitopes represent different antigenic structures. They are recognized as peptides of about 9 to 11 amino acids in length, which are complexed to the protein class of major histocompatibility complex antigens (MHC) on the surface of cells. Those MHC-proteins have a high variability in their regions interacting with the peptides processed from intercellular synthesized polypeptides (in case of MHC class I proteins) or from proteins taken up by the cells using the endosomal pathway (MHC class II). During the latter process, the proteins are proteolyticallly cleaved into peptides which interact with MHC class II proteins followed by presentation of the complex on the cell surface. Peptides presented in this way as T cell epitopes are recognized as complex with the MHC proteins by the variable parts of T cell receptors.

Due to the fundamentally different ways of presentation — B cell epitopes as parts of structurally and functionally intact proteins, T cell epitopes in the form of peptides in complex with cellular surface proteins — prediction of both classes of antigenic sites follows totally different parameters.

II. PREDICTION OF B CELL EPITOPES

The optimal way to predict B cell epitopes is based not only on the knowledge of the amino acid sequence of a protein, but also on structural parameters. In cases where the protein structure is known by X-ray crystallography or NMR analysis, the areas of the amino acid sequence present on the surface of the individual protein molecule can be determined. Since antibody molecules do not interact with processed peptides derived from the protein in question but with the structurally and functionally intact molecules, they preferentially interact with regions on the surface of the polypeptide. Amino acid sequences in interior parts of the protein and the protein core are only accessible to antibody binding when the structure of the molecule is disrupted, e.g., by treatment with detergents or other procedures structure disrupting activities such as treatment with chaotropic agents which may in a mild form occur also during polyacrylamid gel electrophoresis or Western blotting. In these cases also areas of the molecule not regularly exposed to the protein's surface may become accessible for antibody binding.

This is, however, not possible during the normal way of epitope recognition by antibodies interacting exclusively with surface-exposed regions of the antigen.

0-8493-8967-4/95/$0.00+$.50

Surface-exposed regions representing B cell epitopes in a protein molecule may further be divided into two subgroups: discontinuous, structurally defined antigenic sites and epitopes which are specifically defined by short sequences of 4 to 10 amino acids, frequently known as sequential epitopes (Atassi and Smith, 1978). Parts from different regions of a protein may contribute to the formation of discontinuous epitopes which are therefore highly dependent on the structure of the protein and the folding of the amino acid chain into special regions at the surface of the polypeptide. Those discontinuous epitopes can only be defined or predicted when data on the three-dimensional structure of the protein are available and accessible. Without that knowledge, prediction of discontinuous epitopes is only possible with some success when structural data of a related protein with defined homologies in the amino acid sequences and similar functions are known. In these cases it may be possible to calculate the structure of the surface-exposed areas of the related protein by molecular dynamic simulation and thus identify the residues in different parts of the polypeptide forming a related structurally defined antigenic site. Since X-ray crystallography and NMR analysis are only applied to a relatively small number of proteins, this possibility to predict discontinuous antigenic sites will be rarely given.

In the majority of cases, sequential epitopes have to be predicted from the amino acid sequence without additional knowledge of three-dimensional structures. Sequential epitopes consist of small regions of amino acids accessible for antibody interaction on the surface of a protein. Minimal length has been described as four residues, most frequently it is assumed that those epitopes are defined by about 6–10 neighboring amino acids (Geysen et al., 1985; Tang et al., 1988). Of course, not only the sequence of the amino acids is specifically recognized by an antibody, but the three-dimensional structure of this small region with its particular distribution of electrochemical properties (Barlow et al., 1986), which is based on the defined atomic composition in this area of the protein, is also recognized. Accumulated knowledge concerning the antigenic structure of a small number of well-studied proteins has stimulated the possibility to predict the location of continuous, sequentially defined epitopes from certain features of the primary structure (Van Regenmortel, 1988, 1989). Many studies have shown that the location of sequential epitopes tends to correlate with features of segments of an amino acid chain such as their hydrophilicity (Hopp and Woods, 1981; Kyte and Doolittle, 1987), atomic mobility (Westhof et al., 1984; Tainer et al., 1984, 1985) and flexibility (Karplus and Schulz, 1985), surface accessibility (Janin et al., 1978; Emini et al., 1985) and sequence variability in cases where the sequence of several, functionally identical proteins, e.g., in related species, is known (Bittle et al., 1982; Modrow et al., 1987). Even if there is considerable disagreement about which structural parameter correlates best with antigenicity, it is obvious that those parameters do not represent independent variables but are generally related to each other. These regions are preferentially present in loops forming surface projections in the protein structure and therefore either contain a high number of hydrophilic residues or are in the neighborhood of glycosylation sites and have a high degree of atomic mobility and flexibility.

In addition to parameters such as hydrophilicity and flexibility, which are mainly based on the chemical properties of the individual residues present in a defined protein region, features of structural propensities contribute to the antigenicity of an amino acid region. Based on the knowledge of the X-ray structure of a small number of proteins, the probability of secondary structures may be calculated from the primary sequence of a protein; certain factors have been found to be common for antigenic sites (Chou and Fasman, 1978; Garnier et al., 1978).

Therefore, one possible way to predict continuous B cell epitopes is to analyze the amino acid sequence and to predict antigenic sites based on computer programs linking theoretical and empirical data of chemical properties of the amino acids with structural conformation parameters. It is important, that individual parameters such as hydrophilicity, flexibility and surface accessibility are averaged over segments of 6 to 7 residues and are combined in a weighted manner with the probabilities for secondary structures for successful prediction of continuous antigenic sites (Wolf et al., 1988). These predictions of epitopes using an algorithm for the antigenic index of an amino acid sequence were successfully applied to a high number of viral and various other proteins (Jameson and Wolf, 1988; Modrow and Wolf, 1990). As an example for successful use of B cell epitope prediction, we show in Fig. 1 an amino acid analysis of the protein encoded by the open reading frame BMRF2 of the Epstein-Barr virus (EBV) (Modrow et al., 1992). By combination of parameters for secondary structure, hydrophilicity, surface probability and flexibility into the values of the antigenic index, one B cell epitope in the center of the protein sequence could be predicted as preferentially recognized by antibodies. Areas potentially associated with cellular membranes are indicated as shaded and display rather low values for antigenicity. To identify the viral protein encoded by BMRF2 we synthesized an oligopeptide derived from the sequence of the predicted B cell epitope and used it for specific antibody induction in rabbits. This approach led to the identification of the respective protein in EBV infected cell lines. In addition to those successful examples, it has,

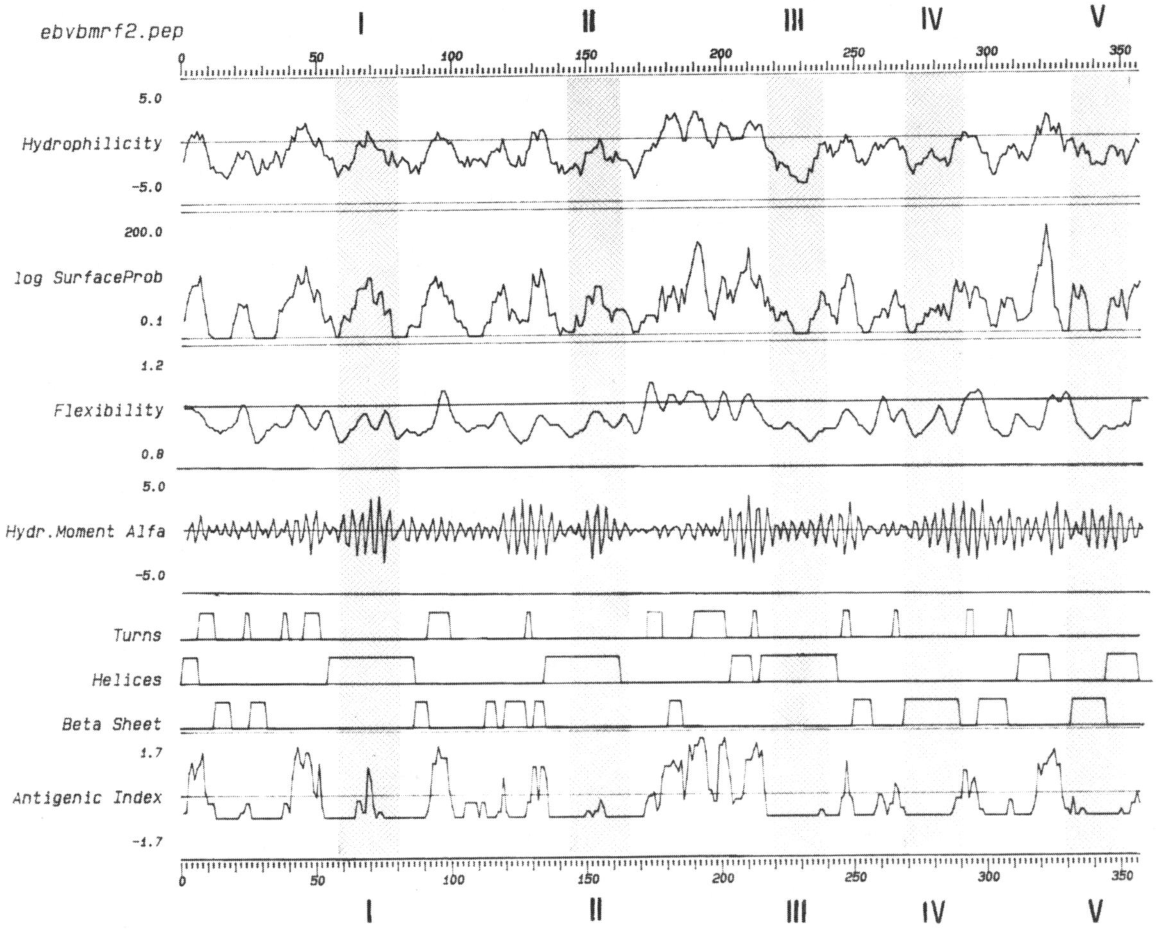

Fig. 1. Linear graph of an amino acid analysis of the BMRF2 encoded open reading frame. Values for hydrophilicity, surface probability, flexibility, antigenicity and secondary structures are shown. Potential transmembrane spanning regions are indicated as shaded areas.The amino acid region between residues 180-220 shows very high values for antigenicity, and a synthetic peptide derived from residues 186-197 was amino terminally esterified with palmitic acid and inoculated into rabbits for antibody induction. The anti-peptide immunoglobulins were successfully applied to characterize the respective EBV protein in cell lysates.

however, to be kept in mind that special functions of the proteins and additional parameters introduced by protein modification may have an enormous influence on the antigenicity of a defined protein area. Glycosylation and phosphorylation have been shown to mask antigenic regions of a polypeptide (Skehel et al., 1984; Klenk, 1990); regions in transmembrane proteins localized not outside the membrane but in the cytoplasm are not accessible to antibodies even when they exhibit high values for surface accessibility and antigenicity.

Therefore, in addition to the values predictable by computer assisted methods, the function of the protein under study has to be kept in mind. Very often the functionally active regions of a polypeptide, i.e., interaction sites with other proteins or nucleic acids in case of DNA-binding proteins, are extremely hydrophilic and are localized at the surface. They are, however, often not accessible for interaction with immunoglobulins due to their participation in other functional important processes.

Therefore, the more knowledge on the function of an individual polypeptide that is contributed to the amino acid sequence, the more successful prediction of continuous antigenic epitope may be performed.

III. PREDICTION OF T CELL EPITOPES

In contrast to B cell epitopes which are recognized as defined parts or structures in native, functionally intact proteins, T cell epitopes are represented by short peptides derived from the individual proteins. These peptides interact with MHC molecules on the surface of antigen-presenting cells, and the complex of the MHC protein/peptide is recognized either by T-helper cells, leading to proliferation of antibody producing B cells, or by cytotoxic T cells, with the consequence of destruction of the antigen-presenting cell (for review, see Townsend and Bodmer, 1989; Parnes, 1989; Bjorkman and Parham, 1990; Rothbard and Gefter, 1991). T-helper cells interact with a complex of peptide and MHC class II proteins, and cytotoxic T cells are specifically recognizing peptides bound to MHC class I antigens. The molecular basis of T cell epitope presentation in complex with cellular membrane proteins renders prediction much more complex than the previously described B cell epitopes.

A very important contribution for the definition of peptide sequences and their respective properties necessary for interaction with MHC molecules came from the publication of the crystallographic structures of MHC class I and class II antigens (Bjorkman et al., 1987; Brown et al., 1993; Zhang et al., 1992; Madden et al., 1992). From those structures, it became obvious that MHC molecules bind peptides derived either from cellular or foreign (viral or bacterial) proteins in the antigen-binding groove formed by two alpha-helices topped onto a beta-pleated sheet plateau. In this antigen-binding groove, pockets could be identified, in which mainly hydrophobic residues have to be present at defined positions of the peptide to be fitted in. Additional interaction sites could be characterized between the charged amino and carboxy terminal end of the peptide and defined regions or pockets in the MHC part (Saper et al., 1991; Garrett et al., 1989). Further very important knowledge for the prediction of amino acid sequences with the potential to interact with individual MHC proteins was contributed by the work of Rammensee and co-workers, which developed methods to elute the peptides complexed with one specific MHC antigen and to sequence the pool of naturally bound peptide fragments (Falk et al., 1991; Rötzschke et al., 1990; Falk and Rötzschke, 1993; Rammensee et al., 1993a,b). Despite the high diversity of amino acids at the respective positions of the peptides, anchor residues represented by hydrophobic amino acids could be identified which are present in the majority of the eluted fragments (Falk et al., 1991; Jardetzky et al., 1991; Hunt et al., 1992). These hydrophobic residues enable the peptide to interact with the hydrophobic pockets present in the MHC proteins. Since the position of the, mostly two, anchor residues in the peptide have to be at defined locations with a defined distance in between, protein sequences may be used as a basis for prediction and peptides with the correct pattern of the distribution of hydrophobic residues may be found by commonly used algorithms for sequence and pattern recognition. Since each MHC molecule has a different distance between the hydrophobic pockets present in the structure of their antigenic grooves, peptides with the potential to interact with a specific MHC protein have to show patterns and distributions of anchor residues characteristic for this individual type of MHC antigen. Peptides binding to the human MHC class I protein HLA-A2 lack the potential to bind to other types since they to not have the right pattern of residues which enables interaction (Rötzschke et al., 1992).

By sequencing pools of peptides eluted from individual MHC class I proteins, the pattern of peptides with the potential to interact with a series of MHC antigens (mouse H2-D^d, H-2K^d, H-2K^b, human HLA-A2, HLA-B27, Table 1) could be described. A further criterium for the potential of a peptide to interact with the antigen-presenting proteins is the correct length. For MHC class I protein, interaction sequencing data indicated a size of 8 to 9 residues as optimal. Longer peptides therefore can interact via their anchor residues when present at the correct positions, but they do not have the possibility to further interact via their charged amino and carboxy terminal ends since they do not have the precise locations necessary for binding (Saper et al., 1991; Garrett et al., 1989). Peptides spanning 15 to 20 residues therefore may bind. The affinity is, however, much lower in comparison to shorter peptides exhibiting the correct length.

MHC class II antigens have the capacity to bind longer peptides up to 20 residues and more due to a different structure of the antigenic peptide-binding groove (Brown et al., 1993; Gorga et al., 1991; Gorga, 1992). Until now, a characteristic pattern of the distribution of individual amino acids had not been identified. There are some indications that the interaction is also performed by hydrophobic interactions. The high restriction in binding as characterized with MHC class I antigens has, however, not been found for MHC class II proteins (Kropshofer et al., 1992; Rudensky et al., 1991; Hunt et al., 1992; Chicz et al., 1992).

In order to predict T cell epitopes specifically binding to individual MHC proteins, it has to be kept in mind that the distribution of hydrophobic anchor residues and the defined length of a peptide are not the only criteria for complex formation. Peptide interaction with individual MHC class I proteins may be easily tested using mutant cell lines, which do not contain a detectable MHC protein concentration on their cell surface due to a defect in the processing of endogenously synthesized proteins and transport of the peptides into the lumen of the endoplasmetic reticulum. Therefore, only "empty" MHC proteins, which are not stable and have a very short half life, are transported to the cell surface. If peptides are added to the culture media with the capacity to interact with the

Table 1
Known motifs for peptide interaction with MHC antigens

	Position								
Class I allele	**1**	**2**	**3**	**4**	**5**	**6**	**7**	**8**	**9**
H-2K[d*]	X	**Y**	X	X	X	X	X	X	**L** **I**
H-2K[b*]	X	X	X	X	**F** **Y**	X	X	**M** **I** **L**	
H-2D[d*]	X	X	X	X	**N**	X	X	X	**I** **L** **M**
HLA-A2[*,**]	X	L	X	X	X	X	X	X	**V**
HLA-B27[***]	**X**	R	X	X	X	X	X	X	**X°**

°Predominant use of R or K.
*Falk et al., 1991; **Hunt et al., 1992; ***Jardetzky et al., 1991.
X stands for any residue; side chains of the amino acids typed in bold contact allele-specific pockets in class I molecules.

individual MHC class I antigen, the complex is stabilized at the cell surface and detectable by immunofluorescence and FACS analysis. The value of the measured signal is linear to the affinity of the interaction capacity shown by the diverse peptides used. Well fitting T cell antigens therefore render very high values in this test system (Stuber et al., 1992). Despite most of the peptides show the correct pattern of anchor residues and length, clearly indicate binding to the respective MHC molecule a lot of others do not. Therefore, it has to be assumed that additional factors are involved that are necessary for peptides to be complexed with MHC proteins (Stuber et al., 1994; Zeh et al., 1994). These additional properties are not well defined, and therefore, precise predictions of peptides with the potential to bind to MHC antigens are rather difficult to perform. In addition to undefined structures and residues involved in complex formation, it has been shown that the length of the peptide bound to MHC class I proteins may be somewhat variable also. For peptides derived from the V3-region of gp 120 of HIV-1 (human immunodeficiency virus) we could show that besides the hydrophobic anchor position a positively charged arginine is essential for the complex formation to HLA-A2.1 and that the minimal size of the peptide has to be 12 residues (Borbe et al., 1995). This protein region of the gp 120 exterior surface proteins has a very stable beta-turn in the native protein structure according to NMR-structured analysis (Czisch et al., submitted; Chandrasekhar et al., 1991). Even in models of the complex between the peptide and HLA-A2, this peptide region seems to form a bulge with makes it necessary for the peptide to have a length of more than the otherwise optimal nine residues. Interaction studies of various shortened forms of the V3-peptide and the mutant T1 cellline specific for HLA-A2 allowed the definition of the minimal size to be 12 residues and showed that a positively charged arginine has to be present at the position 6 of the peptide (Fig. 2).

Interaction and complex formation with MHC molecules is an essential condition for a peptide to serve as T cell epitope; however, it is, only one factor in a complex process. Other factors reflecting the functionality of the T cell antigen are the interaction of the peptide/MHC complex with the T cell receptor on the surface of the respective effector cells, cytotoxic T cell or T-helper cells (Davis et al., 1990). It is an unsolved question: which factors are involved in this protein interaction process if every complex formed between a peptide derived from a foreign protein is automatically recognized and which genetically inherited factors and processes may play a role. Therefore, prediction of T cell epitopes is still, despite the enormous efforts due to the fundamental understanding of the molecular basis of the immunological recognition, an open field ready for many surprises.

ACKNOWLEDGEMENT
The authors thank Mrs. S. Heindl for typing the manuscript.

ABBREVIATIONS

EBV	Epstein-Barr virus
FACS	Fluorescence activated cell sorter
HIV	Human immunodeficiency virus
MHC	Major histocompatibility complex
NMR	Nuclear magnetic resonance
V3-region	Variable domain 3 of the outer glycoprotein gp120 of HIV

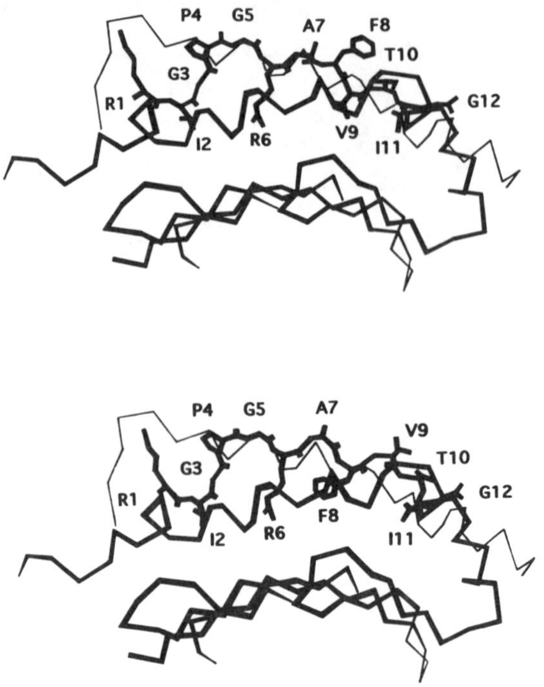

Fig. 2. Proposed complex formations between gp 120/V3 consensus peptide and MHC class I molecules. Light: V3-consensus-peptide, amino acid sequence is given in one letter code beginning at the amino terminal end of the peptide. Dark: Backbone structure of the MHC protein part, antigenic peptide-binding groove. A: Complex of gp 120/V3 and HLA-A2.1. B: Complex of gp 120/V3 and H-2D^d.

IV. REFERENCES

Atassi MZ and Smith JA. A proposal for the nomenclature of antigenic sites in peptides and proteins. Immunochemistry 15:609, 1978

Barlow DJ, Edwards MS and Thornton JM. Continuous and discontinuous proteins antigenic determinants. Nature 322:747, 1986

Bittle JL, Houghten RA, Alexander H et al. Protection against foot-and-mouth disease by immunization with a chemically synthesized peptide predicted from the viral nucleotide sequence. Nature 298:30, 1982

Bjorkman PJ, Saper MA, Samraoui B et al. Structure of the human class I histocompatibility antigen HLA-A2. Nature 329:506, 1987

Bjorkman PJ and Parham P. Structure, function and diversity of class I major histocompatibility complex molecules. Ann Rev Biochem 59:253, 1990

Borbe HC, Stuber G, Wagner R et al. Structural and immunological reactivity of the principal neutralizing determinant V3 of gp 120/HIV-1. J Peptide Science, in press, 1995

Brown JH, Jardetzky TS, Gorga JC et al. Three-dimensional structure of the human class II histocompatibility antigen DR 1. Nature 364:33, 1993

Chandrasekhar K, Proty AT and Dyson J. Solution conformational preferences of immunogenic peptides derived from the principal neutralizing determinant of the HIV-1 envelope glycoprotein gp 120. Biochemistry 30:9187, 1991

Chicz RM, Notan RG, Lam WS et al. Predominant naturally processed peptides bound to HLA-DR1 are derived from MHC-related molecules and are heterogeneous in size. Nature ☺

Chou PY and Fasman GD. Prediction of the secondary structure of proteins from their amino acid sequence. Adv Enzymol 47:45, 1978

Czisch M, Borbe HC, Habazettl J et al. Conformation of the V3 domain of the outer envelope glycoprotein of HIV-1 derived from NMR spectroscopy. Eur J Biochem, in press, 1995

Davis MM. T cell receptor gene diversity and selection. Ann Rev Biochem 59:475, 1990

Emini EA, Hughes JV, Perlow DS et al. Induction of Hepatitis A virus - neutralizing antibody by a virus-specific peptide. Virol 55:836, 1985

Falk K, Rötzschke O, Stevanovic S et al. Allele-specific motifs revealed by sequencing of self-peptides eluted from MHC molecules. Nature 351:290, 1991

Falk K and Rötzschke O. Consensus motifs and peptide ligands of MHC class I molecules. Sem Immunol, 5:81, 1993

Garnier J, Osynthorpe OJ and Robson B. Analysis of the accuracy and implication of simple methods for predicting the secondary structure of globular proteins. J Mol Biol 120:97, 1978

Garrett TPJ, Saper MA, Bjorkman PJ et al. Specifity pockets for the side chains of peptide antigens in HLA-Aw 68. Nature 343:692, 1989

Geysen HM, Barteling SJ and Meloen RH. Small peptides induce antibodies with a sequence and structural requirement for binding antigen comparable to antibodies raised against native protein. Proc Natl Acad Sci USA 82:178, 1985

Gorga JC, Brown JH, Jardetzky T et al. Crystallization of HLA-DR antigens. Res Immunol 142:401, 1991

Gorga JC. Structural analysis of class II major histocompatibility complex proteins. CRC Crit Rev Immunol 11:305, 1992

Hopp TP and Woods KR. Prediction of protein antigen determinants from amino acid sequences. Proc Natl Acad Sci USA 78:3824, 1981

Hunt DF, Henderson RA, Shabanowitz J et al. Characterization of peptides bound to the class I MHC molecule HLA-A2.1 by mass spectrometry. Science 255:1261, 1992

Jameson BA and Wolf H. Predicting antigenicity from protein primary structure: A new algorithm for the prediction of antigenic sites. CABIOS 4:181, 1988

Janin J, Wodak S, Levitt M et al. Conformation of amino acid side-chains in proteins. J Mol Biol 125:357, 1978

Jardetzky TS, Lane WS, Robinson RA et al. Identification of self peptides bound to purified HLA-B27. Nature 353:326, 1991

Karplus PA and Schulz GE. Prediction of chain flexibility in proteins. Naturwissenschaften 72:212, 1985

Klenk H-D. Influence of glycosylation on antigenicity of viral proteins. In: Immunochemistry of Viruses II (v. Regenmortel and Neurath, eds). Elsevier Science Publishers B.V., Amsterdam, p 25–37, 1990

Kropshofer H, Max H, Müller CA et al. Self-peptide released from class II HLA-DR1 exhibits a hydrophobic two-residue contact motif. J Exp Med 175:1799, 1992

Kyte J and Doolittle RF. A simple method for displaying the hydropathic character of a protein. J Mol Biol 157:105, 1987

Madden DR, Gorga JC, Strominger JL et al. The three-dimensional structure of HLA-B27 at 2.1 Å resolution suggests a general mechanism for tight peptide binding to MHC. Cell 70:1035, 1992

Modrow S, Hahn, BH, Shaw GM et al. Computer-assisted analysis of envelope protein sequences of seven human immunodeficiency virus isolates prediction of antigenic epitopes in conserved and variable regions. J Virol 61:570, 1987

Modrow S and Wolf H. Use of synthetic peptides as diagnostic reagents in virology. In: Immunochemistry of Viruses II. (v. Regenmortel and Neureth, eds). Elsevier Science Publishers, p 83–101, 1990

Modrow S, Höflacher B and Wolf H. Identification of a protein encoded in the EB-viral open reading frame BMRF2. J Arch Virol 127:379, 1992

Parnes JR. Molecular biology and function of CD4 and CD8. Adv Immunol 44:265, 1989

Rammensee H-G, Falk K and Rötzsche O. MHC molecules as peptide receptors. Curr Opinion Immunol 5:35, 1993a

Rammensee H-G, Falk K and Rötzschke O. Peptides naturally presented by MHC class I molecules. Ann Rev Immunol 11:213, 1993b

Rothbard JB and Gefter ML. Interactions between immunogenic peptides and MHC proteins. Ann Rev Immunol 9:527, 1991

Rötzschke O, Falk K, Deres K et al. Isolation and analysis of naturally processed viral peptides as recognized by cytotoxic T cells. Nature 348:252, 1990

Rötzschke O, Falk K, Stevanovic S et al. Peptide motifs of closely related HLA class I molecules encompass substantial differences. Eur J Immunol 22:2453, 1992

Rudensky AY, Preston-Hurlburt P, Hong S-C et al. Sequence analysis of peptides bound to MHC class II molecules. Nature 353:622, 1991

Saper MA, Bjorkman PJ and Wiley DC. Refined structure of the human histocompatibility antigen HLA-A2 at 2.6 Å resolution. J Mol Biol 219:277, 1991

Skehel JJ, Stevens DJ, Daniels RS et al. A carbohydrate side chain on hemagglutinins of Hong Kong influenza viruses inhibits recognition by a monoclonal antibody. Proc Natl Acad Sci USA 81:1779, 1984

Stuber G, Modrow S, Höglund P et al. Assessment of major histocompatibility complex class I interaction with Epstein-Barr Virus and human immunodeficiency virus peptides by elevation of H-2 and HLA in peptide loading deficient cells. Eur J Immunol 22:2697, 1992

Stuber G, Storkus W, Leder G et al. Binding capacity of wild-type and mutated p53 nonamer peptides predicted to include HLA-A2.1 binding motifs assessed by peptide-induced stabilization of A 2.1 complex on the peptide loading deficient cells and by its reconstruction after acid denaturation. Eur J immunol 24:765, 1994

Tainer JA, Getzoff ED, Alexander H et al. The reactivity of anti-peptide antibodies as a function of the atomic mobility of sites in a protein. Nature 312:127, 1984

Tainer JA, Getzoff ED, Paterson Y et al. The atomic mobility component of protein antigenicity. Ann Rev Immunol 3:501, 1985

Tang XL, Tregear GW, White DO et al. Minimum requirements for immunogenic and antigenic activities of homologs of a synthetic peptide of influenza virus hemagglutinin. J Virol 62:4745, 1988

Townsend A and Bodmer H. Antigen recognition by class I-restricted T lymphocytes. Ann Rev Immunol 7:601, 1989

Van Regenmortel MHV. Which value of antigenic valency should be used in antibody avidity calculations with multivalent antigens? Mol Immunol 25:565, 1988

Van Regenmortel MHV. Structural and functional approaches to the study of protein antigenicity. Immunol Today 10:266, 1989

Westhof E, Altschuh D, Moras D et al. Correlation between segmental mobility and the location of antigenic determinants in proteins. Nature 311:123, 1984

Wolf H, Modrow S, Motz M et al. An integrated family of amino acid sequence analysis programs. CABIOS 4:187, 1988

Zeh HJ, Lotze GH, Salter RD et al. Flow-cytometric determination of peptide class I complex formation. Identification of p53 peptides that bind to HLA-A2. Hum Immunol 39:79, 1994

Zhang W, Young ACM, Imarai M et al. Crystal structure of the major histocompatibility complex class I H-2K^b molecule containing a single viral peptide: Implications for peptide binding and T cell receptor recognition. Proc Natl Acad Sci USA 89:8403, 1992

Chapter 10

ANALYSIS OF PEPTIDE AND PROTEIN STRUCTURE

H. Jane Dyson

TABLE OF CONTENTS

I. INTRODUCTION

This chapter provides a brief introduction to the methodology of structure determination of peptides and proteins in solution together with a brief review of some of the more exciting recent discoveries using both these and more traditional crystallographic methods to probe the nature of the immune system. Not only do we now have an idea of the structures of antibodies and MHC proteins, but the complexes of these molecules with their peptide and protein ligands bound are beginning to show how these intermolecular interactions work and to give insights useful in the modeling of other protein-protein interactions for drug design purposes. Peptides can be used as immunogens to elicit antibodies that recognize the cognate sequence in folded proteins. In addition, peptides have traditionally been used as probes in epitope mapping of proteins. The nature of peptide fragments of proteins in solution is becoming better understood. Methods for detection of structured conformers in peptide solutions are now available and are proving useful in drug design applications when used with care. In this chapter, I will focus on the techniques for elucidation of structure of molecules in solution and the pitfalls and problems that may arise during these studies. At the end of the chapter, I will describe some of the recent published results in the field of peptide structure in solution, as well as some crystal structures showing peptides bound to molecules of immunological interest.

I.1. Methods for obtaining three-dimensional structures

While a considerable amount of information about local environments within molecules can be obtained from a number of spectroscopic methods such as fluorescence and UV-visible spectroscopy, only a few techniques yield the density of information that is required to map out the three-dimensional structure of a biological molecule. Presently, two methods are used to obtain this information: crystallography, using either with X-ray or neutron diffraction, and nuclear magnetic resonance (NMR) spectroscopy. In both of these methods, there is a large amount of raw data that must be analyzed to give the structural information, and the analysis usually requires the use of large computers and sophisticated computer programs. It is beyond the scope of this review to give an exhaustive treatment of the methodologies for either of the methods. Instead, I will focus on recent advances that have allowed us to

0-8493-8967-4/95/$0.00+$.50

obtain information on the structure of peptides and proteins in solution.

I.2. The nature of peptides in solution

Peptides were long believed to be uniformly disordered in aqueous solution. Early spectroscopic studies using relatively insensitive techniques such as circular dichroism (CD) and optical rotatory dispersion (ORD) spectroscopy failed to detect any evidence of structure in unconstrained peptides in aqueous solution (Epand and Scheraga, 1968; Taniuchi and Anfinsen, 1969; Hermans and Puett, 1971). The C-peptide of ribonuclease A was an apparent exception to this (Brown and Klee, 1971), and, later, a β-turn conformation was detected in an immunogenic peptide by NMR (Dyson et al., 1985). Nevertheless, most unconstrained linear peptides in aqueous solution do not show conformational preferences for structured forms. Even when a strong preference for a folded conformation is observed, it is not usually advisable or necessary to calculate structures for the peptide in solution, as is possible for folded proteins. These methods will be briefly reviewed below, and the reasons that they are unsuitable for peptides will also be discussed.

II. PROTEIN STRUCTURE

Until very recently, the only methods for obtaining a three-dimensional structure of a macromolecule were diffraction methods - X-ray crystallography, neutron diffraction and electron diffraction. Since 1985, new methods have been developed utilizing information from NMR spectroscopy. Since these methods are not widely known, I decided to include in this chapter a brief review of the means whereby structural information is obtained from NMR spectra. Reviews of this subject have appeared in the recent literature (Clore and Gronenborn, 1991a, 1991b).

II.1. Structural information from the NMR spectrum

The NMR spectrum of a macromolecule contains all of the information necessary to describe the three-dimensional structure, since the resonances of the NMR-active nuclei (^{1}H, ^{13}C, ^{15}N, etc.) reflect the chemical environment within the molecule. The problem that arises is of course the identification of the information in structural terms. Much of the considerable progress made in recent years has been in the direction of experimental design for deconvolution and decoding of the available information (Clore and Gronenborn, 1991b). The most commonly used parameter is the nuclear Overhauser effect (NOE), (Neuhaus and Williamson, 1989) which is a measure of the magnetization transfer between nuclei in close spatial proximity, whether connected through chemical bonds or not. One of the seminal developments that allowed NOE information to be translated directly into structural information was the distance geometry algorithm (Havel et al., 1983; Havel and Wuthrich, 1984). Other NMR parameters such as coupling constants can also be used to define backbone geometry in structure calculations (Case et al., 1994). In principle, the resonance frequency itself (the chemical shift) also contains structural information, since it represents a summation of all the local influences on a particular nucleus. However, since many of these influences are subtle and incompletely understood, few attempts are presently being made to utilize chemical shifts as input in structure calculations (Case et al., 1994).

II.2. Assignment of resonances in the NMR spectrum

The first task in the determination of a protein structure by NMR spectroscopy is to identify the resonances in the spectrum with specific nuclei in the protein, known as resonance assignment. Very few NMR spectra contain resonances that are completely resolved, and for larger macromolecules resonance overlap becomes an increasing problem. In addition, the resonances are broader for larger molecules, due to the longer correlation times. Resolution of NMR spectra can be increased by using instruments with higher magnetic fields. Other methods, notably multi-dimensional methods, have also been developed to resolve and identify resonances (Clore and Gronenborn, 1991b).

For reasons of sensitivity, the proton is the most commonly used nucleus for NMR spectroscopy. It is also one of the most abundant nuclei in biological macromolecules. Carbon and nitrogen isotopes with the appropriate nuclear spins (^{13}C and ^{15}N, known as heteronuclei) can also be used, but their sensitivity is much reduced. Recently, methods have been developed for the detection of heteronuclei via the resonance of an attached proton (known as inverse methods) (Live et al., 1984), where considerable additional information from the spectra of the heteronuclei can be obtained with the sensitivity of the proton.

The complete assignment of all of the resonances of a protein would not have been possible without the development of multi-dimensional NMR techniques. Two- and higher dimensional spectra are obtained by adding incrementable delays into the sequence of radiofrequency pulses that are used to give the NMR spectrum (Ernst et al., 1987). Depending on the types of pulses used, two-dimensional spectra may give information on through-bond or through-space connectivities between nuclei. Peptides and proteins are convenient in this respect since the carbonyl group of the peptide bond, with no attached protons, acts as a separator between amino acid residues on the polypeptide chain. The resonances of the nuclei within a given amino acid can thus be identified by using through-bond (correlated) spectroscopy. Connection of these spin systems may then be done in a number of ways, with the most common using the NOE (Wüthrich, 1986). Specific patterns of NOEs between amino acid residues identify secondary structures in proteins (Wüthrich, 1986) and, as we shall see in a later section, conformational preferences for secondary structures in peptides (Dyson and Wright, 1991). A series of NOE connectivities between the $C^{\alpha}H$ of one residue and the NH of the following residue ($d_{\alpha N}(i,i+1)$ indicates the presence of a β-strand, while a helix is indicated by a series of NOEs between the NHs of consecutive residues ($d_{NN}(i,i+1)$, together with medium-range NOEs ($d_{\alpha N}(i,i+2)$, $d_{\alpha N}(i,i+3)$, $d_{\alpha N}(i,i+4)$, and $d_{\alpha\beta}(i,i+3)$ across the strands of the helix. These connectivities are illustrated schematically in Fig. 1. Fig. 2 shows a schematic diagram of the two simplest types of NMR spectra, the two-dimensional correlated spectrum (COSY) below the diagonal, and the two-dimensional NOE spectrum (NOESY) above the diagonal. The diagonal of a two-dimensional spectrum represents a projection of the one-dimensional spectrum, and off-diagonal cross peaks show connectivities between protons with these resonance frequencies. In the COSY spectrum, the connectivities are those between protons that are coupled to each other, usually on adjacent C or N atoms (e.g., $C^{\alpha}H$-$C^{\beta}H$ in Fig. 2). In the NOESY spectrum, the connectivities are those between protons close together in space, but not necessarily coupled (e.g., $C^{\alpha}H(1)$-$NH(2)$ in Fig. 2). In a folded protein, there are NOE connectivities between distant parts of the polypeptide chain; the global fold of the protein can be deduced from these, and the NOEs form the majority of the information used to calculate three-dimensional structures of macromolecules from NMR data.

For larger proteins, the overlap problem becomes increasingly severe, while the resonance lines are also broader. These problems can be addressed by the use of higher dimensional NMR spectra, especially those of proteins which have been uniformly labeled with stable carbon and nitrogen isotopes (Clore and Gronenborn, 1991b). These methods have two advantages. One advantage is that the nuclei can now be identified with reference not only to other protons on adjacent carbon or nitrogen atoms, but also to the carbon or nitrogen to which they are attached, giving an extra dimension for the resonances to be dispersed in. This is illustrated schematically in Fig. 3. The other advantage is that the couplings between adjacent carbon and nitrogen atoms are much larger than those between protons on adjacent atoms, so that spectroscopic methods that rely on correlation between coupled nuclei will work much better. This is particularly

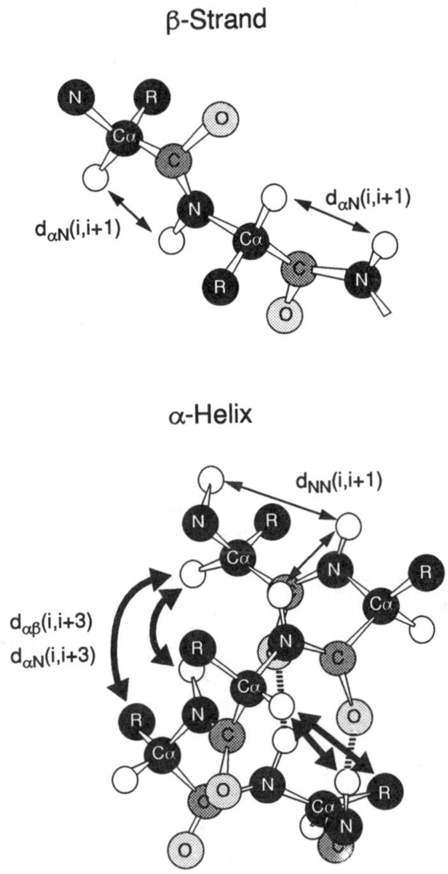

Fig. 1. Schematic diagram showing α-helix and β-strand peptide conformations and the NOE connectivities that are likely to arise due to the close approach of the peptide protons.

important for larger proteins, where coherence transfer through bonds is attenuated due to relaxation. The utility of the heteronuclear-edited three-dimensional techniques in resolving ambiguities in resonance assignment is illustrated in Fig. 4, which shows in part A a portion of a two-dimensional proton NOESY spectrum of a small protein, thioredoxin. Even in this spectrum, there is considerable overlap of proton resonances. Fig. 4B shows the same region of a single plane, corresponding to a particular ^{13}C resonance frequency, of a ^{13}C-edited NOESY spectrum. The overlap has been resolved, and it is clear which cross peaks belong to the $C^{\alpha}H$ of Pro76. Ambiguities are not always completely resolved even in three-dimensional spectra, and the same principles allow extension of the method to four and even higher dimensions.

II.3. Structure calculations

Once the resonances have been assigned to particular nuclei in the protein and the NOE cross peaks have been identified, the information is loaded into a computer program to calculate the three-dimensional structure. The information is usually encoded as a list of upper and lower bounds on internuclear distances rather than as explicit distances, since the precision of the experimental data does not usually justify this. The distance geometry program is run many times with the same set of data to ensure that conformational space is adequately sampled. As a consequence, the output of a distance geometry (or other similar) program is in the form of a series of structures which are overlaid by the computer to show an estimate of the three-dimensional structure of the protein. Such a series is shown in Fig. 5. It is noticeable that some regions of the protein, especially those in hydrogen-bonded secondary structure, form a very close family within the structures, while the conformations of loops may vary widely between structures. This variation is in many cases due to conformational flexibility of certain portions of the polypeptide chain, although it may also arise from a low density of NOE and other constraints in the region, without necessarily implying flexibility.

Thus, we have a picture of the three-dimensional structure of the protein molecule in solution. The solution structures so far obtained for proteins have mainly been identical in folding topology to those obtained by crystallographic means, when such are available. Structures obtained by the two methods can, however, differ in detail. In general, the conformational flexibility observed or implied from solution structures of proteins has not been reflected so far in crystal structures, for example, in the size of the temperature factors in these regions. Part of the explanation may be that the presence of the protein molecule in the crystal itself limits the conformational freedom - the process of crystallization must necessarily regularize the protein or it would not crystallize at all. Since conformational changes and the accessibility of low-energy states different from those observed in the solid state may well be of importance in enzyme reactions and in many examples of protein-protein interaction, the knowledge of the structure and behavior of proteins in solution is necessary for rational design of drugs, antagonists and immunogens.

III. STRUCTURE OF PEPTIDES IN SOLUTION

Peptide structure in solution has been the focus of much recent work, which has been reviewed recently (Montelione and Scheraga, 1989; Dyson and Wright, 1991, 1993).

The covalent chemistry of peptides is the same as that of proteins — both consist of amino acid residues linked by peptide bonds. However, the physical chemistry of peptides and folded proteins differs in fundamental ways. Most peptides, even those constrained in their conformations by disulfides, metal coordination or cyclization, populate a number of conformations of roughly equal energy in solution. A generalized confor-

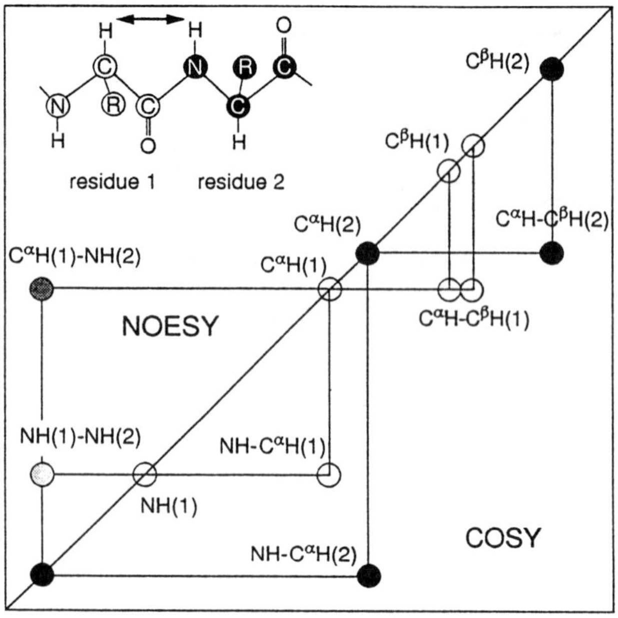

Fig. 2. Composite schematic diagram illustrating cross peak patterns in two-dimensional NMR spectra. The COSY spectrum (connectivities between protons connected through chemical bonds) is shown below the diagonal, and the NOESY spectrum (connectivities between protons close together in space) is shown above the diagonal. (A normal two-dimensional spectrum, COSY, or NOESY, would be symmetrical above and below the diagonal). COSY connectivities with amino acid spin systems are color coded, and the sequential NOE connectivities to be expected between residue 1 (white) and residue 2 (black) are shown in grey. Note that the $C^{\alpha}H(1)$-$NH(2)$ and $NH(1)$-$NH(2)$ would not normally *both* be observed between the same two residues.

mational energy diagram (Ramachandran plot) (Ramachandran et al., 1963; Wright et al., 1988; Dyson and Wright, 1991) is shown in Fig. 6. For linear, unconstrained peptides, it has long been known from experiment and theory that the conformational ensemble in the absence of other preferences consists of a large number of different conformers whose dihedral angles populate largely the β-region of conformational space (Brant et al., 1967). The observation of $d_{\alpha N}(i,i+1)$ NOE connectivities in the NMR spectrum of such a peptide should not therefore be taken to indicate the presence of β-strand, but rather that unfolded states of the peptide are present. Similarly the presence of a series of sequential $d_{NN}(i,i+1)$ NOEs in a peptide may not indicate that the peptide forms a true helix. The conformational ensemble in this case could merely include a number of backbone conformations that populate the α-region of conformational space. Helical conformations do occur in short peptides, but their presence must be verified in a number of other ways before it can be definitively stated that they are present (Wright et al., 1988). A more detailed discussion of these methods is given in the following sections.

Because the conformational ensemble of a linear peptide rarely, if ever, consists of only one or even of several closely related conformations, the calculation of structures using distance geometry, as described in the previous section for proteins, is usually neither practical nor advisable. Instead, several spectroscopic methods can be used to estimate the conformational preferences of the peptide for folded conformations such as reverse turns or helical structures. Since these conformations represent a threshold population of closely related structured forms, conclusions can be made about the likely conformations seen, for example, by receptors on B cells.

Several complementary spectroscopic methods can be used to give information on the conformations present in the ensemble of a peptide in solution.

III.1. Peptide conformational information from circular dichroism

Circular dichroism (CD) is a simple measurement, often employed as a first indication of the presence of structured forms in peptides. It gives an estimate of the extent of secondary structure formation in the peptide (or protein) averaged over the entire sequence (Adler et al., 1973; Brahms and Brahms, 1980). Computer programs exist for deconvolution of CD spectra into components representing helix, sheet, turn and "random coil" (reviewed in Johnson, 1985), but it is difficult to interpret these quantities for peptides because the structures present in solution are constantly changing and may be significantly different from regular secondary structure.

The CD spectrum of a peptide helix shows a pair of strong negative ellipticities at 208 and 222 nm and a positive band near 190 nm (Adler et al., 1973). The CD spectrum can provide valuable qualitative information on the presence of ordered helix in a peptide in solution. The carbonyl groups of the peptide must be strictly aligned in order for the CD signal characteristic of the helix to be observed (Manning et al., 1988), and even small deviations from ideal helical geometry cause attenuation or alteration of the characteristic CD signal. For peptides that adopt relatively small populations of helical conformations with dynamically frayed ends and deviations from ideal helix geometry, the presence of helix may be difficult to detect by CD. Such structures can often be detected by NMR spectroscopy (see Section III.2).

The presence of significant populations of β-turn conformers in solution can be detected by observing a positive ellipticity at 200–210 nm (Woody, 1974), although a type 1 β-turn generally exhibits a CD spectrum similar to that of α-helix (Bandekhar et al., 1982). CD spectra characteristic of β-turns have been observed in short linear peptides in aqueous solution (Dyson et al., 1988a; Shin et al., 1993). A population of β-turns is most readily detected by CD in a small peptide, since the ellipticity measured per residue would be diluted out in longer peptides by the signals from the rest of the peptide. The CD spectrum expected for a β-sheet has been observed for some model peptide systems in non-aqueous solvents (Adler et al., 1973).

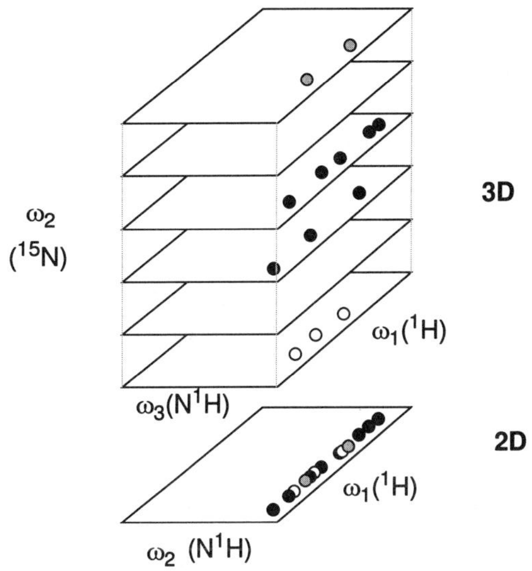

Fig. 3. Schematic representation of the increase in resolution obtained from a heteronuclear three-dimensional spectrum. Cross peaks that are ambiguous in the two-dimensional spectrum due to overlap of the NH resonances of four different amino acid residues are resolved in different planes of the three-dimensional spectrum because the ^{15}N chemical shifts are different.

III.2. NMR spectroscopy

When used carefully, proton NMR spectroscopy provides a powerful technique to detect the presence of folded conformers of a peptide in solution (Dyson and Wright, 1991). The information gained from NMR experiments on peptide conformation has the advantage that it is site specific, but because of the slow timescale of the experiments, all information is a population-weighted average over all structures in the conformational ensemble. This makes the interpretation of NMR information to give peptide conformation difficult.

The most commonly used NMR structural probe for peptide conformation is the NOE. Like all NMR parameters, the measured NOE intensity is a population-weighted average over all conformations of the peptide. However, the observation of a direct NOE between a given pair of protons indicates the presence of at least a threshold population of conformers in which the distance between these protons is relatively short (< 3.5 Å for flexible peptides in solution) (Dyson et al., 1988a,b). The conformational energy diagram (Fig. 6) provides an important guide to the interpretation of NOE data. The conformations for a given peptide generally fluctuate with a range of ϕ- and ω-angles lying within the broad α- and β-minima of Fig. 6. Different patterns of NOE connectivities will be observed for the peptide in each of these minima in ϕ,ω conformational space (Billeter et al., 1982; Wüthrich et al., 1984). Strong NOE connectivities between the $C^{\alpha}H$ of one amino acid residue and the NH of the following residue ($d_{\alpha N}(i,i+1)$ NOEs) in the absence of observable NOEs between successive NH protons ($d_{NN}(i,i+1)$ NOEs) show that backbone dihedral angles are predominantly in the β (extended chain) region of ϕ,ω space. The observation of $d_{NN}(i,i+1)$ NOE connectivities indicates that the conformational ensemble of the peptide includes local structures with dihedral angles in the α_L- or α_R-minima (Fig. 6). Ordered helix is indicated by the presence of relatively strong $d_{NN}(i,i+1)$ NOEs and of medium-range $d_{\alpha\beta}(i,i+3)$ and $d_{\alpha N}(i,i+3)$ NOEs across successive turns of helix (see Fig. 1). Extended regions of $d_{\alpha N}(i,i+1)$ and $d_{NN}(i,i+1)$ NOE connectivities occurring for the same sequence of amino acids are a good indication of conformational averaging between both the α_R- and β-regions of ϕ,ω space. If $d_{\alpha N}(i,i+2)$ NOEs are observed for such a sequence as well, the conformational ensemble may be termed "nascent helix" (Dyson et al., 1988b). A single strong $d_{\alpha N}(i,i+2)$ NOE, together with $d_{NN}(i,i+1)$ connectivities (Fig. 7), is usually indicative of a reverse turn conformation.

Sequential NOEs provide information about the relative populations of backbone conformations in the β- and α_R-minima for each residue in the peptide (Dyson and Wright, 1991). To identify conformational preferences for turn or helical conformations, additional information is needed, either from NMR or from other forms of spectroscopy such as CD. The best diagnostics of folded structures are medium-range NOEs such as the $d_{\alpha\beta}(i,i+3)$ characteristic of helix or the $d_{\alpha N}(i,i+2)$ characteristic of reverse turn, which immediately identify a significant population of non-random folded structures in the conformational ensemble. Additional information on folded structures may come from unusual $^3J_{HN\alpha}$ coupling constants or from evidence that particular protons are involved in hydrogen bonding interactions, indicated by lowered amide proton temperature coefficients or reduced amide proton exchange rates.

Coupling constants are usually one of the less sensitive NMR parameters for the detection of folded conformations in flexible linear peptides because the differences in values for folded and unfolded forms are generally quite small. Exceptions are seen in an analog of the C-peptide of ribonuclease (Osterhout et al., 1989) and in a type VI turn conformation in a short *cis*-proline containing peptide (Dyson et al., 1988a).

The chemical shift of the amide proton resonance is dependent upon the temperature; the temperature coefficient provides information on the interaction of the amide NH with hydrogen-bond-accepting solvent molecules. Observation of a lowered temperature coefficient indicates protection of the amide from solvent, most commonly due to intramolecular hydrogen bonding (Rose et al., 1985). When used in conjunction with NOE or CD data that confirm the presence of a folded structure such as a reverse turn or helix, temperature coefficient measurements can be very informative, but used alone they are not reliable indicators of structure, since local interactions involving the backbone amide and Glu or Asp side chains can affect them (Rose et al., 1985). Unusually low temperature coefficients are an indication of additional solvent protection by hydrophobic cluster conformations (Dyson et al., 1992a) or by interactions between the amide proton and nearby aromatic rings (Kemmink et al., 1993).

Amide proton exchange rates have been extensively used for proteins as a measure of structural integrity and polypeptide chain flexibility (Englander and Kallenbach, 1984). These measurements have not been much employed so far for peptides, probably due to the perception that peptides are unstructured. A

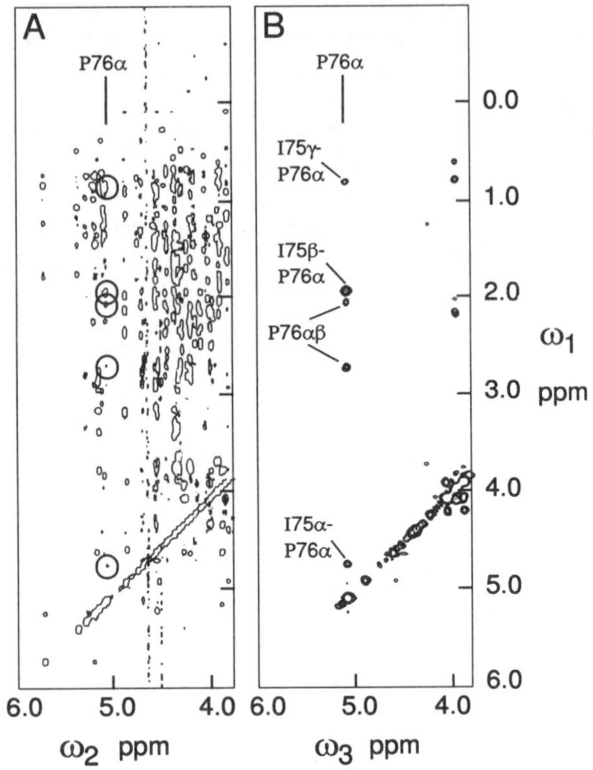

Fig. 4.(A). Portion of a 500 MHz phase-sensitive NOESY spectrum in 99% ^2H_2O of oxidized *E.coli* thioredoxin at 308K, pH 5.7, mixing time 120 ms. (B). Identical region of a single slice (^{13}C = 56.1 ppm) of a NOESY-HMQC spectrum of uniformly $^{15}N/^{13}C$-labeled oxidized thioredoxin under the same conditions, showing NOE connectivities to the *cis*-Proline-76.

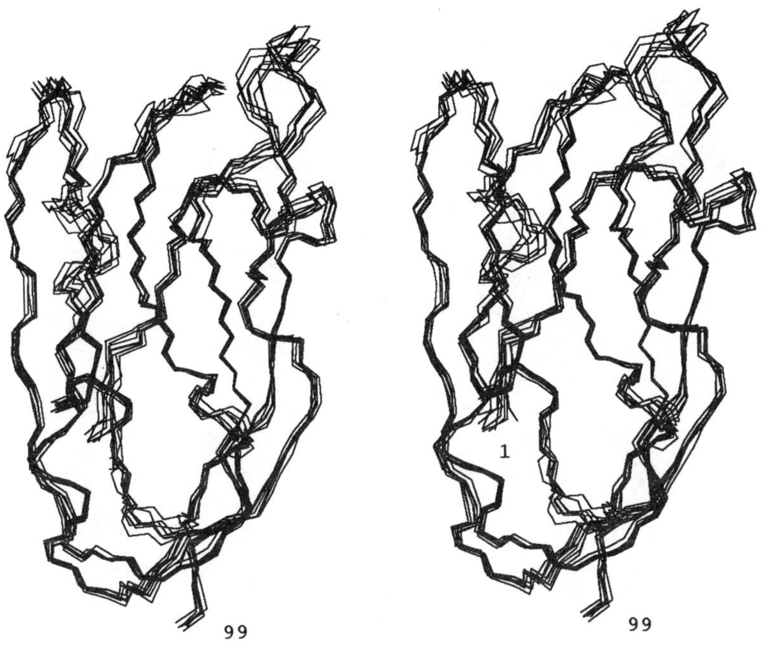

Fig. 5. Stereoview showing the best fit superposition of the backbone N, C^a, C' and O atoms of 16 MD-refined structures of Cu(I) French bean plastocyanin obtained from NMR distance and dihedral angle constraints (Moore et al., 1991). The position of the Cu atom is indicated by a circle.

recent pioneering study (Rohl et al., 1992) showed that amide proton exchange is slowed in the helical conformers of peptides.

III.3. Other spectroscopic methods

Secondary structure estimates for proteins (Dong et al., 1990) and peptides (Williams et al., 1990; Tadesse et al., 1991), similar to those obtained from CD and NMR measurements, have recently been derived using vibrational spectroscopy. The frequency of the amide vibration depends on the nature of the hydrogen bonding at the site, which is determined largely by the secondary structure in which it is involved. Since bands at different frequencies arise from different secondary structures, measurement of relative intensities can give direct information on the relative populations of the various structures in the conformational ensemble of a linear peptide in aqueous solution. Isotopically enhanced Fourier transform infrared (FTIR) spectroscopy has also been used (Tadesse et al., 1991) to provide independent information on conformer populations.

Electron spin resonance using spin-labeled peptide analogs (Miick et al., 1991) has been used in combination with CD and FTIR studies to study helix formation in short alanine-based peptides, which may form conformations which more nearly resemble 3_{10} helix (Miick et al., 1992) than the α-helix that had previously been assumed.

IV. EXAMPLES

IV.1. Immunogenic peptides

The conformational preferences of short linear peptides in solution do not necessarily reflect the conformation of the sequence in the native folded protein and may be quite different (Dyson et al., 1985). The majority of peptide sequences do not have preferences for structured conformers in aqueous solution. For example, a series of peptides representing the entire sequence of the β-sandwich protein plastocyanin displayed almost no preferences for structured conformations in solution, even in regions which contained well-defined turns and helix in the folded protein (Dyson et al., 1992b). Interestingly, no peptide-specific antibodies could be raised in mice to this series of peptides coupled to KLH (T. M. Fieser, personal communication). By contrast, a similar series of fragments from the four-helix bundle protein myohemerythrin showed extensive formation of secondary structure in the conformational preferences of every peptide (Dyson et al., 1992a). The difference in the behavior of the peptide sequences derived from the two protein structural motifs is profound and was interpreted as an indication of a fundamental difference in the mechanism of initiation of the folding of the two protein types (Dyson et al., 1992a,b). Again, interestingly, myohemerythrin peptides are highly immunogenic, as evidenced from the results of exten-

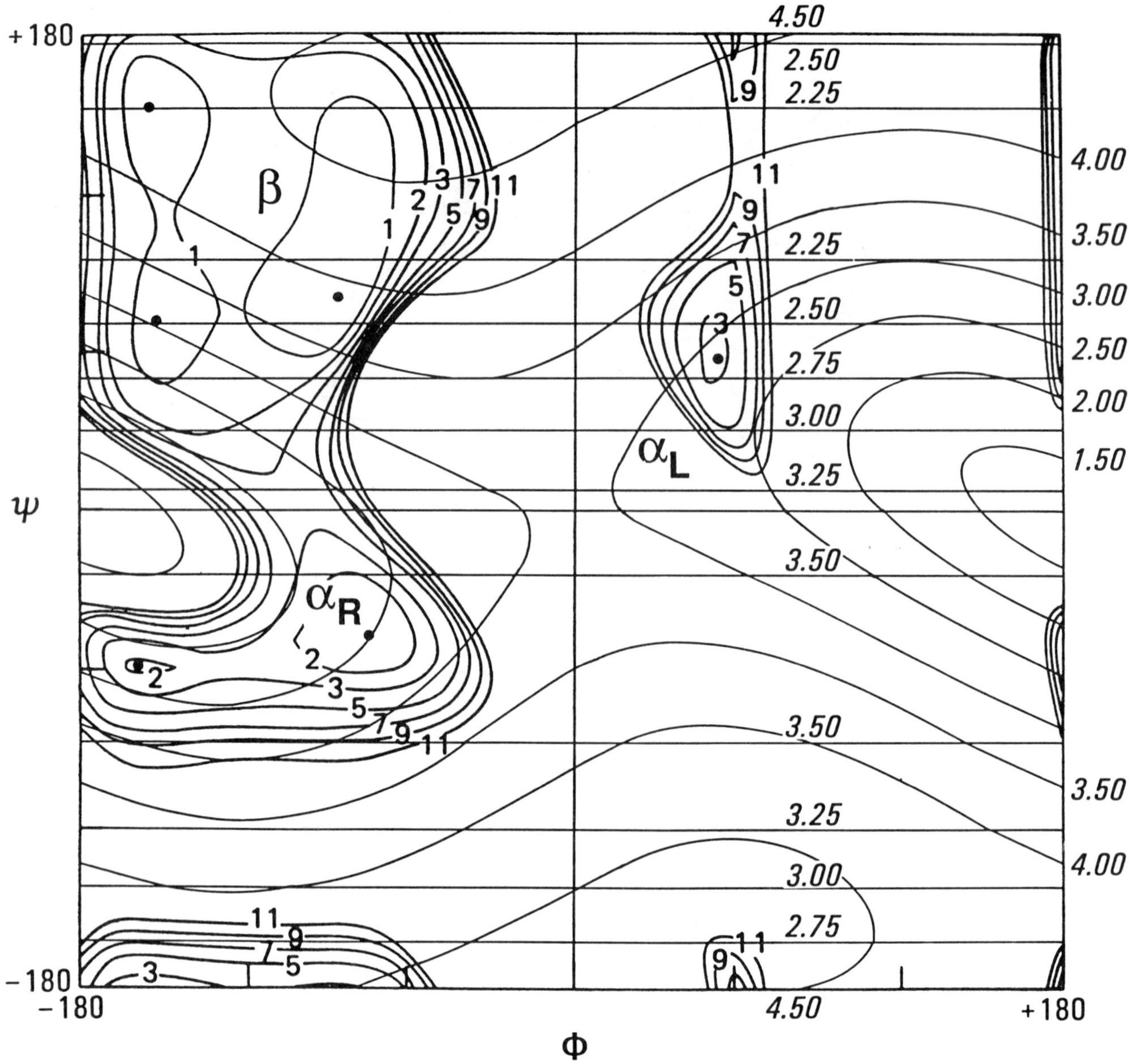

Fig. 6. Conformational energy diagram for the alanine dipeptide (adapted from Ramachandran et al., 1963). Energy contours are drawn at intervals of 1 kcal.mol^{-1}. The potential energy minima for β, α_R and α_L are labeled. The dependence of the sequential $d_{NN}(i,i+1)$ distance (in Å) on the ϕ and ω dihedral angles is shown as a set of contours labeled according to interproton distance at the right of the figure. The $d_{\alpha N}(i,i+1)$ distance depends only on ω for *trans* peptide bonds (Wuthrich, 1986) and is represented as a series of contours parallel to the ϕ-axis. (Reproduced with permission from Dyson and Wright, 1991.)

sive immunological work done on this system (Tainer et al., 1984, 1985; Getzoff et al., 1987; Geysen et al., 1987a, 1987b).

Correlations have been observed between conformational preferences of peptides in solution and their immunogenicity in a number of systems. The behavior of peptide immunogens from mouse LDH-C_4 is an example. An 18-residue random-coil peptide and an 18-residue peptide showing some preference for helix were weakly immunogenic; a 40-residue peptide with a highly populated helical hairpin conformation in solution proved to be highly immunogenic, giving a high proportion of antibodies that recognized the native protein (O'Hern, 1991). A synthetic peptide derived from the myoglobin H-helix is also immunogenic without conjugation to a carrier, presumably due to the presence of a T cell epitope as well as a B cell epitope in the same sequence (Berkower et al., 1986).

This peptide also forms a high proportion of helical structure in aqueous solution even as a monomeric species (Waltho et al., 1990, 1993).

More commonly, peptides are conjugated to carrier molecules for immunization. Fragments of the coiled-coil 24M protein of *Streptococcus pyogenes* were shown to be increasingly immunogenic as the polypeptide chain length was increased; this was correlated with an increased tendency toward helix formation, as indicated by CD spectroscopy, in the longer peptides (Gras-Masse et al., 1988). Conformational preferences for a reverse turn and some (probably nascent) helix were found for a synthetic peptide from *Herpes simplex* antigenic domain (Williamson et al., 1986). Helical turns were observed in a peptide corresponding to the dominant antigenic region of SIV (Dyson et al., 1992c). Folded conformations were found for the repeating tetrapeptide sequences of the malaria circumsporozoite protein using NMR and CD spectroscopy (Dyson et al., 1990), and these results have been used as a basis for the design of conformationally restricted immunogens (Satterthwait et al., 1990).

A number of immunogenic sequences from HIV envelope proteins have been studied by NMR and CD spectroscopy. A folded conformation was predicted for the principal neutralizing determinant of the HIV-1 envelope glycoprotein gp120 (LaRosa et al., 1990), which included two antiparallel β-strands and a short C-terminal α-helix. The sequence Gly-Pro-Gly-Arg in the middle of the sequence is conserved over a number of different isolates of the virus and was predicted to form a β-turn between the two strands. A conformational analysis by NMR (Chandrasekhar et al., 1991) showed the presence of the β-turn in a 40-residue peptide cyclized by a disulfide bond between the N- and C-termini and in a 24-residue linear peptide. No evidence for the α- or β-structure was observed in aqueous solution, although the C-terminal helix appeared to be formed upon addition of trifluoroethanol (TFE) to the 40-residue peptide (Chandrasekhar et al., 1991). This result illustrates the fundamental difference between peptide and protein structure. The structure of this fragment was predicted on the basis of structures found in folded proteins, and it may well be a good representation of that present in the folded gp120 protein. Nevertheless, in solution, only a small section of the peptide fragment was at all structured.

The importance of the correct conformation of the antigen was shown by studies of an anti-peptide antibody and two different forms of the peptide antigen (Oldstone et al., 1991). A 7-residue peptide crosslinked between N- and C-terminal cysteines to form a disulfide forms a turn conformation in aqueous solution and is highly antigenic. The reduced peptide, with no disulfide crosslink, forms no turn conformation and has reduced antibody binding. These studies point to a method for engineering immunogens by crosslinking segments of the polypeptide chain.

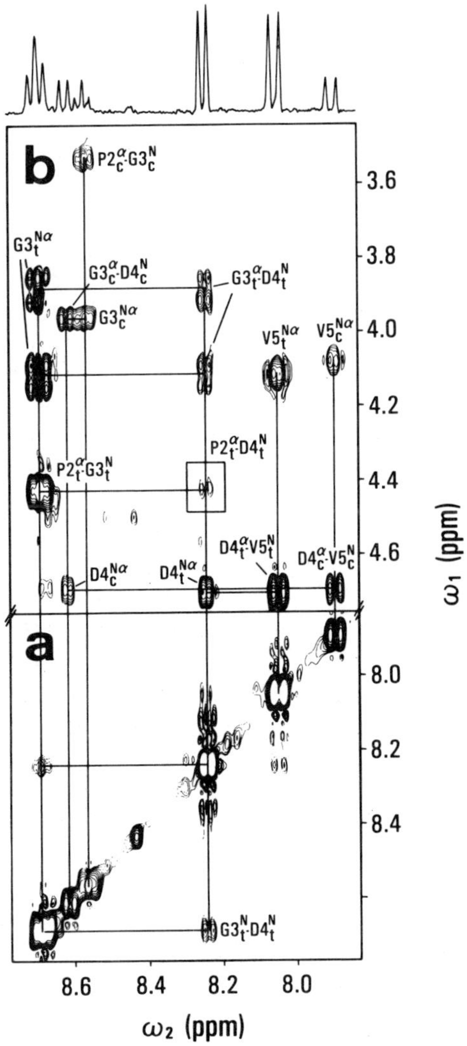

Fig. 7. Portion of the 300 MHz ROESY spectrum of the peptide YPGDV in 90% ^1H_2O/10% ^2H_2O at pH 4.1 and 5°C. A spin-locking mixing period of 400 ms was used. (a) $C^{\alpha}H$-NH region (b) NH diagonal region. The presence of a significant population of conformers containing a β-turn in the sequence YPGD is shown by the $d_{\alpha N}(i,i+2)$ NOE (shown in box) between the $C^{\alpha}H$ of Pro 2 and the NH of Asp 4. (Reproduced with permission from Dyson et al., 1988a.)

IV.2. Structure of a peptide bound to a carrier molecule

The structure of a peptide covalently attached to a protein molecule in aqueous solution has been studied by NMR spectroscopy. The peptide is derived from the sequence of influenza virus hemagglutinin and is present in the trimer interface of the hemagglutinin molecule according to the crystal structure (Wilson et al., 1981). It is highly immunogenic when bound to carrier molecules, and its immunological behavior has been well characterized (Wilson et al., 1984). The conformational preferences of the peptide in aqueous solution have been determined (Dyson et al., 1985): a reverse turn conformation is present in the first four residues of the peptide. This conformation is not present in the folded protein. Instead two turn-like conformations are present at the other end of the sequence (Wilson et al., 1981). The peptide was coupled in a site-specific manner (Ebina et al., 1989) to a small model carrier molecule, bovine pancreatic trypsin inhibitor (BPTI), which was chosen for ease of interpretation of the NMR spectra and also because a single coupling site could be engineered for the peptide (Ebina et al., 1989). When coupled to the model carrier molecule, the peptide still showed the conformational preference for the reverse turn at the N-terminal end. However, in addition, a much stronger preference for turn was observed in the residues VPDY at the other end of the molecule, identical to the site of one of the turn-like structures in the folded protein (S. Ebina, H.J. Dyson and P.E. Wright, manuscript in preparation). These studies show that, for the hemagglutinin peptide system at least, the presence of a folded protein molecule in close proximity to the peptide is capable of inducing a conformation in the peptide similar to that found in the folded protein. Work is continuing on this system using other peptide immunogens.

A crystal structure of a portion of the same peptide bound to an antibody specific for the determinant at the C-terminus of the peptide (Rini et al., 1992) shows a β-turn structure for the bound peptide very similar to that found for the sequence in the native protein. In this case, where the peptide in solution shows no sign of β-turn formation at these sites, we must conclude that the presence of protein (antibody or carrier or the remainder of the native protein) is capable of inducing the conformation with little energy cost.

IV.3. Structure of peptides bound to antibodies

The crystal structures of a number of antigen-antibody complexes have recently been published (reviewed in Davies and Padlan, 1992). An emerging theme when the structures of antibodies and antigens free and complexed are compared is the variety of different adjustments that both antibody and antigen make in the complex. Even the complexation of small peptides causes changes in the complementarity-determining regions (CDR) loops of antibodies (Rini et al., 1992).

The structure of the complex of a peptide derived from the C-helix of myohemerythrin with a protein-reactive anti-peptide antibody (B13I2) to a determinant at the N-terminus of the peptide (Stanfield et al., 1990) shows a well-defined type II β-turn conformation for the bound peptide that is quite different from the helix found in the native protein for this sequence (Sheriff et al., 1987). The sequence was predicted to form a β-turn by the method of Chou and Fasman (Chou and Fasman, 1978), but no evidence was found for the presence of a high population of turn conformations in the peptide in solution (Dyson et al., 1988b). Since one of the most tightly bound of the residues in the complex, His 73, is also one of the ligands of the iron prosthetic group in the native protein, Stanfield et al. (1990) concluded that the protein reactivity of the antibody was probably with the apoprotein, which would permit the conformational changes required for competitive binding of the antibody.

The C-helix peptide of myohemerythrin contains two major epitopes for protein-reactive anti-peptide antibodies (Tainer et al., 1984; Fieser et al., 1987). Besides the N-terminal sequence previously discussed, the C-terminal portion of the peptide is also an epitope for anti-peptide antibodies that bind to the native protein. The complex of the Fab' fragment of an antibody (B13A2) with a peptide consisting of the C-terminal 12 residues of the myohemerythrin C-helix peptide has been studied by isotope-edited NMR spectroscopy (Tsang et al., 1988, 1992). Such a large system is difficult to study by NMR. Nevertheless, sufficient NOE and dynamic information was obtained to characterize the conformation of the bound peptide as helical (Tsang et al., 1992), similar to the structure found in the native protein for this sequence (Sheriff et al., 1987) and to that suggested by the nascent helical conformation found in solution (Dyson et al., 1988b). In this case, unlike that of the hemagglutinin peptide discussed above, the conformations of the peptide in solution, bound to the antibody and in the native folded protein, are all similar.

IV.4. Structure of peptides bound to MHC class 1

Peptide structures in solution and bound to antibodies and carrier molecules may be similar to the structures observed for these sequences in the native proteins from which they were derived. The binding of peptides to the major histocompatibility antigens is apparently quite different. The original studies of the structure of MHC proteins (Bjorkman et al., 1987a, 1987b; Garrett et al., 1989) were not able to distinguish the conformations of any bound peptides. Recent studies have elucidated the structures of specific peptides bound to MHC proteins in an effort to identify principles for the recognition of peptide antigens by MHC molecules (Madison and Schellman, 1972; Fremont et al., 1992; Guo et al., 1992; Madden et al., 1992; Matsumura et al., 1992; Silver et al., 1992). The peptides appear to bind exclusively in extended conformations, consistent with the general dimensions of the binding pocket and regardless of the structure of the peptide sequence in the cognate protein. The extended conformation of the peptide is thus apparently dictated solely by the shape of the binding pocket of the MHC, and tight binding is achieved by specific hydrogen bonding interactions between the backbone of the peptide and side chains of the MHC. Specificity of the stable complex for peptides of foreign origin thus appears to lie in the specific chemical nature of the side chains of the peptide that are exposed at the surface, a hypothesis that is consistent with the observed ability of the immune system to discriminate self from non-self even at the level of single amino acid changes (Matsumura et al., 1992).

V. CONCLUSIONS - WHAT CAN STRUCTURE TELL US?

A knowledge, even if imperfect, of the structures of peptides and proteins in solution is a useful first step in any design process. The use of crystal structures and, more recently, solution structures derived from NMR data has allowed many groups to begin thinking of the design of therapeutics, antagonists, immunogens and so forth. The conformational preferences of peptides in solution are a degree less useful: even when a conformational preference for a folded conformation is observed for a peptide, this may not be the active or physiologically relevant conformation. Such a situation may also arise for proteins, particularly when the structure has been determined in the solid state. Solution structures show much more apparent flexibility, especially in loop regions. Structural information may thus be viewed as a basis for further work, such as the conformational restriction of peptides, rather than as the final word.

ACKNOWLEDGEMENT

This work was supported by grant CA27498 from the National Institutes of Health, Bethesda, MD.

ABBREVIATIONS

BPTI	Bovine pancreatic trypsin inhibitor
CD	Circular dichroism
COSY	Two-dimensional correlated spectroscopy
$d_{\alpha N}(i,i+1)$,$d_{NN}(i,i+1)$	Internuclear distance between the $C_\alpha H$ (or NH) of an amino acid residue in the polypeptide chain and the NH of the next residue, etc.
HIV	Human immunodeficiency virus
FTIR	Fourier transform infrared
KLH	Keyhole limpet hemocyanin
MHC	Major histocompatibility complex
NMR	Nuclear magnetic resonance
NOE	Nuclear Overhauser effect
NOESY	Two-dimensional nuclear Overhauser effect spectroscopy
ORD	Optical rotary dispersion
SIV	Simian immunodeficiency virus

VI. REFERENCES

Adler AJ, Greenfield NJ and Fasman GD. Circular dichroism and optical rotatory dispersion of proteins and polypeptides. Methods Enzymol 27:675, 1973

Bandekar J, Evans DJ, Krimm S et al. Conformations of cyclo(L-alanyl-L-alanyl-epsilon-aminocaproyl) and of cyclo(L-alanyl-D-alanyl-epsilon-aminocaproyl); Cyclized dipeptide models for specific types of beta-bends. Int J Pept Prot Res 19:187, 1982

Berkower I, Buckenmeyer GK and Berzofsky JA. Molecular mapping of a histocompatibility-restricted immunodominant T-cell epitope with synthetic and natural peptides: Implications for T-cell antigenic structure. J Immunol 136:2498, 1986

Billeter M, Braun W and Wütrich K. Sequential resonance assignments in protein ^{1}H nuclear magnetic resonance spectra. Computation of sterically allowed proton-proton distances and statistal analysis of proton-proton distances in single crystal protein conformations. J Mol Biol 155:321, 1982

Bjorkman PJ, Saper MA, Samraoui B et al. The foreign antigen binding site and T cell recognition regions of class I histocompatibility antigens. Nature 329:512, 1987a

Bjorkman PJ, Saper MA, Samraoui B et al. Structure of the human class I histocompatibility antigen HLA-A2. Nature 329:506, 1987b

Brahms S and Brahms J. Determination of protein secondary structure in solution by vacuum ultraviolet circular dichroism. J Mol Biol 138:149, 1980

Brant DA, Miller WG and Flory PJ. Conformational energy estimates for statistically coiling polypeptide chains. J Mol Biol 23:47, 1967

Brown JE and Klee WA. Helix-coil transition of the isolated amino terminus of ribonuclease. Biochemistry 10:470, 1971

Case DA, Dyson HJ and Wright PE. Use of chemical shifts and coupling constants in NMR structural studies on peptides and proteins. Methods Enzymol 239:392, 1994

Chandrasekhar K, Profy AT and Dyson HJ. Solution conformational preferences of immunogenic peptides derived from the principal neutralizing determinant of the HIV-1 envelope glycoprotein gp120. Biochemistry 30:9187, 1991

Chou PY and Fasman GD. Prediction of the secondary structure of proteins from their amino acid sequence. Adv Enzymol 47:45, 1978

Clore GM and Gronenborn AM. Applications of three and four-dimensional heteronuclear NMR spectroscopy to protein structure determination. Prog NMR Spectrosc 23:43, 1991a

Clore GM and Gronenborn AM. Two-, three-, and four-dimensional NMR methods for obtaining larger and more precise three-dimensional structures of proteins in solution. Ann Rev Biophys Biophys Chem 20:29, 1991b

Davies DR and Padlan EA. Twisting into shape. Curr Biol 2:254, 1992

Dong A, Huang P and Caughey WS. Protein secondary structures in water from second derivative amide I infrared spectra. Biochemistry 29:3303, 1990

Dyson HJ, Cross KJ, Houghten RA et al. The immunodominant site of a synthetic immunogen has a conformational preference in water for a type-II reverse turn. Nature 318:480, 1985

Dyson HJ, Rance M, Houghten RA et al. Folding of immunogenic peptide fragments of proteins in water solution. I. Sequence requirements for the formation of a reverse turn. J Mol Biol 201:161, 1988a

Dyson HJ, Rance M, Houghten RA et al. Folding of immunogenic peptide fragments of proteins in water solution. II. The nascent helix. J Mol Biol 201:201, 1988b

Dyson HJ, Satterthwait AC, Lerner RA et al. Conformational preferences of synthetic peptides derived from the immunodominant site of the circumsporozoite protein of *Plasmodium falciparum* by ^{1}H NMR. Biochemistry 29:7828, 1990

Dyson HJ and Wright PE. Defining solution conformations of small linear peptides. Ann Rev Biophys Biophys Chem 20:519, 1991

Dyson HJ, Merutka G, Waltho JP et al. Folding of peptide fragments comprising the complete sequence of proteins. I. Myohemerythrin. J Mol Biol 226:795, 1992a

Dyson HJ, Sayre JR, Merutka G et al. Folding of peptide fragments comprising the complete sequence of proteins. II. Plastocyanin. J Mol Biol 226:819, 1992b

Dyson HJ, Norrby E, Hoey K et al. Immunogenic peptides corresponding to the dominant antigenic region Ala597 to Cys619 in the transmembrane protein of simian immunodeficiency virus have a propensity to fold in aqueous solution. Biochemistry 31:1458, 1992c

Dyson HJ and Wright PE. Peptide conformation and protein folding. Curr Opinion Struct Biol 3:60, 1993

Ebina S, Lerner RA and Wright PE. Chemical modification of bovine pancreatic trypsin inhibitor for single-site coupling of immunogenic peptides for NMR conformational analysis. J Biol Chem 264:7882, 1989

Englander SW and Kallenbach NR. Hydrogen exchange and structural dynamics of proteins and nucleic acids. Quart Rev Biophys 16:521, 1984

Epand RM and Scheraga HA. The influence of long-range interactions on the structure of myoglobin. Biochemistry 7:2864, 1968

Ernst RR, Bodenhausen G and Wokaun A. Principles of Nuclear Magnetic Resonance in One and Two Dimensions. Clarendon Press, Oxford, 1987

Fieser TM, Tainer JA, Geysen HM et al. Influence of protein flexibility and peptide conformation on reactivity of monoclonal anti-peptide antibodies with a protein α-helix. Proc Natl Acad Sci USA 84:8568, 1987

Fremont DH, Matsumura M, Stura EA et al. Crystal structures of two viral peptides in complex with murine MHC class I H-2Kb. Science 257:919, 1992

Garrett TPJ, Saper MA, Bjorkman PJ et al. Specificity pockets for the side chains of peptide antigens in HLA-Aw68. Nature 342:692, 1989

Getzoff ED, Geysen HM, Rodda SJ et al. Mechanisms of antibody binding to a protein. Science 235:1191, 1987

Geysen HM, Rodda SJ, Mason TJ et al. Antigenicity of myohemerythrin. Science 238:1584, 1987a

Geysen HM, Tainer JA, Rodda SJ et al. Chemistry of antibody binding to a protein. Science 235:1184, 1987b

Gras-Masse H, Jolivet M, Drobecq H et al. Influence of helical organization on immunogenicity and antigenicity of peptides. Mol Immunol 25:673, 1988

Guo H-C, Jardetzky TS, Garrett TPJ et al. Different length peptides bind to HLA-Aw68 similarly at their ends but bulge out in the middle. Nature 360:364, 1992

Havel TF, Kuntz ID and Crippen GM. The theory and practice of distance geometry. Bull Math Biol 45:665, 1983

Havel TF and Wuthrich K. A distance geometry program for determining the structures of small proteins and other macromolecules from nuclear magnetic resonance measurements of intramolecular ^{1}H-^{1}H proximities in solution. Bull Math Biol 46:673, 1984

Hermans J Jr and Puett D. Relative effects of primary and tertiary structure on helix formation in myoglobin and α-lactalbumin. Biopolymers 10:895, 1971

Johnson WC. Circular dichroism and its empirical application to biopolymers. Methods Biochem Anal 31:61, 1985

Kemmink J, van Mierlo CPM, Scheek RM et al. Local structure due to an aromatic-amide interaction observed by ^{1}H-nuclear magnetic resonance spectroscopy in peptides related to the N-terminus of bovine pancreatic trypsin inhibitor. J Mol Biol 230:312, 1993

LaRosa GJ, Davide JP, Weinhold K et al. Conserved sequence and structural elements in the HIV-1 pricipal neutralizing domain. Science 24:9932, 1990

Live DH, Davis DG, Agosta WC et al. Observation of 1000-fold enhancement of ^{15}N NMR via proton-detected multiquantum coherences: Studies of large peptides. J Am Chem Soc 106:6104, 1984

Madden DR, Gorga JC, Strominger JL et al. The three-dimensional structure of HLA-B27 at 2.1 Å resolution suggests a general mechanism for tight peptide binding to MHC. Cell 70:1035, 1992

Madison V and Schellman JA. Optical activity of polypeptides and proteins. Biochemistry 11:1041, 1972

Manning MC, Illangasekare M and Woody RW. Circular dichroism studies of distorted α-helices, twisted β-sheets and β-turns. Biophys Chem 31:77, 1988

Matsumura M, Fremont DH, Peterson PA et al. Emerging principles for the recognition of peptide antigens by MHC class I molecules. Science 257:927, 1992

Miick SM, Todd AP and Millhauser GL. Position-dependent local motions in spin-labeles analogues if a short α-helical peptide determined by electron spin resonance. Biochemistry 30:9498, 1991

Miick SM, Martinez G, Fiori WR et al. Short alanine-based peptides may form 3_{10} helices in aqueous solution and not a-helices. Nature 359:653, 1992

Montelione GT and Scheraga HA. Formation of local structures in protein folding. Acc Chem Res 22:70, 1989

Moore JM, Lepre CA, Gippert GP et al. High-resolution solution structure of reduced French bean plastocyanin and comparison with the crystal structure of poplar plastocyanin. J Mol Biol 221:533, 1991

Neuhaus D and Williamson MP. The Nuclear Overhauser Effect in Structural and Conformational Analysis. VCH Publishers Inc, New York, 1989

O'Hern PA. Immunogenicity of peptides having pre-determined alpha-helical and alpha-alpha fold topologies. Mol Immunol 28:1047, 1991

Oldstone MBA, Tishon A, Lewicki H et al. Mapping the immunodominant domain of the human immunodeficiency virus gp41 transmembrane protein: Peptide conformation analysis using monoclonal antibodies and proton nuclear magnetic resonance spectroscopy. J Virol 65:1727, 1991

Osterhout JJ, Baldwin RL, York EJ et al. ^{1}H NMR studies of the solution conformation of an analogue of the C-peptide of ribonuclease A. Biochemistry 28:7059, 1989

Ramachandran GN, Ramakrishnan C and Sasisekharan V. Stereochemistry of polypeptide chain conformations. J Mol Biol 7:95, 1963

Rini JM, Schulze-Gahmen U and Wilson IA. Structural evidence of induced fit as a mechanism for antibody-antigen recognition. Science 255:959, 1992

Rohl CA, Scholtz JM, York EJ et al. Kinetics of amide proton exchange in helical peptides of varying chain length. Interpretation by the Lifson-Roig equation. Biochemistry 31:1263, 1992

Rose GD, Gierasch LM and Smith JA. Turns in peptides and proteins. Adv Protein Chem 37:1, 1985

Satterthwait AC, Chiang L-C, Arrhenius T et al. The conformational restriction of synthetic vaccines for malaria. Bull WHO 68:17, 1990

Sheriff S, Hendrickson WA and Smith JL. Structure of myohemerythrin in the azidomet state at 1.7/1.3Å resolution. J Mol Biol 197:273, 1987

Shin H-C, Merutka G, Waltho JP et al. Peptide models of protein folding initiation sites. 2. The G-H turn region of myoglobin acts as a helix stop signal. Biochemistry 32:6348, 1993

Silver ML, Guo H-C, Strominger JL et al. Atomic structure of a human MHC molecule presenting an influenza virus peptide. Nature 360:367, 1992

Stanfield RL, Fieser TM, Lerner RA et al. Crystal structures of an antibody and its complex with peptide antigen. Science 248:712, 1990

Tadesse L, Nazarbaghi R and Walters L. Isotopically enhanced infrared spectroscopy: A novel method for examining secondary structure at specific sites in conformationally heterogeneous peptides. J Am Chem Soc 113:7036, 1991

Tainer JA, Getzoff ED, Alexander H et al. The reactivity of antipeptide antibodies is a function of the atomic mobility of sites in a protein. Nature 312:127, 1984

Tainer JA, Getzoff ED, Paterson Y et al. The atomic mobility component of protein antigenicity. Ann Rev Immunol 3:501, 1985

Taniuchi H and Anfinsen CB. An experimental approach to the study of the folding of staphylococcal nuclease. J Biol Chem 244:3864, 1969

Tsang P, Fieser TM, Ostresh JM et al. Isotope-edited NMR studies of Fab'-peptide complexes. Peptide Res 1:87, 1988

Tsang P, Rance M, Fieser TM et al. Conformation and dynamics of an Fab'-bound peptide by isotope-edited NMR spectroscopy. Biochemistry 31:3862, 1992

Waltho JP, Feher VA and Wright PE. The H-helix of myoglobin as a potential independent protein folding domain. In: Villafranca JJ ed: Current Research in Protein Chemistry. Academic Press, San Diego, p 283, 1990

Waltho JP, Feher VA, Merutka G et al. Peptide models of protein folding initiation sites: 1. Secondary structure formation by peptides correponding to the G- and H-helices of myoglobin. Biochemistry 32:6331, 1993

Williams RW, Weaver JL and Lowrey AH. Relation between calculated amide frequencies and solution structure in Ala-X peptides. Biopolymers 30:599, 1990

Williamson MP, Hall MJ and Handa BK. ^{1}H-NMR assignment and secondary structure of a Herpes simplex virus glycoprotein D-1 antigenic domain. Eur J Biochem 158:527, 1986

Wilson IA, Skehel JJ and Wiley DC. Structure of the haemagglutinin membrane glycoprotein of influenza virus at 3 Å resolution. Nature 289:366, 1981

Wilson IA, Niman HL, Houghten RA et al. The structure of an antigenic determinant. Cell 37:767, 1984

Woody RW. Studies of theoretical circular dichroism of peptides. In: Blout ER, Bovey FA, Goodman M and Lotan N eds: Peptides, Polypeptides and Proteins. Wiley, New York, p 338, 1974

Wright PE, Dyson HJ and Lerner RA. Conformation of peptide fragments of proteins in water: implications for initiation of protein folding. Biochemistry 27:7167, 1988

Wütrich K, Billeter M and Braun W. Polypeptide secondary structure determination by nuclear magnetic resonance observation of short proton-protn distances. J Mol Biol 180:715, 1984

Wütrich K. NMR of Proteins and Nucleic Acids. John Wiley and Sons, New York NY, 1986

Chapter 11

SYNTHESIS, ANALYSIS AND CONFORMATION OF GLYCOPEPTIDES AND PHOSPHOPEPTIDES

L. Otvos, Jr., G.I. Szendrei, L. Gorbics and L. Urge

TABLE OF CONTENTS

I. INTRODUCTION

Many proteins contain modified amino acids for which no codons exist in the genome. The two most frequent forms of post-translational modifications are the phosphorylation and glycosylation of hydroxy amino acids and the glycosylation of the asparagine residue. The phosphate group is attached to the side chain of serine, threonine, or tyrosine with an ester bond. In most cases, the first sugars are GlcNAc with a β-glycosidic bond or GalNAc with an α-glycosidic bond attached to asparagine or serine/threonine, respectively. Both the oligosaccharide antennae and the negatively charged phosphoryl groups are located on the surface of the molecule and are known to play a crucial role in recognition processes at both macromolecular and cell levels. The structure and biological significance of glycosylated and phosphorylated proteins have been discussed in a great variety of review articles (Frank, 1984; Thomas et al., 1980; Hemmings et al., 1989; Gottschalk, 1972; Kornfeld and Kornfeld, 1985; Montreuil, 1980).

II. ARE CERTAIN NEIGHBORING AMINO ACIDS, SEQUENTIAL CODES, OR SPATIAL ARRANGEMENTS REQUIRED FOR GLYCOSYLATION OR PHOSPHORYLATION OF A GIVEN RESIDUE?

The sequence Asn-Xxx-Ser/Thr was observed to occur at the site of N-glycosylation (Mononen and Karjalainen, 1984), (where Xxx can be any amino acid except Pro and sometimes Cys) but only approximately one third of these sequences are indeed glycosylated. X-ray structural analysis of tripeptide models confirmed the hypothesis that H-bonds between the side chain carbonyl group of Asn and the hydroxyl of Ser or Thr facilitate the attachment of sugar to Asn by decreasing the dissociation constant of the amide group (Pichon-pesme et al., 1988). Exact patterns for O-glycosylation and phosphorylation have not yet been established. Serine, threonine, and proline were found to occur near to O-glycosylation sites by statistical analysis of natural glycoproteins (Wilson et al., 1991). Phosphorylation seems to require both the presence of basic amino acid residues (i.e., Lys, Arg, or His) and a β-turn within or adjacent to the sequence containing the Ser or Thr residue to be phosphorylated (Williams, 1976; Small et al., 1977). In real proteins, established phosphorylation and glycosylation sites appear to be on both type I (III) or type II predicted or spectroscopically determined β-turns (Welphy et al., 1983; Chessa et al., 1983).

If neighboring amino acids or certain secondary structures influence the effectiveness of the incorporation of the phosphate group or carbohydrate moiety *in vivo*, the same variability can be expected when post-synthetic phosphorylation or glycosylation is attempted on both free or resin-bound synthetic peptides. In fact, the yield of the "global" phosphorylation reaction (see later in this chapter) of synthetic neurofilament fragments dramatically dropped when the size of the peptide increased to 10–15 amino acids (Otvos et al., 1989a); highly variable rates of phosphorylation of

0-8493-8967-4/95/$0.00+$.50

medium-sized peptides have constantly been observed (Lang et al., 1992a; Lacombe et al., 1990; Paquet and Johns, 1990). Steric effects on the growing peptide chain were considered as significant parameters in the incorporation rate of unmodified amino acids during standard solid-phase synthesis in general (Krchnak and Vagner, 1992) and in chemical phosphorylation of peptides (Kitas et al., 1991) in particular.

III. SYNTHESIS OF PHOSPHOPEPTIDES

Solid-phase peptide synthesis is the method of choice to prepare a large number of synthetic protein fragments and model peptides for most biochemistry laboratories today (Barany et al., 1987). Accordingly, we will review only the literature of solid-phase phosphopeptide and glycopeptide synthesis. Figure 1 shows the reaction scheme of the currently used strategies.

The following section of a recent report of Petterson and co-workers (1991) illustrates the current status of the literature of phosphopeptide synthesis: "Several methods for the synthesis of O-phosphorylated peptides have been devised. However, none of the methods are generally useful and part of the published information is contradictory, indicating that optimal conditions have not been found for the reliable synthesis of phosphopeptides."

The incorporation of the phosphate group can be accomplished by one of the following three strategies:

1. The post-synthetic phosphorylation protocol uses polyphosphoric acid to phosphorylate the serine residues of cleaved, purified peptides. The final products are available after neutralization and extensive dialysis.

2. The side chain hydroxyl groups of the serine and threonine residues to be phosphorylated are incorporated into the peptide unprotected during the global phosphorylation protocol. Phosphochloridates or phosphoramidites are used to phosphorylate the hydroxyl groups left unprotected after the peptide chain assembly is completed, but before the cleavage. The protecting groups of the phosphate are removed by acidic treatment with concomitant release of the peptide from the solid support.

3. The appropriately protected, phosphorylated serine, threonine, or tyrosine residues are incorporated into the peptides, like any "regular" amino acid during the phosphopeptide syntheses employing the building block strategy. Table 1 summarizes these methods and the advantages or disadvantages of them.

No consensus has been reached yet on the stability of the phosphate esters of the hydroxy amino acids during the repetitive and final acidic or basic cleavage steps. It is generally accepted that the phosphotyrosine survives the treatment with piperidine (used to remove the N-terminal Fmoc protecting group) and most phosphoserines and phosphothreonines withstand the TFA-mediated cleavage from the resin, but every other method is debated. Table 2 lists a small fraction of these controversies.

Several more controversies include (but are not limited to) the disadvantageous acid stability of certain phosphate-protecting groups (Andrews et al., 1991), the too strong acidic properties of some of the oxidants (used in the phosphoramidite-employing strategies) (Otvos et al., 1990b) and the unwanted oxidation of methionine by the oxidants (Andrews et al., 1991), the commercial unavailability of phosphorylated Aloc-amino acids (Staerkaer et al., 1991), the unidentified byproducts formed during the phosphoroxychloridate chemistries (Staerkaer et al., 1991), and the shortness of the model phosphopeptides.

The major factor that is overlooked is the conformation of medium-sized peptides during incorporation of activated phosphoamino acids or the phosphorylation reaction. A systematic study of the involvement of the primary and secondary structure of peptides may reveal sequence-dependent effects that help or delay the phosphorylation reactions. Since the solvents and the temperature may also interfere with the conformation or accessibility of regions of peptides, these parameters need to be scrutinized, as these studies are still missing from the literature. Microwave irradiation during the acylation considerably increased the coupling efficiency of sterically hindered unmodified amino acids (Ile, Val) (Yu et al., 1992) and may also be helpful during the incorporation of the protected phosphate group.

IV. ANALYSIS OF PHOSPHOPEPTIDES

Analysis of unmodified peptides is most often accomplished by amino acid analysis, mass spectroscopy (MS) and NMR spectroscopy, and sequencing. Purification of synthetic peptides and phosphopeptides is traditionally achieved by high-performance liquid chromatographic methods (Mant and Hodges, 1991). As we documented, phosphorylated peptides usually elute from the reversed-phase column before their non-phosphorylated analogs (Otvos et al., 1990a) and can be purified without contamination with the parent peptides (Larson et al., 1992). All the previously listed methods (generally used for unmodified peptides) are appropriate to verify the presence of the phosphate group on synthetic peptides, but all have drawbacks. First, the instrumentation for FAB-MS and NMR is not common in peptide synthetic laboratories, while amino acid analysis and sequencing are common. Second, mass spectroscopy, NMR, and sequencing tend to be expensive for the analysis of series of phosphopeptides. It has been proposed that phosphate esters

Figure 1

Strategies for solid-phase synthesis of posttranslationally modified peptides

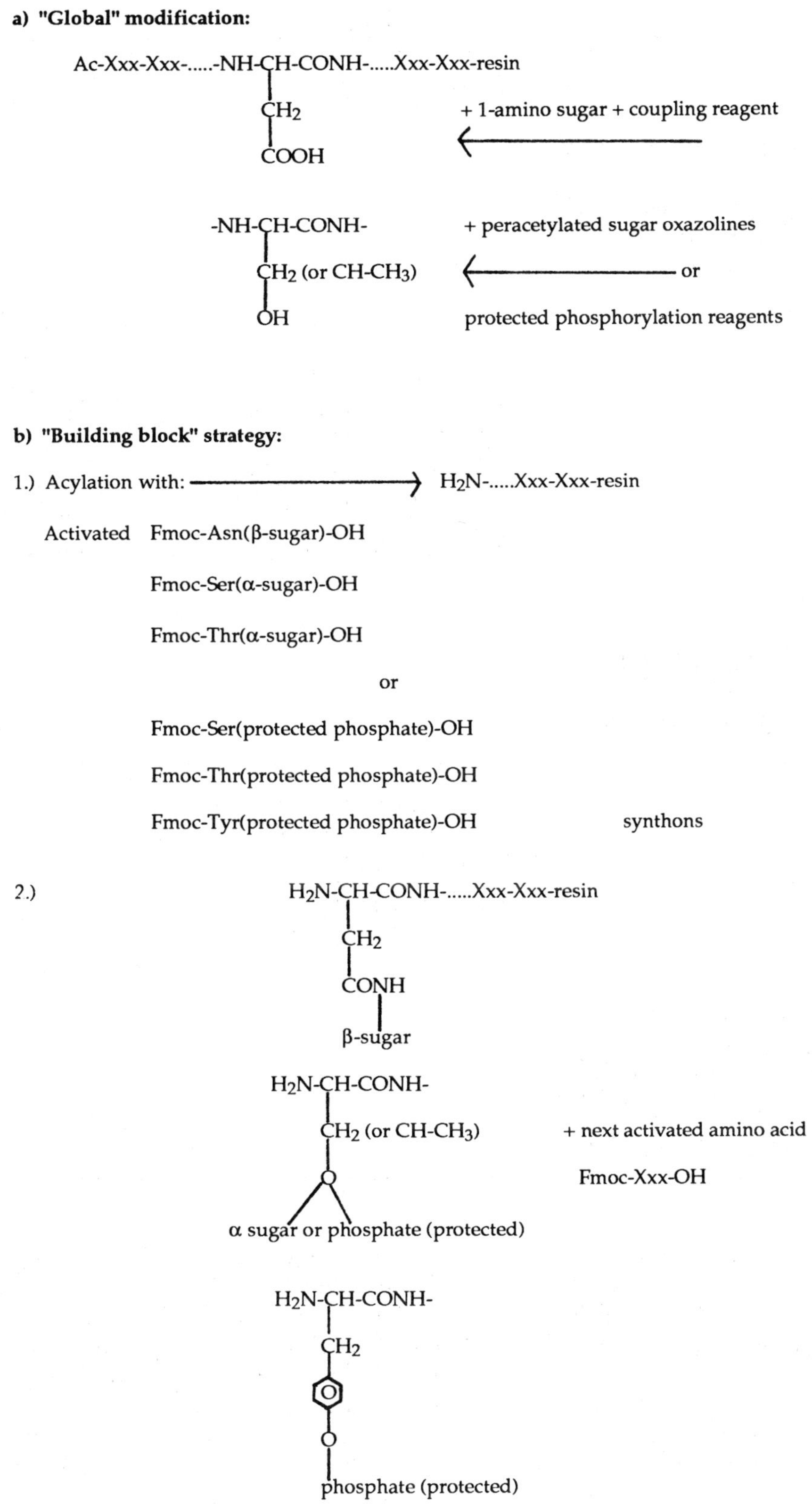

TABLE 1
Chemical phosphorylation of synthetic peptides

Introduction of phosphate group		**Reagent**	**References**
Post-synthetic		Polyphosphoric acid	Otvos et al., 1990a
Advantage:	One synthesis and cleavage - many phosphopeptides		
Disadvantage:	The phosphopeptide isomers need to be separated; virtually no reaction with Thr and Tyr		
"Building block"		Boc-Ser[PO(OPh)$_2$]-OH	Perich et al., 1986 Arendt et al., 1989
		Boc-Ser[PO(OBzl)$_2$]-OH	Perich et al., 1986
		Aloc-Ser[PO(^{t}Bu)$_2$]-OH	Lacombe et al., 1990
		Boc-Thr[PO(OPh)$_2$]-OH	Arendt et al., 1989
		Boc-Tyr[PO(OBzl)$_2$]-OH	Valerio et al., 1984
		Fmoc-Tyr[PO(OBzl)$_2$]-OH	Andrews et al., 1991
Advantage:	Customizable for automated synthesis		
Disadvantage:	Sterically extremely hindered acylating agents; repetitive N-terminal cleavage is difficult		
"Global"		Dibenzyl-phosphochloridate	Otvos et al., 1989a
		Diethylamino-dibenzyl-phosphoramidite	DeBont et al., 1990
		Diethylamino-di^{t}butyl-phosphoramidite	Perich and Johns, 1988 Staerkaer et al., 1991
Advantage:	One synthesis - peptides and phosphopeptides; the same hydroxy amino acids can also be glycosylated		
Disadvantage:	Highly active, chemically labile reagents; problems in resin swelling		

of serine be converted by an elimination/addition reaction into a stabile product (Meyer et al., 1986) which can be detected by standard sequencing methods, but these modifications do not apply to phosphothreonine and phosphotyrosine. Currently, MS (Schroeder et al., 1990; Johns et al., 1986; Moore, 1993) and ^{31}P-NMR (Andrews et al., 1991) are considered to be of the highest value. The conventional organic phosphate analysis (McClare, 1971) is a good alternative, but it requires purified phosphopeptides in the several hundred microgram range. The standard procedure of amino acid analysis of synthetic peptides follows hydrolysis, post- or precolumn derivatization, and HPLC analysis (Allen, 1989). Both PITC- and DABS-Cl-derivatized phosphoamino acids can be baseline separated from the rest of the derivatized amino acids (Gorbics et al., 1994; Murthy and Iqbal, 1991) and offer an easy laboratory protocol to verify

TABLE 2
Cleavage methods during solid-phase synthesis of phosphopeptides

Repetitive	From the resin	For	Against
TFA	HF	Arendt et al., 1989 Tsukamoto et al., 1991	Otvos et al., 1989a Petterson et al., 1991 Perich and Johns, 1989
TFA	$Pd(OAc)_2$ HBr/TFA	Perich et al., 1986 Valerio et al., 1984	
Piperidine		Andrews et al., 1991 (only for Tyr)	Otvos et al., 1989a (for Ser)
Hydrostannolic acid		Lacombe et al., 1990	

the presence of the phosphate group on synthetic peptides. The sensitivity of the phosphoamino acids to strong acids complicates the procedure, but our preliminary results show that a limited acidic hydrolysis liberates the phosphoserine, phosphothreonine, or phosphotyrosine from the peptides and slightly modified conditions of standard DABS- or PITC-amino acid analysis (available in most peptide synthetic laboratories) are suitable for the quantitative analysis of the "regular" amino acids and qualitative analysis of the phosphorylated ones.

V. CONFORMATION OF PHOSPHOPEPTIDES

Both supporting and dissenting papers have been published about the possible influence of an incorporated phosphate group on the conformation of proteins and their fragments (Hider et al., 1985; Chaplin et al., 1988; Kaplan et al., 1984; Ratanabanangkoon et al., 1983). The supporting studies indicate that intrapeptide salt bridges result from ionic forces between the negatively charged Ser- or Thr-linked phosphoryl and the positive Lys, Arg, or His side chain groups. CD spectroscopy was successfully employed to document transitions between different types of β-turns when synthetic peptides corresponding to human neurofilament and τ-proteins were phosphorylated (Otvos et al., 1988; Lang et al., 1992b). Significantly, while acetylation of the single positive charge of the N-terminus of the τ-fragment (amino acids 408–421) (numbering from Goedert et al., 1989) restored the secondary structure of the parent non-phosphorylated peptide, acetylation failed to restore the conformation after N-terminal addition of a histidine residue (the amino acid at the 407 position of the protein). We also detected intermolecular β-pleated sheet formations due to interpeptide salt bridges between chains of these protein fragments (τ: 389–402) carrying phosphate groups and correlated this feature with the ability of these cytoskeletal proteins to form the neurofibrillary tangle aggregates (Lang et al., 1992a, 1992b; Hollosi et al., 1992) of Alzheimer's disease.

An increase in ordered conformation was not the only effect found upon phosphorylation of τ-peptides. While phosphopeptides corresponding to a third region of τ-protein (amino acids 192–204) exhibited CD spectra very similar to the non-modified peptides (Szendrei et al., 1993), the β-pleated sheet-forming tendency of a fourth τ-sequence (bovine τ 92–108) was diminished after phosphorylation of a Ser in mid-chain position of a tetradekapeptide (Lang and Otvos, 1992). Similar to the last case, others have recently reported a decreased α-helix-forming tendency of a synthetic phosphopeptide excised from the cytoplasmic domain of phospholamban (Terzi et al., 1992). Such decreases in ordered structures are in agreement with the general ability of the synthetic phosphopeptides to elute earlier from the column during RP-HPLC than their more ordered glycosylated analogs (Otvos et al., 1992b). Earlier reports assigned a higher retention coefficient for phosphopeptides than for those glycopeptides that were isolated from natural sources (Browne et al., 1982). Based on our data, it currently appears that the decrease of the ordered secondary structure is a

TABLE 3
Aspects of the solid-phase synthesis of glycopeptides

Resins applied		**References**
TFA-cleavable		Bardaji et al., 1991
Super-acid sensitive		Luning et al., 1989a
Pd^0-cleavable		Kunz and Dombo, 1988
Strategy	**Reagent**	
"Building block"	Fmoc-Thr (α or β-Gal or Glc) - OH	Filira et al., 1990; Polt et al., 1992
	Fmoc-Asn (GlcNAc) - OH	Otvos et al., 1989b
	Fmoc-Asn (GlcNAc-1-4-GlcNAc) - OH	Otvos et al., 1990c
	Fmoc-Thr (Ac_4-Man-1-2-Ac_3 Man) - OPfp	Jansson et al., 1990
	Fmoc-Asn (Ac_3GlcNAc) - OPfp	Meldal and Bock, 1990
	Fmoc-Asn (any reducing sugar) - OH	Urge et al., 1992b
	Fmoc-Thr (Ac_4 Gal-1-3-Ac_2GalN_3) - OPfp	Peters et al., 1992b
	Fmoc-Thr/Ser (Ac_4Gal-1-3-Ac_2GalNAc) - OH	Luning et al., 1989b
"Global"	3,4,6, tri-0-acetyl-glucose oxazoline (for O-linked glycopeptides)	Hollosi et al., 1991
	1-amino-GlcNAc (for N-linked glycopeptides)	Anisfeld and Lansbury, 1990
Multiple column	Fmoc-Thr or Ser (Ac_3GalNAc) - OPfp	Peters et al., 1992a

general phenomenon after phosphorylation of a peptide, and increases of ordered conformations are characteristic for certain disease states.

VI. SYNTHESIS OF GLYCOPEPTIDES

The synthesis of glycopeptides has become one of the fastest growing areas of peptide chemistry. The first review was published by Garg and Jeanloz (1985), and the recent synthetic glycopeptide papers were collected by us (Otvos and Hollosi, 1992). The synthesis of glycopeptides (and phosphopeptides) is a new section of the American and European Peptide Symposia and the Solid-Phase Conference. Laboratories specialized for custom peptide synthesis report an exponentially increased demand for post-translationally modified peptides.

The conventional liquid-phase protocols are slowly being replaced by solid-phase techniques. Table 3 represents some aspects of the solid-phase synthesis of glycopeptides.

Due to its mild reaction conditions, Fmoc-chemistry is the method of choice for the synthesis of glycopeptides. In contrast with the phosphopeptide synthetic approaches, the "building block" strategy is preferred to the "global" protocols. The glycosylated amino acid synthons are activated by DCC, DCC + HOPfp, or HBTU or they are directly synthesized as active esters (Meldal and Bock, 1990; Meldal and Jensen, 1990). As far as the stereochemistry of the glycosydic bond is concerned, coupling of the easily available β-1-amino-1-deoxy sugars to the properly protected aspartic acid yields the natural β-N-glycoside connection. Many alternative methods exist for the preparation of the natural α-glycosides of serine and threonine, but the stereoselectivity of these protocols vary in a wide range. The fear of acidic hydrolysis of the sensitive O-glycosidic, especially fucosidic, bonds (Kunz et al.,

1989) during the deprotection step(s) led to the development of the enzymatic cleavage of the side chain protecting *tert*-butyl esters (Schultz et al., 1992).

Early applications of the conventional carbohydrate chemistry to glycopeptides carried over the acetyl- or benzyl-protection of the hydroxyl groups of the sugars, but the protecting groups considerably slowed down the peptide coupling reactions (Otvos et al., 1989b) and were difficult to remove (Nakabayashi et al., 1988). The novel techniques were boosted when it was shown that the commonly used acylating agents leave the sugar side chain hydroxyl groups uncoupled (Otvos et al., 1989b) (similar to the serine- and threonine-hydroxyls), and most of the O-glycosidic bonds, including many of the fucosidic ones (Otvos et al., 1992a; Urge et al., 1992b), withstand the conditions of acidic cleavage from the solid support. Our procedure (based on the original protocol of Likhosherstov and co-workers, 1986) aminating any mono- and oligosaccharide that contains a reducing end, and the subsequent coupling to aspartic acid and incorporation into synthetic peptides (Otvos et al., 1990c; Urge et al., 1991, 1992b), allows us to study the length of the sugar needed to mimic the conformational, chromatographic, and immunological properties of naturally N-glycosylated protein fragments. Figure 2 shows the reaction scheme and the structure of the sugars that we used in our laboratory. As the starting oligosaccharides become more and more complex, the price of the glycopeptides increases dramatically. New, sugar-saving protocols are needed to make the glycopeptide synthesis faster and more economic. The full development of a "global" solid-phase N-glycopeptide synthetic route, proposed by Anisfeld and Lansbury (1990) and Arsequell and co-workers (Arsequell et al., 1993) (these authors presented the data for glutamic acids instead of the natural aspartic acid) to small peptides may offer a solution to this problem (Wong et al., 1993; Kates et al., 1994). We established our "global" O-glycosylation strategy for serine-bound GlcNAc and GalNAc. This technique needs to be generalized to Thr and oligosaccharides as well.

VII. ANALYSIS OF GLYCOPEPTIDES

Due to the remarkable expansion of interest in glycoproteins and glycopeptides, strategies for the structural analysis of carbohydrate moieties (cleaved from the glycoproteins) are widely developed (Beely, 1985). All these methodologies, however, require instrumentation and technology that are specific for carbohydrate laboratories. Standard RP-HPLC is only rarely used for the separation of saccharides, but its abundant availability in the peptide synthetic laboratories underlines the importance of the development of glycopeptide analytical protocols that utilize RP-HPLC. Since the retention times of glycopeptides on RP-HPLC often fall close to the non-glycosylated analogs due to conformational orientation on the surface of the bonded phase (Otvos et al., 1992b), an easy laboratory protocol for the determination of acetamido-sugars in synthetic glycopeptides is of great interest. Similar to the phosphopeptides, utilization of the conditions of standard amino acid analysis fulfills this important need and offers a useful addition to the mass spectroscopy for the verification of the presence of acetamido-sugars in the glycopeptides. Acidic hydrolysis of peptides containing GlcNAc and GalNAc residues (the starting sugar structures in most N- and O-linked glycoproteins, respectively) results in the formation of 2-amino-sugars (Gottschalk, 1972; Foster and Horton, 1959). This intermediate can be derivatized, and the dabsylated derivative originated from GlcNAc elutes between the dabsylated Pro and Val in solvent systems used for standard DABS-amino acid analysis (Gorbics et al., 1993) and GalNAc can be distinguished from GlcNAc. We demonstrated the utility of this methodology on the amino acid analysis of glycosylated analogs of peptide families T and RGD (Gorbics et al., 1993). A similar approach, using PITC as derivatization reagent, was successfully used for the amino acid analysis of a glycosylated soluble intercellular adhesion molecule (Palladino et al., 1992). The amino sugars are still quite labile during the conditions of limited hydrolysis. Mono- and oligosaccharides are more sensitive than amino acids to acidic hydrolysis, and intact reducing sugars or their degradation products may react with amino acids during the hydrolysis (Gottschalk, 1972). Due to this instability, only semi-quantitative analysis of the sugar part of the glycopeptides can be achieved (Gorbics et al., 1993), and it is essential to further modify the protocols of hydrolysis and chromatography to achieve higher yields and separations that will allow fully quantitative determination of the glycosylated residues in synthetic glycopeptides.

VIII. CONFORMATION OF GLYCOPEPTIDES

CD and NMR studies performed on O-glycosylated dipeptide models containing GlcNAc and GalNAc revealed turn-like structures (glyco-turns) fixed by intramolecular H-bonds between the peptide backbone and the first sugar moiety (Mimura, et al., 1989; Hollosi et al., 1990). In this model an intramolecular H-bond is formed between the amide proton of the carbohydrate and the carbonyl oxygen of the Thr. This model suggests that the conformation is already stabilized after the incorporation of the first sugar moiety and this coincides with the observation that enzymatic shortening of the carbohydrate chain of natural glycoproteins results in a less dramatic confor-

Figure 2

Reaction scheme for the preparation of glycosylated asparagine derivatives, suitable for the solid-phase synthesis of N-glycopeptides

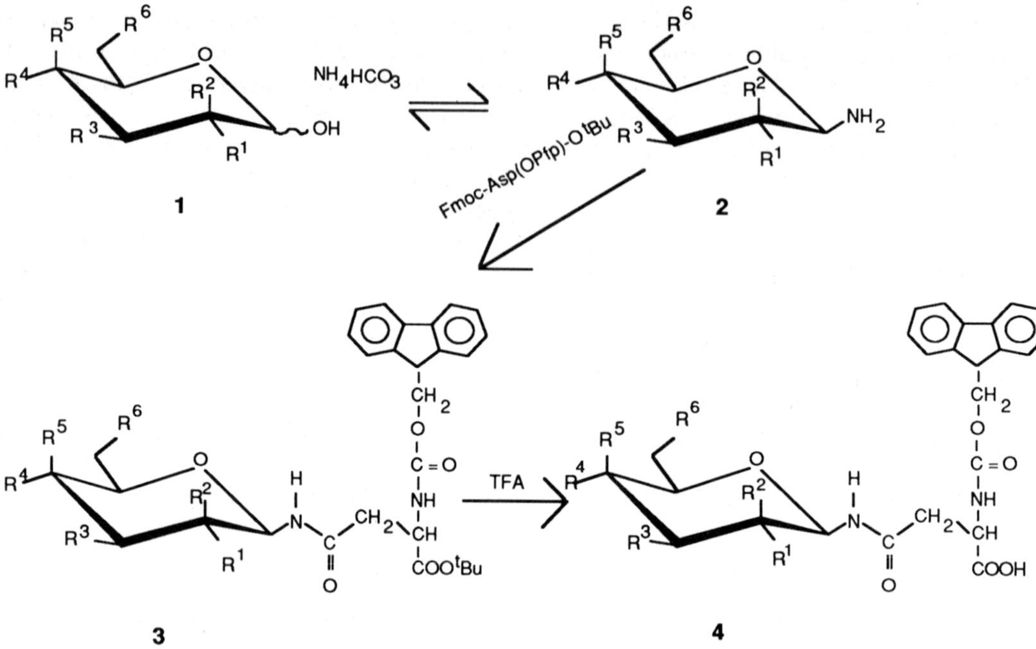

This panel lists the sugar structures that we prepared in our laboratory.

Abbreviated name of the sugar moiety	R1	R2	R3	R4	R5
Gal(β1-3)GlcNAc	NHAc	H	Gal(β1→O)	OH	H
D-GalpNAc	NHAc	H	OH	H	OH
D-Galp	OH	H	OH	H	OH
Gal(β1-4)Glc	OH	H	OH	Gal(β1→O)	H
D-Fucp	OH	H	OH	H	OH
D-Manp	H	OH	OH	OH	H
2dGlcp	H	H	OH	OH	H
D-GlcNAc	NHAc	H	OH	OH	H
GlcNAc(β1-4)GlcNAc	NHAc	H	OH	GlcNAc	H
GlcNAc(β1-4)GlcNAc(β1-4)GlcNAc	NHAc	H	OH	GlcNAc(β1-4)GlcNAc(β1-O)	H
Glc(β1-4)Glc	OH	H	OH	Glc(β1-O)	H
Glc(α1-4)Glc	OH	H	OH	Glc(α1-O)	H
D-Glc	OH	H	OH	OH	H
Glc4	OH	H	OH	[Glc(α1-4)]2Glc(α1-O)	H
Glc6	OH	H	OH	[Glc(α1-4)]4Glc(α1-O)	H
Glc7	OH	H	OH	[Glc(α1-4)]5Glc(α1-O)	H

mational change than the removal of the first one or two carbohydrate moieties (Rose et al., 1984; Gerken et al., 1989). We observed a considerable decrease in the helicity of dodeka- and tridekapeptides after N-glycosylation with a monosaccharide (GlcNAc), and the effect was only slightly enhanced by elongation of the sugar (Otvos et al., 1991). A breakage in the dominant α-helical peptide structure was also found after C-glycosylation (a chemically stable analog of the O-glycosides) (Bertozzi et al., 1992). Nevertheless, we also proved that the H-bond system, which originated from the glyco-turns, is not the only determining factor in the conformation of glycopeptides.

Electron microscopic studies on glycoproteins indicate that, in contrast to the stabilization of β-turns, the peptide chain stiffens around the sugar-binding residue and increases the apparent dimension of the protein (Gerken et al., 1989). This finding indicates that β-pleated sheets (extended structures) can also be generated or stabilized by glycosylation and the indication has been verified by CD and NMR studies of O-glycosylated and non-glycosylated fragments of glucoamylase from *Aspergillus niger* (Williamson et al., 1992). This type of conformational change (in contrast with the β-turn stabilization) is more clearly expressed in O-glycoproteins than in N-glycosylated ones (Gerken et al., 1989). By walking GlcNAc through penta- and octapeptides derived from the glycoprotein of the HIV, we found a stronger stabilization of reversed-turns (detected by CD) when the β-GlcNAc was attached to Asn than when α-GalNAc was attached to Thr of the same sequences (Urge et al., 1992a). Preliminary FT-IR and NMR studies of these glycosylated peptide T analogs support our observations. Stabilization of β-turns of another fragment (the V3 loop) of this protein was also found after glycosylation (Laczko et al., 1992).

The extent of the conformational transition is dependent upon the conformational stability of the peptide and the location of the sugar on the sequence. Most recently, we observed no secondary structural change on strong α-helical T cell epitopic peptides when 2–7 sugar moieties were attached to an asparagine that was located at the N-termini (Urge et al., 1994).

Recent reports of dramatic changes in the activity, stability, and metabolism of peptide drug leads upon glycosylation (Fisher et al., 1991; Bundy et al., 1990) have indicated the applicability of the glycopeptides in rational drug design. In our experiences, the glycosylated T peptides showed an increased resistance to serum proteases (Powell et al., 1993). When the digestion of the glycopeptides went through an endopeptidase-mediated pathway, the increase of the serum half-life was proportional with the increase of the β-turn propensity. Glycosylation of peptide-drug candidates that exhibit a β-turn as the biologically active conformation offers a viable alternative to the "peptide mimetics" approach in drug design. The increased serum half-life and oral availability of the glycopeptides are features that make these compounds even more attractive.

ACKNOWLEDGEMENTS

Our work that constitutes part of this review was supported by NIH grants GM45011 and AG10670 and Alzheimer's Disease Research, a grant from the American Health Assistance Foundation (to L. Otvos).

ABBREVIATIONS

CD	Circular dichroism
DABS	4-Dimethylamino azobenzene-4'-sulfonyl
DCC	Dicyclohexylcarbodiimide
FAB-MS	Fast atom bombardment mass spectroscopy
FT-IR	Fourier-transform infrared spectroscopy
Gal	D-Galactose
Glc	D-Glucose
GalNAc	2-Acetamido-2-deoxy-b-D-galactose
GlcNAc	2-Acetamido-2-deoxy-b-D-glucose
HBTU	2-(1H-benzotriazol-1-yl)-1,1,3,3-tetramethyluronium
HIV	Human immunodeficiency virus
HOPfp	Pentafluorophenol
NMR	Nuclear magnetic resonance spectroscopy
PITC	Phenylisothiocyanate
RP-HPLC	Reversed-phase high performance liquid chromatography
TFA	Trifluoroacetic acid

IX. REFERENCES

Allen G. Laboratory Techniques in Biochemistry and Molecular Biology. Elsevier, New York, 1989

Andrews DM, Kitchin J and Seale PW. Solid-phase synthesis of a range of 0-phosphorylated peptides by post-assembly phosphetylation and oxidation. Int J Peptide Protein Res 38:469, 1991

Anisfeld ST and Lansbury PT Jr. A convergent approach to the chemical synthesis of asparagine-linked glycopeptides. J Org Chem 55:5560, 1990

Arendt A, Palczewski K, Moore W et al. Synthesis of phosphopeptides containing O-phosphoserine or O-phosphothreonine. Int J Peptide Protein Res 33:468, 1989

Arsequell and co-workers Abstract #416, 22nd European Peptide Symposium, Interlaken, 1992

Barany G, Kneib-Cordonier N and Mullen DG. Solid-phase peptide synthesis: A silver anniversary report. Int J Peptide Protein Res 30:705, 1987

Bardaji E, Torres JL, Clapes P et al. Synthesis and biological activity of O-glycosylated morphiceptin analogues. J Chem Soc Perkin Trans I:1755, 1991

Beely JG. Laboratory Techniques in Biochemistry and Molecular Biology. Elsevier, New York, 1985

Bertozzi CR, Hoeprich PD Jr and Bednarski MD. Synthesis of carbon-linked glycopeptides as stable glycopeptide models. J Org Chem 57:6092, 1992

Browne CA, Bennett HPJ and Solomon S. The isolation of peptides by high-performance liquid chromatography using predicted elution positions. Anal Biochem 124:201, 1982

Bundy GL, Pals DT, Lawson JA et al. Potent resin inhibitory peptides containing hydrophilic end groups. J Med Chem 33:2276, 1990

Chaplin LC, Clark DC and Smith LJ. The secondary structure of peptides derived from casein - a circular dichroism study. Biochim Biophys Acta 956:162, 1988

Chessa G, Borin G, Marchiori F et al. Synthetic peptides reproducing the site phosphorylated by Camp-dependent protein kinase in protein phosphatase inhibitor-1: Effect of structural modifications on the phosphorylation. Eur J Biochem 135:609, 1983

DeBont HBA, van Boom JH and Liskamp RMJ. Automatic synthesis of phosphopeptide by phosphorylation on the solid-phase. Tetrahedron Lett 31:2497, 1990

Filira F, Biondi L, Cavaggion F et al. Synthesis of O-glycosylated tuftsins by utilizing threonine derivatives containing an unprotected monosaccharide moiety. Int J Peptide Protein Res 36:86, 1990

Fisher JF, Harrison AW, Bundy GL et al. Peptide to glycopeptide: Glycosylated oligopeptide resin inhibitors with attenuated in vivo clearance properties. J Med Chem 34:3140, 1991

Foster AB and Horton D. Aspects of the chemistry of the amino sugars. Adv Carbohydr Chem 14:213, 1959

Frank AW. Synthesis and properties of N-, O-, and S-phospho derivatives of amino acids. In: Peptides and Proteins, CRC Press - Crit Rev Biochem, Boca Raton, FL, p 51, 1984

Garg HG and Jeanloz RW. Synthetic N- and O-glycosyl derivatives of L-asparagine, L-serine, and L-threonine. Adv Carbohydr Chem Biochem 43:135, 1985

Gerken T, Butenhof KJ and Shogren R. Effects of glycosylation on the conformational dynamics of O-linked glycoproteins: Carbon-13 NMR studies of ovine submaxillary mucin. Biochemistry 28:5536, 1989

Goedert M, Spillantini MG, Jakes R et al. Multiple isoforms of human microtubule-associated protein tau: Sequences and localization in neurofibrillary tangles of Alzheimers disease. Neuron 3:519, 1989

Gorbics L, Urge L, Otvos-Papp E et al. Determination of amino sugar in synthetic glycopeptides during the conditions of amino acid analysis utilizing precolumn derivatization and HPLC analysis. J Chromatogr 637:43, 1993

Gorbics L, Urge L, Lang E et al. Successful and rapid verification of the presence of a phosphate group in synthetic phosphopeptides using the conditions of standard DABS-Cl amino acid analysis. J Liquid Chromatogr 17:175, 1994

Gottschalk A. Glycoproteins: Their Composition, Structure and Function. Elsevier, New York, 1972

Hemmings HC Jr, Nairn AC, McGuinness TL et al. Role of protein phosphorylation in neuronal signal transduction. FASEB J 3:1583, 1989

Hider RC, Ragnarsson U and Zetterqvist O. The role of the phosphate group for the structure of phosphopeptide product of adenosine 3-5-cyclic monophosphate-dependent protein kinase. Biochem J 229:485, 1985

Hollosi M, Perczel A and Fasman GD. Cooperativity of carbohydrate moiety orientation and β-turn stability is determined by intramolecular hydrogen bonds in protected glycopeptide models. Biopolymers 29:1549, 1990

Hollosi M, Kollat E, Laczko I et al. Solid-phase synthesis of glycopeptides: Glycosylation of resin-bound serine-peptides by 3,4,6-tri-O-acetyl-D-glucose oxazoline. Tetrahedron Lett 32:1531, 1991

Hollosi M, Urge L, Perczel A et al. Metal ion induced conformational changes of phosphorylated fragments of human neurofilament (NF-M) protein. J Mol Biol 223:673, 1992

Jansson AM, Meldal M and Bock K. The active ester N-Fmoc-3-O-[Ac_4-α-D-Manp-(1→2)-Ac_3-α-D-Manp-1-]threonin ϵ-O-Pfp as a building block in solid-phase synthesis of an O-linked dimannosyl glycopeptide. Tetrahedron Lett 31:6991, 1990

Johns RB, Alewood PF, Perich JW et al. Fast-atom bombardment mass spectrometry of seryl-containing and O-phosphoseryl containing peptides. Tetrahedron Lett 27:4791, 1986

Kaplan LJ, Bauer R, Morrison E et al. Structure of chromatin reconstituted with phosphorylated H-1: Circular dichroism and thermal denaturation studies. J Biol Chem 259:8777, 1984

Kates SA, De la Torre BG, Eritja R et al. Solid-phase N-glycopeptide synthesis using allyl side-chain protected Fmoc-amino acids. Tetrahedron Lett 35:1033, 1994

Kitas EA, Knorr R, Trzeciak A et al. Alternative strategies for the Fmoc solid-phase synthesis of O^4-phospho-L-tyrosine-containing peptides. Helv Chim Acta 74:1314, 1991

Kornfeld R and Kornfeld S. Assembly of asparagine-linked oligosaccharides. Annu Rev Biochem 54:631, 1985

Krchnak V and Vagner J. Innovation and Perspectives in Solid Phase Synthesis. Epton R. ed: Intercept, Andover, UK, p 419, 1992

Kunz H and Dombo B. Solid-phase synthesis of peptides and glycopeptides on polymeric supports with allylic anchor groups. Angew Chem Int Ed Engl 27:711, 1988

Kunz H, Waldmann H and Marz J. Synthese von N-glycopeptid-partialstructuren der verknupfungsregion sowohl der trans-membrane-neuraminidase eines influenza-virus als auch des factors B das menschlichen komplementsystems. Liebigs Ann Chem p 45, 1989

Lacombe JM, Andriamanapisoa F and Pavia AA. Solid-phase synthesis of peptides containing phosphoserine using phosphate tert.-butyl protecting group. Int J Peptide Protein Res 36:275, 1990

Laczko I, Hollosi M, Urge L et al. Synthesis and conformational studies of N-glycosylated analogues of the HIV-1 principal neutralizing determinant. Biochemistry 31:4282, 1992

Lang E and Otvos L Jr. A serine → proline change in the Alzheimers disease associated epitope Tau 2 results in altered secondary structure, but phosphorylation overcomes the conformational gap. Biochem Biophys Res Commun 188:162, 1992

Lang E, Szendrei GI and Elekes I. Reversible β-pleated sheet formation of a phosphorylated synthetic τ peptide. Biochem Biophys Res Commun 182:63, 1992a

Lang E, Szendrei GI, Lee VMY et al. Immunological and conformational characterization of a phosphorylated immunodominant epitope on the paired helical filaments found in Alzheimers disease. Biochem Biophys Res Commun 187:787, 1992b

Larson JK, Otvos L Jr and Ertl HCJ. Post-translational side chain modification of a viral T cell epitope results in a diminished recognition by specific T cells. J Virol 66:3996, 1992

Likhosherstov LM, Novikova OS, Derevitskaja VA et al. Simple synthesis of amino sugar β-D-glycosylamines. Carbohydr Res 146:c1, 1986

Luning B, Norberg T and Tejbrant J. Solid phase synthesis of mono- and disaccharide-containing glycopeptides. J Chem Soc Chem Commun p 1267, 1989a

Luning B, Norberg T and Tejbrant J. Synthesis of mono- and disaccharide amino acid derivatives for use in solid phase peptide synthesis. Glycoconj J 6:5, 1989b

Mant CT and Hodges RS. High-performance Liquid Chromatography of Peptides and Proteins: Separation, Analysis, and Conformation. CRC Press, Boca Raton, FL, 1991

McClare CWI. An accurate and convenient organic phosphorous assay. Anal Biochem 39:527, 1971

Meldal M and Bock K. Pentafluorophenyl esters for the temporary carboxyl protection in solid phase synthesis of N-linked glycopeptides. Tetrahedron Lett 31:6987, 1990

Meldal M and Jensen KJ. Pentafluorophenyl esters for the temporary protection of the a-carboxy group in solid-phase glycopeptide synthesis. J Chem Soc Chem Commun p 483, 1990

Meyer HE, Hoffmann-Posorske E, Korte H et al. Sequence analysis of phosphoserine-containing peptides: Modification for picomolar sensitivity. FEBS Lett 204:61, 1986

Mimura Y, Inouye Y, Maeji NJ et al. NMR study on conformation of Ac-Thr(a-GalNAc)-Ala-Ala-OMe as a model for mucin type glycosylation. Int J Peptide Protein Res 34:363, 1989

Mononen I and Karjalainen E. Structural comparison of protein sequences around potential N-glycosylation sites. Biochim Biophys Acta 788:364, 1984

Montreuil J. Primary structure of glycoprotein glycans: Basis for the molecular biology of glycoproteins. Adv Carbohydr Chem Biochem 37:157, 1980

Moore WT. Integration of mass spectrometry into strategies for peptide synthesis. Biol Mass Spectrometry 22:149, 1993

Murthy LR and Iqbal K. Measurement of picomoles of phosphoamino acids by high performance liquid chromatography. Anal Biochem 193:299, 1991

Nakabayashi S, Warren CD and Jeanloz RW. The preparation of a partially protected heptasaccharide-asparagine intermediate for glycopeptide synthesis. Carbohydr Res 174:279, 1988

Otvos L Jr, Hollosi M, Perczel A et al. Phosphorylation loops in the synthetic peptides of human neurofilament protein middle-sized subunit. J Protein Chem 7:365, 1988

Otvos L Jr, Elekes I and Lee VMY. Solid-phase synthesis of phosphopeptides. Int J Peptide Protein Res 34:129, 1989a

Otvos L Jr, Wroblewski K, Kollat E et al. Coupling strategies in solid-phase synthesis of glycopeptides. Peptide Res 2:362, 1989b

Otvos L Jr, Tangoren IA, Wroblewski K et al. Reversed-phase high performance liquid chromatographic separation of synthetic phosphopeptide isomers. J Chromatogr 512:265, 1990a

Otvos L Jr, Tangoren IA, Elekes I et al. Solid-phase synthetic methods leading to selectively phosphorylated neurofilament fragments. In: Innovation and Perspectives in Solid Phase Synthesis. Epton R. Ed:, SPCC, Birmingham, U.K., p 421, 1990b

Otvos L Jr, Urge L, Hollosi M et al. Automated solid-phase synthesis of glycopeptides: Incorporation of unprotected mono- and disaccharide units of N-glycoprotein antennae into T cell epitopic peptides. Tetrahedron Lett 31:5889, 1990c

Otvos L Jr, Thurin J, Kollat E et al. Glycosylation of synthetic peptides breaks helices: Phosphorylation results in distorted structure. Int J Peptide Protein Res 38:476, 1991

Otvos L Jr, Thurin J, Urge L et al. Solid phase synthesis and analysis of mid-sized N- and O-linked glycopeptide models. In: Innovation and Perspectives in Solid Phase Synthesis. Epton R. Ed:, Intercept Ltd, Andover, U.K., p 185, 1992a

Otvos L Jr, Urge L and Thurin J. Influence of different N- and O-linked carbohydrates on the retention times of synthetic peptides in reversed-phase high performance liquid chromatography. J Chromatogr 599:43, 1992b

Otvos L Jr and Hollosi M. Development of chemically modified peptides. In: Biologically Active Peptides: Design, Synthesis and Utilization, Williams WV and Weiner DB Eds:, Technomics, Lancaster, PA, p 155, 1992

Palladino DEH, House RM and Cohen KA. Measurement of amino acid compositions of glycoprotein systems by gas-phase hydrolysis and reversed-phase high-performance liquid chromatography. J Chromatogr 599:3, 1992

Paquet A and Johns M. Synthesis of oligophosphoseryl sequences occurring in casein. Int J Peptide Protein Res 36:97, 1990

Perich JW, Valerio RM and Johns RB. Solid-phase synthesis of an O-phosphoseryl-containing peptide using phenyl phosphotriester protection. Tetrahedron Lett 27:1377, 1986

Perich JW and Johns RB. Di-t-butyl N,N-diethylphosphoramidite and dibenzyl N,N-diethylphosphoramidite. Highly reactive reagents for the phosphite-triester phosphorylation of serine-containing peptides. Tetrahedron Lett 29:2369, 1988

Perich JW and Johns RB. Synthesis of O-phosphotyrosine-containing peptides. 1. Synthesis of PTyr-Leu-Gly via benzyl phosphate protection. J Org Chem 54:1750, 1989

Peters S, Bielfeldt T, Meldal M et al. Multiple column solid-phase glycopeptide synthesis. J Chem Soc Perkin Trans I:1163, 1992a

Peters S, Bielfeldt T, Meldal M et al. Solid-phase synthesis of mucin glycopeptides. Tetrahedron Lett 33:6445, 1992b

Petterson E, Luning B, Mickos H et al. Synthesis, NMR and function of an O-phosphorylated peptide, comprising the RGD-adhesion sequence of osteopontin. Acta Chem Scand 45:604, 1991

Pichonpesme V, Aubry A, Abbadi A et al. Crystal molecular structures of 2 tripeptides related to the sequence coding for N-glycosylation. Int J Peptide Protein Res 32:175, 1988

Polt R, Szabo L, Treiberg J et al. General methods for α or β-O-Ser/Thr glycosides and glycopeptides. J Am Chem Soc 114:10249, 1992

Powell MF, Stewart T, Otvos L Jr et al. Peptide stability in drug development. II. Effect of single amino acid substitution and glycosylation on peptide reactivity in human serum. Pharm Res 10:1268, 1993

Ratanabanangkoon K, Keutmann HT, Kitzmann K et al. Properties of the phosphorylated beta-subunit of human choriogonadotropin. J Biol Chem 258:4527, 1983

Rose MC, Voter WA, Saage H et al. Effects of deglycosylation on the architecture of ovine submaxillary mucin glycoprotein. J Biol Chem 259:3167, 1984

Schroeder W, Covey T and Hucho F. Identification of phosphopeptides by mass spectrometry. FEBS Lett 273:31, 1990

Schultz M, Hermann P and Kunz H. Enzymatic cleavage of tert-butyl esters: Thermitase-catalyzed deprotection of peptides and O-glycopeptides. Synlett p 37, 1992

Small D, Chou PY and Fasman GD. Occurrence of phosphorylated residues in predicted β-turns: Implications for β-turn participation in control mechanism. Biochem Biophys Res Commun 79:341, 1977

Staerkaer G, Jakobsen MH, Olsen CE et al. Solid-phase peptide synthesis of selectively phosphorylated peptides. Tetrahedron Lett 32:5389, 1991

Szendrei GI, Lee VMY and Otvos L Jr. The recognition of the minimal epitope of monoclonal antibody Tau-1 depends upon the presence of the phosphate group but not its location. J Neurosci Res 34:243, 1993

Terzi E, Poteur L and Trifilieff E. Evidence for a phosphorylation-induced conformational change of phospholamban cytoplasmic domain by CD analysis. FEBS Lett 309:413, 1992

Thomas G, Podesta EJ and Gordon J. Protein Phosphorylation and Bio-Regulation. S Karger, Basel, 1980

Tsukamoto M, Kato R, Ishiguro K et al. Improved groups for phosphate of O-phosphoserine useful for the solid-phase peptide synthesis. Tetrahedron Lett 32:7083, 1991

Urge L, Kollat E, Hollosi M et al. Solid-phase synthesis of glycopeptides: Synthesis of Na-fluorenylmethoxycarbonyl L-asparagine Nb-glycosides. Tetrahedron Lett 32:3445, 1991

Urge L, Gorbics L and Otvos L Jr. Chemical glycosylation of peptide T at natural and artificial glycosylation sites stabilizes or rearranges the dominant reverse turn structure. Biochem Biophys Res Commun 184:1125, 1992a

Urge L, Otvos L Jr, Lang E et al. Fmoc-protected, glycosylated asparagines potentially useful as reagents in the solid-phase synthesis of N-glycopeptides. Carbohydr Res 235:83, 1992b

Urge L, Jackson DC, Gorbics L et al. Synthesis and conformational analysis of N-glycopeptides that contain extended sugar chains. Tetrahedron 50:2373, 1994

Valerio RM, Alewood PF, Johns RB et al. Synthesis of protected derivatives of O-phosphotyrosine and incorporation in a heptapeptide. Tetrahedron Lett 25:2609, 1984

Welphy JK, Shenbagamurth P, Lennarz WJ et al. Substrate recognition by oligosaccharyltransferase: Studies on glycosylation of modified Asn-X-Thr/Ser tripeptides. J Biol Chem 258:1856, 1983

Williams RE. Phosphorylated sites in substrates of intracellular protein kinases: A common feature in amino acid sequences. Science 192:473, 1976

Williamson G, Belshaw NJ and Williamson MP. O-glycosylation of Aspergillus glucoamylase. Conformation and role in binding. Biochem J 282:423, 1992

Wilson IBH, Gavel Y and von Heijne G. Amino acid distributions around O-linked glycosylation sites. Biochem J 275:529, 1991

Wong SYC, Guile GR, Rademacher TW et al. Synthetic glycosylation of peptides using unprotected saccharide β-glycosylamines. Glycoconj J 10:227, 1993

Yu HM, Chen ST and Wang KT. Enhanced coupling efficiency in solid-phase peptide synthesis by microwave irradiation. J Org Chem 57:4781, 1992

Chapter 12

LIPOPEPTIDES AS ADJUVANTS AND INTEGRATED ADJUVANTICITY

Günther Jung and Wolfgang G. Bessler

TABLE OF CONTENTS

I. SUMMARY

Synthetic lipopeptide analogs derived from the N-terminus of bacterial lipoprotein, tripalmitoyl-S-glyceryl-cysteine (P_3C), constitute potent immunoadjuvants *in vitro* and *in vivo* when administered in combination with antigens. These immunological properties can be demonstrated in various species. When added to Enterobacteriaceae vaccines, lipopeptides markedly enhance the vaccine effect. After the coupling of lipopeptides to haptens or low molecular weight antigens which are not immunogenic per se, a marked specific antibody response is induced, often after only one application of the conjugate. An induced humoral response can be further enhanced by introducing haplotype specific T helper cell epitopes into the conjugate. Lipopeptide-antigen conjugates can also be applied as synthetic vaccines, e.g., they give protection against foot-and-mouth disease. Finally, by the coupling of lipopeptides to epitopes presented by MHC class I molecules, the *in vivo* priming of virus specific cytotoxic T cells can be achieved. The novel, chemically well-defined lipopeptides described here can be synthesized in gram amounts with high purity and reproducibility; they are nontoxic and can be stored for a long time even at room temperature.

II. INTRODUCTION

Immunoadjuvants are substances that enhance the immune response when given simultaneously with antigens. These compounds also improve the efficiency of vaccines, which is important in decreasing the amount required. Adjuvants should show high effectivity and should lack side effects such as inducing autoimmune or allergic reactions, cytotoxicity, or tumor growth. Many immunoadjuvants consist of two components. One component acts as a deposit substance to protect the antigen from degradation. The most commonly used deposit substances are aluminum compounds (phosphate or hydroxide) and Freund's incomplete adjuvant in which the antigen is incorporated into the aqueous phase of a stabilized water in paraffin oil emulsion. Alternative deposit systems

0-8493-8967-4/95/$0.00+$.50

include liposomes or synthetic surfactants. The second components present in many effective adjuvants are substances which stimulate the humoral or cellular immune response or which activate monocytes/macrophages. These stimulants either consist of killed bacteria or fragments thereof. Thus, cell wall components of Gram-negative bacteria such as lipoprotein (Melchers et al., 1975; Bessler and Braun 1975; Bessler et al., 1977), lipopolysaccharide (LPS) (Anderson et al., 1972), and protein I (porin omp-F; Bessler et al., 1979) constitute highly active immunological components; e.g., they activate B lymphocytes and monocytes/macrophages. Biologically active synthetic analogs of these bacterial cell wall constituents have been prepared by various laboratories. The lipid A part of LPS from Gram-negative bacteria, responsible for most of its biological activity, has been synthesized (Galanos and Chedid, 1985). Synthetic muramyl dipeptide (MDP) derived from the bacterial peptidoglycan constitutes an effective immunoadjuvant (Riveau and Chedid, 1987). We could show that synthetic segments of the porin OmpF from *E. coli B* act as polyclonal B lymphocyte activators (Vordermeier et al., 1990). We have also synthesized the N-terminal lipopeptide of the bacterial lipoprotein, which acts as an immunoadjuvant and as a leukocyte activator (Wiesmüller et al., 1983).

III. SYNTHETIC LIPOPEPTIDES AS ADJUVANTS AND LEUKOCYTE ACTIVATORS

The synthetic lipopeptide N-palmitoyl-S-(2,3-bis(palmitoyloxy)-(2RS)-propyl-(R)-cysteinyl-(S)-seryl-(S)-seryl-(S)-asparaginyl-(S)-alanine (P_3C-SSNA) is identical to the lipoprotein N-terminus with respect to its peptide composition, but it contains solely palmitic acid residues. This synthetic lipopeptide (Fig. 1) and a multitude of synthetically prepared analogs thereof constitute potent immunoadjuvants either in combination with antigens (Lex et al., 1986; Reitermann et al., 1989) or after covalent linkage to haptens (Reitermann et al., 1989; Jung et al., 1985; Bessler et al., 1985; Böltz et al., 1988). The newly synthesized lipopeptide analogs Pam_3Cys-Ser-$(Lys)_4$ and Pam_3Cys-Ser-PEG are very soluble in aqueous media. Lipopeptide adjuvants are biologically the most active substances. Most lipopeptide analogs constitute B lymphocyte and macrophage activators for a wide variety of murine inbred strains; no nonresponder strain could be detected so far (Bessler et al., 1985; Prass et al., 1987; Kleine et al., 1987). Lipopeptides induce B cell differentiation into Ig secreting plasma cells, as determined by a hemolytic Jerne plaque assay. On days 3, 4, and 5 after the addition of lipopeptide to splenic cell cultures, predominantly IgM, $IgG_{2(a+b)}$, and IgG_3 were induced, whereas no significant IgA responses could be detected. The amount of Ig produced was comparable to LPS, with the isotype pattern being slightly different (Kleine et al., 1987). Lipopeptides also induce lymphokine production, phagocytosis, induction of superoxide dismutase, or tumor cytotoxicity in human and murine monocytes/macrophages (Bessler et al., 1984; Hoffmann et al., 1988, 1989; Hauschildt et al., 1990). Furthermore, lipopeptides constitute potent activators of human neutrophil granulocytes (Seifert et al., 1992).

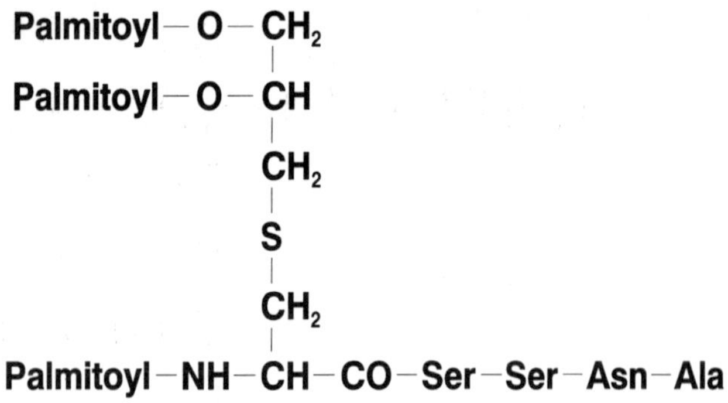

Fig. 1. Structural formula of the N-terminal lipopeptide Pam_3Cys-Ser-Ser-Asn-Ala from the lipoprotein of *Escherichia coli*.

IV. CHEMICAL SYNTHESIS OF LIPOPEPTIDE ADJUVANTS

The synthesis of P_3C antigen conjugates, including the T helper cell epitopes, was performed either on a peptide synthesizer (Applied Biosystems 430A) or on a multiple peptide synthesizer using our own programs developed for Fmoc/tBu strategy. Fmoc-protected amino acids and a p-benzyloxybenzyl alcohol resin were used. The coupling reactions were performed using a 2.5 molar excess of Fmoc-amino acid, HOBt and DIC (diisopropylcarbodiimide). The lipoamino acid P_3C was introduced using the same procedure as for Fmoc-amino acids (for more details and references, see Wiesmüller et al., 1992). The lipopeptides were removed from the resin and deprotected by treatment with 55% trifluoroacetic acid (TFA) for 2.5 h in the presence of thioanisol and resorcinol. Complete removal of the Mtr-protecting group from the arginine residues was obtained after an additional treatment with 95% TFA, 3% thioanisol, 1% resorcinol, and 1% ethanedithiol for 1 h at 50°C. The lipopeptides were purified by repeated precipitation from acetic acid with diethylether and lyophilization. The lipopeptides were analyzed by electrospray-mass spectrometry (Metzger et al., 1992) and amino acid analyses. The racemization was determined by gas chromatography on Chirasil-Val and was found to be below 2% for each of the amino acids.

V. LIPOPEPTIDE ADJUVANTS WITH ALTERED MOLECULAR STRUCTURES

We investigated how alterations of the lipopeptide molecule influence its biologic properties. Studies using mostly leukocyte stimulation assays indicated that various replacements in both the lipid and peptide moieties of the lipopeptide are possible without significant loss of activity provided that particular structural features of the molecule are maintained (Prass et al., 1987; Bessler et al., 1985). For example, protease resistance was enhanced by the incorporation of alpha-methylserine residues into the peptide chain (Metzger et al., 1987) or the sulfur atom was replaced by a methylene group (Metzger et al., 1991). As a rule, (R,R)-configurated P_3C derivatives are immunologically more active than those containing one of the other possible stereoisomers (Wiesmüller et al., 1990). The lipopeptide P_3C-SSNA, with an amino acid sequence identical to the N-terminus of native lipoprotein, and two other lipopeptides, P_3C-SK_4 and P_3C-SE_4, were compared for their stimulatory activity toward Balb/c splenocytes. All three compounds exhibited similar activities (Reitermann et al., 1989); however, P_3C-SK_4 and P_3C-SE_4, due to their increased solubility, were active at lower concentrations than the native P_3C-SSNA. We concluded therefore that the peptide composition, once the minimal dipeptide chain length has been reached (Lex et al., 1986), is not the major factor determining biologic activity.

VI. LIPOPEPTIDE ADJUVANTS AS REPLACEMENT FOR FREUND'S ADJUVANT

In mixture with antigens synthetic lipopeptides were able to act as strong immunoadjuvants *in vitro* and *in vivo* (Fig. 2). P_3C-SK_4 was able to markedly increase the specific *in vivo* Ig response against DNP-BSA: 14 days after the injection of a mixture of antigen and lipopeptide adjuvant into mice, a pronounced rise of anti-DNP-IgG antibodies was found; antibody titers in animals given only the hapten or adjuvant were marginal. Boostering with a mixture of DNP-BSA and P_3C-SK_4 resulted in very high DNP-specific IgG responses (120–180 μg/ml) comparable to the response obtained by immunization with DNP-BSA in combination with complete Freund's adjuvant (Reitermann et al., 1989). The novel lipopeptides offer distinct advantages as adjuvants. They can be reproducibly synthesized in large quantities and high purity. They are thus far known to be neither toxic nor pyrogenic (C. Galanos and W.G. Bessler, unpublished). No toxic side effects or tissue damage have been observed in animals (Wiedemann et al., 1991). Being totally synthetic, lipopeptides do not contain endotoxin contaminations (Hauschildt et al., 1990). The substances are chemically

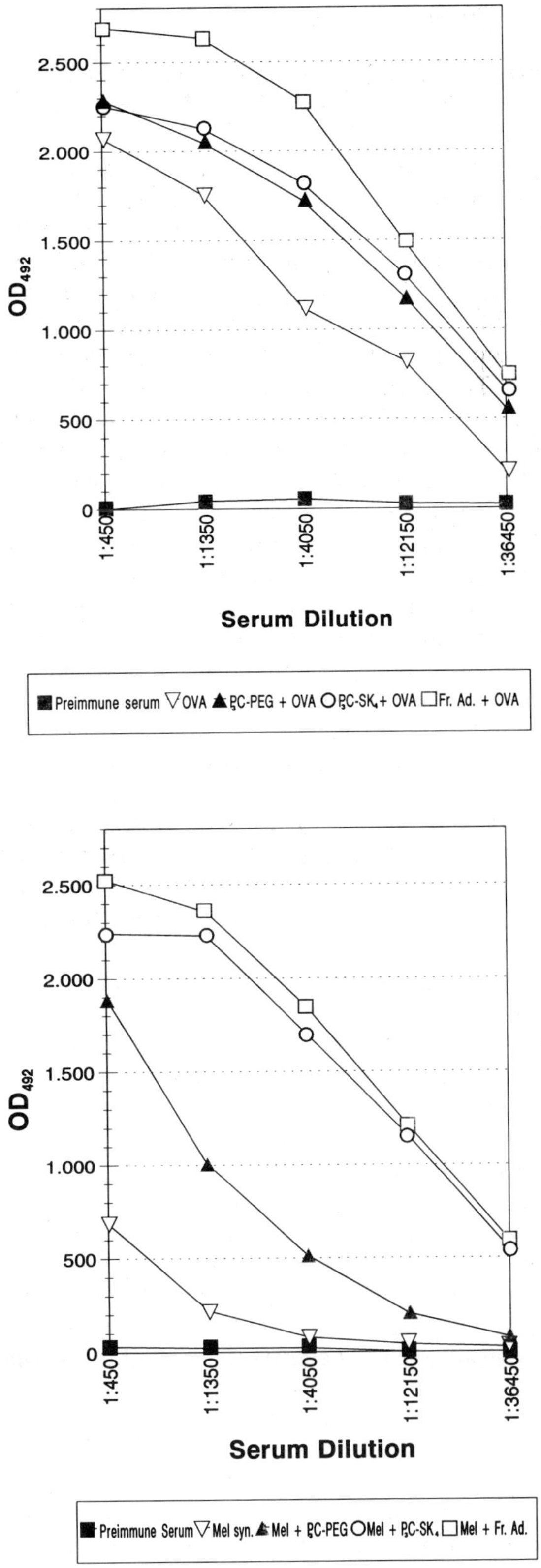

Fig. 2. Immunoglobulin response in mice after the application of (Upper graph) ovalbumin and (Lower graph) synthetic melittin in mixtures with Pam_3Cys-Ser-$(Lys)_4$ or Pam_3Cys-Ser-PEG (PEG = polyethyleneglycol) or Freund's adjuvant.

stable, can be shipped at room temperature, and do not decay during storage.

VII. LIPOPEPTIDES ADMIXED TO VACCINES ENHANCE VACCINE EFFICIENCY

When added to *Enterobacteriaceae* vaccines, lipopeptides were found to provide protection for mice against lethal *Salmonella* infections. As demonstrated in Fig. 3, lipopeptides could substitute for 90% of a vaccine consisting of acetone-killed *S. typhimurium*. For P_3C-SSNA, a pronounced effect was found at an adjuvant concentration of 175 μg (Schlecht et al., 1988). The water soluble lipopeptide adjuvant P_3C-SK_4 exhibited an optimum around 90 μg/mouse. Thus, the supplementation of *Salmonella* vaccines with these synthetic lipopeptides significantly enhances their vaccine efficiency in mice. Variations of the native lipopeptide structure regarding chain length and amino acid sequence of the peptide moiety, as well as modifications of the lipoamino acid, lead to reduction or even complete loss of the adjuvant activity. The immunoadjuvant properties of the lipopeptides as described here are mediated by an enhancement of the humoral immune response without involving the cellular immune response (Schlecht et al., 1988).

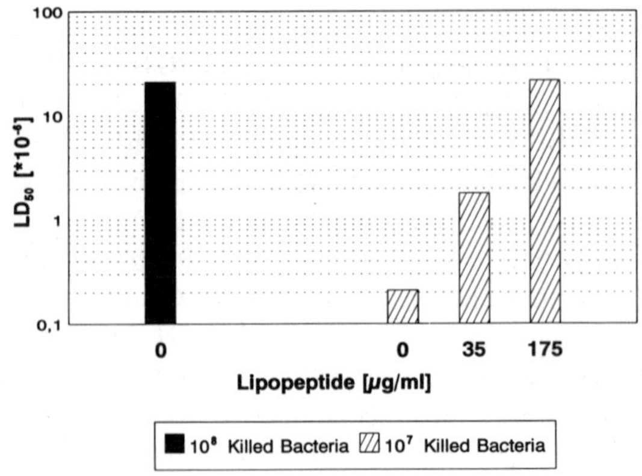

Fig. 3. Lipopeptides enhance the effect of a *Salmonella typhimurium* vaccine. Protection was judged by LD_{50} determinations in mice immunized with either 10^8 or 10^7 acetone-killed *Salmonella* bacteria injected in mixture with various doses (μg/ml) of the lipopeptide Pam_3Cys-Ser-Ser-Asn-Ala.

VIII. LIPOPEPTIDES AS INTEGRATED ADJUVANTS: INDUCTION OF A HUMORAL IMMUNE RESPONSE AGAINST HAPTENS AND LOW MOLECULAR MASS ANTIGENS

Lipopeptides also act as potent immunoadjuvants when coupled covalently to low molecular weight antigens. After being covalently linked to haptens, for example to digoxin or DNP, the lipopeptides rendered these molecules immunogenic, as shown by *in vivo* immunization of mice (Böltz et al., 1988; Reitermann et al., 1989). Furthermore, the toxic effects of digoxin or amanitin were no longer observed after coupling the toxins to lipopeptides (G. Jung et al., in preparation). In a similar way, various peptides coupled to lipopeptide anchors could induce pronounced peptide-specific immune responses *in vivo* and *in vitro*. As shown by us, a partial sequence from the epidermal growth factor receptor (EGF-R), coupled to P_3C-S, induced an epitope-specific anti-EGF-R response (Jung et al., 1985; Bessler et al., 1985). Because the EGF-R tetradecapeptide constitutes most probably a murine T helper cell epitope, the corresponding lipopeptide stimulated boost effects and memory T helper cells. Lipopeptide-antigen conjugates were also used to prepare HIV-specific antibodies (Fig. 4). We prepared oligopeptide segments of the HIV-1 and HIV-2 proteins p17, gp41, and gp120 by chemical synthesis. After covalently coupling the segments to lipopeptides, antigen-specific antibodies were induced *in vivo* in mice and *in vitro* in murine and human cell culture systems. The kinetics of antibody induction and the isotypes were determined. These and other HIV-specific antibodies prepared by our laboratories constitute tools which can be used to characterize viral components of virus-infected cells.

IX. LIPOPEPTIDE BASED VACCINES GIVE PROTECTION AGAINST FOOT AND MOUTH DISEASE

Lipopeptide based vaccines against foot-and-mouth disease were able to protect guinea pigs against lethal virus infections (Wiesmüller et al., 1989). We were able to design and synthesize an effective vaccine against foot-and-mouth disease (FMDV) by coupling the lipopeptide adjuvants to a known B and T helper epitope of the viral protein VP1. Having less than 3,500 Da, the resulting conjugate constitutes the smallest totally synthetic vaccine known so far that gives full protection to guinea pigs. We could show that a single injection of P_3C-SS-(VP1 135–154) was sufficient to protect seven out of seven guinea pigs against challenge with the homologous virus type (Fig. 4). Remarkably, in comparison to a mixture of vaccine diastereomers, only half the amount of pure diastereomer (R,R)-lipopeptide adjuvant-antigen conjugate is needed to induce full protection. High titers of FMDV virus-neutralizing antibodies were also obtained in cattle and swine.

X. LIPOPEPTIDES AS INTEGRATED ADJUVANTS: INDUCTION OF IMMUNOLOGICAL MEMORY BY THE ADDITION OF T HELPER CELL EPITOPES

The response against synthetically prepared antigens can be further enhanced by the additional introduction of a T helper cell epitope into the lipopeptide-hapten-conjugate. In mice, the suitable helper cell epitope applied, which is specifically presented by the BALB/c H-2^d haplotype, consists of a 16 amino acid oligopeptide derived from sperm wale myoglobin. Antibodies obtained after three immmunizations with conjugates containing melittin fragments recognized the synthetic and the native melittin molecule (Fig. 5). The immune enhancing effect was most pronounced for the peptide fragments Mel1–16 and Mel17–26 conjugated to lipopeptide-T_h-peptide conjugates. Our results show that it is possible to markedly enhance a weak hapten-specific immune response by coupling the haptens to lipopeptide conjugated to a haplotype-specific T helper cell epitope. These findings are of importance for the optimization of immunization procedures and for the development of novel synt vaccines.

XI. LIPOPEPTIDE-ANTIGEN CONJUGATES OR MIXTURES AS TOOLS FOR THE PREPARATION OF MONOCLONAL ANTIBODIES

Lipopeptide-antigen conjugates can be effectively used for the immunization of mice (see above). Immunization can be followed by the fusion of the murine splenocytes to myeloma cells yielding antibody producing hybridomas (P. Wernet et al., P. Hoffmann et al., unpublished). Lipopeptide-antigen conjugates can also be successfully applied for the preparation of human monoclonal antibodies. Thus, we immunized, *in vitro*, human peripheral blood lymphocytes with HIV-peptide-lipopeptide conjugates or toxin-lipopeptide conjugates. The immunized cell cultures were fused to cells of the heteromyeloma line CB-F7 (kindly obtained from Dr. v. Baehr, Charité Berlin), and we obtained stable hybridomas producing antigen-specific IgM and IgG (Hoffmann et al., 1990). In these experiments, screening of the hybridoma supernatants for antigen-specific immunoglobulin was done in microtiter plates to which the lipopeptide antigen conjugates had been bound.

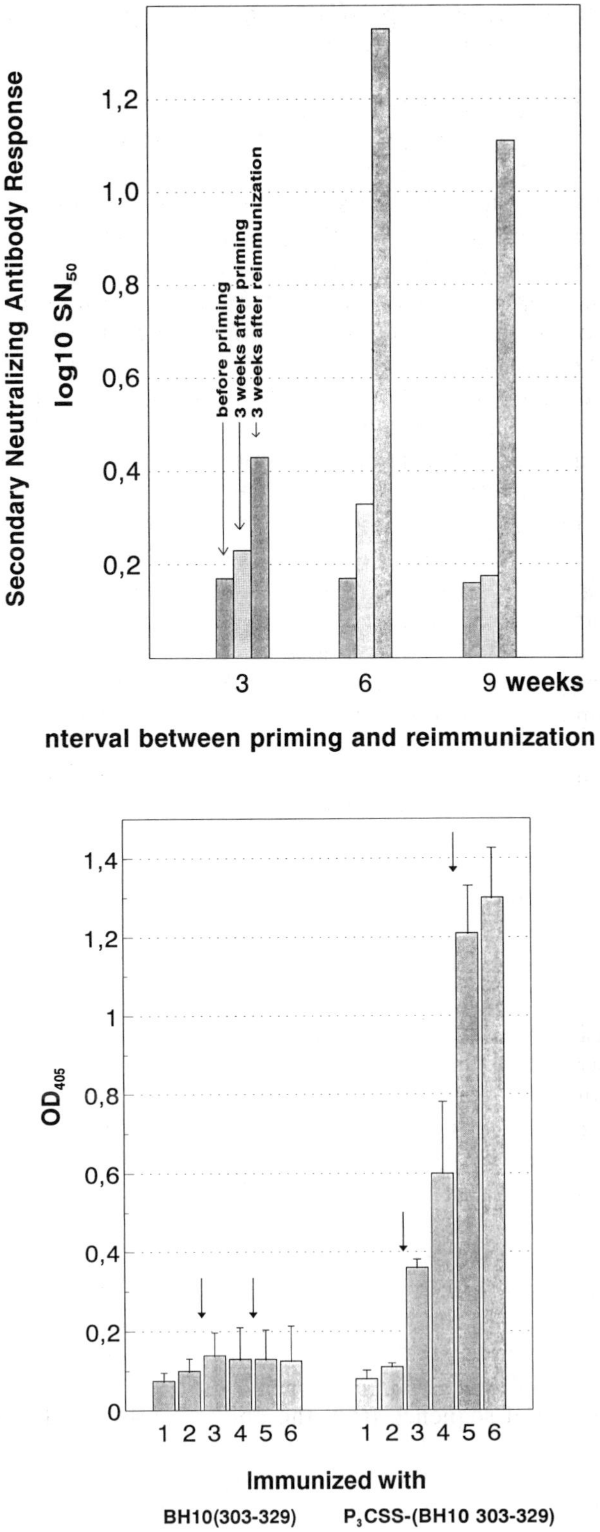

Fig. 4. Enhancement of the humoral immune response by lipopeptide conjugates. (Upper graph). Increase of virus-neutralizing immunoglobulin titers after the immunization of mice with a synthetically prepared viral oligopeptide (containing a B and a T helper cell epitope of FMDV VP1 protein) coupled to Pam_3Cys-Ser. (Lower graph). Increase of HIV peptide-specific immunoglobulin titers after the repeated immunization of mice with a synthetically prepared viral oligopeptide coupled to Pam_3Cys-Ser.

XII. LIPOPEPTIDES AS INTEGRATED ADJUVANTS: INDUCTION OF CYTOTOXIC T LYMPHOCYTES *IN VIVO*

Conjugates of lipopeptides with viral oligopeptides were able to induce peptide-specific cytotoxic T lymphocytes *in vivo* (Deres et al., 1989). Especially for the construction of vaccines, it is important that the lipopeptide immunoadjuvants can be used not only to improve humoral immune reactions, but also, after coupling to viral peptide segments, to induce cytotoxic T lymphocytes that kill virus-infected cells (Fig. 6). A long lasting cytotoxic response against influenza virus has been induced in mice (Schild et al., 1991), and corresponding experiments to induce, using lipopeptide-adjuvant-HIV peptide-conjugates, cytotoxic T cells recognizing HIV-infected targets are in progress (G. Haas, M. Loleit et al., unpublished). The viral oligopeptide sequence corresponding to a potential cytotoxic T lymphocyte (CTL) epitope can be predicted from the allele-specific peptide motifs determined by sequencing of self-peptide pools eluted from MHC class I molecules (Falk et al., 1991; Rötzschke et al., 1990, 1991; Zimmermann et al., 1992).

XIII. MECHANISM OF LIPOPEPTIDE INDUCED ADJUVANTICITY

Lipopeptide adjuvants are derived from a natural immunostimulating bacterial protein which is present in large amounts in the Enterobacteriaceae of our intestine. As outlined above, lipopeptides are potent stimulants for B lymphocytes and for cells of the endoplasmatic reticulum (monocytes, macrophages). Lipopeptide-antigen conjugates display a powerful builtin adjuvanticity in the form of the lipopeptide membrane anchor compound. The enzymatic degradation of the lipopeptides in plasma and tissue is diminished by their rapid uptake into cell membranes, or by the formation of micellar deposits, and as a result the lipopeptide-antigen conjugates will be presented to the immune system for a long period of time (long-term delivery).

The subsequent mechanism of leukocyte activation by the lipopeptide adjuvants is still unknown. After their addition to cell cultures, lipopeptides bind to defined membrane components including proteins of the major histocompatibility complex (Scheuer et al., 1986). A binding protein of 34 kDa could be isolated by lipopeptide-affinity chromatography (Biesert et al., 1987). Separated by two-dimensional electrophoresis the binding proteins consisted of three slightly different polypeptides (Biesert, Kleine, Bessler, unpublished). It is not yet known whether the 34 kDa protein possesses functional activity in the signal transduction process during lipopeptide activation. The

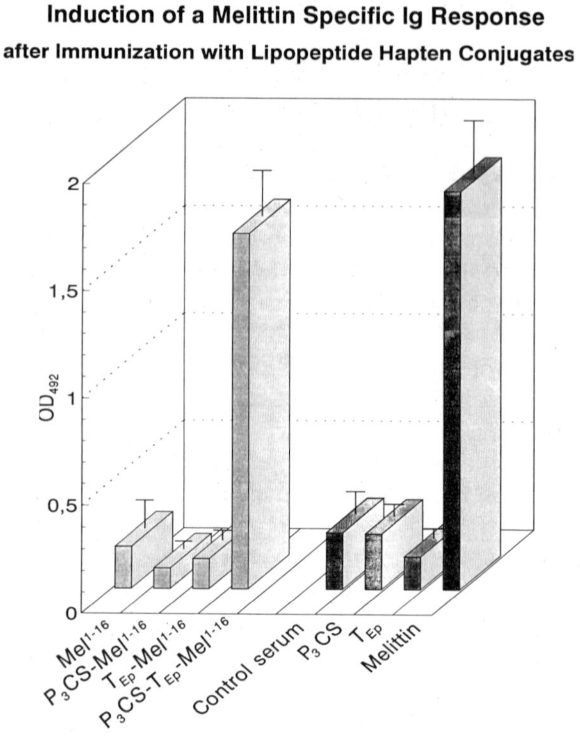

Fig. 5. Enhancement of the humoral immune response against a synthetic melittin fragment (Mel1-16) after covalently linking the oligopeptide to the lipopeptide Pam_3Cys-Ser and/or the T cell epitope from sperm whale myoglobin. Groups of five mice were immunized i.p. in two week intervals for three times with 0.018 μM of each substance. Two weeks after the last immunization, the mice bled from the plexus retroorbitales, and the melittin-specific serum titers were determined by ELISA. Data of three independent ELISA determinations +/- SD (W.G. Bessler, W. Beck, G. Jung et al., unpublished).

analysis of already known second messenger pathways during B cell stimulation by lipopeptides showed that neither cAMP, cGMP, nor phosphatidylinositol metabolism are changed after applying the stimuli. Changes in the intracellular calcium concentrations obtained using lipopeptides as stimuli suggest that these substances might act on intracellular calcium stores or plasma membrane calcium channels without involving the above-mentioned signal transducing mechanisms (Hauschildt et al., 1990). Remarkably, already 2 minutes after their addition to the cells, lipopeptides are found in different compartments within the cell including the nucleus (Uhl, Wolf and Bessler, unpublished; Wolf et al., 1989).

The MHC linked antigen presentation of lipopeptide-antigen conjugates eliciting humoral or cellular immune responses is presently under investigation. The low molecular mass lipopeptide-antigen conjugates are recognized by the antigen-specific B lymphocytes. They are taken up by them or by other antigen presenting

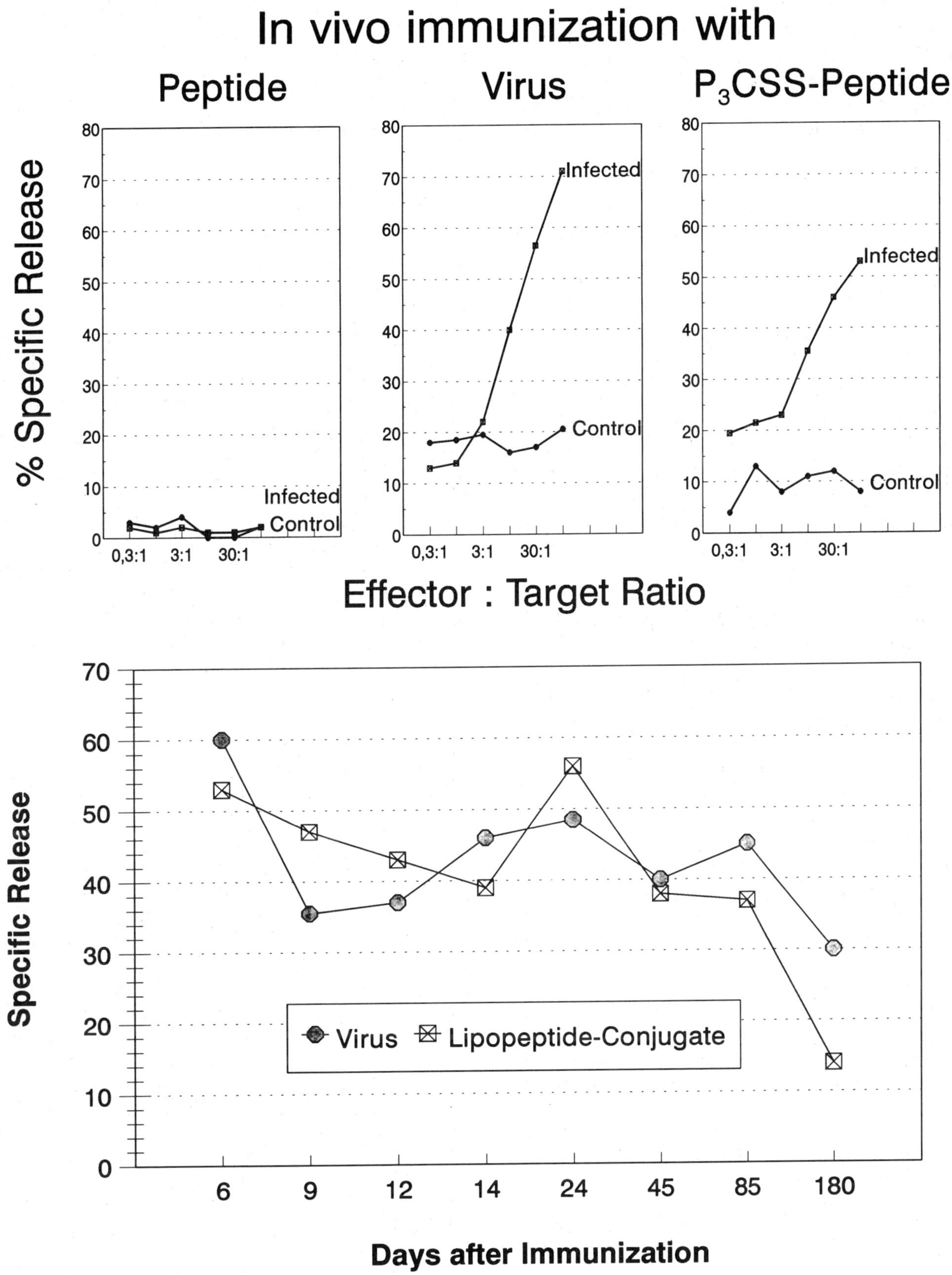

Fig. 6. (Upper graph). CTL response of spleen cells after i.v. immunization of Balb/c mice with Pam_3Cys-Ser-Ser-(NP147-158 (R$^-$)), a partial sequence of the influenza virus nucleoprotein. (Lower graph). Duration of CTL activity after *in vivo* priming of cytotoxic T cells with Pam_3Cys-Ser-Ser-(NP147-158 (R$^-$)) (Schild et al., 1991).

cells (APCs), thereby delivering T cell epitopes to the natural processing pathway and presentation in context with MHC class I or class II molecules, thus inducing - via T helper lymphocytes and B lymphocytes - a humoral immune response or - via T helper and cytotoxic T lymphocytes - a cytotoxic response.

XIV. ADVANTAGES OF THE NOVEL ADJUVANTS

In summary, synthetic lipopeptide analogs derived from bacterial lipoprotein constitute potent immunoadjuvants *in vitro* and *in vivo*, either in combination with or covalently bound to antigen. When using lipopeptide-antigen conjugates, these novel immunogens differ from other carrier-hapten conjugates in the following ways (Wiesmüller et al., 1992):

- Lipopeptide-antigen conjugates display a powerful builtin adjuvanticity in the form of a membrane anchor compound derived from a natural immunostimulating bacterial protein which is present in large amounts in the Enterobacteriaceae of our intestine.
- The low molecular mass lipopeptide-antigen conjugates can be recognized by the antigen-specific B lymphocytes. They are taken up by them or by other antigen presenting cells (APCs), thereby delivering T cell epitopes to the natural processing pathway and presentation in context with MHC class I or class II molecules.
- By the use of lipopeptide adjuvants covalently linked to antigen epitopes presented on MHC class I molecules, it is possible to activate cytotoxic T lymphocytes without using viral or other viable systems.
- Enzymatic degradation in plasma and tissue is diminished by the rapid uptake of the immunogen into cell membranes or by the formation of micellar deposits. As a result, a long-term delivery of the antigen to the immune system is realized.
- Administration of the lipopeptide requires no further adjuvant or additive. In particular, the harmful Freund's adjuvant can be avoided.
- No multiple antigen attachment, e.g., to a carrier protein, is required as in conventional immunogen preparations.
- No toxic side effects or tissue damages have been observed so far in animals for many years (Wiedemann et al., 1991).
- All routine methods of peptide synthesis including scaled up synthesis and multiple peptide synthesis (Jung and Beck-Sickinger, 1992) can be used for the preparation of lipopeptides (Metzger et al., 1992; Wiesmüller et al., 1992). This facilitates complete epitope mapping and reproducible preparation of large amounts of immunogens.
- Lipopeptide antigen conjugates can be analyzed by mass spectrometry (Metzger et al., 1992).
- Lipopeptides exhibit much higher stability against heat, light, or solvent changes in comparison to conventional immunogens or ISCOMB preparations.
- Lipopeptides are fully biodegradable into harmless amino acids and fatty acids, and S-glyceryl-cysteine is a metabolite from *E. coli* lipoprotein found in urine and feces.

XV. PREPARATION OF THE LIPOPEPTIDE-ANTIGEN CONJUGATES

XV.1. Antigen adjuvant mixtures

Lipopeptides can be obtained commercially (Boehringer Mannheim, Penzberg, FRG) or can be prepared by chemical synthesis, as published by us previously (comp. Bessler and Jung, 1992). For the intraperitoneal, intra-, or subcutaneous application in combination with antigens, the water soluble lipopeptide P3C-SK4 is suspended in aqueous solution with antigen by gentle shaking at a 1:1 (w/w) ratio. Water insoluble lipopeptides like P3C-SSNA are suspended together with antigen at a 1:1 (w/w) ratio in aqueous solution by sonification. For oral administration in animals, food pellets are soaked in the lipopeptide antigen solution/suspension. From our experience, the amount of antigen and lipopeptide has to be increased about tenfold compared to parenteral immunization to obtain a sufficient serum antibody titer.

XV.2. Antigen adjuvant conjugates

For covalently coupling antigens to lipopeptide adjuvants, the following procedures are recommended:

1. Conjugation of peptides or proteins soluble in dimethylformamide (DMF): 2 μmol of peptide or protein binding determinants are dissolved in 0.5–1 ml of DMF, and 8 μmol (9.2 mg) of solid P3C-SG-OSu are added. A homogeneous solution is obtained by gentle heating and sonication, and 4 μmol of organic base (N-ethylmorpholine) are added. After stirring for 12 h, 1–2 ml of chloroform:methanol (1:1) are added, and the mixture is cooled in an ice bath for 2 h. The sediment is washed with 1 ml of cold chloroform:methanol (1:1), dissolved in tert. butanol:water (3:1), sonicated if necessary, and freeze dried.
2. Peptides and proteins soluble in water: 2 μmol of peptide/protein are dissolved in 0.8 ml of water, and 4 μmol (6.5 mg) P3C-SK4 are added. The mixture is thoroughly sonicated and a pH of 5.0–5.5 is set up. After 5 mg of EDC ((1,3-dimethyl-aminopropyl)-3-ethylcarbodiimide hydrochloride) dissolved in 100 μl of water has been added, the mixture is stirred at room temperature for 18 h and then dialyzed twice against 1 l of distilled water. The content of the dialysis tubing is freeze dried.

ABBREVIATIONS

APC	Antigen presenting cell
CTL	Cytotoxic T lymphocyte
FMDV	Foot-and-mouth disease virus
MHC	Major histocompatibility complex
LPS	Lipopolysaccharide
P_3C	N-Palmitoyl-S-(2,3-bis(palmitoyloxy)-(2RS)-propyl-(R)-cysteinyl
P_3C-SSNA	N-palmitoyl-S-(2,3-bis(palmitoyloxy)-(2RS)-propyl-(R)-cysteinyl-(S)-seryl-(S)-seryl-(S)-asparaginyl-(S)-alanine
TFA	Trifluoroacetic acid

XVI. REFERENCES

Andersson J, Sjöberg O and Möller G. Induction of immunoglobulin and antibody synthesis *in vitro* by lipopolysaccharides. Eur J Immunol 2:349, 1972

Bessler W and Braun V. Mitogenität von Lipoprotein aus der äußeren Membran von E. coli gegen Lymphozyten verschiedener Spezies. Z Immun-Forsch 150:193, 1975

Bessler WG and Ottenbreit BP. Studies on the mitogenic principle of the lipoprotein from the outer membrane of *Escherichia coli*. Biochem Biophys Res Comm 76:239, 1977

Bessler WG and Henning U. Mitogenic bacterial surface components: B lymphocyte stimulatory activity *in vitro* of Protein I from the outer membrane of E. coli. In: Abstracts of the 13th International Leukocyte Culture Conference. Ottawa, 1979

Bessler WG, Flad HD, Schade U et al. Lymphokine induction in murine and human leukocytes by bacterial lipoprotein and a synthetic analogue and by bacterial protein I. Lymphokine Res 3:234, 1984

Bessler WG, Suhr B, Bühring HJ et al. Specific antibodies elicited by antigen covalently linked to a synthetic adjuvant. Immunbiology 170:6, 1985

Bessler W, Cox M, Lex A et al. Synthetic lipopeptide analogues of bacterial lipoprotein are potent polyclonal activators for murine B-lymphocytes. J Immunol 135:1900, 1985

Bessler WG and Jung G. Lipopeptides as novel adjuvants. Forum Immunol, 1992

Biesert L, Scheuer W and Bessler WG. Interaction of mitogenic bacterial lipoprotein and a synthetic analogue with mouse lymphocytes. Isolation and characterization of binding proteins. Eur J Biochem 162:651, 1987

Böltz T, Hummel RP, Tröger W et al. Distinction between HIV-1 and HIV-2 infection using novel synthetic lipopeptide conjugates as antigens in enzyme immunoassays. J Virol Meth 22:173, 1988

Böltz T, Jung G, Wiesmüller KH et al. Enhancement of the immune response using a synthetic B-lymphocyte mitogen covalently linked to antigens. In: Mani J and Dornand J Eds: Lymphocyte Activation and Differentiation. Walter de Gruyter & Co., Berlin-New York, p 379, 1988

Deres K, Schild H, Wiesmüller KH et al. *In vivo* priming of virus-specific cytotoxic T-lymphocytes with synthetic lipopeptide vaccine. Nature 342:561, 1989

Falk K, Rötzschke O, Stevanovic S et al. Allele-specific motifs revealed by sequencing of self-peptides eluted from MHC-molecules. Nature 351:290, 1991

Galanos C, Lüderitz O, Rietschel ET et al. Synthetic and natural *Escherichia coli* free lipid A express identical endotoxic activities. Eur J Biochem 148:1, 1985

Hauschildt S, Hoffmann P, Beuscher HU et al. Activation of bone marrow-derived macrophages stimulated with bacterial lipopeptide. Cytokine production, phagocytosis and Ia expression. Eur J Immunol 20:63, 1990

Hoffmann P, Heinle S, Schade UF et al. Stimulation of human and murine adherent cells by bacterial lipoprotein and synthetic lipopeptide analogues. Immunobiology 177:158, 1988

Hoffmann P, Wiesmüller KH, Metzger J et al. Induction of tumor cytotoxicity in murine bone marrow-derived macrophages by two synthetic lipopeptide analogues. Hoppe-Seyler's Z Physiol Chem 370:575, 1989

Hoffmann P, Jimenez-Diaz M, Loleit M et al. Preparation of human and murine antibodies: Antigens combined with or conjugated to lipopeptides constitute potent immunogens for *in vitro* and *in vivo* immunizations. Hum Antibod Hybridomas 3:137, 1990

Jung G, Wiesmüller KH, Becker G et al. Verstärkte Produktion spezifischer Antikörper durch Präsentation der antigenen Determinanten mit kovalent verknüpften Lipopeptid-Mitogenen. Angew Chem 97:883, 1985 and Angew Chem Int Ed Engl 24:872, 1985

Jung G and Beck-Sickinger AG. Multiple peptide synthesis methods and their applications. Angew Cem Int Ed Engl 31:367, 1992

Kleine B, Sprenger R, Martinez-Alonso C et al. Analysis of the polyclonal B cell activation by a synthetic analogue of bacterial lipoprotein - qualitative and quantitative approach and comparison with bacterial lipopolysaccharide (LPS). Immunology 61:29, 1987

Kleine B, Biesert L and Bessler WG. The membrane protein composition of lipopolysaccharide (LPS) or lipoprotein activated B lymphoblasts from C57BL LPS responder and nonresponder mice are qualitatively indistinguishable. Res Immunol 140:685, 1989

Lex A, Wiesmüller KH, Jung G et al. A synthetic analogue of *Escherichia coli* lipoprotein, tripalmitoyl pentapeptide, constitutes a potent immune adjuvant. J Immunol 137:2676, 1986

Melchers F, Braun V and Galanos C. The lipoprotein of the outer membrane of *E. coli*: A B-lymphocyte mitogen. J Exp Med 142:473, 1975

Metzger J, Olma A, Leplawy MT et al. Synthesis of a B-lymphocyte activating α-methylserine containing lipopentapeptide. Z Naturforsch 42b:1195, 1987

Metzger J, Wiesmüller K-H, Schaude R et al. Synthesis of novel immunologically active tripalmitoyl-S-glycerylcysteinyl peptides as useful intermediates for immunogen preparations. Int J Peptide Protein Res 37:46, 1991

Metzger J, Wiesmüller K-H and Jung G. Synthesis of N_{alpha}-Fmoc protected derivatives of S-(2,3-dihydroxypropyl)-cysteine and their application in peptide synthesis. Int J Peptide Protein Res 38:545, 1991

Metzger J, Jung G, Bessler WG et al. Lipopeptides containing 2-palmitoylamino-6,7-bis(palmitoyloxy)-heptanoicacid:synthesis, stereospecific stimulation of B-lymphocytes and adjuvanticity. J Med Chem 34:1969, 1991

Metzger J, Beck W and Jung G. Ion-spray mass spectrometry of lipopeptide vaccines. Angew Chem Int Ed Engl 31:226, 1992

Prass W, Ringsdorff H, Bessler W et al. Lipopeptides of the N-terminus of *E. coli* lipoprotein: Synthesis, mitogenicity and properties in monolayer experiments. Biochim Biophys Acta 900:116, 1987

Reitermann A, Metzger J, Wiesmüller KH et al. Lipopeptide derivatives of bacterial lipoprotein constitute potent immune adjuvants combined with or covalently coupled to antigen or hapten. Biochem Hoppe-Seyler 370:343, 1989

Resch K and Bessler WG. Activation of lymphocyte populations with Concanavalin A or with lipoprotein or lipopeptide from the outer cell wall of *Escherichia coli*: Correlation of early membrane changes with induction of macromolecular synthesis. Eur J Biochem 115:247, 1981

Riveau G and Chedid L. Comparison of the immuno- and neuropharmacological activities of MDP and murabutide: Immunopharmacology modulators of non-specific resistance. In: Majde JA Ed: Progress in Leukocyte Biology. Alan R. Liss, New York, p 213, 1987

Rötzschke O, Falk K, Deres K et al. Isolation and analysis of naturally processed viral peptides as recognized by cytotoxic T cells. Nature 348:252, 1990

Rötzschke O, Falk K, Stevanovic S et al. Exact prediction of a natural T cell epitope. Eur J Immunol 21:2891, 1991

Scheuer WV, Biesert L and Bessler WG. Binding a synthetic analogue of mitogenic bacterial lipoprotein to murine major histocompatibility complex (MHC) gene products. Biol Chem Hoppe-Seyler 367:1085, 1986

Schild H, Deres K, Wiesmüller K-H et al. Efficiency of peptides and lipopeptides for *in vivo* priming of virus specific cytotoxic T-cells. Eur J Immunol 21:2649, 1991

Schlecht S, Jung G and Bessler WG. Immunoenhancing activity of heterologuous bacteria and some of their cell wall components on the protective capacity of S. typhimurium vaccines. Adv Biosci 68:317, 1988

Seifert R, Serke S, Huhn D et al. Incomplete functional differentiation of HL-60 leukemic cells by synthetic lipopeptides. Eur J Biochem 203:143, 1992

Uhl B, Wolf B, Schwinde A et al. Intracellular localization of a lipopeptide macrophage activator: Immunocytochemical investigations and EELS analysis on ultrathin cryosections of bone marrow-derived macrophages. J Leukocyte Biol 50:10, 1991

Vordermeier H-M, Hoffmann P, Gombert FO et al. Synthetic peptide segments from the Escherichia coli porin OmpF constitute leukocyte activators. Infection Immunity 58:2719, 1990

Wiedemann F, Link R, Pumpe K et al. Histopathological studies on the local reactions induced by complete Freund's adjuvant (CFA), bacterial lipopolysaccharide (LPS), and synthetic lipopeptide (P_3C) conjugates. J Pathol 164:265, 1991

Wiesmüller KH, Bessler WG and Jung G. Synthesis of the mitogenic S-(2,3-bis(palmitoloxy)propyl)-N-palmitoyl pentapeptide from the *Escherichia coli* lipoprotein. Hoppe Seyler's Z. Physiol Chem 364:593, 1983

Wiesmüller KH, Jung G and Hess G. Novel low-molecular-weight synthetic vaccine against foot-and-mouth-disease containing a potent B-cell and macrophage activator. Vaccine 7:29, 1989

Wiesmüller K-H, Metzger J, Schaude R et al. Synthetic low-molecular weight foot-and-mouth disease virus vaccine and novel lipopeptides for the synthesis of immunogenic lipopeptide-antigen conjugates. In: Innovation and Perspectives in Solid Phase Synthesis; First International Symposium. R Epton Hrsg SPCC (UK) Ltd, Birmingham, p 601, 1990

Wiesmüller KH, Bessler WG and Jung G. Solid phase peptide synthesis of lipopeptide vaccines eliciting epitope-specific B-, T-helper and T-killer cell response. Int J Peptide Protein Res 40:255, 1992

Wolf B, Uhl B, Hauschildt S et al. Interaction of the lymphoid cell line BCL_1 with a lipopeptide analogue of bacterial lipoprotein: Distribution of the activator within the cells detected by electron energy loss spectroscopy (EELS). Immunobiology 180:93, 1989

Wolf B, Hauschildt S, Uhl B et al. Localization of the cell activator lipopeptide in bone marrow-derived macrophages by electron energy loss spectroscopy (EELS). Immunol Lett 20:121, 1989

Wolf B and Bessler WG. Elektronen-Energieverlust-Spektroskopie (EELS) als Methode zur Lokalisierung von Antigenen und anderen Substanzen in Zellen und Geweben. Naturwiss 77:110, 1990

Zimmermann N, Rötzschke O, Falk K et al. Molecular modelling of the class I human histocompatibility molecule HLA-A2 presenting an allele-specific nonapeptide from influenza matrix protein. Angew Chem Int Ed Engl 31:886, 1992

Chapter 13

OPTIMIZATION OF SPECIFIC ANTI-PEPTIDE RESPONSES: IMMUNIZATION AND ADJUVANT

Wim J.A. Boersma and Eric Claassen

TABLE OF CONTENTS

I. INTRODUCTION

The history of vaccination is a story of scientific trial and error. Initially, knowledge with respect to specific pathogens was developed. In addition, the experience of the administration modalities that left the specific effects of vaccination (protection) intact but were designed to minimize the risk for the vaccinee were analyzed. This led to the development of attenuated pathogens. However, an intrinsic drawback of most attenuated pathogens is that upon application in a vaccine the immunogenicity in general is decreased as compared to the intact wild-type pathogen. This in turn led to the development of methods to enhance the response to compensate for the immunogenicity that was lost by attenuation.

One of the reasons for developing peptide or subunit vaccines is to avoid pathogenicity of intact pathogens. As with attenuation, this approach is hampered by a relatively lower immunogenicity of the vaccine as compared to the intact pathogen. As a result, various adjuvants and formulation techniques

0-8493-8967-4/95/$0.00+$.50

TABLE 1
Main categories of enhancers of immune responses

Category	Properties
* Mineral salts	Vehicle, macrophage stimulation, depot Calcium phosphate does not stimulate IgE
* Water-in-oil emulsions	Vehicle, macrophage stimulation, depot
* Cell wall components (lipopolysachariden from bacteria)	Direct B cell stimulation, macrophages and T cell stimulation
* Water-in-oil emulsions to which bacteria are added	Vehicle, macrophage stimulation, depot Direct B cell stimulation (lipopolysacharide) T cell stimulation
* Liposomes	Vehicle, localization IgM responses with haptens or B cell epitopes IgG responses with proteins
* Saponins (ISCOM)	Vehicle, surfactant T cell stimulation (cellular immunity)
* Synthetic compounds	
- Lipophilic quaternary amines	Macrophage stimulation, activation complement system, Induction cytokines (IL-1, interferon-γ), stimulation of delayed type hypersensitivity
- Non-ionic block polymers	Macrophage stimulation, activation complement system
* Body derived adjuvants Cytokines or cytokine fragments	Direct cellular interactions
* Enhancement of immune responses by specific cell targeting antibodies	MHC specific, receptor specific Anti-APC antibodies

have been developed to enhance peptide immunogenicity using a proper formulation of the peptide vaccine.

Adjuvants are widely used not only with vaccines, but even more in experimental immunology for the generation of monoclonal or polyclonal antibodies for application in a wide variety of disciplines. Though the effects of adjuvants depend heavily on the nature, the origin, the structure and physico/chemical properties of the antigen of interest, for the adjuvants a similar list of properties can be given which determine the differences in the outcome of immune responses to the antigen of interest (Table 1).

Adjuvant support for anti-peptide responses is similar to the approach for other antigens. However, for peptides, spatial orientation is made independent of the original structure from which the peptide determinant sequence is chosen. Orientation, exposition and traffic for peptides require an alternative supportive micro-environment to be developed.

In this chapter, we will discuss general aspects of immunization using peptide antigens such as peptide length, presentation, vehicle, routing, localization and doses. The presently known mode of action of adjuvants and the results as presented in the literature and from our own studies with respect to optimization of antigen presentation modalities for peptides based on the selection of specific adjuvants are reviewed. Based on these results, we will suggest a general applicable approach for peptide immunization. As far as peptide immunization is discussed in relation to the production of research reagents, the effect of protocols on animal welfare will be taken into account.

II. ASPECTS OF IMMUNOGEN FORMULATION

Immunization requires in general the introduction of an antigen at a certain site of the body. The form in which the antigen is presented may be either a purified protein, a virus particle or a part of a liposome or immune stimulating complex construct (ISCOM) etc. which to a large extent determines the outcome of the response. To the relatively low immunogenic peptide antigens in general a non-specific immune response enhancer (adjuvant) is added. The character of the antigen, the transport vehicle and routing determine the localization of the antigen. Immunization of peptide conjugates results in responses to the peptide, the coupling agent or neodeterminants formed, the carrier protein and all combination determinants. The immune responses to peptide conjugates therefore require differential analysis. In addition, antigen presentation and immune response can be influenced by the vehicle it is attached to, as well as by the adjuvant formulation.

III. FUNCTION OF ADJUVANTS

We will discuss the nature of the adjuvants, their vehicle and depot aspects, the mode of immune enhancement and the results presently obtained with synthetic peptides in both humoral and cellular immune responses.

Obviously, the first generation of adjuvants, developed without thorough knowledge of the immune system, consisted of micro-organisms or parts thereof that enhanced the immune response to co-administered vaccines. The analysis of the active components of the micro-organisms led to the development of defined and synthetic adjuvants. In this respect, not only the chemical composition of an adjuvant is of importance, but also the structure which is chosen in the formulation of the adjuvant.

An adjuvant that functions as a vehicle is highly effective in that it presents a high dose of the active substances locally. Adjuvants should preferentially enhance the immune response of all cell types involved in the process of generation of protection with a vaccine. To this end, the stimulation of the antigen presentation, the T regulator cells (Th1 and Th2) as well as the B cell and T cell effector cells, is most important.

For the generation of antibodies as reagents, the requirements of immune stimulation are less stringent than for a vaccine which should preferentially in addition to antibody formation also stimulate delayed type hypersensitivity and specific cytotoxic responses. Especially for the enhancement of cellular immune responses, additional criteria have to be fulfilled with respect to the vehicle aspects of the adjuvant.

Functions of adjuvant that contribute to enhancement of immunological function are depot formation, which essentially is a doses effect and vehicle function, improved antigen presentation, which in part is an immunological carrier function, and immunomodulation, i.e., the true adjuvant function. The way in which various adjuvants contribute to these functions is discussed later.

IV. NATURE OF VARIOUS ADJUVANTS

The nature of the adjuvants in general is a balance between (immuno)-toxicity and immune stimulation. Harmless adjuvants have a low stimulative enhancement capacity. Adjuvants like Freunds complete adjuvant stimulate the immune response very effectively, but have dramatic side effects. As a result, not all adjuvants are allowed for human use or for application in veterinary medicine. In addition, in many countries laws on animal experimentation have limited the use of the most potent adjuvants for ethical reasons.

A broad review on the surface chemistry of vaccine adjuvants by Woodard (1990) describes in detail emulsifying and surfactant properties of adjuvants which allow a design of adjuvant combinations which is most suitable for presentation of antigen to professional and other antigen presenting cells.

As has been discussed elsewhere (Boersma et al., 1992a,b), in the case of small protein antigens, i.e., peptides, the exposition of determinants can be dramatically influenced by choice of the adjuvant used. As a result the humoral response may be directed to different parts of the protein amino acid sequences. Electrostatic interaction or adsorption to amorphic or crystalline precipitates (alum) will surely lead to a different tertiary structure and a different mode of antigen presentation (exposition) of a peptide or protein sequence as compared to the same antigens presented emulsified in a water-in-oil emulsion (WIO). Most immunogenic sequences of proteins are hydrophillic, but as a result of this the neighboring sequences often have a less hydrophillic character. The length of hydrophobic and hydrophillic sequences as well as the character of the emulsion (water-in-oil or double emulsions) will determine which part of the peptide is exposed (Fieser et al., 1987). In addition, the structure and folding of an immunogenic sequence is influenced by the micro-environmental conditions. In fact, the development of synthetic determinants with hydrophobic tails are developed to facilitate preferential exposition of the determinant to which the immune response has to be directed preferentially (Neurath et al., 1989; Deres et al., 1989).

Mineral salts consist of hydrated gels of aluminum hydroxide or phosphate and calcium phosphate which form efficient vehicles to which proteins readily adsorb dependent on ion strength and pH. The depot function is most prominent, whereas the immune stimulative capacity is weak. For peptides and sub-unit vaccines, these salts in general are considered insufficient. Protein structure is only mildly distorted. The relatively harmless character led to Food and Drug Administration and World Health Organization approvement of these adjuvants in humans. Local, i.e., vaginal application led to humoral response induction without harmful side effects seen with other adjuvants. The effective depot function has led to combined use of mineral salts with real adjuvants like micro-organisms (*Bordetella pertussis*) (Bomford, 1980)

Micro-organisms and microbial products can exhibit strong adjuvant activity. In general, the bacteria are applied in heat-killed form and co-administered with the immunogen under study. The adjuvanticity can be enhanced by admixing the bacteria with depot preparations of the immunogen (e.g., mineral salts, oil emulsions). As stated above, detoxification (removal of lipid A) usually leads to a loss in adjuvanticity. This has led to the development of effective synthetic analogs of the active components in most cases cell wall components. The most frequently used as adjuvants are mycobacteria (a.o. Bacille Calmette Guerin), *Bordetella pertussis*, *Corynebacterium parvum*. Obvious adverse effects are: sterile inflammations, granuloma's and arthritis (adjuvant arthritis) (Claassen et al., 1992). Intraperitoneal, subcutaneous, intradermal or intramuscular administration is performed to enhance the depot function.

With the micro-organism a whole lot of proteins, membranes and DNA is administrated together with the antigen, and unwanted immune responses to these sometimes rather immunodominant antigens may be the result (anti-DNA, anti-housekeeping proteins).

Toxins have been widely applied as immunological carriers with a high immune stimulating potential. Heat-labile toxin of *E.coli*, tetanus toxoid, as well as cholera toxin are examples best known. Simple coupling of peptides does lead to efficient immune response generation (Arnon, 1990).

Water-in-oil, reversed and double emulsions with vaccines have been in use for a long time (1916). Mineral oil, and also plant and fish derived degradable oils are applied. Freund (1956) applied refined paraffin oil with vaccines, which since then have generally been used as an adjuvant (Freund's incomplete adjuvant, FIA) for experimental purposes mainly. Together with killed mycobacteria, FIA is used as Freunds complete adjuvant (FCA) of which the side effects led to its banning in various countries (Claassen and Boersma, 1992; Claassen et al., 1992). Freunds complete adjuvant (FCA) is probably the most widely applied adjuvant for experimental studies, since it is not allowed for regular use in animals or man. The depot function of oil emulsions is enhanced by the intraperitoneal, subcutaneous, intradermal or intramuscular administration.

The antigen to which an immune response has to be elicited is taken up in general into the hydrophillic phase. In practice, the oil phase of most adjuvants contains over 40% (v/v) oil phase. The oil phase consists mainly of three types of constituents: emulsifiers (±5%), metabolizable oils and non-metabolizable and mineral oils. A proper balance between metabolizable and non-metabolizable oils establishes the right level of enhancement at a well tolerated level of irritation. Metabolizable oils are plant (e.g., soybean, peanut) or fish-derived. Non-metabolizable oils are derivatives of paraffins, stearins, etc. which have similar adjuvant activity as mineral oils but cause less irritation. In some admixtures., squalene is the main oil component. Saponins (e.g., QuilA) may be added to increase adjuvant activity. Efficient emulsions generally contain droplets of 0.1 –1.0 μm diameter. This allows emulsions to generate a surface contact area of about 10–30 square meters with 1 ml of oil.

To form a stable water-in-oil emulsion, surfactants have to be applied. However, an intrinsic property of all surface active compounds is that they will cause damage to almost any tissue to which they are brought into contact. For water-in-oil emulsions, lipophilic surfactants are required, since the oil phase forms the continuous phase (Bokhout et al., 1981; Woodard and Jasman, 1985; McKercher, 1986; Woodard, 1990). Immune stimulating oils have to be refined to such a degree that carcinogenic derivatives (polycyclic aromats) are removed (McKercher, 1986).

Specol has advantageous properties certainly with respect to animal care, easiness and reproducibility of preparation and administration and the use of constituents which generally are regarded as safe for use in animals. Bokhout et al. (1981) first described specol as a selected water-in-oil emulsion of which the application has been evaluated in the prevention of future diseases by vaccination as well as in the enhancement of natural *in vivo* immune responses. This adjuvant was developed based on their observation that vaccinated pigs were protected against pathogens not present in the vaccine.

Specol is an adjuvant admixture which has immunostimulating properties to almost the same level as has FCA when the antigen is mixed into the adjuvant. Even separately administered, significant immune stimulation is obtained. However, specol does have far less adverse reactions than those observed for FCA. In mice the level of adverse reactions seems to be dependent on the genetic background of the mice. In C3H/He mice no or only mild irritation was observed upon i.p. injection of specol. In contrast, NZB, Balb/c and C57Bl mice show chronic mild to severe peritonitis. The organs in the abdomen showed fixtures with connective tissue in a similar way as is observed when FCA/IFA is used in mice intraperitonially. Though proliferative T cell responses and DTH reactions could be demonstrated with use of specol, formal proof of the support of a cytolytic T cell response has yet to be produced. In contrast to most adjuvant admixtures, specol can be prepared easily and is highly reproducible, since all constituents can be obtained from the suppliers in quality controlled batches. The low viscosity of specol as compared to many blends (IFA) facilitates administration by injection (Boersma et al., 1992b).

Liposomes are balloons or envelopes of concentric multi-lamellar layers of phospholipids which in themselves are not immunogenic. The antigen is encapsulated in the water or lipid phase or can be coupled to the surface (for review, see Buiting et al., 1992). Van Rooijen and Nieuwmegen (1980) showed that surface exposition of antigens is most efficient in antigen presentation. For peptides the elongation with a hydrophobic tail (lipopeptides) enhances the uptake and correct surface orientation of the determinants of interest in a given antigen. As such, a liposome can be envisaged as a hapten vehicle with a large surface. The repetitive presence of the determinant may lead to B cell activation (IgM responses). T cell dependent responses are found only when a T cell determinant is included in the antigen. The fusion of liposomal and endosomal membranes delivers the antigen in the MHC class II processing pathway. Therefore, peptide presentation in a context of liposomal membranes is expected to generate complexes with MHC class II, but not with class I (Germain, 1986; Bevan, 1987).

Saponins are a plant derived family of rather complex glycoside containing compounds which are amphiphilic in nature and tend to form micelles. Together with cholesterol, some of these compounds tend to form particulate matrix structures with a high adjuvanticity. By hydrophobic interaction, proteins can strongly complex with these structures to form (ISCOM) immune stimulating complexes. Peptides can be coupled to the matrix proteins and elicit even cellular immune responses. This property is of advantage for the immunization with antigens having amphipathic features or transmembrane sequences. Most saponins are toxic and local, and depot-like administration may lead to necrosis. Their hemolytic action is due to the surfactant activity.

Synthetic adjuvants are non-ionic block polymers that have a low intrinsic toxicity. They are built from repetitive combinations of poly-oxy-ethylenes and poly-oxy-propylenes. Due to their partly hydrophillic and partly hydrophobic nature, they are readily used with liposomes and water-in-oil emulsions. Dimethyldioctadecylammonium bromide (DDA) is a representative of a specific category of adjuvants, called the quaternary amines. Muramyl dipeptides (MDP) are synthetic analogs of membrane components of mycobacteria.

Lipopeptides present a novel intrinsic adjuvant model for peptide immunizations. Lipopeptides not only can serve to efficiently immunize peptides, but in addition they can serve as a carrier with adjuvanticity (for review, see Bessler and Jung, 1992).

Since cytokines are body derived adjuvants which function as the primary immune response regulators. The analysis of adjuvant effects, apart from antibody production, DTH responses and cellular cytolysis, is based on demonstration of cytokine production. As a consequence, immunomodulation is aimed at mimicking adjuvant effects by administration of the cytokines itself.

Body derived adjuvants can be administrated or locally produced as a secondary effect of an immunization procedure using various other adjuvants (microorganisms, microbial membranes). Some cytokines have been evaluated for their adjuvant activity: interleukins 1 and 2, interferon-γ. These cytokines influence growth and differentiation of T and B cells as well as the antigen presentation by various cells. However, the mode of action of cytokines is complex, such that the application of cytokine subunits with specific response enhancing modalities was developed (IL-1).

Cytokines seem promising in that they mimic the actions that are generated with adjuvants. Especially, subunit cytokines may enhance immune responses without expressing the adverse side effects. Various modes of administration have been investigated. Experimentally intraperitoneal administration is most effective (Staruch and Wood, 1983; Boraschi et al., 1988; Tagliabue and Boraschi, 1993).

V. ADJUVANT MODE OF ACTION

Adjuvants in general have a multitude of functions. When the antigen is administered together with the adjuvant, in general the adjuvant also functions as the vehicle which takes care of localization of the antigen. Antigen administration in water-in-oil or oil-in-water emulsions, antigen bound to or enclosed in liposomes, antigen bound or absorbed onto amorphous or crystalline material as in alum hydroxide/phosphate gels, and with nitrocellulose or solid beads of various compositions all have in common that specific localization (intrasplenal, intraperitoneally, intra- or subcutaneous) can be obtained such that administration results in local immune responses. The surface area of the vehicle to which the antigen is attached determines the load and density of the antigen during antigen presentation.

Apart from localization and vehicle capacity, the various adjuvants also offer a depot function. When deposited at a given site the antigen leaks slowly from the adjuvant compound and gradually becomes available for stimulation of the immune system.

Adjuvants are foreign to the body. Therefore, if appropriately administered, they lead to sterile inflammations that attract the various cells of the a-specific defence system — macrophages — which in turn produce immunomodulating factors that increase the level of immune surveillance. Therefore, macrophages as well as other antigen presenting cells play an important role in the generation of the adjuvant effect. Most adjuvants directly or indirectly stimulate not only the generation of IL-1, but also other factors that support growth and differentiation.

Mineral salts have a major drawback that they do not support cell mediated immunity very well. The mild irritation which is caused does suggest that this adjuvant functions by local production of cytokines to activate antigen presenting cells (APC).

Water-in-oil emulsions thus far have not been shown to be effective in the elicitation of specific local cytokine production. In draining lymph nodes and spleen, the production of IL-1, -2, -4 and interferon-γ were demonstrated (Zegers et al., unpublished observations). From these results no skewing in either Th1 or Th2 response direction may be expected using this type of adjuvant. Nevertheless, their effective mode of action seems mostly to rely on the enhancement of hydrophobicity of antigens, targeting to antigen presenting cells and on the presentation of antigens on a large surface area. The oil microspheres carry antigen to the surface of antigen presenting cells. Endocytosis and presentation in the context of MHC class II is therefore expected. The activation of macrophages suggests that the cytokine which may be involved is most likely IL-1 (for review, see Boersma et al., 1992b). Recent experiments suggest induction of IL-2, IL-4 and interferon-γ together with expression of the T cell activation marker gp39 (Zegers et al., unpublished results).

Microbial products may lead to even mitogenic activation of B cells (lipopolysaccharide), activation of macrophages, T cells and the complement system. Lipid A as well as the lipopolysaccharide are hydrophobic and amphiphilic (lipid A). Administrated separately or together with antigen, they generate cytokine responses. Most prominent is the production of interferon-γ by natural killer (NK) cells. This cytokine in turn activates antigen presenting cells like macrophages. Activated macrophages generate tumor necrosis factor-α. The result of this adjuvant is switching of the humoral immune response to IgG2a in the mouse. Combination of lipid A or lipopolysaccharide with non-ionic block polymers or alum leads to a shift to IgG1 antibodies. The same effect can be obtained with rec. interferon-γ, which is much more expensive than lipopolysaccharide or lipid A and in addition cannot be applied to the sites where the local immune response is generated.

Cholera toxin (CTX) is investigated mostly for its mucosal immunity stimulating character. Cholera toxin was shown to enhance the IL-1 production by macrophages. However, it inhibits T cell IL-2 production but enhances proliferation. *in vitro* enhanced IgA synthesis in lipopolysaccharide stimulated B cells was shown (Lycke et al., 1989)

Preparations consisting of a mixture of components from a diversity of micro-organisms have been shown to severely influence the traffic of lymphocytes within the mucosal immune system, thereby enabling immunization on site to support responses also in other mucosal locations (Ruedl et al., 1993).

Saponins and ISCOM promote a strong immunogenic enhancement, most probably by direct interaction with cellular membranes. Apart from humoral responses, saponins also seem to support CD8 (class I) mediated cellular cytotoxicity (Newman et al., 1992; Campbell and Peerbaye, 1992).

Their effect as true adjuvants in activation of APCs is best demonstrated by the transfer of cells from animals pretreated with saponins only. CD8 effected cytolysis has been demonstrated and it might be dependent on interferon-γ production. Also the generation of IgG2a humoral responses in mice is indicative for this mode of action. How the induction of interferon-γ is established is not known. Certainly, the toxic and hemolytic activities of some saponins are not required for adjuvant activity. For the application via the oral route, the induction of interferon-γ is thought to be the most important. This is in agreement with the antibody isotypes observed (Finkelman et al., 1988a).

The group of synthetic adjuvants is rather heterogenous. The mode of action of non-ionic block polymers (NBPs) is not fully understood, but the activation of the complement system as well as the activation of macrophages seems to play an important role therein. NBPs usually have an amphipathic character. They stabilize water-in-oil emulsions and thereby may stabilize the tertiary structure of antigens as well. In contrast to most adjuvants studied, NBPs have a strong adjuvant effect, especially for poly- and oligosaccharides, whereas for most proteins they are not as potent as many other adjuvants. The mode of action may be similar to lipid A since the predominant antibody response is Ig2a in the mouse which is enhanced by interferon-γ production. The isotype distribution is influenced by the balance between the proportions of hydrophobic and hydrophillic moieties in the NBP composition. CTL responses have not been documented, while it has been demonstrated that DTH responses are supported (Zigterman et al., 1988; for review, see Verheul and Snippe 1992).

Liposomes *in vivo* act by facilitation of antigen uptake, ingestion, fragmentation, processing and presentation mainly by macrophages (for review, see Claassen, 1992). Liposomes are used together with other immune stimulating agents (MDP, saponins, etc.). Targeting of liposomes using antibodies anchored in the membrane is controversial (Buiting et al., 1992). Secretory humoral immunity (IgA) has been observed after oral application of pH stable liposomes (Jackson et al., 1990).

Humoral responses elicited with antigens presented in liposomes in general show profiles similar to responses elicited with water-in-oil emulsions. Like these emulsions, liposomes seem to support primarily CD4 dependent T cell responses (DTH, B cell help). For this class II associated immune responses, the lysosomal degradation of antigen is required. Recently, activation of CD8 cells in cytolysis (class I dependent) after application of antigen with pH sensitive liposomes has been documented. These liposomes introduce antigen in the cytoplasmic phase of antigen processing. However, questions with respect to the quantitative contribution of this route of processing based on liposomal delivery of antigen remains to be clarified.

Lipopeptides are non-toxic and non-immunogenic analogs of bacterial cell walls. They are amphipathic in nature which seems to be a favorable asset of adjuvants. *In vivo* as well as *in vitro*, their ability to generate CTL responses even to single CTL-T-epitopes has been demonstrated (Martinon et al., 1992; Nair et al., 1992). Conjugation as well as simultaneous and separate administration of adjuvant and antigen elicits immune responses. A drawback is that relatively high doses (tenfold) of antigen are needed in most cases, but then high antibody titers have been obtained. In a number of cases valuable assets of lipopeptides were demonstrated. Toxicity of toxins decreased without loosing immunogenicity, whereas non- or low immunogenic substances showed enhanced immunogenicity (Bessler and Jung, 1992). Lipopeptides are very easily applied and are commercially available (see Chapter 12).

The lipophilic character of DDA (forms liposome-like structures in aqueous solutions) might be responsible for its capacity to enhance T cell responses. However, most responses documented are DTH responses, which is a CD4 T cell function, whereas CTL responses have not been described in detail yet. Adjuvant and antigen need to be administrated together following the same route, which suggests that in general binding of antigen and adjuvant is required to generate optimal responses. Binding of antigen to DDA serves the purpose of increased hydrophobicity of the antigen and neutralization of negatively charged moieties. In contrast, DDA is not an optimal adjuvant to elicit antibody responses. For sufficient antibody responses, combination with, e.g., alum or water-in-oil emulsions is required. The mild side effects of DDA and some other quaternary amines make them promising adjuvants for veterinary application. For application in humans, new less irritative analogs need to be developed (for a review, see Hilgers and Snippe, 1992).

Cytokines are pleiotropic immunomodulatory agents which together form a partly degenerate system. The mode of action of various adjuvants has been unravelled to the level of immunomodulatory substances among which are cytokines. No specific patterns of cytokine expression as a result of the use of adjuvants have been reported. In addition, cytokines mutually influence each other function. Fragments of cytokines have been selected which only induce immune responsiveness. The application of the active component together with antigen has led to promising results for thymus dependent as well as for thymus independent antigens.

As has been summarized above, most adjuvants which support CMI responses, especially CTL responses, require an increase in the hydrophobicity of the antigen. It remains unclear what might be the mode of action of this function. The chemical nature suggests that binding or association with membranes of antigen presenting cells is facilitated. With interferon-γ the conjugation of antigen enhanced the immune response (Heath and Playfair, 1990).

VI. PEPTIDE IMMUNIZATION

Upon immunization of antigenic peptides which were all of sufficient length, ±20 amino acids, and which were used as free immunogens, according to some authors the success rate was unexpectedly high (Lerner et al., 1981; Attassi and Webster, 1983; Briand et al., 1985; Boersma et al., 1989a, 1992a). Routing, doses of immunogen and adjuvant applied were different in these experiments. Bixler et al. (1985) for lysozyme and Francis et al. (1987a,b) for the foot-and-mouth disease virus found that only a selection of all the peptides of a protein was able to generate an antibody response with the free peptide. Using a rather large number of peptides selected by computer prediction methods from various proteins, we found that only a minor proportion, ± 20%, is able to generate an anti-peptide response using the unconjugated peptide (Boersma et al., unpublished results). McCormick et al. (1988) showed that irrespective of the T cell epitope character of the peptide the response was dependent on the adjuvant, doses and timing as well as on the sex of the mice used. No general rules were derived. The generation of antibodies even with long peptides does not warrant the generation of anti-native protein antibodies as was shown for the insulin β-receptor. Four out of four peptides did not produce anti-native receptor antibodies (Pessino et al., 1989).

From our own experience, we derived the conclusion that short peptides, in general, tend to elicit antibodies which recognize the denatured form of the protein only (Boersma et al., 1988a). Application of short peptides, therefore, may be an advantage for the elicitation of antibodies to be used in assays where the antigen is encountered in denatured form, e.g., in sodium dodecyl sulphate poly-acrylamide gel-electrophoresis and fixed tissue specimen. In contrast, Dyson et al. (1985) showed that an immunogenic nonapeptide which showed a conformational preference in aqueous solutions could elicit anti-native antibodies to heamagglutinin of influenza virus. Ultrastructural analysis showed that longer peptides indeed tend to imitate the native structure of a protein much better since these peptides maintain a specific space filling structure (Horiuchi et al., 1986).

A short overview of various possible applications of peptides with various adjuvants is given in Table 2.

VII. POLYPEPTIDE IMMUNIZATION

For non-immunogenic peptides, (i.e., true B cell epitopes), coupling to a carrier protein to provide T cell help is essential. This is not a matter of molecular mass since immunogenicity of peptides in general is not increased by homo-polymerization with, e.g., glutaraldehyde (Boersma, unpublished results). Polymerization using carbodiimides led to encouraging results, though in part it was based on the coupling of T-determinants (Borras-Cuesta et al., 1988). Straightforward covalent elongation with a selected T cell determinant is a more effective option (Hackett et al., 1985).

When Broekhuizen et al. (1987) used a fusion protein of β-galactosidase elongated with an immunogenic sequence of foot-and-mouth disease Virus which was incorporated once, twice or four times, they clearly showed that the multi-copy determinants were more effective in the protection of infected animals than were fusion proteins which had only one determinant incorporated. The same fusion proteins were excellent for the production of specific monoclonal antibodies (Kottenhage, personal communication). Similar experiments of Van der Ploeg (1989) confirmed these results for herpes simplex virus 1 glycoprotein determinants. Single determinant antigens led to lower immune responses than did antigens which had two or four determinants incorporated.

In an experiment performed by Jacob et al. (1986) using cholera toxin delineated peptides (10–15 AA long) that were conjugated to tetanus toxoid or bovine serum albumin, it was found that in rabbits (1 mg conjugate in complete freunds adjuvant [CFA]) four out of six selected peptides elicited anti-peptide responses of which two showed a cross-reactivity with the native protein. A similar ratio was observed by Lord et al. (1989), producing anti-peptide antibodies cross-reactive with the human interferon-γ. Recently, a method was developed for the synthesis of branched peptides (Tam and Zavala, 1989). This multiple antigen peptide method was developed for sensitive detection of anti-peptide antibodies, but in addition the multimeric peptides show enhanced immunogenicity as compared to free peptides, peptide conjugates or peptides still attached to their solid support (McLean et al., 1991).

VIII. PEPTIDE CARRIER CONJUGATES

For peptides not including a T cell determinant sequence, the properties of a carrier protein determine to a large extent the outcome of immunizations. Size, structure, homology with mammalian proteins and relative frequency of certain amino acids are important points to consider.

Larger proteins are preferred as carriers because they contain sufficient reactive groups, -NH2, -SH, arginine, etc., for coupling to generate satisfactory peptide/protein ratios. In principle, most proteins will do if derived from a non-self source. However, the more genetically distant a protein is to the animal to be immunized, the better the chances for a potent immunogenicity. Keyhole limpet hemocyanin (KLH), a large aggregated protein from a gastropod, is therefore often used as a carrier protein in mice, rabbits,

TABLE 2
Immune response modulation using various adjuvants for peptide immunization

Adjuvant	Routes Applicable	Dose	Isotype	T cells Restriction Class I	T cells Restriction Class II	Species	References
Alum	ip	10 μg	Ig	NI	NI	Mouse	Hu and Kitagawa, 1990
		200 μg					
	ip, sc, id		IgG1	NO	Th1	Animals	Nicklas, 1992
	im					Human	
Specol	ip, sc, id	50 μg	IgG1	NO	Th1	Mouse	Boersma et al., 1992b
		200 μl			DTH*	Cattle*	
						Pigs*	
MLA	im	30–40 μg	IgG2a	NO	NI	Human	Alving et al., 1990
	sc, im, ip	40 μg			DTH	Mouse	Altman and Dixon, 1989
					Th1	Animals*	Gustafson and Rhodes, 1992
DDA	sc,	10–50 μg	Ig	NO	DTH	Mouse	Snijders et al., 1992
		10–1000 μg					
	sc		NI	NO	DTH*	Animals*	Hilgers and Snippe, 1992
						Human	
NBP	sc, ip	0.1–1 mg	all	NO	DTH	Mouse	Geerlings et al., 1989
		100 μg	except				Kalish et al., 1991
			D, E, A				
Saponin	sc, im,	10–50 μg	IgG2a	NI	CTL	Mouse	Campbell and Peerbaye, 1992
	id, ip	*50–100 μg*					
Liposomes	iv, id, ip	1–50 μg	M, G	NO	DTH	Animals	Buiting et al., 1992
		5 mg lipid	A		CTL		
				CTL		Mouse	Martin et al., 1993
Cytokines	iv, ip, sc.	5–20 μg	M, G	NO	Th	Animals	Tagliabue et al., 1992
(IL-1)		*0.1–50 μg*					
ISCOM	sc, id, ip	0.1–50 μg	G, A	NO	DTH	Animals*	Lövgren et al., 1987
	in, im			CTL	CTL*		Claassen and Osterhaus, 1992
Syntex AF	sc, ip	1–50 μg	IgG2a	NO	Th1	Animals	Allison and Byars, 1992
					DTH		
					CTL		
Immuno-	sc	10–100 μg	G1, G2a		NI	Animals	Skea and Barber, 1992
targeting			G2b, M				
MHC class II							

* Protection to challenge with pathogens. NI: not investigated; NO: not observed. Amounts of antigen in normal print, amount of adjuvant in *italics*.

goats and sheep. Other widely used carriers are from bacterial origin: tetanus toxoid, purified protein derivative of tuberculin and diphtheria toxoid. Less efficient in mammals are ovalbumin and bovine serum albumin (BSA) (Geerligs et al., 1989).

It is important to keep in mind that anti-carrier protein antibodies especially in polyclonal reagents may cross-react with protein determinants in the test systems applied for evaluation of immune responsiveness (Boersma et al., 1992a). In addition, a carrier protein which is very immunogenic may contain immunodominant epitopes such that the response to attached peptides is negligible (Neurath and Kent, 1985). In practice, however, this problem only very seldom is encountered or recognized. It turns out that the immunogenicity of the carrier is changed by coupling of new determinants. Apparently, a new ranking of immunogenicity of B cell determinants is reached which results in antibodies to the newly introduced (peptide) determinants.

The size of a carrier is important since a larger protein can accommodate more synthetic peptides. However, coupling of relatively large numbers of synthetic peptides, depending on the protein, can have a negative influence on the properties of a conjugate. Overloading may lead to precipitation and decreased immunogenicity (Peeters et al., 1989). Also cross-linking agents like diazo compounds and glutaraldehyde may drastically decrease the solubility and immunogenicity of a carrier-peptide conjugate.

Large aggregated proteins like KLH (>600 kDa) and toxoids tend to have an adjuvant-like activity. Empty "ISCOM-cages" are not suitable to be applied directly as a peptide carrier. ISCOM with immunogenic proteins to which peptides are easily coupled have a relatively large carrier potential and in addition show a marked adjuvant effect (Claassen and Osterhaus, 1992). Similarly, peptides may be coupled to bacterial surfaces like those of *Lactobacillus* for immunization along various routes (Boersma et al., 1994).

Recently, the use of virus particles (hepatitis B surface antigen) as a carrier for peptides with a hydrophobic tail that dissolves in the capsid has been proposed as an alternative for both carrier and adjuvant (Neurath et al., 1989). Empty capsides for rotavirus and hepatitis B virus have been reported as peptides carriers.

Using SP, the antigen usually is an assembly which consists of a carrier protein, a peptide and bridging coupling reagents. We are of the opinion that the actual role of a determinant can best be judged from the results of immunizations, i.e., the anti-peptide antibody responses in sera, the elicitation of delayed type hypersensitivity (DTH) and cytotoxic T cell lysis (CTL) responses.

IX. THE IMMUNOGEN VEHICLE: ANTIGEN LOCALIZATION

The non-specific vehicle to which the antigen (peptide or peptide conjugate) is coupled to, or is dissolved in, determines the localization and often the dose of an antigen that can be applied. Sepharose beads, pieces of nitrocellulose sheaths, aluminium hydroxide gels and oil emulsions can be used for the purpose of delivery in the peritoneum, under the skin or even in the spleen. Some of these vehicles have both adjuvant and carrier properties, alum, oil emulsions, and viral capsules, respectively, while others, such as sepharose beads and nitrocellulose are relatively inert and have a vehicle function only.

Comparison of effectiveness of a peptide either conjugated to an immunogenic protein carrier or an inert matrix (beads, resins) showed that in most cases the coupling of the peptide to a protein carrier is most efficient (Fischer et al., 1989; Boersma et al., 1992a).

Mineral salts, emulsions, liposomes and ISCOM, all by their semi-particulate nature, have an adequate vehicle function to transport sufficient amounts of antigen to the desired location. There they take care of depot function with a specific release rate. Inert carriers allow the use of low doses of antigen, but require a repetition of immunization. The major function of the inert carriers is the vehicle aspect. The antigen requires an integral T cell epitope.

Biodegradable polymers in general (polylactic acid based) are used as antigen vehicles, which in addition have as a major mode of action that they allow local presentation of antigen on a large surface area with increased hydrophobicity. When applied to elicit responses at the mucosal immune system interface, the size of the particles (1–10 μm) determined the effectiveness of this mode of antigen presentation. In principle, biphasic release of antigen can be obtained which leads to priming and booster with a single administration (Eldridge et al., 1990, 1991; Jani et al., 1989, 1990). Cellular immune responses are not very well documented. However, the chemical nature of the polymers suggests that addition of a specific adjuvant is necessary to promote T cell immunity. For this, admixture of DDA might be worthwhile trying (Hilgers and Snippe, 1992).

The location where the antigen is deposited, in part, determines the lymphoid organs activated and the isotype of the antibody response. Routine immunizations, intraperitoneally and subcutaneously, which deposit antigen mainly in the spleen and peripheral lymph nodes result in IgG responses. The routing is limited by the form of the antigen. Insoluble antigens cannot be immunized intravenously. In order to waste little of the antigen in the periphery, the antigen can be directly applied locally in the spleen either free or attached to a vehicle (nitrocellulose, sepharose beads). Doses 50–100x lower than normal could be used on intrasplenic immunization (Spitz, 1986; Hong et al., 1989). However, intrasplenic deposition of a short (10 AA) peptide conjugated to tetanus toxoid administered as free conjugate, or as the conjugate precipitated to alum or as the conjugate bound to sepharose beads; all led to similar anti-peptide and anti-conjugate antibody responses in serum. The doses immunized were the same for all conjugates based on the carrier weight. Peptides directly bound to sepharose did not elicit antibody responses. In these experiments, the intrasplenic deposition did not show major advantages compared to intraperitoneal administration (Boersma et al., 1992a). Application of antigen for nitrocellulose has been described by Forrest and Ross (1993).

A peptide with a sequence homology with a genetic variant of apolipoprotein E was applied to generate specific antibodies. Upon intraperitoneal immunization with a peptide-carrier conjugate in a water-in-oil emulsion, in this case, the anti-peptide responses were extremely low. Subcutaneous and intracutaneous immunization of the same peptide conjugates finally led to low but significant responses. No cross-reactivity with native protein was observed. After intracutaneous immunization (inguinal, axial) or subcutaneously in the foot pad, only weak responses were determined. The cross reactivity with the native protein apolipoprotein E was low but significant (Gerritse et al., 1991).

Generally, for serum immune responses the immunization via the subcutaneous route (administration in the upper part of the hind foot or in the neck) is very effective and can be performed in a simple manner. Immunization subcutaneously on the upper side of the foot, however, does not allow large quantities of adjuvant to be administered (50 μl). The efficiency is most likely based on the concentration of antigen in the local depot and the gradual breakdown and release into the immune system.

X. TARGETING

Targeting of antigens to certain molecular structures on cells is a form of specific routing of an antigen. For peptides, enhanced responsiveness has been shown to occur by targeting of these peptides to surface immunoglobulin or class II molecules (Casten, 1988; Skea and Barber, 1992, 1993). Immunological targeting to, e.g., MHC molecules acts as a replacement of adjuvant in that it targets the antigen and at the same time acts as an activation for the antigen presenting cells. To this end, MHC-specific antibodies were applied (Carayanniotis and Barber, 1988, 1990). Memory induction has been demonstrated with anti-MHC class II-antigen conjugates.

Using β2-micro-globulin as an adjuvant peptide, priming for class I restricted cytolysis has been demonstrated with peptides related to protein antigens (ovalbumin) as well as with viral determinants (Sendai, vesicular stomatitis virus; Rock et al., 1993).

Wyss-Coray et al. (1992) showed that anti-CD4 and anti-CD2, but not anti-CD8 antibodies were able to enhance anti-peptide responses by targeting to T effector and antigen presenting cells, respectively.

Targeting to constitutive structures on dendritic cells like CD45 was effective (Nussenzweig et al., 1982), but targeting to differentiation antigens was not (Springer et al., 1978). For BSA, Lees et al. (1990) showed that especially targeting with anti-IgD led to strong potentiation of antibody responses. Analysis of these responses suggest that essentially polyclonal activation is the basis for the response enhancement.

Targeting of liposomes to specific receptors, e.g., the ganglioside GM1 receptor, mimic a situation such as that found in the localization of AB5 type toxins like shigella toxin, cholera toxin, etc. Targeting can be established by the uptake of GM1 in the liposomal membrane (Liu et al., 1991, 1992).

Targeting to specific cellular structures as a mode of immune response enhancement in vaccines or other clinical applications will be implemented in an outbred situation only when antigens are used for targeting to antigens which share the determinant in all individuals treated. Therefore, MHC targeting will be of use in a limited number of cases only.

XI. HUMORAL RESPONSE ISOTYPE

Antibodies are dual function molecules. Fab is specific for the antigen, while the Fc portion mediates the effector function of the molecule. Therefore, isotypes elicited in humoral immune responses are of great importance. Specific T cell subsets control the isotype switching by production of response modulators. Therefore, isotypes have different roles in protective immune responses. Binding to Fc receptors may facilitate the opsonization of micro-organisms and mediates antibody dependent cellular cytolysis. Complement binding to the Fc part is important for activation of the complement system.

Frequently, with specol in Balb/c mice IgG1 responses were observed. In the mouse IL-4 stimulates the generation of IgG1 and IgE, while interferon-γ is involved in the IgG2a responses (Finkelman et al., 1988a,b).

Liposome associated proteins elicited IgG2a and IgG2b antibodies in the mouse (Phillips and Emili,

1992). Whereas MDP stimulated preferentially IgG1 in their experiments with a combination of liposomes and MDP, a shift to IgG2 was observed.

With SAF, the combination adjuvant which includes a muramyl dipeptide analog, a non-ionic block polymer in a water-in-oil emulsion (Byars et al., 1990) a skewing of the response to IgG2a antibodies was demonstrated in mice which was attributed to the induction of interferon-γ production (Th1 responses) (Kennedy et al., 1989; Byars et al., 1990).

In mice, saponins stimulate production of IgG1, IgG2a and IgG2b. The major part of the response is formed by IgG2a antibodies. IgE was not observed (Kensil et al., 1991). Quil-A, a purified fraction of saponin, elicited mainly IgG1 in mice (Kennedy et al., 1989), whereas IgE responses have been reported also (Allison and Byars, 1992). For those responses where elicitation of IgE has to be evaded (vaccination) the choice of adjuvants should be carefully made. For aluminium phosphate, it has been documented that no detectable IgE production was stimulated.

Using MHC class II, targeting enhanced responsiveness in humoral immunity was mainly IgG1 restricted (70%), whereas about 25% was of the IgG2a isotype in mouse (Skea and Barber, 1993).

The function of adjuvants and vehicles in vaccines and in experimental reagent production is therefore the induction of cytokines which modulate the response to an antigen in such a way that it is most protective.

XII. IMMUNIZATION AND TOLERIZATION

Immunization should preferentially lead to immune responsiveness to the antigen administered. In auto-immune diseases, the response is directed to antigens to which in normal situation tolerance exists. To investigate the development of pathogenesis, immune response modulation, etc., various disease models have been developed. When these diseases have to be induced into normal experimental animals, the protective barrier of immune tolerance has to be overcome.

Many (auto-immune) disease inducing protocols are based on the administration of a specific peptide T cell determinant, and a strong adjuvant protocol is necessary to initiate the inflammatory or immune-silencing response (Aichele et al., 1994). Models for induction of adjuvant arthritis, experimental allergic encephalomyelitis, auto-immune uveitis, auto-immune diabetes, etc. all are based on the application of FCA in a protocol inducing severe harm to the experimental animal. Also, for the induction of cytotoxic T cells, often FCA or adjuvant with a similar effect are applied (Claassen and Boersma, 1992b). The skewing of immune responsive T cell balance (Th1/Th2) probably is based on the specific immuno regulating factors of which the production is induced in various cell types by the different components present in the adjuvant.

XIII. NEW DEVELOPMENTS AND TRENDS

Modern approaches also in immunology require the use of well defined components in the systems used for application in a pharmaceutical setting. From investigations of the mode of action of components of less well defined adjuvants like FCA, it has become apparent that the rational design of adjuvants built from well defined components might lead to optimal results.

Three major types of approaches can be distinguished which in all cases the aim is the support of immune responses of the cellular and of the humoral immune system:

- Composite adjuvants. Clearly defined components based on classical adjuvants
- Empty viral envelopes. Provides vehicle, carrier and localization aspects
- Liposomes, virosomes, ISCOM

One of the commercial preparations based on this assumption is SAF. Its components support vehicle function (squalene microsphere emulsion), induction most likely of cytokine production (MDP) and the increase of hydrophobicity (NBP) of the antigen involved. T cell functions (including CD8 dependent cytotoxicity) are very well supported, which is not the case with most other adjuvants. SAF, most importantly, probably does not have the negative side effects of adjuvants like CFA, of which it has all positive activities.

ISCOM, like SAF, combine various functions of adjuvants. They provide a vehicle not only for hydrophillic but also for hydrophobic antigens. They support CD4 and CD8 dependent cellular immune reactions and local application, e.g., use as an oral vaccine can easily be envisaged. The particle size is very much like the size of emulsion droplets and micro-organisms which can be used as vaccine carriers. When toxicity of the components can be brought under control without loss of adjuvant activity, ISCOM probably are the most promising vaccine carriers with adjuvant activity. For experimental purposes, e.g., the elicitation of antibody responses, ISCOMs seem to be less appropriate in that they require a relatively high amount of antigen for production of the particles. For large-scale production, however, this is less of a problem. Nevertheless, ISCOM may be expected to be rather expensive as vaccine carriers when compared to live bacterial vaccines.

XIV. APPLICATIONS OF ADJUVANTS WITH PEPTIDES

Most applications of peptides are found in experimental immunization for antibody production or testing the efficacy of peptide vaccines. Peptides are thought to be relatively low immunogenicly. That is why in most experimental protocols the most potent of adjuvantia, complete freunds adjuvant (CFA), is used. Conjugates are emulsified in CFA and immunized in experimental animals like mice and rabbits. For the production of polyclonal antibodies, however, this has the drawback of contamination with antibodies to bacterial antigens and DNA. Therefore, adjuvantia without bacterial antigens were developed which also provide milder treatment for the laboratory animal (Bokhout et al., 1981; RIBI adjuvant system, Ribi et al., 1987). SP might be applied as (subunit)-vaccines for application in the veterinary situation and in humans. Under those conditions, the use of adjuvantia is restricted since most of them exhibit adverse side effects. A large number of substances (lipid derivatives, liposomes: Tamauchi et al., 1983; non-ionic block polymers: Zigterman et al., 1988; Hunter et al., 1981; virosomes, micelles and the like: Morein and Simons 1985) have been promoted as promising adjuvant for the use in vaccine formulations and certainly are worthwhile of being used in the generation of diagnostic reagents (Warren et al., 1986; Geerligs et al., 1989).

In most adjuvant formulations, certainly water-in-oil or double emulsions, antigen presentation is influenced by the hydrophobic character of various components. For FMDV vaccine, the role of various adjuvant formulations leads to different outcomes of vaccine trials (Barteling and Vreeswijk, 1991). Certainly for peptides with hydrophobic and hydrophillic regions, antigen presentation may be dependent on the vehicle and adjuvant formulation chosen (Fieser et al., 1987). For generation of diagnostic antibodies, we favor the use of gentle adjuvant, i.e., water-in-oil emulsions, without addition of micro-organisms.

Mineral salts have the reputation of being only mildly supportive for responses to weak immunogens. The precipitation of peptide conjugates dependent on the carrier protein applied on aluminum hydroxide gels leads to good anti-peptide responses (Geerligs et al., 1989; Boersma et al., 1988a, 1989b,c). Comparison of results of immunization of conjugates of the same peptide using six different adjuvant protocols showed insignificant differences only when tested for anti-peptide reactivity or reactivity to the coupling reagent dependent neo-determinant. Though it has been claimed that alum precipitated antigens may sub-optimally stimulate cell mediated immunity, they do elicit delayed type hypersensitivity reactions. Till now, alum hydroxide gels are the only substance generally approved for use in humans and primates (Bomford, 1980; Allison and Gregoriadis, 1990). Only a few communications were found with aspects referring to application or optimization of aluminum hydroxide gels in anti-peptide immune responses (Nicklas, 1992). For viomycin coupled to BSA (10 μg antigen with 200 μg of the alum gel) (Hu and Kitagawa, 1990), significant specific immune responses in the mouse were reported. Side effects observed with aluminium gels were similar to those seen with immunization of proteins or intact pathogens without adjuvants (Nicklas, 1992).

We have used specol for numerous experiments in mice, and for the purpose of eliciting anti-peptide antibodies that were cross-reactive with the native proteins from which these peptides were derived. The antigen/adjuvant emulsion was in general administered intraperitoneally. The ratio of cross-reactive antibodies over total anti-peptide antibodies was such as to allow the selection of particular antibodies (Boersma et al., 1988b; Deen et al., 1990; Zegers et al., 1990). In almost all the cases (over 150), we succeeded in eliciting anti-peptide antibody responses, and high levels of IgG in secondary responses were obtained. With use of specol in Balb/c mice, we mainly evoked IgG1 responses similar to most water-in-oil emulsion's (Allison and Byars, 1991). However, in Balb/c mice also with alum as an adjuvant, in our hands preferentially IgG1 responses were elicited (Haaijman et al., 1988). Following this approach, highly specific and often unique antibodies intended for clinical diagnostic applications were developed for human immunoglobulin isotypes IgG3 (Boersma et al.,1988a, 1989b) and IgG2 (Boersma et al., 1989a, 1991) to the human androgen hormone receptor (Van Laar et al., 1989; Zegers et al., 1991a), to human fatty acid binding proteins (Kleine et al., 1990a), to human apolipoprotein E2 (Gerritse et al., 1992), to human α_1-antitrypsin and a genetic variant of α_1-antitrypsin (Zegers et al., 1991b), and to human HIV-1 variant proteins (Laman et al. 1990, 1992). In most cases the peptides were not selected for optimum immunogenicity, but predominantly to include differences between the target protein from which the peptide was derived and other strongly homologous proteins. Differences as slight as one amino acid substitution could be discriminated by these antibodies. Peptides derived from α_1-antitrypsin, APO-E2 and the human androgen receptor are to a high degree homologous to mouse (i.e., self-) proteins. Most responses were generated using peptide-carrier conjugates, but on various occasions we encountered peptides which contained a T cell and a

B cell epitope and which were able to elicit antibody response with specol as the adjuvant even when non-conjugated.

Subcutaneous injection of peptides with a T and B cell epitope in specol in the footpad allowed *in vivo* pre-activation of lymph node cells to demonstrate the presence of T cell epitopes in peptides. We could demonstrate the ability of peptides to stimulate isolated T cells to IL-2 production and proliferation. In addition, a DTH reaction could be elicited by subcutaneous administration of the peptide antigens in the footpad (Boersma et al., 1991). The nature of an antigen to a large extent may determine the effect of an antigen. In a typical experiment using a model peptide of 11 amino acids coupled to a KLH as the carrier protein, we did not observe significant differences in secondary humoral reactions measured as anti-peptide antibody titers in serum, when alum, alum and *Bordetella pertussis*, administration in specol, pretreatment with specol (day -2) or the regular FCA/IFA combination was used as an adjuvant. This might be explained by the nature of KLH which itself has irritative properties (Boersma et al., 1992a,b).

The role of bacterial cell wall components has been reviewed recently by Gustafson and Rhodes (1992). Tsujimoto et al. (1986) demonstrated enhancement of cellular (class II) and humoral responses with MDP derivatives. Lipopolysaccharides seem to play an important role in activation of NK cells. They suggested that the production of interferon-γ might play an important role in the enhancement seen with a.o. monophosphoryl lipid A (MLA), a derivative of lipopolysaccharide (Odean et al., 1990). Also for liposomes including MLA the induction of long-lasting IgG anti-peptide responses efficiently could be elicited (Friede et al., 1993). The induction of interferon-γ in addition explains the production of IgG2a in responses supported by bacterial cell wall products. Richards et al., (1989) on intramuscular administration of alum absorbed R32tet$_{32}$, an important malaria determinant, elicited significant immune responses in monkeys (30 μg antigen, 32 μg MLA). Alving et al. (1990) found similar results in a combination with this determinant taken up in liposomes and MLA in the liposomal membrane for administration in humans (5 μg antigen, 50 μg MLA). No serious side effects of MLA have been reported. Pertussis toxin subunits were administrated to mice in a water-in-oil emulsion containing, apart from MLA, trehalose dimycolate also. Alum precipitated p72 HBV peptides with tetanus toxoid as a carrier upon subcutaneous as well as intramuscular administration (40 μg antigen with 100 μg MLA) showed specific immune responses in mice without side effects (Altman and Dixon, 1989).

A peptide covalently linked to a multi-meric lipid backbone resulted in CD8 mediated cellular immune responses in mice (Nardelli and Tam, 1993). A similar construct was shown to be a macrophage activation composite structure supportive for foot-and-mouth disease specific peptide responses (Wiesmuller et al., 1989).

DDA, a lipophilic quarternairy amine, is one of the most active compounds of this kind in the stimulation of immune responses in animal models. It is readily dispersible in water and therefore can easily be applied for immunization purposes following various routes. In various hapten-carrier systems, as well as with carrier-peptide systems, an efficient support of humoral immune responses and specific DTH reactions has been demonstrated (Hilgers and Snippe, 1992). From experiments with hapten carrier systems by Snippe et al. (1980, 1982), it could be envisaged that DDA could be a suitable adjuvant also for peptides. This was confirmed mainly with peptides used for antibody responses to peptide hormones to elicit detection reagents as well as in models for immunological manipulation of fertility (Silversides et al., 1984; Goubau, 1989). Snijders et al. (1992) showed that peptides coupled to β-galactosidase specific DTH responses were elicited in mice. Adjuvant activity was shown in protection experiments with various intact pathogens (Hilgers and Snippe, 1992).

SAF is an adjuvant formulation (Syntex, Palo Alto, CA) which combines the adjuvant effects of MDP, a synthetic homolog of mycobacterial membrane components and of non-ionic block polymer surfactants. Squalene replaces the oil phase of classical WIO to produce water-in-oil emulsions. The emulsion is rather stabile and has a relatively low viscosity which enables administration through small bore needles. SAF supports DTH and humoral immune responses. Class I mediated responses have not been documented, but protection studies with a number of vaccines have been conducted with promising results (Byars et al., 1990). However, the use of influenza hemagglutinin (HA) as a demonstration project is not too convincing in that most influenza hemagglutinins are sufficiently immunogenic to elicit protective antibodies without adjuvants. Nevertheless, it supports the notion that SAF is a very effective adjuvant with a broad application range (Allison and Byars, 1992). One of the advantages of SAF is that it combines efficient adjuvanticity with low levels of side effects.

ISCOM in general have been applied with large proteins incorporated in the matrix. Using these preformed ISCOM as carriers successful immunizations were performed with epitopes of foot-and-mouth disease virus and with a small peptide hormone

(Lövgren et al., 1987). The high potential of stimulation is demonstrated by the low doses (1–3 μg) which need to be applied.

Inert carriers as well as biodegradable micro-spheres have been applied for immobilization of soluble antigens and the delivery of peptides only (Nilsson and Larsson, 1992; Eldridge et al., 1992). Side effects are minimal and increased effectiveness in antigen uptake plays a role in immune response augmentation. Biodegradable micro-spheres of specific diameters have access to the mucosal immune system (Peyers patches), and incorporated antigens may elicit IgA responses (Eldridge et al., 1992; O'Hagan et al., 1989, 1991). In general these vehicles have been applied to generate humoral immune responses. Dependent on the antigen construct, cellular responses may be expected. One of the specific assets is that biodegradable micro-spheres deliver the antigen with time, which means that a continuous or even a discontinuous repetitive delivery of antigen (prime and boost in one delivery) is feasible (O'Hagan et al., 1991).

Cytokines as immunological adjuvants were recently reviewed by Heath and Playfair (1992). They concluded from the presently available data that the adjuvant effect of cytokines may sometimes not be powerful, such as the most potent experimental/research adjuvants, but they rival adjuvants presently allowed for use in humans. For IL-1, subunits have been defined with different immunomodulating capacity. Comparison to alum or MDP showed equal effects with immunogenic proteins (Tagliabue et al., 1992). Since cytokines are potent only when present in relatively high local concentrations, the routing of cytokines as immunomodulators needs further improvement before routine use as adjuvants can be envisaged. As long as these have not been identified, the induction of local production of cytokines using the adjuvants discussed above may give a better effect (Odean et al., 1990).

XV. ROUTE OF ADMINISTRATION FOR ADJUVANTS

For most routes applied, e.g., the skin (s.c./i.c.), muscle (i.m.), and blood (i.v./i.s.), the adjuvants described above will function relatively well and their mode of action is at least partly understood. Their function is a combination of depot, vehicle, carrier and immunomodulator. Enhanced antigen presentation immunomodulation by truncated cytokines allows a different approach. For most of these adjuvants, due to their toxicity, their applicability for oral administration has not been investigated in depth.

Since most pathogens have their porte d'entree through the mucosa, the investigation of protective mucosal immune responses is of utmost importance. However, relatively little attention has been paid to the elicitation of IgA responses, especially with peptides.

For the antigen presentation of peptide antigens along the mucosal surfaces, however, localization as well as presentation is less well investigated. It is still dependent on the application of inert particles or sheets, e.g., polylactic acid (Eldridge et al., 1990, 1991, 1992; Jani et al., 1989, 1990), nitrocellulose, nylon membranes, ion-exchanger membranes, etc. of specific size (localization) (Nilsson and Larsson, 1992); the application of mucosal adherence of toxins (e.g., cholera toxin, heat-labile toxin of *E. coli* (see a.o. Vajdy and Lycke, 1992; Jackson et al., 1993; Arnon, 1990) or their subunits that are required for mucosal interaction (Holmgren et al., 1993) and the use of envelopes of viruses that harbor the mucosa (e.g., polio virus); or the application of semi-virulent pathogenic carriers (*Salmonella*).

Only a few non-pathogens can serve as adjuvant at the mucosal surfaces. Live *Lactobacillus* of various strains (casei, plantarum a.o.) can serve as adjuvants after intraperitoneal and also after oral administration (Boersma et al., 1994). Their efficacy using peptide antigens is relatively low. Peptides need to be coupled or produced at the surface by *Lactobacillus* to generate acceptable levels of immune responsiveness (Boersma et al., 1994).

When HBV envelopes are used as carriers for oral immunization in humans and in animals, safe enhancers like taurine, parotin or relatively large quantities of lithium can be applied to enhance the immune response (Kuriyama et al., 1988). These oral adjuvants, except for lithium, seem to act via IL-1 and mitogenic activation of the humoral immune system. Quillaja saponin is a mixture of potent adjuvants. Crude saponin extract are unreliable. Quil-A, one of the purified derivatives, is used as the basis for the formation of ISCOM (Dalsgaard, 1978; Morein, 1990). Intraperitoneal application of low doses of quillaja saponin (400 μg) are lethal in mice, but the application of 20 mg along the oral route did not cause visible harmful effects. This was successfully utilized in the application of oral rabies vaccines (Campbell and Peerbaye, 1992). The effect of oral application and vaccine administration may be spaced in time without loss of effectiveness.

ISCOMs are generally applicated intramuscularly or subcutaneously. In contrast to many other adjuvants the administration of ISCOM related antigen along these routes is most effective when the interval between two administrations is 6 to 8 weeks. Short intervals give low antibody titers (Lövgren et al., 1990).

Oral immunization of animals with ISCOM only in part has been successful. Relatively high doses of antigen (100 μg) were needed, but then also class I restricted responses were demonstrated (antigen ovalbumin). But the support of mucosal immune responses has been documented upon intranasal application of the antigens/vaccines (Jones et al., 1988). Intravaginal responses have been demonstrated as well, but these required relatively high amounts of antigen (Thapar et al., 1991).

XVI. EFFICACY VERSUS SAFETY OF ADJUVANTS

In a scale of arbitrary units of injury from I (low) to IV (severe), the routes of application of adjuvant/antigen combinations score relatively low by application intravenously (when possible), intraperitoneally and subcutanously; intermediate (II) with intradermal and intramuscular application; whereas intrasplenal and in the foot is in the III category for most adjuvants. Freunds complete adjuvant, liposomes and ISCOM by various routes (i.e., intramuscularly) cause a high degree of injuries and therefore are either better not used (FCA) or applicated via routes that cause only a low degree of harm.

The discussion was in a recent Forum in Immunology (44th, Res Immunol 143, 1992). Snippe et al. (1982) give a useful overview of four categories of adjuvants. Two categories can be distinguished of which category I includes adjuvant applications for which the efficacy aspect is more important than the safety/toxicology aspect and category II includes applications where safety is more important than efficacy.

In the first group, experimental and food animals are distinguished. All adjuvants seem to be applicated in the former category and only four — water-in-oil and oil-in-water emulsions as well as aluminium salts and liposomes — in the latter category. For food animals only, FCA is considered not allowed, whereas some doubts are expressed with respect to application of most other adjuvants. In this category, application may be considered based on the presently available safety records of the adjuvants. The negative effects of FCA are discussed further by Claassen et al. (1992). They suggest that presently sufficient alternatives are available to eliminate FCA from animal experimentation in most fields of biomedical research.

In the second group, companion animals and humans are distinguished. For humans only aluminium salts are allowed. Most other adjuvants are not considered for application except for specific purposes. Water-in-oil and oil-in-water as well as FCA is considered not to be applicable in humans at all. For companion animals the application of ISCOM is thought to be safe enough for application. Water-in-oil emulsions as well as FCA are considered not to be applicated in companion animals. All other adjuvants are worthwhile to consider for application based on the presently available safety records.

From this overview it can be concluded that especially for application in humans and food and pet animals the development of safe adjuvant formulations is of great importance. The efficacy of adjuvants seems in most cases to be the counter balance of safety. For many vaccines needed, an appropriate level of adjuvanticity is required to reach a sufficient level of vaccine efficacy. To obtain hold of formulations which combine the two in a critical equilibrium is a challenge to modern scientists.

XVII. COMMERCIAL ADJUVANTS COMPARED

Various, mainly complex, combinations of adjuvants have been made commercially available for veterinary as well as for research purposes. A selection has been evaluated by various authors. Though we will not go into details, it seems worthwhile to summarize the results as far as peptides are concerned. Most adjuvants were compared to Freunds with respect to immunostimulation and adverse side effects. All adjuvants were designed to decrease side effects, aiming at application for veterinary or human use.

Titermax is a low toxicity adjuvant which contains NBPs as surfactants in a double emulsion (Hunter et al., 1989). Titermax®, Adjuvax®, Ribi-system® and Alhydrogel® or Lipovant® were compared to Freunds complete adjuvant by Bennett et al. (1992) using haptens and peptides. They demonstrated that Titermax, a combination of an inert micro-particulate silica in an emulsion of water, squalene (metabolizable oil) and a copolymer, led to relatively long-lasting high titers. The magnitude of humoral immune responses varied with the experimental animal and route of administration applied (goat, rabbit, rat, mouse). Titermax led to highest titers when anti-peptide antibody titers were compared. In most experiments the Ribi adjuvant was second best.

Comparison of Ribi adjuvant with Freunds complete adjuvant in support of immunization of rabbits with a synthetic peptide antigen showed that the titers obtained with FCA were longer lasting and of higher avidity than those observed with Ribi. Adverse effects granuloma formation and increased creatinine levels were observed especially in animals immunized using FCA but not with Ribi (Deeb et al., 1992). Lipman et al. (1992) evaluated the response to haptens which is rather similar to peptide epitopes. Coupled to BCG as a carrier the response elicited was higher with Ribi's adjuvant than observed with Freund with as a bonus a lower level of adverse side effects.

Recently Leenaars et al. (1994) presented an investigation of various adjuvants with respect to their potential to support mainly humoral immune responses and the level of undesirable side effects cause by these reagents. FCA was compared to specol (Boersma et al., 1992b), Ribi adjuvant and Titermax. Specol is essentially a water-in-oil adjuvant (based on mineral oil and a surfactant/emulsifier). From this study in which anti-peptide responses also were evaluated, it may be concluded that for efficient activation of the humoral immune response in rabbits the use of FCA was concluded to be the adjuvant with the highest record of adverse effects. In addition, apparently the doses of adjuvant which are generally applied can be reduced without losing the stimulatory effect.This reduces the level of adverse effects with all adjuvants.

Specol in primary and secondary subcutaneous or intramuscular immunization experiments gave equal or second best results when compared to FCA, whereas Titermax and Ribi scored relatively low. It was shown that with 1/8 (0.125 ml) of the regular dose of FCA (1.0 ml) the same anti-peptide titers were still obtained at a reduced level of adverse effects. All adjuvants showed undesirable side effects. Specol, an adjuvant with the lowest complexity, led in most immunizations to an acceptable level of adverse effects. For Titermax and specol no side effects were observed after intradermal application. Prediction of side effect as a result of a combination of adjuvant with a specific antigen was not possible. When Rabies virus ISCOMs were applied as adjuvant/carrier, similar effectiveness was demonstrated as that observed with FCA or specol in support of humoral immune responses (intraperitoneal and subcutaneous administration) (Leenaars et al., unpublished results).

XVIII. CONCLUSIONS

We support the conclusion of Perraut et al. (1993) who investigated combinations of adjuvants and antigens, including a peptide antigen related to herpes simplex virus. The quality of the immune response to selected antigens was related to the adjuvant formulation applied. In vaccine and immunogenicity studies only investigation of the antigen with various adjuvants may lead to the selection of the optimal combination. Nevertheless, we are of the opinion that a sufficient number of studies has been published which prove that the use of Freunds complete adjuvant can be avoided in most cases excluding a few disease induction models and more generally the induction of the response to auto-antigens which in fact is an aberrant phenomenon as well.

For the immunization with peptides, water-in-oil adjuvant like specol seems very effective in humoral responsiveness and DTH responses at low levels of adverse effects.

ISCOM are very effective in support of humoral and cellular immune response, including cytotoxic responses, but it must be taken into account that in most studies so far ISCOM has been applied as a carrier in which proteins or viruses with a high intrinsic immunogenicity have been incorporated. Proper controls with those proteins or viruses as a peptide-carrier were not included in most studies. Oral application of ISCOM has not been highly effective, though the risk of toxicity seems to be limited along this route.

DDA is a very elegant adjuvant for analysis of T cell immune responses, but it is defective in that it does not seem to support humoral immune responses equally well.

Most commercial cocktail adjuvants seem to vary broadly with respect to their adjuvant effect. Presently, SAF, which theoretically is the most sophisticated adjuvant in that it is a combination of effective tricks, is the adjuvant which supports the whole scala of immune responses at low adverse effect levels. SAF, however, is a relatively expensive adjuvant.

The application of an efficient adjuvant in combination with effective immunotargeting though productive is rather complicated. The orientation of the antigen with the targeting device is of utmost importance and can be designed by trial and error only.

Most pathogens have their route of entry via the mucosal surfaces. Therefore, new vaccines aim at enhancement of mucosal immunity. Clear deficiencies in modalities for presentation of peptides and proper adjuvant support is found in stimulation of mucosal immunity. New developments are needed to replace immune response enhancement based on carriers with the risk of pathogenic side effects such as toxins or toxoids or other products which clearly have similar disadvantages. Especially for mucosal administration with live (attenuated) pathogenic microorganisms via the respiratory tract (nose, etc.), the efficacy of antigen delivery should be enhanced to generate response via the nose associated lymphoid tissue (NALT) with a limited mucosal surface or the lungs to overcome degradation of the antigen by alveolar macrophages.

For the research scientist, sufficient means are now at hand to perform immunization via various routes without causing too much harm or side effects. For practical application on a large scale in veterinary and human use, choices still have to be made. Peptide vaccines or immunizations in humans will require strong immunostimulation without side effects at all. New ISCOM with decreased toxicity, empty viral envelopes (HBV) or mixed type combination adjuvants like SAF seem most promising, but we await their approval.

XIX. PROCEDURES

XIX.1. Selection of peptide sequences for synthesis

For practical purposes, when there are no restrictions posed by homology, etc., then the antigenicity prediction as described by a computer program is helpful. We used the program of the Genetics Computer Group, Wisconsin (Devereux et al., 1984). Terminal segments may be very useful since they loop out of the globular structure. Peptides are chosen at least 15 (15–25) amino acids long. Amino acids for coupling purposes may be added at strategic terminal positions. The sequence is chosen such that both ends of the peptide do not represent a minimum in hydrophillicity. Peptides can be elongated either N- or C-terminally by a cysteine or arginine residue for coupling purposes.

XIX.2. Peptide derivatization

For an introduction into various amino acid-specific chemical protection techniques, which have been developed for proteins but in general can be applied to peptides as well, see Atassi (1987). For the selection of proper coupling reactions, there are the following rules of the thumb: (a) use preferably bifunctional reagents, (b) use specific coupling procedures for an amino acid that has a low frequency in the peptide, (c) avoid amino acids that are crucial for recognition and (d) elongate peptides with an amino acid for coupling purposes, e.g., cysteine or arginine.

Frequently applied reagents: Sulfo derivatives of N-hydroxysuccinimide compounds are less susceptible to hydrolysis and water soluble. In general, they lead to better conjugate yields (2-4x). Most other coupling reagents are hygroscopic and readily hydrolyze. Store at -20°C and equilibrate to ambient temperature before use. Tris, glycine and other amino group containing reagents (ammonium sulphate) should be avoided. SP-protein conjugates stored at 4°C were not stable (Briand et al., 1985). It is advised to store all conjugates at protein concentrations >1 mg.ml^{-1} at -20°C.

Biotinylation (long chain biotin, LCB): Biotin binds extremely well to avidin. However, to prevent unwanted interactions by the avidin bound, due to the extreme depth of the binding pocket (4-5 amino acids deep), it is best to couple biotin via a long aliphatic chain linker (Suter and Butler, 1986). N-Hydroxysuccinimidyl-6-(biotinamido)-hexanoate (Pierce, Rockford, IL) is bound equimolar to the peptide. The procedure is easiest to perform while the peptide is still bound to the resin with all its protecting groups still in place. Then coupling to the unprotected -NH2 terminal side is easily obtained. Be sure that all protecting groups are removable using TFA. Ingredients: DMF, amylalcohol, acetic acid, diethylether, a saturated solution of phenol in water, long chain biotin and imidazole (10 mg LCB in 10 μl imidazole). Procedure: The resin-bound peptide is washed thoroughly with DMF. Leave the resin submersed in DMF. LCB is added in imidazole to equimolar concentration with the peptide. Incubate overnight (room temperature). Wash with DMF. Subsequently wash with tert.amylalcohol (2x), acetic acid (2x), tert.amylalcohol (2x), diethylether and dry in vacuum. With conjugated avidin the biotinylation can be monitored. Then cleave the peptide from the resin with 95% TFA, 5% phenol in water. Purify peptides according to standard procedures.

Introduction of sulfhydryl groups: When peptides or proteins do not contain free -SH groups, they can be introduced easily using Traut's reagent: methyl-4-mercaptobutyrimidate (iminothiolane) (Traut et al., 1973). This compound reacts preferentially with amino groups. On mild reduction the peptide or protein is tagged with free -SH groups. The -SH groups introduced can be used for coupling with a hetero-bifunctional coupling agent or they can be allowed to form -S-S- bonds (King and Kochoumian, 1978). Procedure: For the best result a threefold concentration of 2-iminothiolane (MW 140) over the peptide concentration is used. The reaction is allowed to take place in a sodium phosphate buffer (NaPi), pH 8.1. Iminothiolane is added dropwise to the peptide solution (5–10 mg.ml^{-1}) in a small reaction vial and allowed to react for 1 h at room temperature. Subsequently, the peptide is separated from iminothiolane by chromatography. (Dialysis membranes with low molecular weight cut off [1 kD] give disappointing results). Recently, a method for the introduction of -SH groups at carboxyl sites has also been described (Lin et al., 1990).

Determination of available -SH groups: Available -SH groups can be determined using Ellmans reagent (Ellman, 1969). Ingredients: NaPi buffer 0.1 M, pH 8.0.; 5,5'dithio-bis-(2-nitrobenzoic acid) (DTNB, Ellman's reagent), 4mg.ml^{-1} in NaPi 0.1, pH 8.0.; solution of cysteine or β-mercaptoethanol as a standard. Procedure: Briefly, add 0.1 ml of the sample to 0.1 ml of the reagent solution and mix with 4.8 ml NaPi, pH 8.0 buffer. After a few minutes the reaction is completed. Read absorbance at 412 nm. $E_{412} = 1.36.10^4 cm^{-1}.M^{-1}$ versus blank.

XIX.3. Homo-bifunctional coupling

Activated -SH: Coupling two active -SH which are incorporated into the protein/peptide. Ingredients: N-succimidyl 3-(2-pyridyldithio)propionate (SPDP) solution 20–40 mM (m.w. 312.4; 1.56–12.5 mg.ml^{-1}) in ethanol or DMSO; protein (carrier) 1–10 mg.ml^{-1} in NaPi buffer 0.1 M, pH 7.5, with 0.1 M NaCl; small reaction vial; sepharose G-25 column (PD-10, Pharmacia, Uppsala, Sweden); acetate buffer 0.1 M, pH 4.5 with 0.1 M NaCl; dithiotreitol (DTT) 0.1 M; n-ethyl-maleimide 10 mM. Procedure: Coupling principle: protein + SPDP (10-fold excess, minimum) ➡ protein with protected -SH; idem for peptide. On reduction of one of the 2-pyridyl disulfides with DTT, preferably the protein, the protecting groups are removed. Mixing the peptide-2-pyridyl disulfide with the thiolated carrier protein yields an -S-S- coupling product. The coupling, i.e., the release of the pyridyl-2-thione, can be followed spectrophotometrically E_{343} = 8.08.10^{-3}M^{-1}.cm^{-1}. Allow the protein to react with SPDP under stirring in a small reaction vial for 30 min at room temperature. For the determination of protein concentrations spectrophotometrically a correction factor for the 2-pyridyl disulfide should be taken into account (E_{280} = E_{280}-[pyridyne-2-thione released, molair] x 5.1.10^{3}; Pharmacia methodology SPDP method). To transfer to acetate buffer and to remove excess SPDP bring protein-2-pyridyl disulfide over a G-25 column (PD-10). Reduce this disulfide with DTT (final concentration 50 mM, 30 min room temperature). Transfer to NaPi pH 7.5, 0.1 M NaCl and remove reagents over G-25 column (PD-10, Pharmacia). Similarly, the peptide is treated with SPDP and excess of reagents is removed by HPLC chromatography. Finally, the 2-pyridyl disulfided peptide and the thiolated protein are allowed to form a disulfide bond (1 h overnight, room temperature). Thiolated protein (pH 7.0) may be stored at -20°C. Remaining free -SH groups if necessary can be blocked with n-ethyl-maleimide (10 mM).

Glutaraldehyde (coupling -NH2): Ingredients: Carrier protein: 10 mg.ml^{-1} in PBS; peptide: 10 mg.ml^{-1} in water; dialysis membrane cut off 10 kD; glutaraldehyde (Baker's); glycine 0.2 M; PBS; small reaction vials. Procedure: Dialyze (4°C, overnight) carrier protein (1 ml) versus 0.2% glutaraldehyde (1.6 ml) in 200 ml PBS. Dialyze two times versus 250 ml PBS (2 x 2 h). Transfer activated carrier into reaction vial. Slowly add SP: 100 mol/mol carrier protein. Incubate overnight (4°C). Then add 0.1 ml 0.2 M glycine to block all active aldehyde groups left (2 h, room temperature). Dialyse extensively to PBS. Store conjugate at -20°C (Zegers et al., 1990, adapted from Boorsma. 1984). For the conjugation of peptides which are not readily soluble a method has been described by Gerritse et al., 1991.

XIX.4. Hetero-bifunctional coupling

Carbodiimide EDC (coupling, e.g., -NH2, -COOH): Ingredients: N-methyl-imidazole 0.5 M, pH 6.0; carrier 10 mg.ml^{-1} in N-methyl-imidazole; peptide 20 mg.ml^{-1} in N-methyl-imidazole; EDC (1-ethyl-3-(3-dimethyl-aminopropyl)cardodiimide hydrochloride, Pierce) 100 mg.ml^{-1} in N-methyl-imidazole; dialysis membrane cut off 10 kD; small reaction vials. Procedure: Mix carrier and peptide solution in reaction vial: SP 100 mol/mol carrier (mol weight carrier 150 kD) for other carriers similar, e.g., BSA (mol weight 60 kD): 50 mol SP/mol carrier protein. Subsequently, add slowly (dropwise) the EDC (mol weight 191.7) solution to 50 mol/mol SP. Mix 30 min. at room temperature. Then dialyze versus PBS overnight at 4°C. Store at -20°C (Deen et al., 1990).

Maleimide/N-hydroxysuccinimide MBS (coupling of -SH and -NH$_2$): Ingredients: NaPi buffer 0.1 M, pH 7.0 (SPB); carrier protein 10 mg.ml^{-1} in SPB; coupling agentm-maleimidobenzoyl-N-hydroxysuccinimideester (MBS, mol weight 314.2, Pierce) 20 mg.ml^{-1} in dimethylformamide (DMF); peptide: 20 mg.ml^{-1} in SPB; 2 small reaction vials; PD-10 column (Pharmacia). Procedure: Carrier solution in vial is stirred on ice. MBS is added dropwise/slowly in 3 portions with 5 min intervals to 200 mol MBS/mol carrier (mol weight 150 kD). MBS with NaPi buffer gives a precipitate. This is discarded by centrifugation (Eppendorf, minifuge). Activated carrier is separated from unreacted MBS via a desalting column (Pharmacia, PD10). A little bit of precipitate can be discarded. The activated carrier is transferred to a clean reaction vial. The peptide is added to 100 mol SP/mol carrier (mol weight carrier 150 kD). Stir at room temperature for 1 h. Store at -20°C.

Azidophenylglyoxal APG (arginine to tyrosine, histidine; photoaffinity labeling): Glyoxal containing compounds react selectively (at pH 7–8) with arginine residues (Ngo et al., 1981) (for a review, see Atassi 1987). Coupling arginine containing peptides to proteins (Zegers et al., unpublished results) and vice versa, arginine containing proteins to any peptides containing tyrosine, histidine or lysine, has been described. Ingredients: SP: 5.10^{-3}M (±10 mg.ml^{-1}) in H_2O; APG: 1.10^{-2} M (1.9 mg.ml^{-1} in borate buffer 0.1 M, pH 7.9), prepared and kept in the dark; UV lamp (λ = 360 nm, socalled black ray 40 W). NaBH4 10^{-2} M in H_2O. KLH 10 mg.ml^{-1} in PBS. Dark glass mini-reaction vial. Procedure: The procedure starts in complete darkness! Incubate 0.2 ml SP with 0.2 ml APG (for one Arg in the peptide; APG: Arg = 2) at 37°C for 2 h. The course of the reaction can be followed spectrophotometrically: APG free has a maximum at 297 nm, whereas the APG-Arg product has a maximum at 250 nm. 0.2 ml of the reaction mixture is mixed with 0.2 ml KLH solution. Mix well and place this solution in a smal petri dish or a 24-well tissue culture plate under the UV lamp for 30 min. Subsequently, the coupling product is reduced for stabilization with 0.4 ml $NaBH_4$ for 2 h on ice. Dialyze to PBS. Coupling conditions: Peptides with limited solubility. Peptides sometimes are not readily soluble in water, water solutions like borate or phosphate buffers, etc. Alternatively, DMSO or DMF can be used as a solute. If peptides still are not dissolved urea may be used. Recently, a modification of a glutaraldehyde coupling of peptides to alkaline phosphatase (Zegers et al., 1990) has been described by Gerritse et al. (1991), using 8 M urea to dissolve the peptides. Procedure: Briefly, dialyze AP (10 $mg.ml^{-1}$) against 0.2% glutaraldehyde in PBS for 16 h at 4°C. Remove excess glutaraldehyde by dialyzation of AP against PBS. Transfer into a small reaction vial. Dissolve peptide (1 $mg.ml^{-1}$) in 8 M urea, pH 7.0 by stirring for 10 min at room temperature. Remove the non-dissolved material by centrifugation. Solubilized peptides are mixed with the activated AP (peptide/Ap = 100/1) for 16 h at 4°C. Block remaining active glutaraldehyde groups by adding 100 μl 0.2 M lysine.HCl, incubate another 2 h. Excess reagents are removed by dialysis to PBS.

Coupling of peptides to enzymes: Peptide-enzyme conjugates can be applied for detection of receptors for peptide ligands (antibodies, peptide-hormone receptors, etc.) (Laman et al., 1990, 1992). In principle, the methods described for peptide coupling in general can be applied. In our hands, GA and MBS-like reagents fulfill our criteria best.

XIX.5. Immunization

For the production of immunological reagents, two routes are most widely used: (1) injection of conjugate or peptide together with adjuvant intraperitoneally in mice or (2) subcutaneously on the back of rabbits or on the upper side of the feet or in the inguinal and axillary regions of the mice. In both cases, serum IgG responses are obtained after a booster immunization. The spleen and lymph node cells can be used for the production of MAbs. With subcutaneous injections the doses should best be distributed over more places (four) for better results.

XIX.6. Doses

Peptides of 15–20 amino acids are immunized in doses between 50 and 100 μg per mouse or 100–200 μg per rabbit. Conjugates contain a number of peptides per carrier and in general a dose range between 50 and 150 μg of the conjugate, based on the carrier content, will be sufficient.

XIX.7. Adjuvants

Water-in-oil emulsions: Water phase (9/20 v/v) including the antigen is slowly pipetted onto the oil phase (Bokhout et al., 1981) under thorough mixing on a Vortex. The emulsion is relatively stable and can be used for administration. Using complete Freund's adjuvant (CFA), thorough mixing of the components (incomplete, bacteria, antigen) until water-in-oil and oil-in-water emulsions are obtained is necessary for good results. Mixing is easily performed when two glass syringes are connected by a double luer lock. In general, 45/55 v/v of antigen solution and CFA are used. (Caution, the mixture of CFA is a skin irritant). In mice, intraperitoneally, 200 μl of the emulsion is administered, while subcutaneously 50 μl per location is injected. In rabbits, 250 μl per location is used for water-in-oil emulsions as well as for CFA.

Alum precipitation: Precipitates of proteins or peptide conjugates onto alum (aluminium-potassium-sulfate gel) act as depot administrations. Ingredients: 10% w/v solution of aluminium-potassium-sulfate in phosphate buffered saline (PBS) is prepared; 1M $NaHCO_3$ (2.1 g/25 ml PBS); mini-reaction vial. Procedure: Dissolve 100 μg of protein to 210 μl PBS. Add dropwise the $NaHCO_3$ (90 μl) and mix. Subsequently, 200 μl of the alum gel is added slowly under vigorous stirring (or vortexing). Leave the mixture overnight at 4°C, freshly prepared alum precipitates may be toxic! Mix before introduction. Use 200 μl of the mixture per mouse.

***Bordetella pertussis*:** Similar to mycobacteria in Freund's complete adjuvant various other bacteria (mainly membrane/cell wall components) show adjuvant activity. For this purpose heat-killed *B. pertussis* bacteria are frequently used. At the time of immunization or 1 day in advance $2x10^9$ bacteria are injected intraperitoneally in mice. For rabbits $8x\ 10^9$ bacteria are administrated subcutaneously. Do not mix *B. pertussis* suspensions with alum precipitates, since in that case large aggregates will develop.

XIX.8. Scheduling

The time course of development of serum antibody responses may in part be dependent on the type of carrier protein used. When KLH is applied, mice serum antibodies (IgG + IgM) increase gradually up to at least 21 days. For an estimate of responses, repeated tapping of the animals at various time points (14 and 21 days) after immunization is advised. In rabbits, taking some serum 14 days after each immunization is sufficient. Immunizations with peptide-carrier conjugates are given with 4 week intervals. In general, after two to three subsequent immunizations the maximum titers are obtained.

Best results following booster immunizations are obtained when priming titers have decreased. However, using KLH, which itself has adjuvant properties, anti-carrier titers will remain high for months. Nevertheless, normal immunization schedules can be completed.

The timing of the immunization preceding spleen cell fusion depends on the routing: soluble antigens (conjugates) can be introduced intravenously 3 days before fusion. Less soluble antigens are given intraperitoneally (i.p.) with IFA, as a water-in-oil emulsion or precipitated on aluminium hydroxide gel 4 days before an intended spleen cell fusion. Though we prefer spleen cells for fusion, for fusion of lymph nodes, regional nodes are activated by immunization with adjuvant (as for i.p. immunization). Mirza et al. (1987) performed fusion of lymph node cells at day 14 after primary immunization for the production of anti-insulin (5.6 kD) MAbs.

XIX.9. Monitoring of antibody responses in ELISA

ELISA is the most easy and rapid method for analysis of experimental antibody reponses. In general, optimal conditions for each ELISA are found by titration of coating and detector reagents. For regular ELISAs, we folowed the procedure as described in detail by Haaijman et al. (1988). For various special applications, other coating procedures were tested and practiced.

Passive coating: Coating of protein and peptides in PBS 5–10 μg.ml^{-1} overnight at 4°C or 1–2 h at room temperature. Peptides are best coated according to this procedure. If the peptides do not adhere (a conjugate is the best control), then coating at lower or higher pH is advised. Citrate (pH 2.5) or borate (pH 9.5) provide conditions which will solve most problems.

Glutaraldehyde: Antigens that do not adhere spontaneously to plastic surfaces can be coated actively onto glutaraldehyde pre-activated polystyrene plates (Nunc 2.69620). Procedure: Glutaraldehyde is diluted to 0.025% in Na_2CO_3 (0.1 M, pH 9.0). The wells of 96-well polystyrene plates are filled with 100 μl. Leave the plates wrapped in aluminium foil for 120 min at 56°C. After the incubation the plates are washed with water and subsequently dried with hot air in a short time (15 min). Immediately, the antigen (5 μg protein.ml^{-1} in Na_2CO_3 (0.1 M, pH 9.0) is added in a volume of maximum 100 μl. The plates with antigen are incubated for 1–2 h at room temperature or at 4°C overnight. Pre-activated plates cannot be stored (Klasen et al. 1983). Include proper controls since glutaraldehyde may introduce some background.

Poly-L-Lysin: Proteins or peptides which do not adhere to polyvinylchloride plates, may adhere to poly-L-lysine. Procedure: Poly-L-lysine (stock: 10 mg.ml^{-1} in PBS) is diluted to 10 μg.ml^{-1} in PBS. To each well, 100 μl is added and incubated at 4°C overnight or 1–2 h at room temperature. After incubation the plates are washed in PBS (3x) and dried in air at room temperature. Plates can be stored at -20°C. This procedure has been shown to be useful for coating DNA to ELISA plates.

Carbodiimide in imidazole: Various compounds can be coated to polystyrene plates using activation with carbodiimides. The procedure is carried out in imidazole (see also EDC coupling, Deen et al., 1990). Procedure: ELISA plates are pre-coated according to the glutaraldehyde method with BSA (5 μl.ml^{-1}). The compounds to be coupled are pre-treated with carbodiimide (EDC, couples to NH2 and COOH groups). For DNA: n ml of DNA 20 μl.ml^{-1} + n ml EDC (0.2 M in N-methylimidazole 0.1 M, pH 6.0) are mixed and incubated for 1 h at room temperature. Of this mixture, a maximum 100 μl per well is incubated at 4°C overnight.

Fixed antigen ELISA: antigen is coated as for standard procedure. Subsequently, the coating is incubated with formalin/HgCL2 for 1–2 h. Longer incubation does not lead to better discrimination. The plates are washed and the ELISA is developed as usual. Ingredients: The fixative is prepared as follows: HgCl2 (42 g) is added to a mixture of acetic acid 35 ml + aqua d. 595 ml + formalin 40% 70 ml. The wells are incubated with this mixture for 1–2 h.

Imunoblotting: Ingredients: Nitrocellulose (Immobilon, Millipore); Whatman 3 filtration paper; PBS; DAB (20 mg DAB + 100 μl H_2O_2 (30%) in 100 ml PBS + 0.05% Tween 20 (PT buffer); 0.1% BSA in PT buffer. Electrophoresis in agar. Procedure: First agar electrophoresis is performed. Immobilon is washed successively in methanol, aqua d. and PBS. Dress immobilon layer over the agar electrophoresis gel. Subsequently, overlay with 3 layers of wet and 7 layers of dry filtration paper. Put on a glass plate and a pressure of 1 kg. Leave for 1 h. The nitrocellulose blot is rolled up and put in a test tube (15 ml). 5 ml PTB buffer is added to block free nitrocellulose. Wash 2x with PT buffer. Apply antibody/serum appropriately diluted (5 ml). Incubate on a roller overnight at 4°C or 1–2 h room temperature. Wash 3x with PT buffer. Apply second antibody conjugate (Peroxidase) diluted in PTB. Incubate on a roller as for the first antibody. Wash 3x with PT buffer. Stain with 5–10 ml substrate solution (for DAB in dark). Take the blot from the incubation tube and wash for 15 min in aqua d. Dry under filter paper and a glass plate. Keep blots away from light. Store data as photo negatives (Radl et al., 1988, Nooij et al., 1990).

The BioRad immunodot apparatus is very easy for direct immunoblotting. In ELISA format the antigen is deposited (1 g) on a pre-treated Immobilon sheath. The pre-treatment now includes a last wash with $NaHCO_3$ (0.5M, pH 8.0). The antigen applied is diluted (30 μg.ml^{-1}) in the same solution. The rest of the procedure is similar as that described for blotting.

XIX.10. Materials

Buffers and media PBS: 0.01 M phosphate buffered saline 0.15 M NaCl, pH 7.4, PT: 0.05% Tween 20 in PBS, PTG: 0.1% gelatine in PT, PBS-GEL: 0.1–0.5% gelatin in PBS, DEA: diethanol amine 10 mM (Merck p.a.) + 1 mM $MgCl_2$, pH 9.8. N-Methylimidazole

(Aldrich, M5, 083-4, Brussels, Belgium); DMF, DMSO, TFE, amylalcohol, acetic acid, diethylether, TFA, ethanol and urea preferably all of p.a. quality.

Adjuvant A water-in-oil emulsion is simply made by thorough mixing of the antigen solution in PBS with specol or span Tween (the Netherlands Veterinary Instute, Lelystad, The Netherlands; Bokhout, 1981). Complete and incomplete Freunds can be obtained from Difco (inc.: 0639-60-6, com.: 06638-60-7).

Coupling reagents: Derivatization MBS (Pierce, 223310), APG, HSAB, Biotin NHS, Iminothiolane can be obtained from Pierce. GA: J.T.Baker, DTT: Aldrich 14,046-0, N-ethylmaleimide: Sigma (E-3876), EDC: Sigma (E-7750). PD10 pre-packed sephadex G25 colums to be obtained from Pharmacia, Uppsala, Sweden.

Antigens: KLH (Calbiochem, 374811, San Diego, CA), CGG: Sigma, G-6516, BSA: Sigma A-9647, St. Louis, MO.; ELISA can be performed in polyvinyl plates (Falcon, 3911, Becton and Dickinson Labware, CA., or Titertek, 77-172-05) or polystyrene plates (NUNC, 2.69620). Conjugates anti-mouse, anti-goat, anti-rabbit conjugates can be obtained from various suppliers: Dakopatts, Sigma KPL-Labs. Alkaline phosphatase can be obtained from Sigma P-6774 Type VII-T; PNPP: p-nitrophenylphosphate: Boehringer (107905); horseradish peroxidase (PO) RZ=3.0 (type VI; P-8375); o-phenylenediamine-di-hydrochloride: OPD (1078054): Kodak, Rochester, USA; DAB: 3.3'-diaminobenzidine: Sigma D-5637; FITC (fluorescein Isothiocyanate, isomer 1) can be obtained from Nordic Immunology, Tilburg, The Netherlands.

ABBREVIATIONS

AA	Amino acids
BSA	Bovine serum albumin
CD	Cluster of differentiation
CTL	Cytolytic T cell lysis
CTX	Cholera toxin
DDA	Dimethyldioctadecylammonium bromide
DTH	Delayed type hypersensitivity
CFA	Complete Freund's adjuvant
IFA	Incomplete Freund's adjuvant
IFN-γ	Interferon gamma
IL-#	Interleukin-#
I.P.	Intraperitoneal
ISCOM	Immune stimulating complex
KLH	Keyhole limpet hemocyanin
MDP	Muramyl dipeptides
MHC	Major histocompatibility complex
MLA	Monophosphoryl lipidA
NBP	Non-ionic blockpolymers
NK	Natural Killer
SAF	Syntex adjuvant formulation
SP	Synthetic peptide
Th1	T helper 1
Th2	T helper 2
WIO	Water-in-oil emulsion

XX. REFERENCES

Aichele P, Kyburz D, Ohashi PS et al. Peptide induced T cell tolerance to prevent autoimmune diabetes in a transgenic mouse. Proc Natl Acad Sci USA 91:444, 1994

Allison AC and Byars NE. An adjuvant formuation that selectively elicits the formation of antibodies of protective isotypes and cell mediated immunity. J Immunol Meth 95:157, 1986

Allison AC and Gregoriadis G. Vaccines: Recent trends and progress. Immunol Today 11:427, 1990

Allison AC and Byars NE. Immunological adjuvants: Desirable properties and side effects. Mol Immunol 28:279, 1991

Allison AC and Byars NE. Immunological adjuvants and their mode of action. In: Vaccines: New Approaches to Immunological Problems. Ed. WR Ellis. Butterworth-Heineman, Boston, p 431, 1992

Altman A and Dixon FJ. Immunomodifiers in vaccines. Adv Vet Sci Comp Med 33:301, 1989

Alving CR, Richards RL, Egan JE et al. Adjuvant effects of liposomal lipid A: Human malaria vaccine. In: Proc Int Endotoxin Soc. Eds. Nowotny, Spitzer and Ziegler. Excerpta Med, Amsterdam, pp 67, 1990

Arnon R. Synthetic peptides as the basis for vaccine design. Mol Immunol 28:209, 1990

Atassi MZ. Chemical modification and cleavage of proteins and chemical strategy in immunochemical studies of proteins. In: Immunochemistry of Proteins I, Ed. Atassi MZ, Plenum Press, New York, p 1, 1987

Attassi MZ and Webster RG. Localization, synthesis and activity of an antigenic site on influenza virus heamagglutinin. Proc Natl Acad Sci USA 80:840, 1983

Barteling SJ and Vreeswijk J. Developments in foot-and-mouth disease vaccines. Vaccine 9:75, 1991

Bennett B, Check IJ, Olsen MR et al. A comparison of commercially available adjuvants for use in research. J Immunol Meth 153:31, 1992

Bessler WG and Jung G. Synthetic lipopeptides as novel adjuvants. Res Immunol 143:548, 1992

Bevan MJ. Antigen recognition. Class discrimination in the world of immunology. Nature 325:192, 1987

Bixler GS, Yoshida T and Atassi MZ. Antigen presentation of lysozyme: T cell recognition of peptide and intact protein after priming with synthetic overlapping peptides comprising the entire protein chain. Immunology 56:103, 1985

Boersma WJA, Deen C, Zegers ND et al. IgG subclass distribution in juvenile human tonsil: IgG3 and IgG4 results of specific antibody production using synthetic peptides. Adv Exp Med Biol 237, Hystophysiology of the Immune Response, Eds. S Fossum and B Rolstad, Plenum Press, New York, p 125, 1988a

Boersma WJA, Claassen E, Deen C et al. Antibodies to short synthetic peptides for specific recognition of partly denatured protein. Analyt Chim Acta 213:187, 1988b

Boersma WJA, Deen C, Gerritse K et al. Anti-peptide antibodies as subclass specific reagents: Epitope mapping of human IgG2. Prot Biol Fluids 36:161, 1989a

Boersma WJA, Deen C, Haaijman JJ et al. Antibodies to a short synthetic peptide related to the hinge segment of human IgG3 recognizes thermally induced or fixative induced conformational changes in the human IgG3 molecule. Immunology 68:427, 1989b

Boersma WJA, Deen C, Laman JD et al. Synthetic peptides: A unique model to analyze the induction and regulation of determinant specific immune responses. In: Lymphatic Tissues and *In Vivo* Immune Responses, Eds. Imhof B, Berrih-Aknin S, Ezine S, Marcel Dekker In, New York, 273, 1991

Boersma WJA, Haaijman JJ and Claassen E. Use of synthetic peptide determinants for the production of antibodies. In: Immunohistochemistry. 2nd Edition, Ed. A C Cuello, John Wiley & sons, New York, p 1, 1992a

Boersma WJA, Bogaerts WJC, Bianchi ATJ et al. Adjuvant properties of stable water in oil emulsions: Evaluation of the experience with specol. Res Immunol 1443: 503, 1992b

Boersma WJA, Zegers ND, Van Den Bogaerdt A et al. Development of safe oral vaccines based on *Lactobacillus* as a vector with adjuvant activity. Proc ICHEM 2nd Internatl Congr Biotech UK, Brighton: Industrial Immunology, p 43, 1994

Bokhout BA, Van Gaalen C and Van der Heijden PhJ. A selected water-in-oil emulsion: Composition and usefulness as an immunological adjuvant. Vet Immunol Immunopathol 2:491, 1981

Bomford R. The comparative selectivity of adjuvants for humoral and cell mediated immunity I effect on the antibody response to bovine serum albumin and sheep red blood cells of Freunds incomplete and complete adjuvants, alhydrogel, corynebacterium parvum, bordetella pertussis, muramyldippetides and saponin. Clin Exp Immunol 39:426, 1980

Boorsma DM. Direct immunoenzyme double staining applicable for monoclonal antibodies. Histochemistry 80:103, 1984

Boraschi D, Nencioni L, Villa L et al. *In vivo* stimulation and restoration of the immune response by the non-inflammatory fragmemt 163–171 of human interleukin-1β. J Exp Med 169:675, 1988

Borras Cuesta F, Fedon V and Petit-Camurdan A. An Enhancement of peptide immunogenicity by linear polymerization. Eur J Immunol 18:199, 1988

Briand JP, Muller S and Van Regenmortel M. Synthetic peptides as antigens: Pitfalls of conjugation methods. J Immunol Meth 78:59, 1985

Broekhuizen MP, Van Rijn JMM, Blom AJM et al. Fusion proteins with multiple copies of the major antigenic determinant of foot and mouth disease virus protect both the natural host and laboratory animals. J Gen Virol 68:3137, 1987

Buiting AMJ, Van Rooijen N and Claassen E. Liposomes as antigen carriers and adjuvants *in vivo*. Res Immunol 143:541, 1992

Byars NE, Allison AC, Harmon MW et al. Enhancement of antibody responses to influenza B virus haemagglutinin by use of a new adjuvant formulation. Vaccine 8:49, 1990

Campbell JB and Peerbaye YA. Saponin. 44th Forum in Immunology. Res Immunol 143:526, 1992

Carayanniotis G and Barber BH. Characterization of the adjuvant-free serological response to protein antigens coupled to antibodies specific for class II MHC determinants. Vaccine 8:137, 1990

Carayanniotis G, Skea DL, Luscher MA et al. Adjuvant independent immunization by immunotargeting antigens to MHC and non-MHC determinants *in vivo*. Mol Immunol 28:261, 1991

Casten LA. Enhanced T cell responses to antigenic peptides targeted to B cell surface immunoglobulin, Ia or class I molecules. J Exp Med 168:171, 1988

Chavali SR, Barton LD and Campbell JB. Immunopotentiation by orally administered Quillaja saponins: Effects in mice vaccinated intraperitoneally against rabies virus. Clin Exp Immunol 74:339, 1988

Claassen E and Boersma WJA. 44th Forum in Immunology Characteristics and practical use of new generation adjuvants as an acceptable alternative for Freund's complete adjuvant. Res Immunol 143:475, 1992

Claassen E, De Leeuw W, De Greeve P et al. Freunds complete adjuvant: An effective but disagreeable formula 44th Forum in Immunology. Res Immunol 143:478, 1992

Claassen E. Detection, localization and kinetics of immunomodulating liposomes *in vivo*. In: Liposomes as an *In Vivo* Tool to Study and Manipulate Macrophage Function. 41st Forum in Immunology. Res Immunol 143:235, 1992

Claassen I and Osterhaus A. The iscom structure as an immune enhancing moiety: Experience with viral systems. 44th Forum in Immunology. Res Immunol 143:531, 1992

Dalsgaard K. A study of the isolation and characterization of the saponin Quil-A. Acta Vet Scand 19:suppl 1, 1978

Dalsgaard K, Hilgers L and Trouve G. Classical and new approaches to adjuvant use in domestic food animals. Adv Vet Science Comp Med 35:121, 1990

Deeb BJ, Digiacomo RF, Kunz LL et al. Comparison of Freunds and Ribi adjuvants for inducing antibodies to the synthetic antigen (TG)-AL in rabbits. J Immunol Meth 152:105, 1992

Deen Carla, Claassen E, Gerritse K et al. An improved and modified carbodiimide coupling method for synthetic peptides: Enhanced antipeptide antibody responses. J Immunol Meth 129:119, 1990

Deres K, Schild H, Wiesmüller K-H et al. *In vivo* priming of virus-specific cytotoxic T lymphocytes with synthetic lipopeptide vaccin. Nature 342:561, 1989

Devereux J, Haeberli P and Smithies O. A comprehensive set of sequence analysis programs for the VAX. Nucleic Acids Res 12:387, 1984

Dyson HJ, Cross KJ, Houghton RA et al. The immunodominant site of a synthetic immmunogen has a conformational preference in water for a type II reverse turn. Nature 318:480, 1985

Eldridge JH, Staas JK, Meulbroek JA et al. Biodegradable microspheres as a vaccine delivery system. Mol Immunol 28:287, 1991

Eldridge JH, Hammond CJ, Meulbroek JA et al. Controlled vaccine relaese in the gut associated lymphoid tissues. I. orally administered biodegradable microspheres target the peyer's patches. J Control Rel 11:205, 1990

Eldridge JH, Staas JK, Tice TR et al. Biodegradable poly (DL-lactide-co-glycolide) microspheres. Res Immunol 143:557, 1992

Ellman GL. Tissue sulfhydryl groups. Arch Biochem Biophys 82:70, 1969

Fieser TM, Tainer JA, Geysen HM et al. Influence of protein flexibility and peptide conformation on reactivity of monoclonal antipeptide antibodies with a protein α-helix. Proc Natl Acad Sci USA 84:8568, 1987

Finkelman FD, Katona IM, Mosmann TR et al., Interferon-χ regulates the isotypes of immunoglobulin secreted during *in vivo* humoral immune responses. J Immunol 140:1022, 1988a

Finkelman FD, Katona IM, Urban JF et al. IL-4 is required to generate and sustain *in vivo* IgE responses. J Immunol 141:2335, 1988b

Fischer PM, Comis A and Howden MEH. Direct immunization with synthetic peptidyl-polyamide resin: Comparison with antibody production from free peptide and conjugates with carrier proteins. J Immunol Meth 118:119, 1989

Forrest JMS and Ross HA. Splenic implantation of nitrocellulose pellets as a means of raising antibodies to nanogram quantities of antigen. J Immunol Meth 159:275, 1993

Francis MJ, Fry MC, Rowlands DJ et al. Immune response to uncoupled peptides of foot-and-mouth disease virus. Immunology 61:1, 1987a

Francis MJ, Hastings GZ, Syred AD et al. Non responsiveness to a foot and mouth disease virus peptide overcome by addition of foreign helper T cell determinants. Nature 33:168, 1987b

Freund J. The mode of action of immunologic adjuvants. Adv Tuberc Res 7:30, 1956

Friede M, Muller S, Briand J-P et al. Induction of immune response against short synthetic peptide antigen coupled to small neutral liposomes containing monophosphoryl lipid A. Mol Immunol 30:539, 1993

Geerligs HJ, Weijer WJ, Welling GW et al. The influence of different adjuvants on the immune respons to a synthetic peptide comprising amino acid residues 9–21 of herpes simplex virus type 1 glycoprotein D. J Immunol Meth 124:95, 1989

Germain RN. The ins and outs of antigen processing and presentation. Nature 322:687, 1986

Gerritse K, Fasbender M, Boersma WJA et al. Conjugate formation in urea to couple insoluble peptides to alkaline phosphatase for ELISA and *in situ* detection of antibody forming cells. J Histochem Cytochem 39(7):987, 1991

Gerritse K, De Knijff P, Van Ierssel G et al. Immunological discrimination between the human apolipoprotein E2(ARG -> Cys) and E3 isoforms. J Lipid Res 33:273, 1992

Goubau S, Silversides DW, Gonzalez A et al. Immunization of cattle against modified peptides of gonadotropin releasing hormone conjugated to carriers: Effectiveness of Freund's and alternative adjuvantia. Thriogenology 32:557, 1989

Gustafson GL and Rhodes MJ. Bacterial cell wall products as adjuvants: Early interferon gamma as a marker for adjuvants that enhance protective immunity. Res Immunol 143:483, 1992

Haaijman JJ, Coolen J, Deen C et al. Monoclonal antibodies directed against human immunoglobulins: Preparation and evaluation procedures. In: Reviews on Immunoassay Technology 1:59. Ed. SB Pal, MacMillan, London, 1988

Hackett CJ, Hurwitz JL, Dietzschold B et al. A synthetic decapeptide of influenza virus hemagglutinin elicits helper T cells with the same fine recognition specificities as occur in response to whole virus. J Immunol 135:1391, 1985

Heath AW and Playfair JHL. Conjugation of interferon gamma to antigen enhances its adjuvanticity. Immunology 71:454, 1990

Heath AW and Playfair JHL. Cytokines ad immunological adjuvants. Vaccine 10:427, 1992

Hilgers LAT and Snippe H. DDA as an immunological adjuvant. Res Immunol 143:494, 1992

Holmgren J, Lycke N and Czerkinsky C. Cholera toxin and cholera B subunit as oral mucosal adjuvant and antigen vector systems. Vaccine 11:1179, 1993

Hong TH, Chen ST, Tang TS et al. The production of polyclonal and monoclonal antibodies in mice using novel immunization methods. J Immunol Meth 120:151, 1989

Horiuchi N, Caufield P, Fischer JO et al. Similarity of synthetic peptide from human tumor to parathyroid hormone *in vivo* and *in vitro*. Science 238:1566, 1986

Hu JG and Kitagawa T. Studies on the optimal immunization schedule of experimental animals. VI. Antigen dose response of aluminium hydroxide aided immunization and booster effect under low antigen dose. Chem Pharm Bull 38:2775, 1990

Hunter RL, Strickland F and Kezdy F. The adjuvant activity of nonionic block polymer surfactants. I. The role of hydrophile-lipophile balance. J Immunol 127:1244, 1981

Hunter RL, Bennett B, Howeron D et al. Non-ionic blockpolymer surfactants as immunological adjuvants: Mechanisms of action and novel formulations. In: Immunological Adjuvants and Vaccines. Eds. G Gregoriadis, AC Allison, G Poste, Plenum Press, New York p 133, 1989

Jackson S, Mestecky J, Childers NK et al. Liposomes containing anti-idiotypic antibodies: An oral vacine to induce protective secretory immune responses specific for pathogens of mucosal surfaces. Infect Immunity 58:1932, 1990

Jackson RJ, Fujihashi K, Xu-amano J et al. Optimizing oral vaccines: Induction of systemic and mucosal B cell and antibody responses to tetanus toxoid by use of cholera toxin as an adjuvant. Infect Immunity 61:4272, 1993

Jacob CO, Grossfeld S, Sela M et al. Priming immune responses to cholera toxin induced by synthetic peptides. Eur J Immunol 16:1057, 1986

Jani P, Halbert GW, Langridge J et al. The uptake and translocation of latex nanospheres and microspheres after oral administration to rats. J Pharm Pharmacol 41:809, 1989

Jani P, Halbert GW, Langridge J et al. Nanoparticle uptake by the rat gastrointestinal mucosa: Quantitation and particle size distribution. J Pharm Pharmacol 42:821, 1990

Jones PD, Tha Hla R, Morein B et al. Cellular immune responses in the murine lung to local immunization with influenza A virus glycoproteins in micelles and immunostimulatory complexes (iscoms). Scand J Immunol 27:645, 1988.

Kalish ML, Check IJ and Hunter RL. Murine IgG isotype responses to the Plasmodium cynomolgi circumsporozoite peptide (NA-GG)5 - I. Effect of carrier, copolymer adjuvants and lipopolysacharide on isotype selection. J Immunol 146:3583, 1991

Kennedy JS, Hughes BW, Masada MP et al. Influence of adjuvants on the quantity, affinity, isotype and epitope specificity of murine antibodies. J Immunol Meth 121:157, 1989

Kensil CR, Patel U, Lennick M et al. Separation and characterization of saponins with adjuvant activity form Quillaja saponaria molina cortex. J Immunol 146:431, 1991

King TP, Li Y and Kochoumian L. Preparation of protein conjugates via intermolecular disulfide bond formation. Biochemistry 17:1499, 1978

Klasen EA, Rigutti A, Bos A et al. Development of a screening system for detection of somatic mutations. II. The use of peptides and insoluble protein fragments in a non-competitive solid phase enzyme immunoasay. J Immunol Meth 59:281, 1983

Kleine AH, Glatz JF, van Nieuwenhoven FA et al. Type specific immunodetection of human heart fatty acid binding protein with polyclonal anti-peptide antibodies. Proc First Int Workshop on Fatty Acid Binding Proteins, Maastricht 4–5 september 1989. Mol Cell Biochem 98:41, 1990

Kuriyama Stsujii T, Ishizaka S, Kikuchi E et al. Enhancing effects of oral adjuvants on anti-HBsAg responses induced by hepatitis B vaccine. Clin Exp Immunol 72:383, 1988

Laman JD, Gerritse K, Fasbender M et al. Double immunocytochemical staining for the *in vivo* detection of epitope specificity and isotype of antibody forming cells against synthetic peptides homologous to human immunodeficiency virus-1. J Histochem Cytochem 38:457, 1990

Laman JD, Schellekens MM, Abacioglu YH et al. Variant specific monoclonal and group specific polyclonal HIV-1 neutralizing antibodies raised with synthetic peptides from the gp-120 third variable domain. J Virol 66:1823, 1992

Leenaars PPAM, Hendriksen CFM, Angulo AF et al. Evaluation of several adjuvants as alternatives to Freunds complete adjuvant in rabbits. Vet Immunol Immunopathol 40:225, 1994

Lees A, Morris SC, Thyphronitis G et al. Rapid stimulation of large antibody responses with conjugates of antigen and IgD antibody. J Immunol 145:3594, 1990

Lerner RA, Green N, Alexander H et al. Chemically synthesized peptides predicted from the nucleotide sequence of the hepatitis B virus genome elicit antibodies reactive with the native envelope protein of Dane particles. Proc Natl Acad Sci USA 78:3403, 1981

Lin C, Mihal KA and Krueger RJ. Introduction of sulfhydryl groups in proteins at carboxyl sites. Biochem Biophys Acta, 1038:382, 1990

Lipman NS, Trudel LJ, Murphy JC et al. Comparison of immune response potentiation and *in vivo* inflammatory effects of Freund's and Ribi adjuvants in mice. Lab Animal Sci 42:193, 1992

Liu D, Mori A and Huang L. Large liposomes containing ganglioside GM1 accumulate effectively in the spleen. Biochem Biophys Acta 1066:159, 1991

Liu D, Wada A and Huang L. Potentiating of the humoral response of intravenous antigen by splenotropic liposomes. Immunol Lett 31, 177, 1992

Lövgren K, Lindmark J, Pipkorn R et al. Antigen presentation of small molecules and peptides conjugated to preformed iscom as carrier. J Immunol Meth 98:137, 1987

Lövgren K, Kaberg H and Morein B. An experimental subunit vaccine (iscom) induced protective immunity to influenza virus infection in mice after a single intranasal administration. Clin Exp Immunol 82:435, 1990

Lord SC, McCullen GL, Marzuki S et al. Functional domains of human interferon gamma probed with antipeptide antibodies. Mol Immunol. 26:637, 1989

Lycke N, Bromander AK, Ekman L et al. Cellular basis of immunomodulation by cholera toxin *in vitro* with possible association to the adjuvant function *in vivo*. J Immunol 142:20, 1989

Maharaj I, Froh KJ, and Campbell JB. Immune responses of mice to inactivated rabies vaccine administered orally: Potentiation by quillaja saponin. Canad J Microbiol 32:414, 1986

Martin S, Niedermann G, Leipner C et al. Intracellular processing of hapten-modified protein for MHC class I presentation: Cytoplasmic delivery by pH sensitive liposomes. Immunol Lett 37:97, 1993

Martinon F, Gras-Masse H, Boutillon C et al. Immunization of mice with lipopeptides bypasses the prerequisite for adjuvant: Immune response of Balb/c mice to human immuno-deficiency virus envelope glycoprotein. J Immunol 149:3416, 1992

McCormick DJ, Hillman K and David CS. Production of antibodies against murine class II antigens using unconjugated synthetic peptides of the β-chain region 63–78. J Immunol Meth 113:165, 1988

McKercher PD. Oil adjuvants: Their use in veterinary biologics In: Advances in Carriers and Adjuvants for Veterinary Biologists. Eds. Nervig RM, Gough PM, Kaeberle ML, Whetstone CA, Iowa State University Press; p1, 1986

McLean GW, Owsiaka AM, Subak-Sharpe JH et al. Generation of antipeptide and anti-protein sera: Effect of peptide presentation on immunogenicity. J Immunol Meth 137:149, 1991

Mirza IH, Wilkin TJ, Cantarini M et al. A comparison of spleen and lymphe node cells as fusion partners for the raising of monoclonal antibodies after different routes of immunization. J Immunol Meth 105:235, 1987

Morein B. The ISCOM an immunostimulating system. Immunol Lett 25:281, 1990

Morein B and Simons K. Subunit vaccines against enveloped viruses; virosomes, micelles and other protein complexes. Vaccine 3:83, 1985

Mowat AM, Donachie AM, Reid G et al. Immune stimulating complexes containing quil A and protein antigen prime class I MHC-restricted T lymphocytes *in vivo* and are immunogenic by the oral route. Immunology 72:317, 1991

Nair S, Zhou X, Huang L et al. Class I restricted CTL recognition of a soluble protein delivered by liposomes containing lipophilic polylysines. J Immunol Meth 152, 237, 1992

Nardelli B and Tam JP. Cellular immune responses induced by *in vivo* priming with a lipid conjugated multimeric antigen peptide. Immunology 79:355, 1993

Neurath AR and Kent SBH. Antigenic structure of hepatitis virus. In: Immunochemistry of Viruses. The basis of Serodiagnosis and Vaccines, Ed. Van Regenmortel MHV and Neurath AR, Elsevier, Amsterdam, p 325, 1985

Neurath AR, Strick N and Girard M. Hepatitis B virus surface antigen (HbsAg) as a carrier for synthetic peptides having an attached hydrophobic tail. Mol Immunol 26:53, 1989

Newman WJ, Wu JY, Gardner BH et al. Saponin adjuvant induction of ovalbumin specific CD8 cytotoxic T lymphocyte responses. J Immunol 148:2357, 1992

Ngo TT, Yam CF, Lenhoff HM et al. p-Azidophenylglyoxal A hetrobifunctional photoactivable crosslinking reagent selective for argininyl residues. J Biol Chem 256:11313, 1981

Nicklas W. Aluminium salts. Res Immunol 143:489, 1992

Nilsson BO and Larsson A. Inert carriers for immunization. Res Immunol 143:553, 1992

Nooij FJM, Van der Sluijs-Gelling, Jol-Van der Zijde CM et al. Immunoblotting techniques for the detection of low level homogeneous immunoglobulin components in serum. J Immunol Meth 134, 273, 1990

Nussenzweig MC, Steinman RM, Witmer MD et al. A monoclonal antibody specific for mouse dendritic cells. Proc Natl Acad Sci USA 79:161, 1982

Odean MJ, Frane CM, Van der Vieren M et al. Involvement of gamma interferon in antibody enhancement by adjuvants. Infect Immunity 58:427, 1990

O'Hagan DT, Palin KJ and Davis SS. Poly(butyl-2-cyanoacrylate) particles as adjuvants for oral immunization. Vaccine 7:213, 1989

O'Hagan DT, Jeffrey H, Roberts MJJ et al. Controlled realease microparticles for vaccine development. Vaccine 9:768, 1991

Peeters JM, Hazendonk TG, Beuvery EC et al. Comparison on four bifunctional reagents for coupling peptides to proteins and the effect of the three moieties on the immunogenicity of the conjugates. J Immunol Meth 120:133, 1989

Perraut R, Hundt E, Garraud O et al. Comparison of the effects of adjuvants and adjuvant doses on the quantitative and qualitative antibody response to selected antigens in New World squirrel monkeys Saimir sciureus. Vaccine 11:730, 1993

Pessino A, Gherzi R, Damiani G et al. Antipeptide antibodies toward the extracellular domain of insulin receptor beta-subunit. Biochem Biophys Res Comm 162:1236, 1989

Phillips NC and Emili A. Enhanced antibody response to liposome associated protein antigens: Preferential stimulation of IgG2a/b production. Vaccine 10:151, 1992

Radl J, Wels J and Hoogeveen CM. Immunoblotting with (sub)class-specific antibodies reveals a high frequency of monoclonal gammapathies in persons thought to be immunodeficient. Clin Chem 34:1839, 1988

Ribi E, Ulrich JT and Masihi KN. Immunopotentiating activities of monophosphoryl lipid A. In: Immunopharmacology of Infectious Diseases: Vaccine Adjuvants and Modulators of Non-Specific Resistance, Ed Majde JA, Alan R. Liss, New York, p 101, 1987

Richards RL, Schwarz GM, Schultz C et al. Immunogenicity of liposomal malaria sporozoite antigen in monkeys: Adjuvant effects of aluminium hydroxide and non-pyrogenic liposomal lipid A. Vaccine 7:506, 1989

Rock KL, Fleischacker C and Gamble S. Peptide priming of cytolytic T cell immunity *in vivo* using β-2 microglobulin as an adjuvant. J Immunol 150:1244, 1993

Ruedl Ch, Albin B, Bock G, Wick G et al. Oral administration of a bacterial immunomodulator enhances murine intestinal lamina propria and the peyer's patch lymphocyte traffic to the lung:

Possible implications for infectious disease prophylaxis and therapy. Int Immunol 5:39, 1993

Silversides DW, Allen AF, Misra V et al. A synthetic LHRH vaccine. II. Temporal apects of titer development and formulation trials in Balb/C mice. J Reprod Immunol 14:47, 1984

Skea DL and Barber BH. Adjuvant independent induction of immune responses by antibody-mediated targeting of protein and peptide antigens. Res Immunol 143:568, 1992

Skea DL and Barber BH. Studies of the adjuvant independent antibody response to immunotargeting. J Immunol 151:3557, 1993

Snijders A, Benaissa-Trouw BJ, Visser Vernooy HJ et al. A DTH inducing T cell epitope of Semliki Forest virus mediates effective T helper activity for antibody production. Immunology 77:322, 1992

Snippe H, Willers JMN and Inman JK. The specificity of antibody formation in mice following immunization with hapten carrier complexes mixed with the surfactant DDA dimethyl dioctadecyl ammonium bromide. Immunology 39:361, 1980

Snippe H, De Reuver MJ, Beunder JW et al. Delayed type hypersensitivity in rabbits. Comparison of the adjuvant dimethyl dioctadecyl ammonium bromide and Freund's adjuvant. Int Arch Allergy 67:139, 1982

Spitz M. Single-shot intrasplenic immunization for the production of monoclonal antibodies. In: Methods in Enzymology, Eds. Langone JJ and Vunakis HV, Academic Press, New York, 121, 33, 1986

Springer T, Galfre G, Secher DS et al. Monoclonal xenogenic antibodies to murine cell surface antigens: Identification of novel leukoccyte differntiation antigens. Eur J immunol 8:539, 1978

Staruch MJ and Wood DD. The adjuvanticity of interleukin 1 *in vivo*. J Immunol 130:2191, 1983

Suter M and Butler M. The immunochemistry of sandwich ELISA's. II. A novel system prevents the denaturation of capture antibodies. Immunology Lett 13:313, 1986

Tagliabue A and Boraschi D. Cytokines as vaccine adjuvants: Interleuklin 1 and its synthetic peptide 163–171. Vaccine 11: 594, 1993

Tagliabue A, Ghiara P and Boraschi D. Non-inflammatory peptide fragments of IL-1 as safe new-generation constructs. 44th Forum in Immunology. Res Immunol 143:563, 1992

Tam JP and Zavala F. Multiple antigen peptide A novel approach to increase detection sensitivity of synthetic peptides in solid phase immuno-assays. J Immunol Meth 124:53, 1989

Tamauchi H, Tadakuma T, Yasuda T et al. Enhancement of immunogenicity by incorporation of lipid A into liposomal model membranes and its application to membrane associated antigens. Immunology 50:605, 1983

Thapar MA, Parr EJ, Bozzaola JJ et al. Secretory immune repsonses in the mouse vagina after parenteral or intravaginal immunization with an immunostimulatin complex (ISCOM). Vaccine 9:129, 1991

Traut RR, Bollen A, Sun TT et al. Methyl-4-mercaptobutyrimidate as a cleavable cross-linking reagent and its application to the Escherichia coli 30s ribosome. Biochemistry 12:3266, 1973

Tsujimoto M, Kotani S, Shiba T et al. Adjuvant activity of 6-O-acyl-muramyldipeptides to enhance primary cellular and humoral immune responses in guinea pigs: Dose-response and local reactions observed with selected compounds. Infect Immun 53(3):517, 1986

Vajdy M and Lycke NY. Cholera toxin adjuvant promotes longterm immunlogical memory in the gut mucosa to unrelated immunogens after oral immunization. Immunology 75:488, 1992

Van der Ploeg JR and Drijfhout JW et al. Immunological properties of multiple repeats of a linear epitope of herpes simplex virus type I glycoprotein-D. J Immunol Meth 124:211, 1989

Van Laar JH, Voorhorst-Ogink MM and Zegers N. Characterization of polyclonal antibodies against the N-terminal domain of the human androgen receptor. Mol Cell Endocrinol 67:29, 1989

Van Rooijen N and van Nieuwmegen R. Liposomes in immunology: Evidence that their adjuvant effect results from surface exposition of the antigens. Cell Immunol 49:402, 1980

Verheul AFM and Snippe H. Non ionic block polymer surfactants as immunological adjuvant. Res Immunol 143:512, 1992

Warren HS, Vogel FR and Chedid LA. Current status of immunological adjuvants. Ann Rev Immunol 4:369, 1986

Wiesmuller K-H, Jung G and Hess G. Novel low molecular weight synthetic vaccine against foot and mouth disease containing a potent B cell and macrophage activator. Vaccine 7:29, 1989

Woodard LF and Jasman RL. Stable oil-in-water emulsions: Preparation and use as vaccine vehicles for lipophillic adjuvants. Vaccine 3:137, 1985

Woodard LF. Surface chemistry and classification of vaccine adjuvants and vehicles. In: Bacterial Vaccines, Alan R. Liss, New York, p 281, 1990

Wyss-Coray T, Brander C, Bettens F et al. Use of antibody/peptide constructs to direct antigenic peptides to T cells: Evidence for T cell processing and presentation. Cell Immunol 139:268,1992

Zegers N, Gerritse K, Deen C et al. An improved conjugation method for controlled covalent coupling of synthetic peptides to proteins using glutaraldehyde in a dialysis method. J Immunol Meth 130:195, 1990

Zegers ND, Claassen E, Neelen C et al. Epitope prediction and confirmation for the human androgen receptor: Generation of monoclonal antibodies for multi-assay performance following the synthetic peptide strategy. Biochem Biophys Acta 1073:23, 1991a

Zegers ND, Claassen E, Gerritse K et al. Detection of genetic variants of human α-1-antitrypsin with site specific monoclonal antibodies. Clinical Chemistry 37(9):1606, 1991b

Zigterman GJWJ, Snippe H, Jansze M et al. Non-ionic blockpolymers NBP as adjuvants for future vaccines. Adv Biosci 68:301, 1988

Chapter 14

EPITOPE SPECIFIC DETECTION OF ANTIBODIES

Eric Claassen and Wim J.A. Boersma

TABLE OF CONTENTS

Note: Parts of this paper have been previously published in J. Immunol. Meth. **150**, 207–216, 1992 by Claassen et al.

I. INTRODUCTION

Analysis of humoral immune responses is best performed by looking at the production of antigen specific antibodies in combination with their isotype (subclasses). For this purpose, a number of *in vitro* and *in vivo* assays have been developed, none of which are generally applicable or without specific drawbacks. Especially in those cases where one might want to investigate antibodies in a tissue context (i.e., *in situ/in vivo* in freshly frozen or archive material), there remains only one method of choice. This method, originally developed by Van Rooijen and Claassen (1986), makes use of the fact that lightly fixed (acetone) antibodies can still recognize and bind the epitope against which they are directed. By coupling the antigen to an enzyme, and subsequent incubation on tissue sections followed by routine immunohistochemistry, one can thus reveal antibodies based on their antigen specificity (Fab). This method has a number of advantages over other methods and those are listed in Table 1. Disadvantages include the fact that quantitative results are rather laborious to procure and the fact that interpretation of cell-cell interactions and tissue compartments in, e.g., the spleen requires some skill. In this chapter, we will point out how the use of synthetic peptides enables the precise study of antibody formation (viz. humoral responses) on an epitope specific level by making use of synthetic-peptide-enzyme conjugates. The method which was originally developed for use in *in situ/in vivo* detection in tissue sections can also be used in ELISA and ELISPOT assays with great sensitivity.

II. THE TECHNIQUE

We have developed a direct double-immunoenzyme approach for the detection of antigen specific antibody forming B-cells (AFC) *in situ*. In animal studies with model antigens (such as albumins and γ-globulins) specificity of the antibodies in tissue sections could be demonstrated after incubation with antigen-enzyme conjugates and their isotype could be simultaneously determined by using an anti-immunoglobulin (Fc-chain specific)-enzyme conjugate followed by appropriate (double) enzyme chemistry (Van Rooijen and Claassen, 1986). By means of hapten-enzyme and hapten-carrier conjugates made with haptens such as trinitrophenyl (Claassen and Van Rooijen, 1984), penicilloyl (Boorsma et al., 1986) and arsenate (Brown and Claassen, 1988), immune responses against thymus-dependent and T-independent (Claassen et al.,1986a, b) and/or soluble versus particulate carriers (cf. Claassen et al., 1987a) could be studied. The method was adapted for use in other lymphoid tissue such as that

0-8493-8967-4/95/$0.00+$.50

TABLE 1
Advantages of epitope specific detection of antibody forming cells in tissue sections

- Positive detection of rare AFC (1 in 10^6–10^7)
- Small sample size possible (human biopsies; Hoefakker et al., 1993)
- Use of archive material, retrospective studies, possible
- Use of frozen tissue is convenient for storage and timing in experimental design
- Detection of AFC having low or blocked secretion
- Detection of membrane antibody positive (viz. memory) B-cells
- Detection of multiple antigen specificities in a single sample
- No induction of antibodies by culturing (low background)
- Determination of tissue compartments, actual clone size and cell-cell interactions
- Combination with *in situ* hybridization
- Mounted slides provide permanent record of results
- Sequential application and analysis of crossreactive (re-)agents is possible
- Simultaneous detection of antigen specificity and isotype by double staining

associated with the gut (Jeurissen et al., 1985; Gerritse et al., 1991a) or the lung (Van der Brugge-Gamelkoorn et al., 1986).

After modification in conjugation and staining procedures, this technique was also used for the simultaneous *in situ* detection of crossreacting idiotype and anti-idiotype AFC. In other assays, the rapid clearance of Id-anti-Id complexes and the technical demands of quantitating auto-anti-Id makes these studies very complicated (cf. Brown and Claassen, 1991). In rabbits, we studied B-cell development (Claassen et al., 1986c) and persistence in adoptive immunity (Claassen et al., 1987b) and we were also able to demonstrate both an experimental and a spontaneous auto-anti-allotype immune response which could not be detected by means of any other assay (Claassen and Adler, 1988).

In all our studies, special emphasis was placed on the development and localization of AFC in lymphoid tissues as a function of the carrier type which was used (cf. Van Rooijen et al., 1989; Laman et al., 1991b). The relation of these AFC with different other splenic cell types was established *in vivo/in situ* by means of several double-staining techniques as reviewed before (Claassen et al., 1992). The use of experimental antigens and animal studies, as described in the above studies, could not solve major questions arising in virus induced pathogenicity (Laman et al., 1989) or autoimmunity (Gerritse et al., 1993a,b). We therefore started out to develop a method for the detection of epitope specific AFC *in situ* in human lymphoid tissues involved in natural (auto)-immune responses. In this chapter, we will outline the principle of a new technique for the detection of such AFC, based on the use of synthetic peptides, with special emphasis on detailed technical demands and properties.

III. PRINCIPLE OF THE METHOD FOR EPITOPE SPECIFIC DETECTION OF ANTIBODIES

As shown schematically in Figure 1, the method works as follows: selected synthetic peptides (9–25 amino acids long) are covalently coupled to enzymes such as horseradish peroxidase (HRP), alkaline phosphatase (AP) or β-galactosidase (GL) (Fig. 1A) To obtain a configuration that most resembles that of the native protein a coupling method which provides a random orientation of the peptides is preferred (Zegers et al., 1990). Cryostat sections (8 μm, -20^{0}C) of lymphoid tissue from experimental animals (Laman et al., 1990; Gerritse et al., 1991a) or human autopsy/biopsy material (Laman et al., 1991a) are incubated with the conjugate. When antibodies crossreactive with the peptide-epitope(s) are present in the section the conjugate is bound (Fig. 1B), and the cell will be colored after appropriate cytochemistry. Antibodies visualized in this way can be both cytoplasmic (AFC) or extracellular, the latter in the form of, e.g., immune complexes (colour pictures in: Laman et al., 1990) or on memory B-cells (Liu et al., 1988). By making use of an isotype specific antibody (labeled with a different enzyme) directed against the Fc portion of the Ig molecule the isotype can be simultaneously determined (Fig. 1C), due to the resulting intermediate violet color in cells binding both conjugates (singly stained cells are specific for either the peptide or the isotype under study).

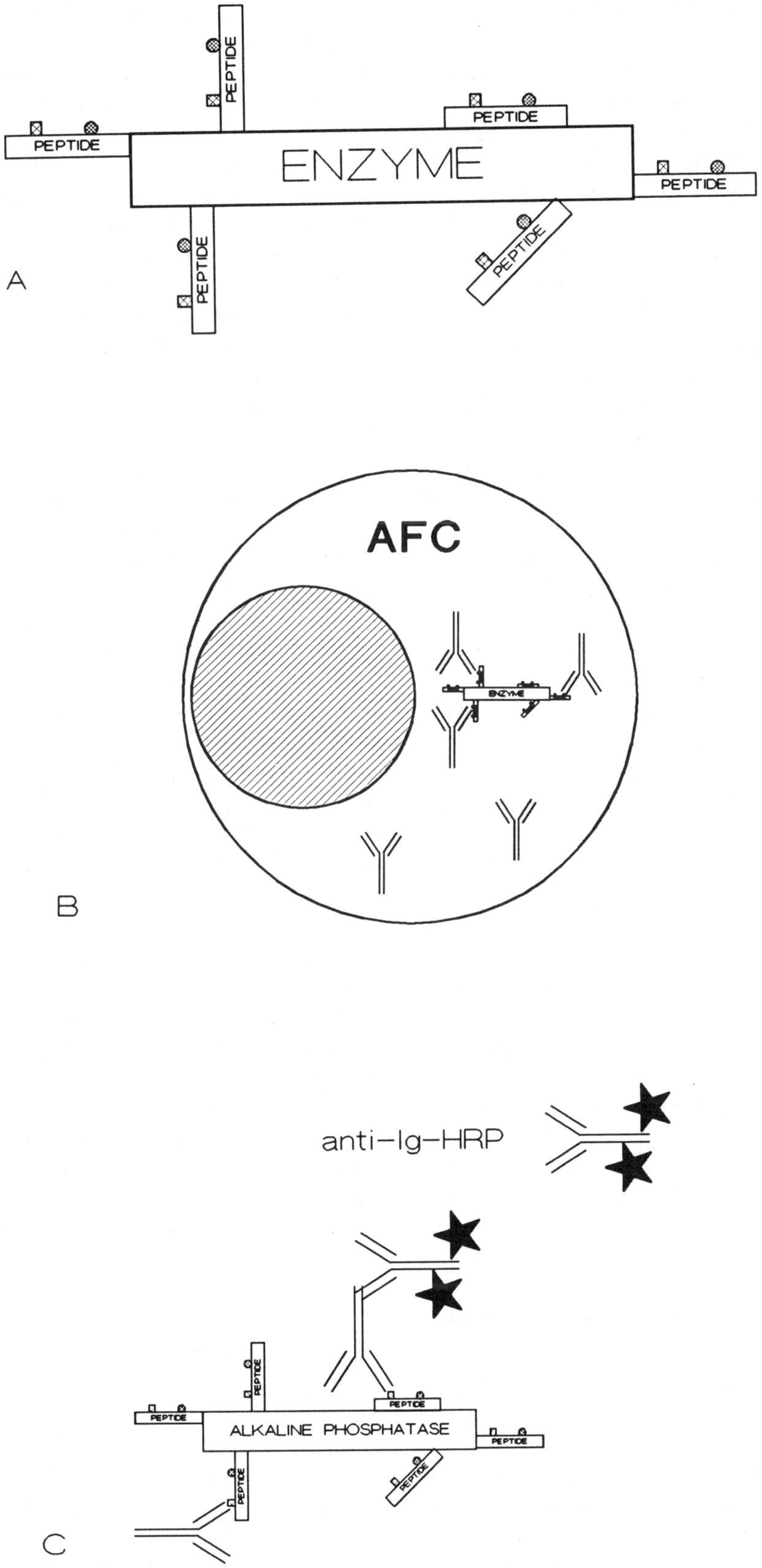

Fig. 1. Synthetic peptides are covalently coupled, in a random orientation, to enzymes such as horseradish peroxidase, alkaline phosphatase or β-galactosidase. Cryostatsections of lymphoid tissue are incubated with the conjugate, and this can be bound by antibodies cross-reactive with the epitopes present in the peptide. By making use of an isotype specific antibody (labeled with a different enzyme) directed against the Fc portion of the Ig molecule the isotype can be simultaneously determined. The number of B-cell epitopes, recognized by the antibodies in the section may vary from one to several and is depending on the length and nature of the peptide.

IV. SELECTION OF PEPTIDES

Peptides are selected primarily based on the biological question under study, e.g., neutralizing epitopes from (putative, subunit) vaccines. Numerous peptide sequences possibly involved in autoimmunity (e.g., from myelin basic protein) have already been put forward and tested in *in vitro* assays and/or animal experiments (Wucherpfennig al., 1991). In the case of a suspected molecular mimicry involved in autoimmunity or pathogenesis (cf. Laman et al., 1989) one should select those epitopes from the protein that resemble native antigenic structures, e.g., parts of a virus that have sequence homology with human proteins (Lernmark et al., 1988) (see also Chapter 20). An educated guess, based on comparison in databases, can in this case be supported by making use of methods for the *in vitro* study and simulation of antigen/protein processing and the release of immunogenic peptides by proteolytic enzymes (cf. Van Noort and Van der Drift, 1989).

For the selection of the exact peptide sequences and the construction of these conjugates we used computer prediction programs and knowledge obtained in a parallel line of research in which we generated diagnostic antibodies by means of selected synthetic peptides, as we reviewed in detail elsewhere (Boersma et al., 1991). A number of new coupling procedures was developed to amplify the proportion of antibodies that recognize the synthetic peptide and also crossreact with the native protein (cf. Deen et al., 1990; Zegers et al., 1990). By adapting these procedures we were able to conjugate synthetic peptides to enzymes such as AP (Laman et al., 1990), HRP and GL (Laman et al., 1991c). Controlled conditions enabled the production of conjugates without loss of enzyme activity and with excellent binding capacities of the synthetic peptides to antibodies *in vitro* (ELISA) and *in situ*. These conjugates were successfully used to detect both AFC and immune complexes directed against selected, B-cell epitopes of HIV-1 proteins such as gp120 (Laman et al., 1991a).

V. SELECTION AND COUPLING OF ENZYMES

Basically, we have employed three enzymes, viz. AP, HRP and GL (see also Table 2).

Alkaline phosphatase. AP was most frequently used, mainly because it can be readily modified with the dialysis glutaraldehyde method to provide simple coupling through amino groups present in immunoglobulins (cf. Claassen and Adler, 1988) and peptides (Zegers et al., 1990). The advantage of this modification of the classical one-step glutaraldehyde method is that no homopolymers (enzyme-enzyme or peptide-peptide) are formed and much less antigen is needed (Zegers et al., 1990). Furthermore, this conjugation can also be performed in 8 M urea with otherwise poorly soluble peptides without significant loss of enzyme activity (Gerritse et al., 1991b). In addition to this, AP can be revealed as a bright transparent blue (cf. Claassen et al., 1986b) or red (cf. Brown and Claassen, 1988) color, both extremely well suited for immunoenzyme cytochemical double and triple stainings (for color pictures, cf. Laman et al., 1991b). The red reaction product obtained through the use of Fast Red TR is also fluorescent and can thus be used for quantitation or fluorescent double/triple labelings (cf. Claassen 1991).

Horseradish peroxidase. In double staining the blue reaction product of AP, we have found HRP in combination with the red immunodecoration by 3-amino-9-ethylcarbazole to yield best results. The more frequently used dye diaminobenzidine is not suited for double staining with AP substrates in our method because the dye is not transparent but (electron)dense. Coupling of peptides to HRP cannot be obtained through amino acid residues on HRP (e.g., by glutaraldehyde) since not enough (only 1–2) are available. Coupling of synthetic peptides should be performed by making use of the periodate oxidation method (cf. Laman et al., 1991c). HRP is more difficult to couple, but on the other hand it is simple to use, cheap (also the substrate), can be easily amplified and the conjugates can be frozen down (-20^{0}C).

β-galactosidase. This enzyme was chosen because its greenish blue substrate provides an opportunity for cytochemical triple and quadruple stainings in combination with AP (blue + red) and HRP (red/brown, NB: GL can be double stained even with diaminobenzidine). Furthermore, after acetone fixation of lymphoid tissue, absolutely no background due to endogenous enzyme activity is observed. This latter fact provides the possibility for prolonged staining times (2–4 hours), and we have found assays with GL conjugates to be far more sensitive than those using HRP or AP (Van den Eertwegh et al., 1992a).

VI. FREEZING AND STORAGE OF TISSUES

After sacrificing the mice, remove immediately (within 5–10 minutes) organs of interest (e.g., lymph nodes, spleen and liver). Organs can be kept in a dry sterile Petri dish on ice for up to 10 minutes (do not immerse or keep in aqueous media!). Tissues are snap frozen in liquid nitrogen directly after isolation.

A few millimetres of water are first put in aluminium tissue containers (Sanbio, Uden, The Netherlands); the water freezes when the containers are placed (floating) in a Dewar vessel with liquid nitrogen and gives a layer of ice which prevents the tissue from drying out during storage (do not freeze the tissues

TABLE 2
Applications of antigen-enzyme conjugates in the detection of antigen specific antibodies

Antigen	Analyte	Sensitivity#	Linearity	Precision	Interferences	Selected references
In situ technique						
Hapten * arsonate	HRP##	10	10–50	100%	Background over 200 μg/ml	Brown and Claassen, 1991
* trinitrophenyl	HRP###	50	50–200	100%	Occasional natural α-TNP antibodies	Claassen and Van Rooijen, 1985
	AP	2	2–25	100%	Endogenous AP in gut and kidney	Claassen and Van Rooijen, 1984
Peptide	AP	8	8–40	100%	Adjuvant induced AFC (controls)	Laman et al., 1990
	HRP/GL	5/10	10–40	100%	Adjuvant induced AFC (controls)	Laman et al., 1991c
Peptide in 8 M urea or DMF	AP	5	10–25	100%	pro-zone effect at high concentrations	Gerritse et al., 1991b
Protein * <160 kd	AP/HRP	5	10–20	100%	—	Van Rooijen and Claassen, 1986
* 160 kd (IgG)	AP/HRP	1	5–20	100%	—	Claassen and Adler, 1988
* >160 kd	HRP	50	50–200	NA	—	Voorbij, 1989
ELISA	HRP/GL	0.01	0.05–2	100%		Laman et al., 1991c
(competitive)	AP	NA	0.2–10	100%		Gerritse et al., 1991b
spot-ELISA	AP	5	5–15	100%	Purity of catching antibody	Vos et al., 1990

Legend see also Table 1. # Sensitivity is defined as the smallest amount of conjugate which still stains 100% of the AFC expressed in micrograms of conjugate per milliliter. Linearity is defined as the range which is used for efficient double staining. ## Conjugate prepared by bridging with bovine serum albumin. ### Conjugate prepared by bridging with poly-L-lysine. DMF = dimethyl formamide; Ig = immunoglobulin G; NA = not assessed.

with water!). The aluminium containers are filled with liquid nitrogen, almost to the brim. Freeze the fresh tissue by simply tossing it into the container. Note three things here. (1) Tissue parts should not be toolarge (>3x3 mm); they may break while freezing. (2) Very small organs (like murine lymph nodes) are hard to section; these can be embedded in a fissure made in a slice of liver. You can use the liver as a support while cutting cryosections. (3) As the morphology of most lymphoid tissues is not uniform throughout its length, we cut them in several pieces, freeze them separately, and cut sections from each part.

Let most of the nitrogen evaporate, then close the container and put it back in the liquid nitrogen. Note two things here: (1) Do not close the container with too much nitrogen in it because the cap may come off with the pressure and tissue might be lost. (2) Do not wait to long with closing the container and putting it into nitrogen because the tissue will thaw and become useless. Containers are stored either at -20 or at -80°C with a well-closed cap, preferably in a freezer that is opened only occasionally. Thawing is fatal: Transport of containers to and from the freezer should be done in liquid nitrogen. Give the tissues enough time to get to -20°C (i.e., from liquid nitrogen or -80°C before cutting!). Eight-micrometer sections are cut at -18 to -20°C (cryostat box at -30°C and the specimen and knife at -20°C) and picked up on glass slides. Try to put the sections as close together as possible (25–35 on a slide) and always include at least one negative and positive control sections on each glass slide. After sectioning, slides are kept overnight in a box with humidified atmosphere! Such a box can be made from a closed plastic container in which wet tissues are placed on the bottom, when the sections become wet with dripping condensed water they are useless! (i.e., test out such a box first). After one night in the "humid" box, air dry sections for 1 hour fix them in 40 ml of fresh acetone (analytical grade) containing 0.02% hydrogen peroxide for exactly 10 minutes at room temperature. Air dry the sections and use them immediately; do not store under any condition.

In those cases where HRP was used 0.02% hydrogen peroxide was added, just before fixation, to inhibit virtually all endogenous peroxidase activity. However, acetone does not effectively inactivate all infectious agents and fixations like 0.37% v/v formaldehyde in PBS or 0.5% paraformaldehyde(w/v) in PBS, for 10 minutes at room temperature are the methods of choice under biohazard conditions (Laman et al., 1991d).

VII. IMMUNOCHEMISTRY AND AMPLIFICATION

Enzymes are revealed as described in detail elsewhere: AP as brilliant blue with Naphthol AS MX phosphate as the substrate and Fast Blue BB Base (in the presence of levamisole to inhibit endogenous AP; Claassen et al., 1986c) or as bright red (also fluorescent) with Fast Red TR Salt (Brown and Claassen, 1988; Claassen, 1991) as the indicator dye. HRP as bright red with 3-amino-9-ethylcarbazole (Claassen et al., 1986c) or brown with 3,3-diaminobenzidine-tetrahydrochloride (Van Rooijen et al., 1989). GL as greenish blue with 5-bromo-4-chloro-3-indolyl-β-D-galactoside (Bondi et al., 1982; Van den Eertwegh et al., 1992a).

The presumed insensitivity of direct techniques does not, in our hands, apply to the detection of AFC with antigen-enzyme conjugates; however, in these rare cases where low staining intensity is observed, a number of amplification steps can be used. Efficient amplification can be performed by prolonged incubation in the substrate medium only with GL; with both HRP and AP it is advisable to change to fresh substrate solution for a second staining after 10 (HRP) and 25 (AP) minutes, respectively. Amplification before (and even after the first staining period) is also possible by a second incubation step with enzyme labeled anti-enzyme antibodies or with enzyme-anti-enzyme-antibody complexes (PAP or APAAP; Mason and Sammons, 1978). Furthermore, specialized conjugates with several enzyme molecules can be prepared, e.g., by the use of poly-L-lysine polymerized HRP (Claassen and Van Rooijen, 1984) or polymerization of AP with glutaraldehyde.

VIII. CONTROLS

Normal controls should include enzyme only, substrate only, and at least one irrelevant conjugate (i.e., same enzyme with different peptide). In animal studies, animals injected with medium and/or adjuvant-only serve as negative controls; in human studies, lymphoid tissue sections of healthy (or unrelated disease) persons must be included on the same glass slide. Furthermore, the staining must be inhibited when free (or conjugated) antigen is added to the incubation medium.

In using synthetic peptides, we found it necessary to incorporate a positive control, this being an enzyme conjugate of the native protein. Upon double staining with a peptide-enzyme-X conjugate and a native protein-enzyme-Y conjugate, all cells staining for peptide reactivity should also stain for the protein. This serves as a 100% control for the three-dimensional configuration of the peptide in relation to the native protein (Laman et al., 1990).

The double staining with an anti-isotype-enzyme conjugate does not only provide detailed information on the isotype distribution of the specific AFC, but it also enables confirmation of the fact that B-cells are stained (positive control). Further experimental and technical controls (when necessary) and additional caveats can be deduced from the paper on limitations of the antigen enzyme approach (Van Rooijen, 1987).

IX. SIMULTANEOUS IDENTIFICATION OF OTHER MARKERS

The main strength in the method for detecting specific AFC lies in the fact that particularly those cells actively involved in a given (experimental or natural) *in vivo* immune response against a defined antigen (or even a single epitope thereof) can be identified. This is in sharp contrast to studies using only anti-Ig stainings in which all immunoglobulin forming cells (also "bystanders") are detected (these cells can also be observed, as a bonus, in the double-label technique we use, as singly stained cells, see no 1C in Fig. 1).

In addition to this, the development and localization of the specific AFC can be studied in relation to, e.g., the localization of the antigen, antigen presenting cells (e.g., macrophages), T-helper and T-suppressor cells, cytokine producing cells and several other cell types involved in induction or regulation of the response (as reviewed by Van den Eertwegh et al., 1992b). In some cases, these double stainings may result in an intermediate, violet color, indicating that both markers under investigation reside in the same cell. Furthermore, this can be done *in situ*, offering an integral picture, complete with cell-cell interactions *in vivo*, of the immune response (see also Laman et al., 1991b; Van den Eertwegh et al., 1993).

X. APPLICATIONS OF THE TECHNIQUE

Assays. Antigen-enzyme conjugates can be effectively used for the detection of specific antibodies *in situ*, as described here (Table 2), but also with superior results in developing of anti-Ig coated, antibody catching, ELISA (enzyme linked sorbent assay; Vos et al., 1990; Laman et al., 1991c) and the even more sensitive variant, the competition-ELISA (Gerritse et al., 1991b). In a spot-ELISA, which is an improved modern version of the classical plaque forming cell assay, these conjugates dramatically enhanced the sensitivity of the assay (10–20 times; Vos et al., 1990).

Theoretically, these conjugates should also perform well in receptor-ligand studies (e.g., hormones; neurotransmitters; cytokines; etc.), but this has not yet been performed.

Experiments. The method we describe here is the only way to analyze the *in situ* interactions which occur during an immune response (cf. Laman et al., 1991b). This is of particular importance when one wants to study the relation of antibody formation with local histopathological effects or pathogenicity (relation with clinical stage; Laman et al., 1991a). In all cases where one cannot efficiently evaluate an immune response due to lack of material (or only frozen tissue available) this method is very suitable, since only little material is needed and the method is sensitive enough to pick up rare AFC. In those cases where antibodies bind directly to the target antigen (e.g., autoimmunity) and are therefore not observed in the blood or liquor (CSF) this is also the method of choice. Furthermore, we use it in tissues where lymphoid cells (and AFC) are scarce such as skin and parts of the lamina propria of the gut.

XI. DISCUSSION

The use of antigen-enzyme conjugates for the detection of specific AFC *in vivo* offers several important advantages (cf. Brown and Claassen, 1988; Claassen and Adler, 1988). The method we describe here for the epitope specific detection of AFC in animals and man with synthetic peptides has additional advantages over other methods. The main advantage is the fact that one can now study the immune response against natural antigens *in situ/in vivo*, in its anatomical context, in relation with other cell types, mediators, or the inducing antigen itself. Also notable is the fact that this can be done without suspending the tissue, thereby destroying cell-cell interactions and possible feedback mechanisms. Furthermore, this is one of the few methods with which one can study antibody formation in stored material, thereby opening up the possibility of retrospective studies on frozen tissue.

ABBREVIATIONS

AFC	Antibody forming cell
AP	Alkaline phosphatase
APAAP	Alkaline phosphatase anti-alkaline phosphatase
ELISA	Enzyme-linked immunosorbent assay
ELISPOT	Enzyme-linked immunospot forming assay
GL	β-galactosidase
HRP	Horseradish peroxidase
PAP	Peroxidase anti-peroxidase
PBS	Phosphate buffered saline

XII. REFERENCES

Boersma WJA, Haaijman JJ and Claassen E. Use of synthetic peptide determinants for the production of antibodies. In: Immunohistochemistry, 2nd Edn., A.C. Cuello, Ed: Wiley & Sons, New York, 1, 1991

Bondi A, Chieregatti G, Eusebi V et al. The use of β-galactosidase as a tracer in immunocytochemistry. Histochemistry 76:153, 1982

Boorsma DM, Claassen E, Kors N et al. Detection of specific anti-hapten-antibody producing hybridoma cells in tissue sections of spleen and liver, by hapten-enzyme conjugates in an immunoenzyme approach. Hybridoma 5:347, 1986

Brown AR and Claassen E. Detection of antibody, idiotype, and anti-idiotype forming cells by *in situ* immunocytochemical staining. J Immunol Meth 109:235, 1988

Brown AR and Claassen E. B lymphocytes in the spleen: Idiotype and anti-idiotype antibody forming cells. Quantitative and qualitative aspects of an *in situ* enzymehistochemical approach. 38th Forum in Immunology (org. Claassen E.), Res Immunol 142:321, 1991

Claassen E and Van Rooijen N. TNP-enzyme conjugate for the detection of anti-TNP antibody producing cells *in vivo*. J Immunol Meth 75:181, 1984

Claassen E and Van Rooijen N. A trinitrophenyl-poly-L-lysine-horseradish peroxidase conjugate for the detection of anti-TNP antibodies *in vivo*. J Histochem Cytochem 33:840, 1985

Claassen E, Kors N and Van Rooijen N. Influence of carriers on the development and localization of anti-trinitrophenyl antibody forming cells in the murine spleen. Eur J Immunol 16:271, 1986a

Claassen E, Kors N, Dijkstra CD et al. Marginal zone of the spleen and the development and localization of specific antibody forming cells against thymus-dependent and thymus-independent type-2 antigens. Immunology 57:399, 1986b

Claassen E, Adler LT and Adler FL. Double immunocytochemical staining for the *in situ* study of allotype distribution during an anti-trinitrophenyl immuneresponse in chimeric rabbits. J Histochem Cytochem 34:989, 1986c

Claassen E, Kors N and Van Rooijen N. Influence of carriers on the development and localization of anti-2,4,6-trinitrophenyl (TNP) antibody forming cells in the murine spleen. II. Suppressed antibody response to TNP-Ficoll after elimination of marginal zone cells. Eur J Immunol 16:492, 1986d

Claassen E, Adler LT, Saito K et al. Functional and histological studies of adoptive immunity in neonatally transplanted rabbits. Immunology 60:37, 1987a

Claassen E, Kors N and Van Rooijen N. Immunomodulation with liposomes: The immune response elicited by liposomes with entrapped dichloromethylene diphosphonate and surface associated antigen or hapten. Immunology 60:509, 1987b

Claassen E and Adler L. Sequential double immunocytochemical staining for the *in situ* identification of an auto-anti-allotype immune response in allotype suppressed rabbits. J Histochem Cytochem 36:1455, 1988

Claassen E. Alkaline phosphatase-fast red fluorescence: Rediscovery of a useful label. (Letter to the Editor) J Immunol Meth 139:149, 1991

Claassen E, Gerritse K, Laman JD et al. New immuno-enzyme-cytochemical stainings for the *in situ* detection of epitope specificity and isotype of antibody forming B-cells in experimental and natural (auto) immune responses in animals and man: A review. J Immunol Meth 150:207, 1992

Deen C, Claassen E, Gerritse K et al. An improved and modified carbodiimide coupling method for synthetic peptides: Enhanced antipeptide antibody responses. J Immunol Meth 129:119, 1990

Gerritse K, Posno M, Schellekens MM et al. Mucosal immune responses and systemic immunological memory after oral administration of TNP-*Lactobacillus* conjugates in mice. In: Lymphatic Tissues and *In Vivo* Immune Responses, Imhof B, Berrih-Aknin S, Ezine S. Eds., Marcel Dekker, New York, 497, 1991a

Gerritse K, Fasbender M, Boersma WJA et al. Conjugate formation in urea to couple insoluble peptides to alkaline phosphatase for ELISA and *in situ* detection of antibody forming cells. J Histochem Cytochem 39:987, 1991b

Gerritse K, Slierendrecht B, Deen C et al. *In situ* detection of specific anti-myelin basic protein antibody forming cells with horseradish peroxidase myelin basic protein conjugates. EOS J Immunol Immunopharmacol 13:63, 1993a

Gerritse K, Deen C, Fasbender M et al. The involvement of specific anti myelin basic protein antibody-forming cells in multiple sclerosis immunopathology. J Neuroimmunol 49:153, 1993b

Hoefakker S, van 't Erve EHM, Deen C et al. Immunohistochemical detection of co-localizing cytokine and antibody producing cells in the extrafollicular area of human palatine tonsils, Clin Exp Immunol 93:223, 1993

Jeurissen SHM, Claassen E, Van Rooijen N et al. Intra-intestinal priming leads to antigen-specific IgA memory cells in peripheral lymphoid organs. Immunology 56:417, 1985

Laman JD, Claassen E, Van Rooijen N et al. Immune complexes on follicular dendritic cells as a target for cytolytic cells in AIDS. AIDS 3:543, 1989

Laman JD, Gerritse K, Fasbender M et al. Double immunocytochemical staining for the *in vivo* detection of epitope specificity and isotype of antibody forming cells against synthetic peptides homologous to human immunodeficiency virus-1, J Histochem Cytochem 38:457, 1990

Laman JD, Rácz P, Tenner-Rácz K et al. Immunocytochemical determination of antigen- and epitope-specificity of HIV-1 specific b-cells in lymph node biopsies from HIV-1 infected persons. AIDS 5/3:255, 1991a

Laman JD, Van den Eertwegh AJM, Claassen E et al. Cell-cell interactions. In: Immune System Accessory Cells, L. Fornusek and V. Vetvicka. Eds. CRC Press, Boca Raton FL, 201, 1991b

Laman JD, van den Eertwegh AJM, Deen C et al. Synthetic peptide conjugates with horseradish peroxidase and -galactosidase for use in epitope-specific immunocytochemistry and ELISA. J Immunol Meth 145:1, 1991c

Laman JD, Kors N, Heeney J et al. Fixation of cryosections under HIV-1 inactivating conditions: Integrity of antigen binding sites and cell surface antigens. Histochemistry 96:177, 1991d

Lernmark A, Dyrberg T, Terenius L et al. Eds. Molecular mimicry in health and disease. Int. Congress Series, Exerpta Medica, Amsterdam, 1988

Liu Y-J, Oldfield S and MacLennan ICM. Memory B-cells in T-cell dependent antibody responses colonize the splenic marginal zones. Eur J Immunol 18:355, 1988

Mason DY and Sammons R. Alkaline phospatase and peroxidase for double immunoenzymatic labeling of cellular constituents. J Clin Pathol 31:454, 1978

Van den Eertwegh AJM, Laman JD, Schellekens MM et al. Complement mediated follicular localization of T-independent antigens: The role of marginal zone macrophages revisited. Eur J Immunol 22:719, 1992a

Van den Eertwegh AJM, Boersma WJA and Claassen E. Immunological functions and *in vivo* cell-cell interactions of T-cells in the spleen. Critical Rev Immunol 11(6):337, 1992b

Van den Eertwegh AJM, Noelle RJ, Roy M et al. *In vivo* CD40-gp39 interactions are essential for thymus dependent humoral immunity. I. *In vivo* expression of CD40 ligand, cytokines and antibody production delineates sites of cognate T-B cell interactions. J Exp Med 178:1555, 1993

Van der Brugge-Gamelkoorn GJ, Claassen E and Sminia T. Anti-TNP-forming cells in bronchus-associated lymphoid tissue (BALT) and paratracheal lymph node (PTLN) of the rat after intratracheal priming and boosting with TNP-KLH. Immunology 57:405, 1986

Van Noort JM and Van der Drift ACM. The selectivity of cathepsin D suggests an involvement of the enzyme in the generation of T-cell epitopes. J Biol Chem 264:14159, 1989

Van Rooijen N. The limitations of immunoenzyme approaches to distinguish between <specific> and <non-specific> antibody forming cells, with particular respect to immunocytochemical studies on the *in situ* immune response. Histochem J 19:369, 1987

Van Rooijen N and Claassen E. Recent advances in the detection and characterization of specific antibody forming cells in tissue sections. A review. Histochem 18:465, 1986

Van Rooijen N, Kors N, Claassen E et al. Binding of different antigen-enzyme and antibody-enzyme conjugates by intracellular antibodies in cytoplasm and Golgi complex of plasma cells. A double immunocytochemical study. Histochemistry 83:61, 1985

Van Rooijen N, Claassen E, Kraal G et al. Cytological basis of immune functions of the spleen. Progress Histochem Cytochem 19/3:1, 1989

Voorbij HAM. Dendritic cells and the development of thyroid autoimmune disease and type 1 diabetes mellitus. Thesis Vrije Universiteit Amsterdam, Thesis Publishers, Amsterdam, 1989

Vos Q, Claassen E and Benner R. Enumeration of anti-insulin antibody-secreting cells using an improved spot-elisa. J Immunol Meth 126:89, 1990

Wucherpfennig KW, Weiner HL and Hafler DA. T-cell recognition of myelin basic protein. Immunol Today 12/8:277, 1991

Zegers N, Gerritse K, Deen C et al. An improved conjugation method for controlled covalent coupling of synthetic peptides to proteins using glutaraldehyde in a dialysis method. J Immunol Meth 130:195, 1990

Chapter 15

MULTIPLE ANTIGEN PEPTIDES FOR IMMUNIZATION

Netty D. Zegers, Eric Claassen and Wim J.A. Boersma

TABLE OF CONTENTS

I. INTRODUCTION

A new approach to generate anti-peptide antibodies has been initially developed by the group of Tam and colleages (Posnett et al., 1988; Tam, 1988). A branched core of lysine residues was constructed by making use of the α and ϵ amino groups of lysine. Subsequently, multiple copies of the peptide epitope of choice could be synthesized onto all arms of the core. The resulting structure has been called multiple antigen peptide (MAP) (Fig. 1).

Two categories of MAP application can be distinguished: (i) as a highly efficient immunogen and (ii) as an efficient antigen for detection in immunoassays. The MAP approach for immunization purposes offers several advantages over the conventional carrier-peptide construct. It is a construct of intermediate molecular mass with a defined orientation of the peptide. Most of the molecule (may be more than 90%) consists of the peptide antigen in contrast to the carrier-peptide conjugate, where the peptide in general is only part of the construct. Ad i. First, probable immunodominant determinants or even suppressor determinants of the carrier protein can be circumvented by applying a MAP. Second, purification is rather easy, and, dependent on the peptide sequence chosen, the construct can be used directly for immunization without coupling to a carrier protein. Third, the peptide dose of the administered MAP is exactly known, in contrast with the peptide dose administered as a peptide-carrier conjugate. In addition, the dendritic peptide chains on the MAP are probably mobile, which may contribute to enhanced immunogenicity. These characteristics make the MAP extremely useful for application as immunogen or as a synthetic vaccine.

Ad ii. MAPs were also shown to be very useful as antigens in solid phase immunoassays. Coating of MAPs to polymer surfaces leads to more efficient epitope exposition than is obtained with free monomeric peptides and thus are very effective for sensitive detection in ELISA (Tam and Zavala, 1989; Habluetzell et al., 1991; Marsden et al., 1992; Marguerite et al., 1992).

Here, we will discuss the MAP immunogen properties only.

II. ALTERNATIVE SYNTHESIS

The basic design of a MAP allows alternative synthesis to offer investigators highly flexible possibilities to explore various abilities of the MAP approach for immunization purposes. By making use of two different protecting groups (Fmoc and tBoc) for lysine at the last level of branching, two different peptide synthesis routes can be applied. Thus, two different peptides can be synthesized onto the core. (Tam and Lu, 1989).

Chemical coupling of a purified peptide with a N- or C-terminal cysteine and a chloroacetylated oligomeric lysine core matrix was described as an alternative for construction of the MAP (Lu et al., 1991). This method offers the advantage to study the influence of orientation of the peptide to the lysine core. It was found that antibodies raised with these MAPs were mainly directed against the distal (and most flexible) part, opposite of the conjugation site, of the peptide (Lu et al., 1991).

Drijfhout and Bloemhoff (1991) described a similar coupling method: the addition of a protected branched lysine core to a functionalized peptide. The lysine core is kept protected by N-(S-acetylmercapto-acetyl)-glutamyl residues. Removal of the S-acetyl

0-8493-8967-4/95/$0.00+$.50

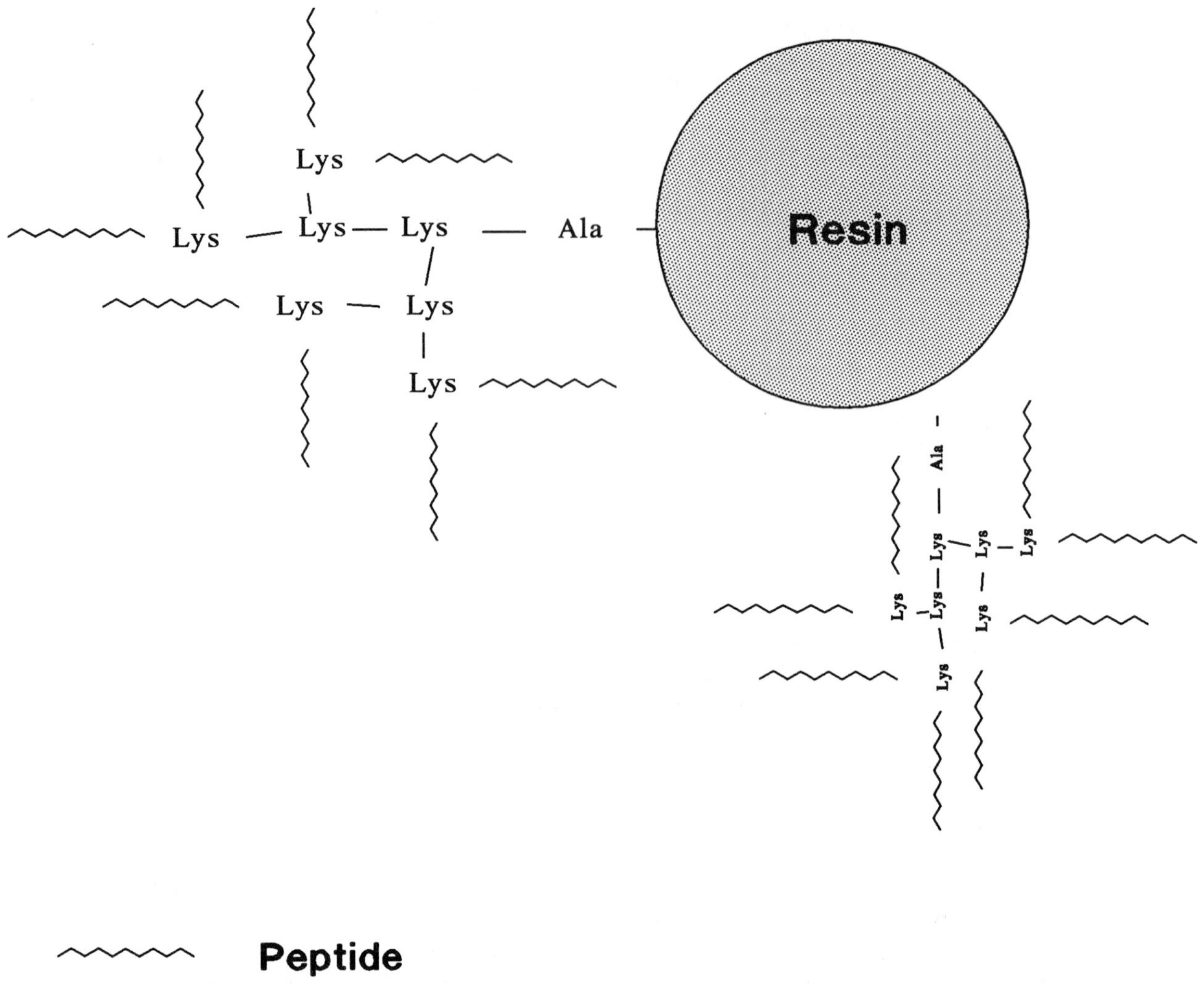

Fig. 1. Schematic drawing of a multiple antigen peptide. To the functional groups of the resin, the first amino acid (β-Ala as an internal standard amino acid) is coupled, and, subsequently, a Lys with two Fmoc protecting groups for the α and ϵ amino groups is coupled. After removing the Fmoc groups, two Lys residues can be coupled. This step can be repeated several times, until the branching of choice has been reached: the core. Four and eight branched cores are applied most frequently. The selected peptide is then synthesized onto the arms of the core. Here, two MAP molecules on a resin bead are shown.

groups just prior to the coupling procedure yields thiol functions which can be used for coupling to, e.g., S-(Npys)-cysteinyl peptides.

Virtually the same approach was followed by Baleux and Dubois (1992). Both the lysine core and the peptide antigen were provided with S-(Npys)-cysteine. Conjugation was carried out by removal of the Npys group of the lysine core with dithiothreitol and subsequent addition of the S-(NPys) peptide.

Coupling of a fully protected peptide, which contained a T cell determinant (TD), and a protected MAP, containing peptides with a B cell determinant (BD), proved to be an effective method for fragment condensation (McLean et al., 1992). The TD peptide can be synthesized in bulk in advance and thus does not require repetitive synthesis of the TD each time a di-epitope MAP is synthesized.

III. MAPS AS IMMUNOGENS

An octameric MAP was found to be highly immunogenic in mice (Balb/c) and rabbits in the first study carried out with MAPs. The peptide (14 residues) was derived from the human T cell antigen receptor β-chain constant region. The antibodies which were raised reacted with the peptide in its monomeric form as well as in its octameric form and with the intact β-chain protein (Posnett et al., 1988). The antibodies were mainly directed against the N-terminal residues of the peptide. No antibodies against the core were detected. The subclass distribution of the subsequently generated monoclonal antibodies was predominantly IgG1 and, to a lesser extent, IgA and IgG2a, which the authors attribute to the T cell dependent nature of the antigen. The authors did not demonstrate that the selected peptide sequence with which the MAP was constructed itself contained a T cell determinant, nor was it investigated whether a new T cell determinant had been formed upon assembly of epitopes by synthesis of the MAP. It is also conceivable that repeating epitopes on the same molecule may cause crosslinking of antigen-specific immunoglobulin receptors on B cells and thus behave as a relatively T cell independent immunogen which can induce isotype switching. However, known T cell independent immunogens as lipopolysaccharide (LPS) or ficoll activate B cells to synthesize predominantly IgG3 besides IgG1. Posnett and colleagues did not investigate the IgG3-specific response.

In a subsequent study, five out of six different MAPs with octameric branched peptides derived from different proteins were highly immunogenic. The antibodies which were induced crossreacted with the native proteins (Tam, 1988). For one determinant investigated the analogous carrier-SP conjugate induced substantial lower levels of antibodies than was observed with the MAP construct.

Co-workers of Marsden compared the efficacy of six different MAPs with conventional carrier-peptide conjugates of the same peptide sequences as immunogens (McLean et al., 1991). The titers of sera from rabbits immunized with branched peptides were higher than those immunized with carrier-peptide conjugates. These findings are confirmed by others (Wang et al., 1991). However, Briand et al. (1992) have tested several peptides sequences, as a carrier-peptide conjugate or as a MAP, in immunization studies. They found that three out of four MAPs were immunogenic, but, most importantly, only one out of these four MAPs raised antibodies crossreactive with the native protein. It appeared that immunization with two MAPs, in which the sequence of the peptides was derived from the C-terminus of a protein, did not result in protein-specific antibodies. The orientation of the determinants with respect to the core matrix may have played a major role. From this, we tend to conclude that there is no advantage in using MAPs with respect to the generation of anti-native protein antibodies with synthetic immunogens.

With the $(NANP)_{10}$ repeat of *Plasmodium falciparum* circumsporozoite protein, it was demonstrated that a MAP construct with 4 branches could overcome genetic MHC restriction. The linear $(NANP)_{10}$ peptide is only immunogenic in H-2^b mice (DelGuidice et al., 1986; Good et al., 1987), while with the $(NANP)_{10}$-MAP construct IgG antibodies were raised in various mouse strains with different MHC background (Pessi et al., 1991). However, MAP constructs with the repeat of other malaria parasites (*P. vivax* and *P. berghei*) did not overcome genetic MHC restriction. It is difficult to explain the finding that the MAP-$(NANP)_{10}$ construct overcame MHC restriction, since it seems a favorable exception. The authors speculated on three hypotheses. (i) A MAP might bind to MHC class II molecules as a multivalent ligand. However, it was shown that not all MAPs could overcome MHC restriction, thus this may only be found with special combinations of MAPs sequences and MHC haplotypes. (ii) (NANP)-specific B cells are activated with a MAP independently of T cell help through Ig crosslinking. However, IgG antibodies were observed and proliferation of lymph node cells from immunized mice was observed. (iii) A MAP follows different proteolytic pathways, as compared to linear peptides, leading to products that bind with higher affinity to Ia molecules. In addition, we think that it is conceivable that new T cell determinants may be formed in the MAP with the $(NANP)_{10}$ sequence.

IV. INCORPORATION OF A T CELL DETERMINANT

Enhancement of the antibody response by addition of a TD peptide in a MAP has been reported (Tam et al., 1990; Munesinghe et al., 1991; McLean et al., 1992; Nardelli et al., 1992a).

It was investigated whether incorporation of a TD in an essentially non-immunogenic BD-MAP would result in an immunogenic construct. MAPs containing 4 or 8 copies of a T cell determinant combined with 0, 1, 4 or 8 copies of a B cell determinant were compared for their efficacy to raise (neutralizing) antibodies (Tam et al., 1990). Also, the complementary MAPs in which the TD and BD were mutually exchanged (4 or 8 copies of the B cell determinant combined with 0, 1, 4 or 8 copies of the T cell determinant) were synthesized and immunized. Moreover, in the same study the effect of orientation of the determinants [$(BD\text{-}TD)_4$, $(BD\text{-}TD)_8$, $(TD\text{-}BD)_4$ or $(TD\text{-}BD)_8$] on the antibody response was investigated. The BD (residues 93–108, repeat domain) and the TD (residues 265–276) were derived from the circumsporozoite protein (CS) of *P. berghei*. The BD monomer and the 4 and 8 branched MAPs containing either the BD or the TD alone did not induce antibodies in mice. The BD-TD monomer induced only a very low level of CS-specific antibodies. In this study, 4 or 8 branched MAPs containing linearly synthesized tandem arrayed BD-TD or TD-BD peptides induced high levels of recombinant CS-specific antibodies and protected mice from chal-

TABLE 1
Peptide sequences

Peptide	Character	Sequence	Derived from
SP149	TD	FERFEIFPKEKC*	TD from influenza (AA 111–120)
SP190	BD	C*DEGKLQHLVNELT	α_1-Antitrypsin (AA 256–268)
SP191	TD-BD	FERFEIFPKEKDEGKLQHLVNELTC*	TD + AA 256–268 from α_1-antitrypsin
SP215	TD and BD	DKTVERKCCVECPPCPAPPVA	Human IgG2 (hinge) (AA 95–115)

* An extra cysteine was added to the sequences for coupling purposes.
TD: T cell determinant; BD: B cell determinant.

lenge (50–80% protection). The (BD-TD)$_4$-MAP scored highest followed by (TD-BD)$_4$-MAP or (TD-BD)$_8$-MAP and (BD-TD)$_8$-MAP. This study showed that incorporation of a T cell determinant in a MAP indeed rendered the construct immunogenic as compared to the MAPs that only contained the BD. Four branches appeared to be sufficient, if not optimal, for eliciting a high immune response in case a TD is incorporated.

After mapping the *P. berghei* circumsporozoite protein for T cell helper activity, Migliorini et al. (1993) chose two apparent T cell determinants (residues 20–39 and 57–70) for constructing multiple antigen peptide constructs comprising only one of the T cell determinants or one of the T cell determinants co-linearly synthesized to a tandem repeat of B cell determinant of the CS protein. Mice immunized with one of the T-B constructs were protected after a challenge with infective sporozoites and a high level of B cell determinant-specific antibodies could be detected. Strikingly, also mice immunized with a MAP construct comprising only the 57–70 T cell epitope were protected, while peptide-specific antibodies were not found. These latter results indicate that protection can be obtained either by effector T cells or by high levels of antibodies. Thus, a double mechanism of protection can be acquired by a synthetic peptide vaccine. With evaluation of the antigen-specific antibody response alone, protection against infection after challenge with a pathogen may be missed.

We have investigated the effect of incorporation of a T cell determinant on the antibody response either in a linear peptide or in a MAP. The immunogens were compared with the linear peptide conjugated to a carrier protein, keyhole limpet hemocyanin (KLH). We determined whether or not a T cell dependent antibody response intrinsically is formed by a MAP without a TD sequence. The T cell responses were assessed in an *in vitro* proliferation assay after priming *in vivo* with the different immunogens.

The peptides used for this study are listed in Table 1. Peptide SP190, a pure B cell determinant, was derived from the human S-variant of α_1-antitrypsin (residues 256–268). SP191 contained the sequence of SP190 which was N-terminally elongated with a TD from the influenza virus hemagglutinin (residue 111–120) (Hackett et al., 1985; Zegers et al., 1993). A third peptide, SP215, was derived from human IgG2 (residues 95–115) and contained an intrinsic TD and BD (Boersma et al., 1989). Peptides SP190 and SP191 were coupled to KLH via an extra added cysteine, and SP215 was coupled via EDC (using carboxyl and amino groups). These peptides were used for immunization in mice (BALB/c) as free peptide, MAPs or conjugated to KLH. Sera were analyzed for specific antibodies in ELISA.

From Fig. 2, it is clear that SP190 (BD) is not immunogenic. The KLH-SP190 conjugate induced peptide-specific antibodies which were crossreactive with α_1AT as expected. When SP190 was offered as a branched peptide (MAP), peptide-specific antibodies were detected and these antibodies were also reactive with the α_1AT protein in ELISA. The level of protein-specific antibodies raised with the MAP was lower than the level raised with the KLH-SP190 conjugate. It can be concluded from these results that a non-immunogenic peptide can be made immunogenic by offering it as a MAP or as a carrier-peptide conjugate.

From Fig. 3, it is clearly seen that incorporation of a TD renders the peptide immunogenic, though the antibody response induced with the free TD-BD peptide (SP191) is low as compared to the antibody responses induced with the MAP or the conjugate including the same sequences. Incorporation of a TD in a MAP therefore specifically enhanced the protein-specific antibody response. The level of the α_1AT-

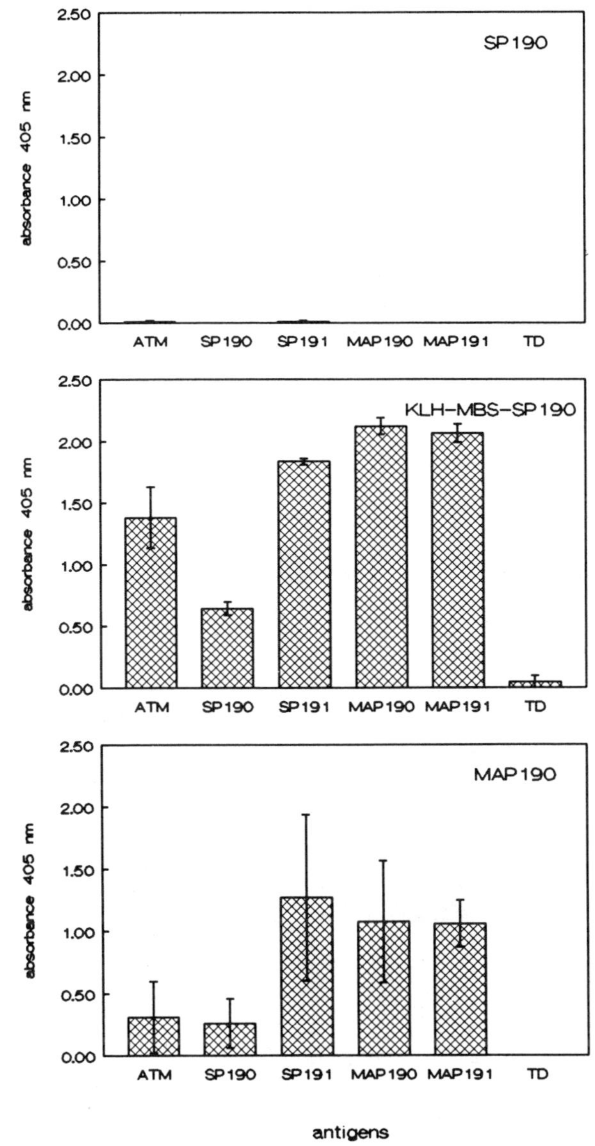

Fig. 2. Comparison of the specific reactivity of mouse sera after immunization with different peptide constructs containing a B cell determinant peptide. Mice were immunized twice with a 4 week interval with the linear peptide SP190 (upper graph), with a carrier-peptide construct KLH-MBS-SP190 (middle graph), or with a branched peptide MAP190 (lower graph). One week after the second immunization, sera were analyzed in ELISA for reactivity with linear peptides [SP190 and SP191, or the T cell determinant (TD)], with branched peptides [MAP190 and MAP191], or with the native protein α_1-antitrypsin [ATM]. Mean serum responses of three mice with standard deviation are shown. See Table 1 for sequences and character of peptides.

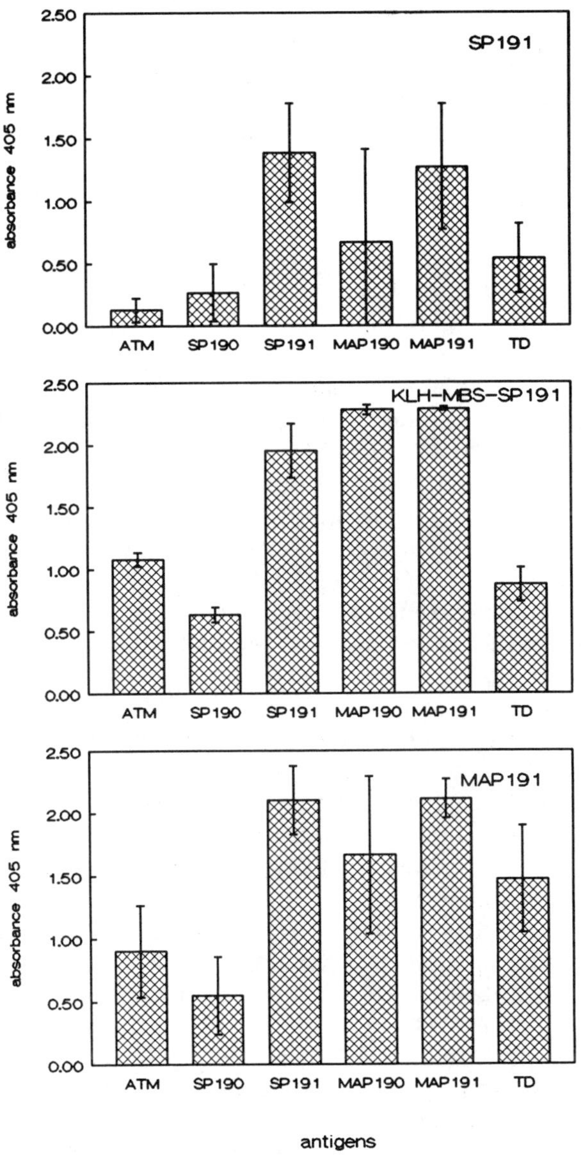

Fig. 3. Comparison of the specific reactivity of mouse sera after immunization with different peptide constructs containing a B and a T cell determinant peptide. Mice were immunized twice (4 week interval) with the linear peptide SP191 (upper graph), with a carrier-peptide construct KLH-MBS-SP191 (middle graph), or with a branched peptide MAP191 (lower graph). Serum immune responses were analyzed in ELISA as described for Fig. 2.

specific antibody response after immunization with MAP191 is similar to the response after immunization with the KLH-SP191 conjugate.

SP215 is immunogenic and induces peptide-specific antibodies crossreactive with the native protein (Fig. 4). Conjugation of this peptide with EDC to a carrier protein decreased the protein-specific response, probably due to modification of the epitope by the coupling procedure. Also with a conjugate made by coupling via the cysteines, protein-specific antibodies were not found (not shown). For this peptide, with many functional groups available for coupling, the MAP approach seems the most appropriate one. Indeed, high levels of protein-specific antibodies were raised with MAP215.

Lymph node cell proliferation assays were performed after *in vivo* priming with the immunogens. Most obviously, *in vitro* restimulation with a peptide or a MAP, which contains a TD, induced a specific enhancement of the proliferation (Fig. 5). Especially, the stimulation indices for restimulation with SP191, MAP191 or SP215 of cells primed with SP191, MAP191, SP215 or MAP215 were of the same order of magnitude as after restimulation with concanavalin A, a polyclonal mitogen (not shown). No proliferation was seen with SP190 and MAP190 (both contain only a BD) restimulation of cells primed with SP190 (BD), MAP190 (BD), SP191 (TD-BD) or MAP191 (TD-BD).

We found subclass distribution patterns (IgG1, IgG2a, IgG2b, IgG3) which were more or less similar for all immunogens used (data not shown). The only apparent difference observed was that the level of BD-specific IgG2b antibodies elicited with MAP190 (BD) was higher as compared to MAP191 (TD-BD). Also the ratio IgG2b/IgG1 was higher after immunization with MAP190 than that found with the other immunogens. It remains to be demonstrated whether a MAP without a TD in the peptide sequence can behave as a T cell independent immunogen capable of inducing isotype switching.

We showed that for strong and native protein-specific responses incorporation of a TD into a MAP can be an advantage. For the sequences used in our study, a MAP with only a BD did not stimulate T cells *in vitro* in a proliferation assay. Despite the absence of T cell activation, specific IgG antibodies (predominantly IgG2b and IgG1) could be detected. The antigen presentation as well as the B cell activation mechanism by MAPs still remains to be elucidated. In the present case, for a peptide which contained several residues sensitive to modification during coupling procedures, the MAP approach is most appropriate.

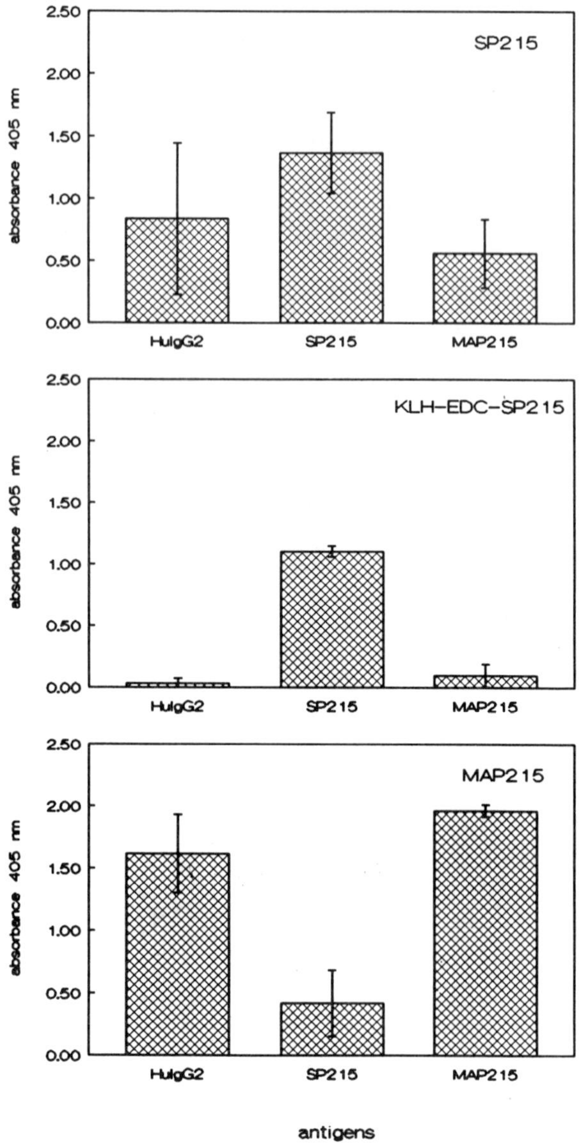

Fig. 4. Comparison of the specific reactivity of mouse sera after immunization with different peptide constructs containing a B cell and a T cell determinant peptide. Mice were immunized twice (4 week interval) with the linear peptide SP215 (upper graph), with a carrier-peptide construct KLH-EDC-SP215 (middle graph), or with a branched peptide MAP215 (lower graph). One week after the second immunization, sera were analyzed in ELISA for reactivity with the linear peptides, SP215, with branched peptide MAP215, or with the native protein human IgG2. See Table 1 for sequences and character of peptides.

V. MAPS AS VACCINES

Major advantages of synthetic peptide vaccines are the avoidance of any pathogenic sequence which is present in the pathogen in its natural or attenuated form and the safe production of the vaccine. Furthermore, the immune response with MAP is directed only against the selected protective peptide sequence. MAPs are expected to be extremely useful for application as vaccines, since they present a multivalent construct with high potential immunogenicity. A MAP is of advantage, as compared to conjugation of a peptide to a carrier protein which is needed for immunogenicity, in that the immune response mounted against the carrier protein as is observed with a conventional carrier-peptide construct can be avoided. With a MAP, the immunodominant B cell determinant, relevant for elicitation of protective antibodies, together with efficient promiscuous T helper or cytotoxic T cell determinants, can be selected.

Several investigators explored the application of MAPs as vaccines in model studies. Mice could be protected from infection with rodent malaria after immunization with a MAP containing a BD and a TD of the circumsporozoite protein of the parasite (Tam et al., 1990; Zavala and Chai, 1990; de Oliveira et al., 1994).

Neutralizing antibodies against the HIV-1 were elicited in guinea pigs after immunization with a MAP containing a rather long sequence (33 amino acids) from the third variable (V3) loop of gp120. The antibody response exceeded the response found with a peptide-bovine serum albumin conjugate prepared with glutaraldehyde with a factor 100. Moreover, the antibody response persisted more than 3.5 years (Wang et al., 1991).

Putative vaccines with intrinsic adjuvant properties were constructed by linking tripalmitoyl-S-glycerylcysteine (P3C) to a MAP with a peptide sequence from the V3 loop of gp120 from HIV-1 which includes a neutralizing B cell epitope, T helper and T-cytotoxic epitopes (Defoort et al., 1992; Nardelli et al., 1992b). The lipophilic membrane-anchoring group (P3C) can be used for inclusion in liposomes. Such complexes induced specific HIV-1 neutralizing antibodies and elicited cytotoxic T lymphocytes. Even with a single administration of the P3C-MAP construct without incorporation in a liposome, the same responses were obtained (Nardelli and Tam, 1994), demonstrating the promising potential use of MAPs as vaccines.

VI. CONCLUDING REMARKS

MAPs offer a very useful alternative for conventional immunogens (linear peptides and peptide-carrier conjugates), provided that a TD is incorporated in the peptide sequence or that a TD is formed by the sequences in the construct. Since a MAP is a spatial structure, conformational determinants also may be formed. MAPS are very flexible in use by various optional synthesis possibilities. In general, the specific antibody response raised with MAPs is higher than the antibody response raised with linear peptides and comparable with or even superior over the antibody response with peptide-carrier conjugate. Moreover, carrier-specific antibodies, immunodominant determinants of the carrier and carrier induced epitope suppression may be prevented by the use of MAPs, which is another important advantage for application in vaccine development.

In case the amino acid sequence of an antigen contains residues which are important for immunogenicity and of which side chains are sensitive to coupling reagents, coupling should be avoided. MAPs then offer an excellent alternative. Besides antibody production, T effector cell mediated protection also can be obtained by applying a MAP as vaccine including a TD.

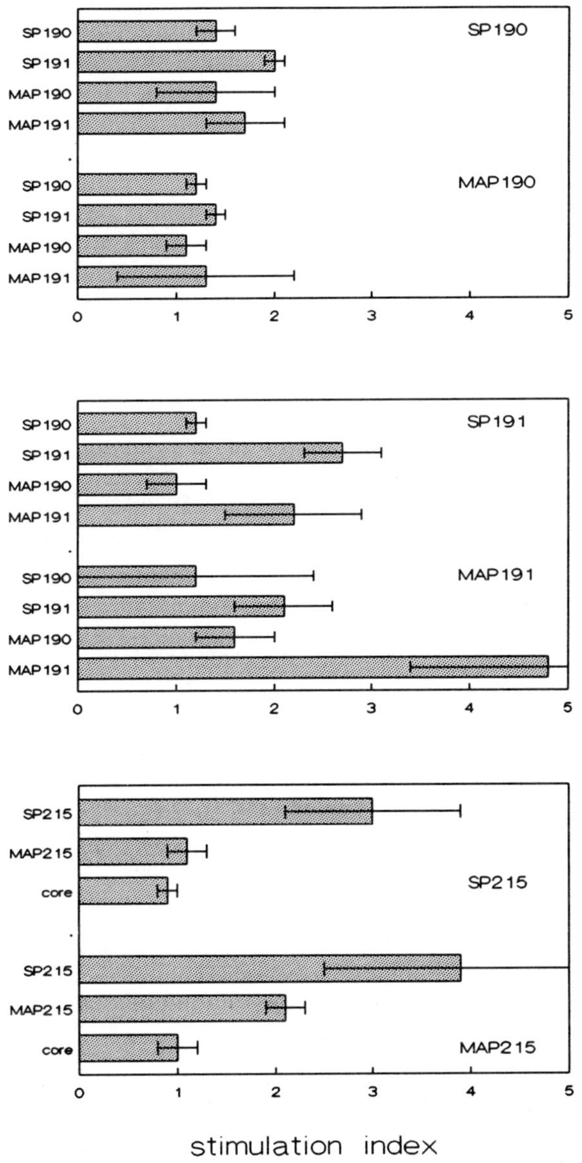

Fig. 5. Proliferation of stimulated lymph node cells primed *in vivo* with a linear peptide or with a MAP. Proliferation of stimulated lymph node cells primed with (Upper graph) SP190 or MAP190, (Middle graph)) SP191 or (Lower graph) MAP191, SP215 or MAP215. Cells were restimulated *in vitro* with different antigens (Y-axis). The mean stimulation index with standard deviation for three mice is shown on the X-axis after ^{3}H-Tdr incorporation and is calculated by dividing the cpm in cultures with antigen by the cpm in cultures without antigen. A stimulation index greater than 2 is significant.

From several studies it has become clear that a MAP with four arms of branching is sufficient to induce an optimal immune response.

ABBREVIATIONS

α_1AT	α_1-Antitrypsin
BD	B cell determinant
CS	Circumsporozoite
EDC	1-Ethyl-3-(3-dimethylaminopropyl)carbodiimide
ELISA	Enzyme-linked immunosorbent assay
Fmoc	9-Fluorenylmethoxycarbonyl
KLH	Keyhole limpet hemocyanin
LPS	Lipopolysaccharide
MAP	Multiple antigen peptides
MBS	m-Maleimidobenzoyl-succinimideester
P3C	Tripalmitoyl-S-glycerylcysteine
t-Boc	*t*-Butoxycarbonyl
TD	T cell determinant

VII. REFERENCES

Baleux F and Dubois P. Novel version of multiple antigenic peptide allowing incorporation on a cysteine functionalized lysine tree. Int J Pept Protein Res 40:7, 1992

Boersma WJA, Deen C, Gerritse K et al. Anti-peptide antibodies as subclass specific reagents: Epitope mapping of human IgG2. Prot Biol Fluids 36:161, 1989

Briand JP, Barin C, van Regenmortel MHV et al. Application and limitations of the multiple antigen peptide (MAP) system in the production and evaluation of anti-peptide and anti-protein antibodies. J Immunol Meth 156:255, 1992

Defoort J-P, Nardelli B, Huang W et al. Macromolecular assemblage in the design of a synthetic AIDS vaccine. Proc Natl Acad Sci USA 89:3879, 1992

Drijfhout JW and Bloemhoff W. A new synthetic functionalized antigen carrier. Int J Pept Protein Res 37:27, 1991

DelGuidice G, Cooper JA, Merino J et al. The antibody response in mice to carrier-free synthetic polymers of *Plasmodium falciparum* circumsporozoite repetitive epitope is IA^b-restricted: Possible implications for malaria vaccines. J Immunol 137:2952, 1986

Good MF, Maloy WL, Lunde MN et al. Construction of synthetic immunogen: Use of new T-helper epitope on malaria circumsporozoite protein. Science 235:1059, 1987

Habluetzell A, Pessi A, Bianchi E et al. Multiple antigen peptides for specific detection of antibodies to a malaria antigen in human sera. Immunol Lett 30:75, 1991

Hackett CJ, Hurwitz JL, Dietzschold B et al. A synthetic decapeptide of influenza virus hemagglutinin elicits helper T cell with the same fine recognition specificities as occur in response to whole virus. J Immunol 135:1391, 1985

Lu Y, Clavijo P, Galantino M et al. Chemically unambiguous peptide immunogen: Preparation, orientation and antigenicity of purified peptide conjugated to the multiple antigen peptide system. Mol Immunol 28:623, 1991

Marguerite M, Bossus M, Mazingue C et al. Analysis of antigenicity and immunogenicity of five different chemically defined constructs of a peptide. Mol Immunol 29:793, 1992

Marsden HS, Owsianka AM, Graham S et al. Advantages of branched peptides in serodiagnosis. Detection of HIV-specific antibodies and the use of glycine spacers to increase sensitivity. J Immunol Meth 147:65, 1992

McLean GW, Owsianka AM, Subak-Sharpe JH et al. Generation of anti-peptide and anti-protein sera. Effect of peptide presentation on immunogenicity. J Immunol Meth 137:149, 1991

McLean GW, Cross AM, Munns MS et al. Rapid attachment of a helper T cell epitope to branched peptides by fragment condensation to give enhanced immunogenicity. J Immunol Meth 155:113, 1992

Migliorini P, Bretschart B and Corradin G. Malaria vaccine: Immunization of mice with a synthetic T cell helper epitope alone leads to protective immunity. Eur J Immunol 23:582, 1993

Munesinghe DY, Clavijo P, Calle MC et al. Immunogenicity of multiple antigen peptides (MAP) containing T and B cell epitopes of the repeat region of *P. falciparum* circumsporozoite protein. Eur J Immunol 21:3015, 1991

Nardelli B, Lu Y, Shiu DR et al. A chemically defined synthetic vaccine model for HIV-1. J Immunol 148:914, 1992a

Nardelli B, Defoort J-P, Huang W et al. Design of a complete synthetic peptide-based AIDS vaccine with a built-in adjuvant. AIDS Res Human Retroviruses 8:1405, 1992b

Nardelli B and Tam JP. Cellular immune responses induced by *in vivo* priming with a lipid-conjugated multimeric antigen peptide. Immunology 79:355, 1994

de Oliveira GA, Clavijo P, Nussenzweig RS et al. Immunogenicity of an alum-adsorbed synthetic multiple-antigen peptide based on B and T cell epitopes of the *Plasmodium falciparum* CS protein: Possible vaccine application. Vaccine 12:1012, 1994

Pessi A, Valmori D, Migliorini P et al. Lack of H-2 restriction of the *Plasmodium falciparum* (NANP) sequence as a multiple antigen peptide. Eur J Immunol 21:2273, 1991

Posnett DN, McGrath H and Tam JP. A novel method for producing anti-peptide antibodies. J Biol Chem 263:1719, 1988

Tam JP. Synthetic peptide vaccine design: Synthesis and properties of a high-density multiple antigenic peptide system. Proc Natl Acad Sci USA 85: 5409, 1988

Tam JP and Zavala F. Multiple antigen peptide. A novel approach to increase detection sensitivity of synthetic peptides in solid-phase immunoassays. J Immunol Meth 124:53, 1989

Tam JP and Lu Y-A. Vaccine engineering: Enhancement of immunogenicity of synthetic peptide vaccines related to hepatitis in chemically defined models consisting of T and B cell epitopes. Proc Natl Acad Sci USA 86:9084, 1989

Tam JP, Clavijo P, Lu Y et al. Incorporation of T and B epitopes of the circumsporozoite protein in a chemically defined synthetic vaccine against malaria. J Exp Med 171:299, 1990

Wang CY, Looney DJ, Li ML et al. Long-term high-titre neutralizing activity induced by octameric synthetic HIV-1 antigen. Science 254:285, 1991

Zavala F and Chai S. Protective anti-sporozoite antibodies induced by a chemically defined synthetic vaccine. Immunol Lett 25:271, 1990

Zegers ND, van Holten C, Boersma WJA et al. Peptide induced memory (IgG) response, crossreactive with native proteins, requires covalent linkage of a specific B cell epitope with a T cell epitope. Eur J Immunol 23:630, 1993

Chapter 16

EFFECTS OF CONFORMATION, AMINO ACID SEQUENCE, AND CARRIER COUPLING ON THE IMMUNOLOGICAL RECOGNITION OF PEPTIDE AND PROTEIN ANTIGENS

Ronald Jemmerson

TABLE OF CONTENTS

I. INTRODUCTION

Applications for anti-peptide antibodies have been found in a number of areas of biology and medicine, including vaccine development and protein purification and identification. Although anti-peptide antibodies as reagents have been very useful in many cases, frequently they are not effective. This is a subject that has not been thoroughly reviewed. In this chapter, limitations in the use of anti-peptide antibodies are discussed along with plausible explanations for those limitations and attempts that have been made to get around them. Many of the examples presented are from studies carried out in the author's laboratory. Examples from the work of others are included to substantiate the author's findings. However, the work cited should not be considered exhaustive of the information available in the literature regarding this subject.

II. MODEL FOR B LYMPHOCYTE ACTIVATION

To fully understand the biochemical basis for limitations in the use of anti-peptide antibodies it is helpful to consider the current model by which B lymphocytes are activated, leading to an antibody response (Fig. 1). B lymphocytes with surface immunoglobulin having the appropriate specificity bind the form of the antigen present in an inoculum (Raff et al., 1970; Pernis et al., 1971). Usually, but not always, this is the native form of the antigen (see later discussion). The antigen is taken up via surface immunoglobulin and processed in endocytic vesicles (Lanzavecchia, 1985; Myers, 1991). Processing involves the breakdown of a protein antigen into smaller peptides by proteases (Brodsky and Guagliardi, 1991). Some of the resulting peptides, usually 13–16 amino acids long (Rudensky et al., 1991), have appropriate chemistries for associating with major histocompatibility complex (MHC) class II molecules (Rothbard and Gefter, 1991). T cells with receptors capable of binding the binary complex of peptide and MHC class II molecule can then associate with the B lymphocyte (Schwartz, 1985). In addition to the T cell receptor, several other molecules on the T lymphocyte interact with ligands on the B lymphocyte. These include adhesion molecules such as LFA-1 and CD2 (Springer, 1990) and other receptors that deliver signals to the T lymphocyte such as CD28 (Liu and Linsley, 1992). Lymphokines produced by the T lymphocyte in response to these signals are delivered to the B lymphocyte (Miyajima et al., 1992). In addition, signaling through surface immunoglobulin and other molecules such as MHC class II (Cambier et al., 1987) and CD40 (Noelle et al., 1992) result in B lymphocyte proliferation and differentiation to the antibody-producing plasma cell stage. Some of the lymphokines produced by the T lymphocyte such as γ-interferon and interleukin-4 affect the heavy chain isotype of the antibody produced by the activated B lymphocyte (Coffman et al., 1988).

The physiology of B and T lymphocyte activation clearly demonstrates a dichotomy by which these cells recognize antigen. While B cells recognize the immu-

0-8493-8967-4/95/$0.00+$.50

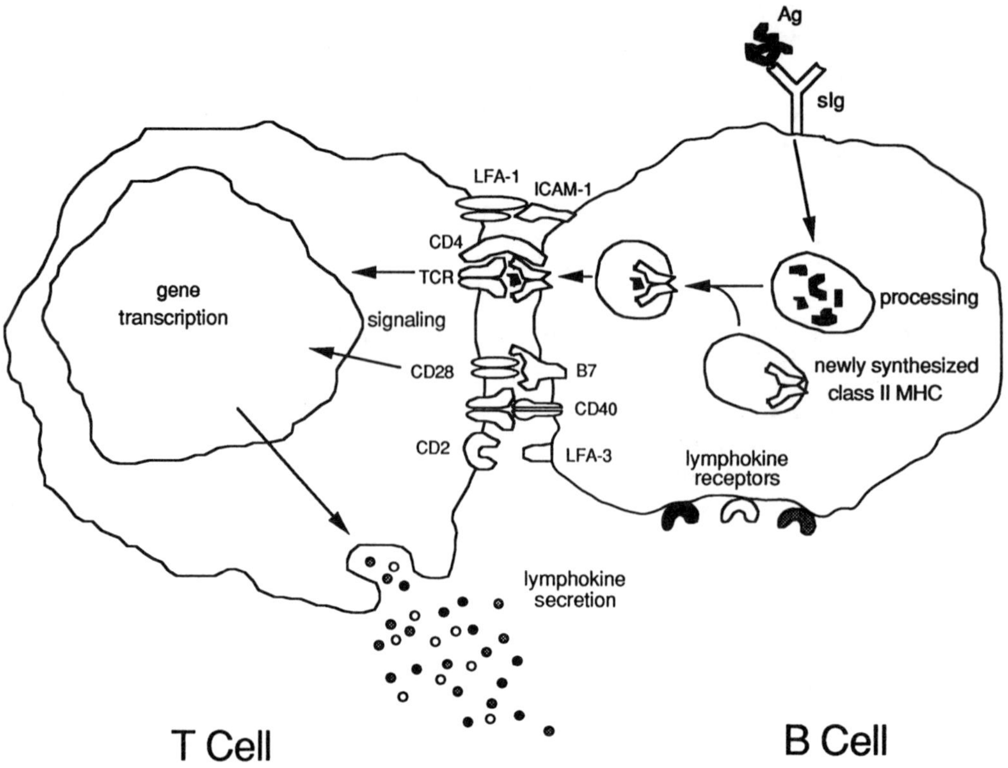

Fig. 1. A model for B lymphocyte activation. The B lymphocyte recognizes the native protein antigen and the T lymphocyte recognizes peptides in association with MHC class II molecules. See text for details.

nizing form of a protein antigen, T cells recognize short peptides in complex with MHC class II molecules. Thus, peptides are invariably effective in substituting for the intact antigen in regard to T lymphocyte recognition (Schwartz, 1985; Rothbard and Gefter, 1991). However, whether antibodies produced by the B lymphocyte bind peptides depends on several factors. These are discussed in later sections. There is no evidence that after injection into an animal, antigen is modified in any way prior to B lymphocyte recognition. Antigen does associate with follicular dendritic cells in the spleen and lymph nodes in the form of immune complexes (Szakal et al., 1989), but the antigen is probably not degraded or denatured by the cells.

The conformational specificity of protein antigens studied in the author's laboratory supports the idea that processing of protein antigens does not occur prior to B lymphocyte recognition of antigens. When mice were injected intraperitoneally with chicken avidin in aqueous solution, essentially all the antibodies elicited reacted with native avidin. This was shown by the ability of soluble, native avidin to inhibit the binding of the antibodies to avidin adsorbed to microtiter wells in a solid-phase enzyme-linked immunosorbent assay (ELISA)(see Fig. 2). In solid-phase immunoassays, some antigen molecules denature as they attach to the plastic (Friguet et al., 1984; Hollander and Katchalski-Katzir, 1986). Thus, antibodies specific for both native and denatured or peptide forms can be detected in such assays. However, only antibodies that bind to the native form are inhibited by the addition of soluble, native antigen to the wells along with antibody (Jemmerson, 1987). When the anti-avidin antibodies were tested for binding peptides of avidin represented in a cyanogen bromide digest of reduced and alkylated avidin, no binding of the antibodies to the peptides in a solid-phase ELISA was observed (results not shown). This supports the notion that when a protein antigen is present in an inoculum in its native form, the B lymphocytes with appropriate surface immunoglobulin receptors recognize the native form and respond to it. Processing of the antigen prior to B lymphocyte recognition probably does not occur.

III. *IN VITRO* DENATURATION OF PROTEIN ANTIGENS

Certain manipulations of protein antigens during the preparation of an inoculum can result in the denaturation of at least a small fraction of the protein molecules. These manipulations include lyophilization of the protein and emulsification of the antigen in mineral oil adjuvants. The presence of denatured forms of the antigen in the inoculum leads to the activation of B lymphocytes specific for these forms. The antibodies that result bind denatured antigen and peptides, but may or may not bind the native protein (see later discussion).

It was previously documented that emulsification of a protein antigen in mineral oil can result in protein denaturation. Scibienski (1973) compared the conformational specificity of antibodies elicited against hen egg-white lysozyme either injected emulsified in adjuvant or as an inoculum in aqueous solution. Antibodies cross-reactive with carboxymethylated, reduced lysozyme (a denatured form) were readily apparent in the sera of guinea pigs injected with the antigen emulsified in adjuvant, but not in sera from animals injected with the antigen in aqueous solution. In another study, human hemoglobin was emulsified in complete Freund's adjuvant (which contains mineral oil) and the electron paramagnetic resonance (EPR) spectrum was recorded and compared to the spectrum of soluble hemoglobin (Berzofsky et al., 1976). Denaturation effects would have been apparent by new signals appearing in the $g = 2$ range and by the broadening of the principle signal at $g = 6$. It is doubtful that a broadening of the EPR signal would have been useful in detecting denatured forms that only amounted to a few percent of the total population and which broadened the peaks ever so slightly. At higher attenuation, methemoglobin in adjuvant did show low spin signals possibly indicative of denatured forms. However, a similar attenuation of soluble methemoglobin was not presented, so the results are difficult to assess. What is clear from this study is that at most only a small fraction of hemoglobin molecules was denatured by emulsification in adjuvant. These denatured molecules can be highly immunogenic and can elicit antibodies in significant amounts that react with denatured antigen but not with native antigen.

Whether lyophilization affects the integrity of protein conformation depends on the nature of the protein. There are documented cases where the specific activity of a protein can be decreased by lyophilization (Townsend and DeLuca, 1988; Goodnough and Johnson, 1992). Bovine pancreatic ribonuclease was shown to aggregate upon lyophilization from aqueous solution, the aggregates amounting to 12% of the total protein (Crestfield et al., 1962). Lyophilization was shown to alter the antigenic sites of apolipoprotein B in a radioimmunoassay where as much as 26% of the antigenic activity was lost (Marcovina et al., 1992). In the same study, it was found that lyophilization did not have a detectable affect on the antigenicity of apolipoprotein A-1.

In our studies of the conformational requirements for antibody binding to cytochrome *c*, we have frequently obtained monoclonal antibodies specific for non-native forms of the protein (Jemmerson, 1987; Jemmerson et al., 1991). To determine if these may have arisen in response to variant conforms of cytochrome *c* resulting from lyophilization of the protein, we examined the effect of lyophilization on the conformation of the protein. To do this, we employed a T cell activation assay that can distinguish between native and non-native forms of a protein antigen. As noted above, T cells require interaction with antigen-presenting cells (for example, the B lymphocyte in Fig. 1) for activation. If the antigen-presenting cells are fixed with paraformaldehyde, then they cannot process native antigens and can only present forms of the antigen that are already in an appropriately unfolded state to bind the MHC class II molecule (Shimonkevitz et al., 1983). Neither native pigeon cytochrome *c*, kept in aqueous solution following its purification, nor the lyophilized protein in conjunction with fixed antigen-presenting cells leads to activation of a T hybridoma which is specific for the carboxyl terminal segment (81–104) of pigeon cytochrome *c* (Fig. 3). This indicates that lyophilization does not have a significant denaturing effect on cytochrome *c*, at least in the region of the protein to which the T cells respond. Since this is the same region recognized by some antibodies that bind denatured cytochrome *c* (Jemmerson et al., 1991), it is apparent that lyophilization is not responsible for denaturation of the protein that led to the elicitation of these antibodies. Interestingly, commercial preparations of pigeon cytochrome *c* were found to be stimulatory in this assay (results not shown), emphasiz-

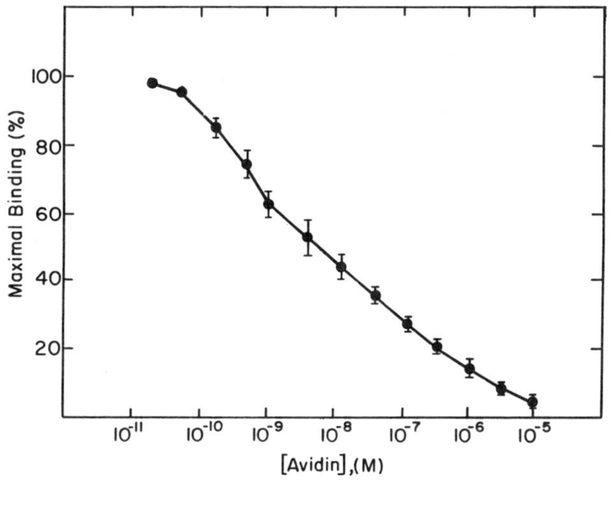

Fig. 2. Competitive ELISA for mouse (C57Bl/6 X C3H/HeJ) antibodies elicited against avidin injected intraperitoneally in aqueous solution. Sera from five mice were diluted 1:2000, the titer found to saturate 50% of the antigen binding sites on the microtiter plate. Soluble avidin at various concentrations was tested for blocking the binding of these antibodies to avidin adsorbed to the microtiter wells. Error bars represent the standard deviations. The results show that essentially all the antibodies elicited against soluble avidin bind the native protein.

ing the need for further purification of the protein before use to eliminate denatured forms.

One should always be aware that even for the most carefully purified protein some of the molecules, perhaps even less than 1%, can be denatured by certain manipulations considered routine. Thus, from an inoculum, B lymphocytes can be exposed to both native and non-native forms of the antigen. The non-native forms lead to the production of antibodies that cross-react with peptides even when antibodies to native forms may not. It is, therefore, important to distinguish between these two types of antibody specificities. This can be accomplished by including appropriate controls in the assays to examine antibody binding.

IV. ARTIFACTS OF IMMUNOASSAYS

Antibodies are extremely sensitive to the conformation of a protein antigen (Benjamin et al., 1984). If the immune system of an animal is presented with a protein antigen in its native form, then the antibodies elicited by the activated B lymphocytes are generally specific for the native form. The antibodies may or may not react with peptides. Similarly, if the immune system is presented with non-native forms of a protein antigen, whether denatured conforms or peptides, the antibodies produced will favor recognition of peptides and denatured protein and may or may not react with the native antigen. The extent of cross-reactivity in both cases depends on the type of protein domain recognized by the B lymphocyte, be it globular or linear like. Antibodies to linear-like or random coil domains of proteins readily interact with peptides, while antibodies to globular domains do not. Although this was the basis for the earlier definition of sequential versus conformational determinants (Sela, 1969) , in recent years this dichotomy has not been appreciated due to artifacts in assays systems which have not been appropriately controlled.

Most immunoassays involve the non-covalent attachment of antigen to plastic (Engvall, 1980). Antibodies then bind to the antigen, and their binding to the solid phase can be detected using a second antibody that is appropriately labeled. During the adsorption of protein antigens to plastic in the solid phase, some molecules are denatured exposing sequential determinants not present on the native protein (Friguet et al., 1984; Hollander and Katchalski-Katzir, 1986). Since protein antigens can be denatured *in vitro* prior to inoculation into an animal, as discussed above, and since both native and denatured forms of antigen are in the solid phase, antibodies to both native and non-native forms which are present in an antiserum will be detected in a typical immunoassay. To distinguish between the two types of antibodies, it is necessary to examine the ability of soluble, native antigen to inhibit the binding of antibodies to the solid phase (Jemmerson, 1987). In Fig. 2, the specificity of anti-avidin antibodies for native avidin was demonstrated in such an assay where increasing amounts of soluble avidin added to the microtiter wells prior to the addition of the antibodies showed an increasing ability to inhibit the binding of the antibodies to avidin adsorbed in the wells. An avidin concentration of 10^{-5} M essentially completely blocked the binding of the antibodies to insolubilized avidin. We have observed similar results in studies of the conformational specificity of cytochrome *c* where 10^{-5} M antigen was sufficient to completely inhibit the binding of anti-native cytochrome *c* antibodies to cytochrome *c* adsorbed in the solid phase (Jemmerson, 1987).

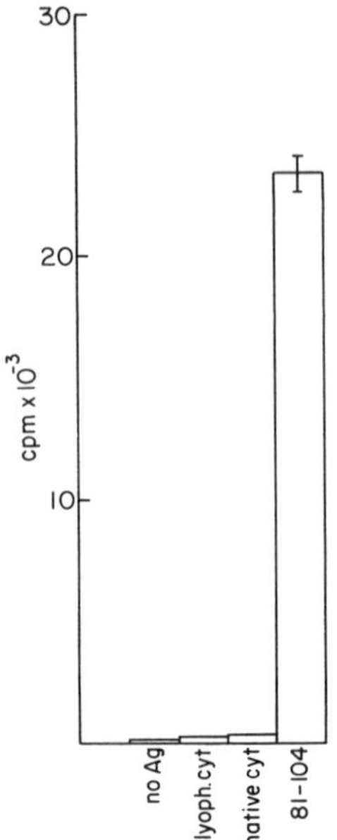

Fig. 3. T cell response to native and lyophilized pigeon cytochrome *c* using fixed antigen-presenting cells demonstrates the absence of denatured antigen molecules in the lyophilized preparation. The T hybridoma, 2B4, specific for the pigeon cytochrome *c* peptide 81-104 was employed (see Jemmerson et al., 1991 for details of the assay). The antigen-presenting B cell hybridoma, LK35.2, fixed with paraformaldehyde is unable to process antigen and can only present through its MHC class II molecules non-native forms and peptides of the antigen provided exogenously. T hybridoma recognition of the complexes was determined by assay of the production of interleukin-2 using as the readout, proliferation (^{3}H-thymidine incorporation) of the interleukin-2 dependent cell line CTLL. The concentration of the antigens was $1x10^{-5}$ M which resulted in the maximal response to the 81-104 peptide. In this system, the T cells responded well to the 81-104 peptide and did not respond to either native pigeon cytochrome *c* or the lyophilized protein. This indicates that lyophilization does not denature cytochrome *c* to any detectable extent, at least in the region recognized by the T hybridoma.

Often anti-protein antibodies are examined for binding peptides in immunoassays. In this case, it is important to demonstrate that the antibodies can be inhibited by the native protein to confirm that the peptide-reactive antibodies do, in fact, bind the native protein. When avidin is emulsified in complete Freund's adjuvant for immunization of mice, some of the antibodies elicited when tested in ELISA do bind peptides derived from the cleavage of the reduced and alkylated protein with cyanogen bromide (Fig. 4A). These peptide-reactive antibodies are present in very low titer (approximately 0.01%) relative to the total anti-avidin antibodies, yet they are readily detected when the peptides are used as antigens in the immunoassay (Fig. 4A). These antibodies cannot be inhibited in binding the peptides by the addition of soluble, native avidin (Fig. 4B). Even at 10^{-5} M soluble avidin, no inhibition is observed. This indicates that the peptide-reactive antibodies do not bind native avidin. They probably arose in response to non-native forms derived from the emulsification of the protein in mineral oil adjuvant since they were not apparent in sera from mice immunized with avidin in aqueous solution.

Similar results are observed when monoclonal antibodies are examined. For example, as shown in Fig. 5, two monoclonal antibodies specific for cow cytochrome *c* were examined for binding native cow cytochrome *c* in a competitive ELISA. These antibodies were prepared using standard hybridoma technology (Jemmerson and Johnson, 1991). One of the monoclonal antibodies (4H9) did bind native cow cytochrome *c* since its binding to cytochrome *c* on the microtiter plate was completely inhibited by soluble cytochrome *c* at a concentration of 10^{-5} M. The other monoclonal antibody (1D3) did not bind native cow cytochrome *c* since its binding to cow cytochrome *c* adsorbed to the plate was not inhibited by the soluble protein. Although both of the monoclonal antibodies were shown to bind the region around residue 44 by testing their binding to cytochromes *c* that varied at this position, they recognized different conformations of the protein. Monoclonal antibody 1D3 that failed to bind the native protein did react with peptides containing sequences around residue 44 of cow cytochrome *c* (Fig. 6). In competitive ELISA, the antibody bound with high affinity the 1–80 fragment that was derived by cyanogen-bromide cleavage, bound the synthetic peptide 40–54 with lower affinity, and did not bind the synthetic peptide 41–48. This antibody may be conformationally dependent, even though it does not bind the native conformation. As the peptides get shorter and less capable of maintaining the appropriate conformation or as contact residues are lost by the shortening of the peptides, the affinity for antibody decreases. The other antibody specific for native cow cytochrome *c* did not bind these peptides.

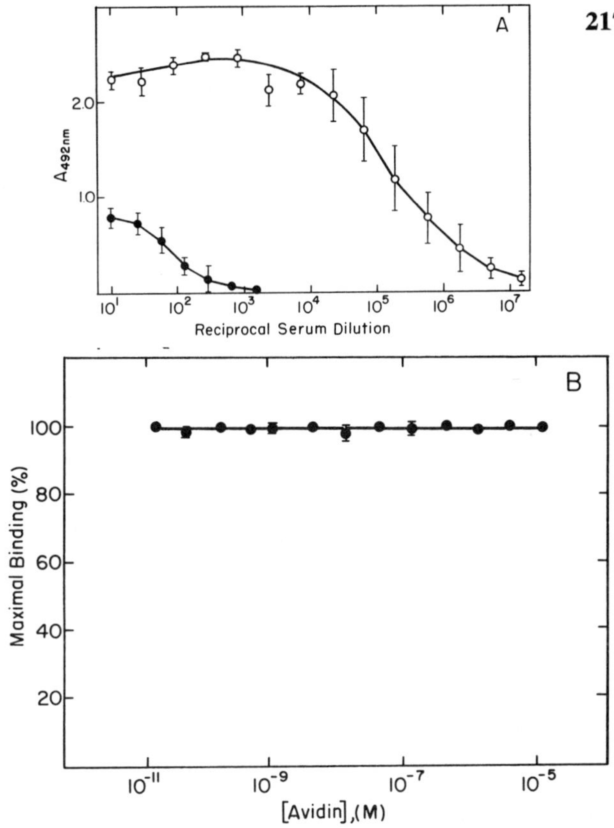

Fig. 4. Conformational specificity of polyclonal mouse antibodies elicited against avidin emulsified in complete Freund's adjuvant. In (A), the antibodies were tested for binding avidin (○) and cyanogen bromide-cleaved peptides of reduced and alkylated avidin (●) by ELISA with the antigens adsorbed to microtiter wells. In (B), an attempt was made to block the binding of the antibodies to the peptides on the assay plate in the presence of increasing amounts of soluble avidin. The peptide-reactive antibodies did not bind the native protein.

V. RECOGNITION OF NATIVE PROTEIN ANTIGENS BY ANTI-PEPTIDE ANTIBODIES

Many of the reported cases of anti-peptide antibodies binding proteins do not actually involve recognition of the native antigen. The control experiments necessary to distinguish between binding of antibodies to native or non-native antigen forms, discussed previously, are often not included in immunoassay protocols. Thus, there is considerable confusion in the literature on this matter, and the frequency of antibodies binding both native proteins and peptides is highly exaggerated. In a study of several hundred monoclonal antibodies elicited against rat cytochrome *c*, the vast majority of antibodies that were inhibited by soluble, native cytochrome *c* in ELISA did not bind cyanogen bromide-cleaved fragments (Jemmerson, 1987). In contrast, most of the antibodies which bound the fragments were not inhibited by soluble, native cytochrome *c*. Thus, it is clear that these antibodies are not specific for the native antigen, but rather recognize denatured forms adsorbed to the ELISA plate. Results of a more recent but similar study support these findings (Schwab et al., 1993). Spangler (1991) examined a series of anti-peptide monoclonal antibodies (seven total) reactive with three different protein antigens (cholera toxin, myohemerythrin, and sickle hemoglobin) for their conformational specificity. She found that while all of the antibodies reacted with the antigens in solid-phase assays (where some antigen molecules denature upon adsorption to the microtiter wells), none of the antibodies bound the soluble proteins. This was shown by size-exclusion high performance liquid chromatography where the elution profile of the antigen was not affected by the presence of the specific antibody and by competitive ELISA. Since the three protein antigens employed in the study are all globular, the findings are consistent with the contention that anti-peptide antibodies cannot bind globular domains on native globular proteins.

Confirmation that one of the monoclonal antibodies in Spangler's study did not bind the native protein came from X-ray crystallography studies (Stanfield et al., 1990). Fab fragments of the monoclonal antibody B13I2, specific for the myohemerythrin peptide containing residues 69–87, were crystallized in complex with the peptide. This peptide contains a histidine at position 73, which in native myohemerythrin is coordinated to an internal iron. In the antibody-peptide complex this histidine is buried in the antibody binding pocket. Thus, this antibody cannot possibly recognize myohemerythrin in its native conformation since the histidine would not be available for antibody binding.

There are bona fide cases where anti-peptide antibodies appear to react at least to some extent with native protein antigens. For instance, there are a number of cases where anti-peptide antibodies neutralize the infectivity of viruses (see Rothbard, 1992 for a review on this subject). As an example, an eight amino acid long peptide of the human immunodeficiencey virus (HIV) gp120 protein when coupled to keyhole limpet hemocyanin elicited antibodies in guinea pigs which neutralized infectivity of HIV *in vitro* and inhibited cell-cell fusion (Javaherian et al., 1989). Neutralization by antibodies does not necessarily imply that the antibodies recognize the virion in its "native" state. If infectivity involves partial unfolding of the viral coat proteins, the anti-peptide antibodies could then bind linear-like or random coil conformations not expressed on the native virion. In studies of anti-peptide antibody binding to poliovirus, none of the nine different anti-peptide antibodies bound the native virus, although several of them bound a form of the virus isolated following cell penetration (Fricks and Hogle, 1990). This suggests that viral neutralization by anti-peptide antibodies could involve the recognition of conformational variants of the viruses that exist during cell infection. Since the molecular basis for neutralization is poorly understood, it is impossible at present to interpret the results of anti-peptide antibody neutralization studies more clearly.

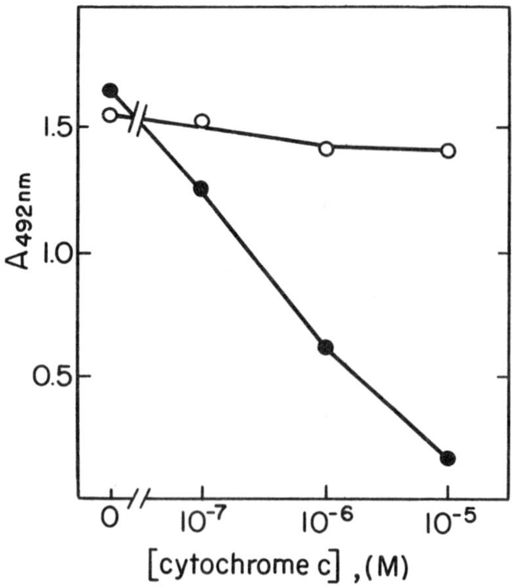

Fig. 5. Two monoclonal antibodies specific for the region around residue 44 of cow cytochrome *c* recognize different conformational states of the antigen. The monoclonal antibody 4H9 (●) recognizes the native protein as shown by the ability of native cytochrome *c* to inhibit the binding of the antibody to cytochrome *c* attached to microtiter plates in competitive ELISA. The monoclonal antibody 1D3 (○) recognizes a denatured form of the protein since soluble cytochrome *c* does not inhibit its binding to plate-adsorbed cytochrome *c*.

There are many other examples where anti-peptide antibodies bind proteins *in situ* or serve as a means for protein isolation and identification (Walter, 1986; see also other chapters of this book). For example, a peptide corresponding to the predicted carboxyl terminal sequence of the superantigen of mouse mammary tumor virus, Mmtv-7, elicited antibodies in hamsters that were useful in precipitating the protein from cells transfected with the gene (Winslow et al., 1992). Since cellular proteins were extracted with non-ionic detergent, they likely maintained their native conformation.

A possible explanation for these successes, in light of the observations that anti-peptide antibodies do not bind the globular proteins that have been carefully examined, is that the protein domains recognized by the antibodies are not globular but are possibly more linear in conformation. Whatever the explanation may be, what is clear is that anti-peptide antibodies do not always bind the native form of a protein. Sometimes this may not be a problem, depending on the particular use to which the antibodies are applied. Even though anti-peptide antibodies may not bind the native protein, their ability to recognize non-native forms make them useful for some applications such as detection of protein bands in Western blots where the protein has been boiled in sodium dodecyl sulfate prior to electrophoresis (Burnette, 1981). Also, anti-peptide antibodies can sometimes be more useful than antibodies to the native protein to detect bacterial clones containing the corresponding DNA. Thus, for example, Millan (1986) produced antibodies to cyanogen bromide-cleaved peptides of reduced and alkylated human placental alkaline phosphatase to detect polypeptide expression in clones of *Escherichia coli* transformed with cDNA. Prior attempts to identify clones using several monoclonal antibodies specific for the native protein were not successful.

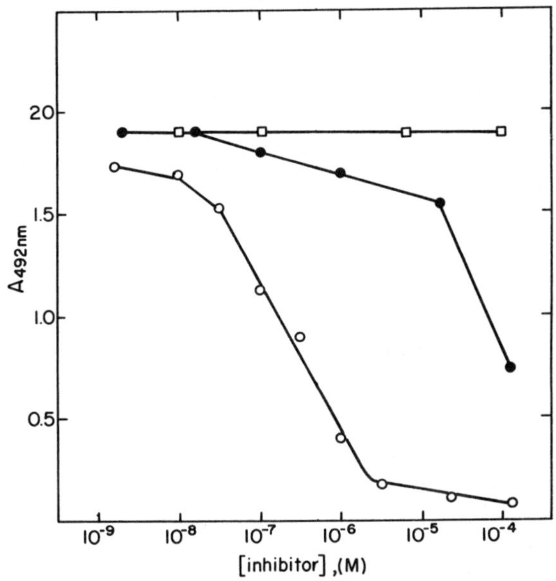

Fig. 6. Binding of the monoclonal antibody 1D3 to peptides of cytochrome *c*: 1-80 (○), 40-54 (●), and 41-48 (□) by competitive ELISA. This antibody does not bind native cytochrome *c* but rather a denatured form of the protein present on the assay plate. The two longer peptides inhibit this binding with different affinities.

VI. A SOLUTION TO THE DISORDER-ORDER PARADOX

Based on the putative high frequency of anti-peptide antibodies reactive with proteins that was observed by some researchers (Niman et al., 1983; Geysen et al., 1985), the disorder-order paradox was proposed (Dyson et al., 1988). It was paradoxical that antibodies elicited against a short peptide lacking a uniquely definable conformation would also react with a native protein containing the same sequence but in a highly ordered conformation. One proposal put forth was that anti-peptide antibodies might induce a conformational change in the protein, thus forcing upon it a conformation similar to that of the peptide against which the antibody was elicited. This idea was supported by the finding that anti-peptide antibodies specific for sequences of myohemerythrin reacted with flexible regions of the polypeptide chain in ELISA (Tainer et al., 1984). However, it was never satisfactorily demonstrated that these anti-peptide antibodies bound the native protein. In fact, as mentioned above, at least three of these antibodies were shown not to react with the native protein in solution-phase assays (Spangler, 1991). The most likely explanation for the disorder-order paradox is that the anti-peptide antibodies do not bind the native protein but rather bind denatured forms that exist in more linear or random coil conformations (see Fig. 7). The failure to perform adequate control experiments to take into account the denaturation of some protein antigen molecules when bound to plastic and/or the presence of denatured protein molecules in the native protein preparations caused confusion of the different conforms of the antigen present in the immunoassays. It should be noted that a likely explanation for the apparent correlation between polypeptide chain flexibility and protein antigenicity (Tainer et al., 1985; Westhof et al., 1984) is that during denaturation of a polypeptide the flexible regions are more likely to deform than more tightly tethered regions, thus exposing linear determinants that peptide-reactive antibodies can then recognize. Any correlation between polypeptide chain flexibility and antigenicity has not held up in examining the epitopes of globular proteins that have been defined by X-ray crystallography (Davies et al., 1988). These epitopes do not occur in particularly flexible regions of the protein molecules.

VII. CONFORMATIONAL RESTRICTION OF PEPTIDES

There are two major reasons why anti-peptide antibodies do not bind globular protein domains. First, the conformational state of the peptide used to elicit the antibodies is in all likelihood distinct from the conformation of the corresponding segment of the native protein. For a long time, it was believed that short peptides did not exist in a particular conformation but rather in equilibrium with a number of random coiled structures (Wetlaufer, 1981). The use of a sensitive technique, two-dimensional nuclear magnetic resonance spectroscopy, provides evidence that preferred conformations may be adopted by certain peptide sequences in aqueous solution (Wright et al., 1988). However, the conformation assumed is probably not identical to that of the corresponding segment of the native protein and may not even be similar in many cases. The sequence 69 to 87 of myohemerythrin discussed previously adopts a β-turn in the short peptide but is α-helical in the native protein (Stanfield et al., 1990). Amino acid sequences some distance from a particular segment in a protein can influence the folding of that segment. Thus, the same sequence of amino acids (4–8 residues in length) in different proteins can exist in different conformations (Wilson et al., 1985).

The second reason why anti-peptide antibodies do not bind globular protein domains is that these domains are composed of discontinuous segments that present a confluent surface. Antibodies to native globular proteins can contact two to four discontinuous segments (Davies et al., 1988). Since anti-peptide antibodies would only recognize one continuous segment if it happened to assume the conformation of the corresponding peptide, neighboring segments would likely hinder the binding of the antibodies either sterically, by charge repulsion, or by hydrophilic/hydrophobic interference (Laver et al., 1990).

There have been attempts to get around these problems but with limited success. The conformational restriction of peptides for antibody recognition has focused mostly on the maintenance of loop structure using an intrachain disulfide bond. The earliest attempt at this approach dealt with a peptide of lysozyme containing amino acids 64–82 (Arnon et al., 1971). This peptide contains cysteine residues at positions 64 and 80. Oxidation of the peptide to form the intrachain disulfide bond resulted in a peptide having a conformation apparently similar, although not identical, to the corresponding segment of the native protein. Thus, antibodies to native lysozyme cross-reacted with the peptide, although with lower affinity, and antibodies to the peptide bound the native protein, also with lower affinity. The performic acid-oxidized synthetic loop (in an open-chain conformation) did not bind either of the antibodies.

Using this same approach, Brown et al. (1984) observed several-fold increases in the affinities of antibodies specific for the hepatitis B surface antigen in binding synthetic peptides. As was the case in the lysozyme study, the sequence of the hepatitis B surface antigen contained cysteines for disulfide bond formation. In other studies, cysteine residues have been appended to natural sequences to stabilize loop structure (Dorow et al., 1985; Schulze-Gahmen et al., 1986; Jemmerson and Hutchinson, 1990). In addition, in one study the effect of including glycine spacers between the natural sequence and the appended cysteines was examined (Jemmerson and Hutchinson, 1990). In all of these studies, effects of cyclization on antibody binding were apparent.

We incorporated cysteine residues at the amino and carboxyl termini of a peptide containing residues

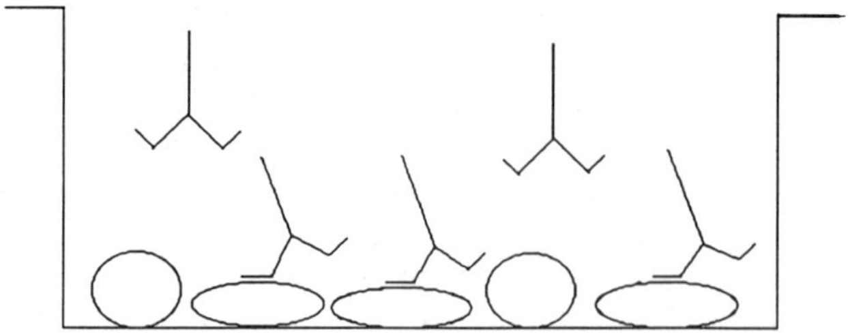

Fig. 7. A cartoon depicting the binding of anti-peptide antibodies to a denatured conform of a protein antigen in solid-phase ELISA. During adsorption of the antigen to the solid phase some molecules denature (ovals), exposing linear-like or random coil determinants that are not present on the native molecule (circles). Anti-peptide antibodies react well with these denatured forms. In the preparation of polyclonal antisera to the intact protein some peptide-reactive antibodies are elicited against denatured forms of the antigen that derive from emulsification of the antigen in mineral oil adjuvants or by lyophilization.

41–48 of horse cytochrome *c*. In the native protein, this segment is part of an ω-loop with a β-turn centered around the Pro-Gly sequence at residues 44 and 45 (Fig. 8). Furthermore, antibodies to native cytochrome *c* are known to bind to this region and are affected by amino acid changes at position 44 and/or 47. These antibodies did not bind the linear sequence 41–48 nor did they bind the cyclized Cys41–48Cys, CysGly41–48GlyCys, or dimeric looped peptides. Antibodies to the 41–48 peptide or cyclized Cys41–48Cys peptide failed to bind the native protein (Jemmerson and Hutchinson, 1990; Jemmerson, unpublished results). It is clear that the cyclized peptides were restricted in conformation since the anti-41–48 peptide antibodies bound the cyclized peptides with significantly lower affinities than they bound the linearized peptides produced by reduction and alkylation. We do not know how the cyclized peptides compared in conformation to the corresponding segment in native cytochrome *c*. The distance between the α-carbons at positions 41–48 in cytochrome *c* as determined by the atomic coordinates of tuna cytochrome *c* (Takano and Dickerson, 1981) is 6.2 Å. The distance between the α-carbons of cysteines involved in disulfide bonds is about 5.5 Å. Thus, the cyclized constructs should approximate the distance between the α-carbons of residues 41 and 48 found in native cytochrome *c*. Assuming that the conformations of the synthetic and natural "loop" are similar (which they may not be), the failure of the anti-cytochrome *c* antibodies to bind the cyclized peptides could reflect the discontinuous nature of the epitopes in this region on native cytochrome *c*. This possibility was addressed in additional experiments that will be discussed in the next section.

Another approach to the stabilization of secondary structure in short peptides involves the substitution of hydrogen bonds with covalent mimics. Since hydrogen bonds provide a major means for stabilizing secondary structure in proteins, it was reasoned that replacement of the hydrogen bonding potentials by covalent bonds would have the same effect (Satterthwait et al., 1990). This tactic was employed by Lerner and co-workers to induce secondary structure in synthetic peptides containing a repeating sequence (AsnProAsnAla)3 of the circumsporozoite protein of *Plasmodium falciparum* which causes malaria (Satterthwait, 1988). It had previously been found that the unconstrained peptide elicited an immune response that was partially protective (Herrington et al., 1987). Secondary structure prediction suggested a preference by the peptide for a reverse turn conformation. Such predictions are only about 50% accurate (Schulz, 1988), but assuming that this segment in the native protein did form a reverse turn and by modeling the peptide in this way it was observed from space-filling models that the two asparagine residues in each tetramer repeat could be involved in an amide-amide hydrogen bond. This hydrogen bond was thus replaced by a hydrazone-hydrocarbon linkage. Antibodies specific for the conformationally restricted peptide reacted with the recombinant circumsporozoite protein as well as with the glutaraldehyde-fixed sporozoite. However, the titers were no different than those of antibodies elicited against the random coil peptide. Since affinity measurements were not carried out, it is impossible to know whether stabilization of secondary structure in the repeating peptide actually resulted in an immunogen which mimicked the native cirumsporozoite epitope more effectively than did the random coil peptide.

Attempts to induce α-helical structure in peptides using helix-forming amino acids have been successful. At one time it was believed that T cell stimulatory peptides were α-helical or had the propensity to adopt α-helical structure (Spouge et al., 1987). Paterson and co-workers (1981) attempted to enhance T cell antigenicity by incorporating helix-promoting amino acids at the termini of a T cell stimulatory peptide (residues 95–104) from pigeon cytochrome *c* (Carbone et al., 1987). This segment is α-helical in the native protein (Takano and Dickerson, 1981; Bushnell et al., 1990). Repeats of alanine-α-aminoisobutyric acid appended at the termini did stabilize α-helical structure in the short peptide and did increase T cell antigenicity, but it was found that the latter was probably due to enhanced ability of the peptides to interact with the lipid bilayer of the antigen-presenting cell due to the increased hydrophobicity of the peptides. What is clear, however, is that the addition of appropriate amino acids at the termini of shorter peptides can stabilize α-helical structure which the peptides may have the propensity to adopt (Paterson et al., 1981; Forood et al., 1993). Insertions of helix-promoting amino acids at non-contact positions may enhance anti-peptide antibody binding to a native protein (Kaumaya et al., 1992).

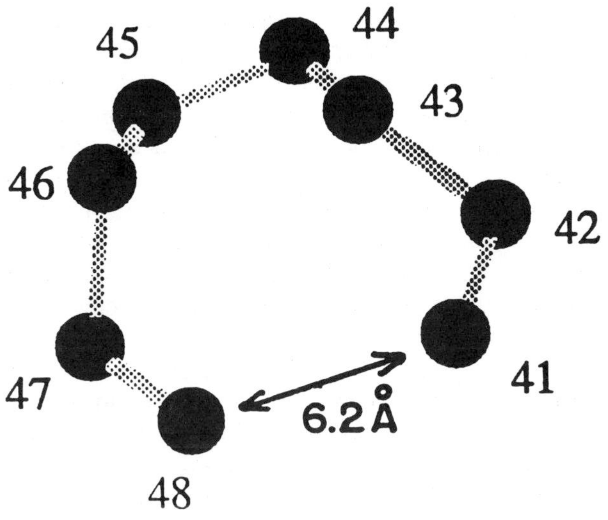

Fig. 8. The α-carbon backbone structure in native cytochrome *c* of residues 41-48. Residue 44 has been identified as an immunodominant residue of native cytochrome *c*. This sequence is part of an ω-loop extending to residue 54 with a β-turn around residues 44 and 45 (Bushnell et al., 1990).

In conclusion, while attempts have been made to stabilize secondary structure in short peptides, the results have not been very impressive. More information needs to be obtained regarding the types of secondary structure innate to short peptides and further approaches to stabilize native-like conformations need to be developed. This is an area of interest to peptide chemists, so advances in this area are likely (DeGrado et al., 1989; Nakanishi et al., 1992). A major problem that must be considered is that any structural change in a peptide within the amino acid sequence of a potential epitope could directly influence antibody binding by lowering the affinity of the antibody for the peptide. Any changes in primary structure that are made to induce or stabilize a particular secondary structure must be well "hidden" from the antibody or remain peripheral to the recognition unit itself.

VIII. PROTEIN SURFACE SIMULATION

Even given the possibility to stabilize a particular secondary structure in a short peptide, because of the discontinuous nature of epitopes on globular proteins, it may be necessary to prepare peptide constructs which mimic the surface contour of a protein and include as many as 15–16 amino acid residues from discontinuous segments (Davies et al., 1988). An attempt at this approach was reported by Atassi 15 years ago in a study of the antigenicity of hen egg-white lysozyme (Atassi, 1978). He called this approach "surface-simulation synthesis." Linear peptides were constructed which included amino acids along the surface of lysozyme from discontinuous segments. Where there were spaces between residues putatively involved in antibody binding, glycine residues were inserted as spacers. Using this approach, Atassi claimed to have modeled three major antigenic sites on native lysozyme for polyclonal antibodies elicited in rabbits and goats. Curiously, none of these sites were observed when mouse monoclonal antibodies were analyzed, including three antibodies whose binding sites have been mapped by X-ray crystallography (Davies et al., 1988; Newman et al., 1992). Since this surface-simulation synthesis approach has not been reproduced, its usefulness remains unconfirmed. Considering the conformation requirements for antibody recognition of native globular proteins, it is doubtful that linear peptides could effectively mimic secondary structures on a protein even if they contained the necessary contact residues in sequence with appropriate spacers.

We employed another approach in an attempt to simulate the surface of a protein antigen (Jemmerson and Paterson, 1986). We had obtained evidence that the antigenic region around residue 44 on horse cytochrome *c* included a discontinuous segment containing residue 26 (see Fig. 9). In native cytochrome *c* there is actually a hydrogen bond between the carbonyl oxygen of residue 44 and the N-ϵ-hydrogen of the side chain of histidine at position 26. We synthesized a branched peptide containing the sequence of residues 21–28LysGlyGly-amide (the three latter residues not part of the native sequence) with the sequence of residues 41–49 extending from the side chain of the lysine residue after residue 28. To facilitate the synthesis, the amino terminus at residue 21 was acetylated. Neither of the linear peptides (21–28 and 41–49) alone nor the branched peptide bound monoclonal antibodies specific for this region of cytochrome *c*. This construct in all likelihood did not place the two sequences in the appropriate relative orientation for antibody recognition. Notwithstanding that limitation, it is also unlikely that either segment adopted the native secondary structure. As noted above, residues 41–48 exist as a β-turn around residues 44 and 45. Incorporation of cysteine residues at the termini could, perhaps, have stabilized a β-turn. Therefore, we next attempted to synthesize a double-branched peptide on a backbone of LysAlaAlaLys, extending the sequence Cys41–48Cys from the first lysine side chain and the sequence CysGly18–28AlaCys from the other lysine side chain. The AlaAla sequence in the backbone served to space the two segments at a similar distance to their spacing in the native molecule, and the Gly and Ala residues appended to the side chains also served as spacer residues to extend the 21–28 sequence farther out from the backbone than the 41–48 sequence (see Fig. 9). Neither the reduced nor oxidized forms of this peptide reacted with antibodies to native cytochrome *c* (Jemmerson, unpublished results). Definition of the epitopes in this region is now in progress using X-ray crystallography. Knowledge of these structures could facilitate additional approaches to mimic the antibody binding sites.

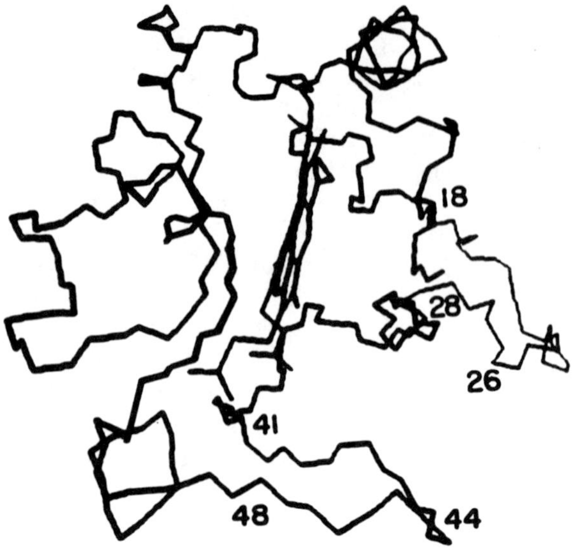

Fig. 9. The backbone structure of cytochrome *c* derived using the atomic coordinates of tuna cytochrome *c* (Takano and Dickerson, 1981). The heme moiety is seen edge-on in the center. Residue 44 is immunodominant for antibodies binding one of the major antigenic regions of the native protein. The carbonyl oxygen of residue 44 is hydrogen bonded to the N-ϵ-hydrogen of the side chain of histidine at position 26. From experiments examining the effects of monoclonal antibodies on the proteolysis of cytochrome *c* by trypsin, it appeared that the segment around the dipeptide 26-27 as well as the segment containing residue 44 were involved in antibody binding (Jemmerson and Paterson, 1986).

IX. CARRIER AND SEQUENCE EFFECTS ON ANTIBODY RECOGNITION

In addition to the effects of conformation of a protein antigen on antibody recognition, changes in primary structure can influence the ability of an antibody to bind an antigen. More than 20 years ago, Margoliash et al. (1970) demonstrated that antibodies elicited against cytochromes *c* in rabbits bound with highest affinity the immunizing protein and with lower affinities cytochromes *c* from other species. The extent of cross-reactivity of these antigens correlated directly with their similarity in amino acid sequence. Furthermore, they found that a population of antibodies specific for human cytochrome *c* was sensitive to a single amino acid change from isoleucine to threonine at position 58 (Nisonoff et al., 1970). Antibodies can be sensitive to even more subtle changes in primary structure. For example, a single amino acid change from threonine to serine, representing the loss of a methylene group in the side chain, dramatically reduces the ability of a monoclonal antibody to bind cytochrome *c* (Fig. 10). The monoclonal antibody 4A9 elicited against horse cytochrome *c*, which has threonine at position 47, binds with lower affinity donkey cytochrome *c*, which has serine at that position. Residue 47 is the only amino acid difference between horse and donkey cytochromes *c*. It should be noted that monoclonal antibodies to other sites on horse cytochrome *c* bind identically to horse and donkey cytochromes *c*, indicating that the single amino acid change at position 47 induces a localized effect only.

These and similar observations made by many others (Benjamin et al., 1984) emphasize the importance of retaining the primary structure of the native protein in the peptide sequence to be used as an immunogen. For example, if the amino acid sequence of a protein is deduced from the nucleotide sequence of its gene, an error in a single base could greatly diminish the usefulness of antibodies raised against a peptide containing the error. Furthermore, these observations suggest that chemical coupling of a peptide to a carrier can alter immune recognition if the coupling site is involved in antibody recognition. In fact, such effects are well substantiated (Briand et al., 1985). We found that antibodies to residues 41–48 of horse cytochrome *c*, elicited against the peptide coupled to ovalbumin, reacted weakly with the free peptide and with higher affinity to peptides that were extended in sequence at the amino and carboxyl termini (Jemmerson and Hutchinson, 1990). This suggested that glutaraldehyde cross-linking to lysine residues on the carrier molecule occurred through one and/or the other terminus and that the antibodies made contact with the covalent bond to the cross-link. When the amino terminus of the free peptide was acetylated, the affinity for the antibody was increased. Apparently, since the antibodies were elicited against a form of the peptide without a free amino terminus, they were sensitive to the presence of the positive charge in the free peptide. Similar effects have been observed by others (Dyrberg and Oldstone, 1986). Effects of coupling peptides to carrier molecules are probably most dramatic when the cross-link occurs at a residue position internal to the peptide or when the cross-linking agent is part of the antibody binding site. These problems can be circumvented by applying different coupling methods or, in some cases, by immunization with the carrier-free peptide (Briand et al., 1985).

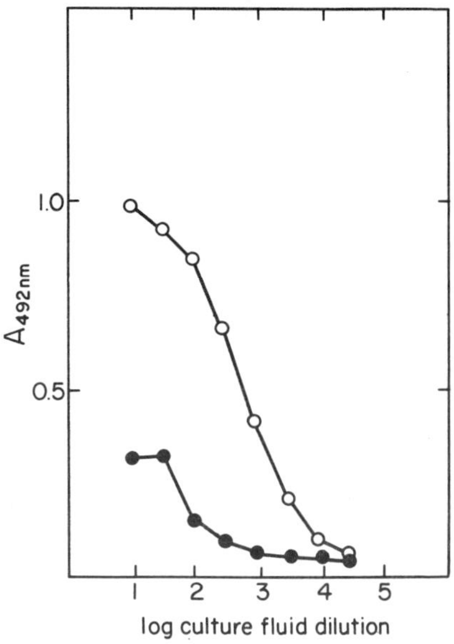

Fig 10. ELISA of the binding of the monoclonal antibody 4A9 to horse (○) and donkey (●) cytochrome *c*. This antibody is sensitive to the single amino acid difference between these two cytochromes *c* at position 47 (threonine changed to serine). Other antibodies to the region around residue 44 bind these cytochromes *c* identically in this assay.

X. SUMMARY

A major limitation to the use of anti-peptide antibodies as tools in biology and medicine is that they frequently do not recognize the native protein antigen. The extent of this problem has not been widely appreciated due to certain artifacts that are often not controlled in antibody binding assays. This limitation is most apparent when the peptide derives from a globular domain of a protein and largely reflects conformational constraints on antibody recognition. Attempts to circumvent this problem by inducing secondary or tertiary structure in peptide immunogens have not been very successful, but further research in this area may lead to solutions to this problem. For some applications where recognition of the native form of a protein is not essential, such as Western blots, anti-peptide antibodies can still be useful even for globular proteins. A more minor problem in the use of anti-peptide antibodies is that they may recognize neo-structures present in the peptide-carrier complex that are not present in the carrier-free peptide or in the intact protein. This problem can be circumvented by applying different coupling methods or, where possible, by immunizing with the carrier-free peptide.

ACKNOWLEDGMENTS

The data presented in the graphs in this chapter have not been previously published. This work was supported by a grant from the United States Public Health Service, National Institutes of Health (AI-23803) and by a grant from the National Science Foundation (DMB-9019181). The author thanks the following students for their efforts in preparing the antibodies and for testing their conformational specificities: Lam Kongpeng, Jon Ness, and Michael Sangster. The artwork of Tim Leonard and editorial assistance of Jeanne Minnerath are greatly appreciated.

ABBREVIATIONS

ELISA Enzyme-linked immunosorbent assay
EPR Electron paramagnetic resonance
MHC Major histocompatibility complex

XI. REFERENCES

Arnon R, Maron E, Sela M et al. Antibodies reactive with native lysozyme elicited by a completely synthetic vaccine. Proc Natl Acad Sci USA 68:1450, 1971

Atassi MZ. Precise determination of the entire antigenic structure of lysozyme: Molecular features of protein antigenic structures and potential of 'surface-simulation' synthesis--a powerful new concept for protein binding sites. Immunochemistry 15:909, 1978

Benjamin DC, Berzofsky JA, East IJ et al. The antigenic structure of proteins: A reappraisal. Ann Rev Immunol 2:67, 1984

Berzofsky JA, Schechter AN and Kon H. Does Freund's adjuvant denature protein antigens? EPR studies of emulsified hemoglobin. J Immunol 115:270, 1976

Briand JP, Muller S and Van Regenmortel MHV. Synthetic peptides as antigens: Pitfalls of conjugation methods. J Immunol Meth 78:59, 1985

Brodsky FM and Guagliardi LE. The cell biology of antigen processing and presentation. Ann Rev Immunol 9:707, 1991

Brown SE, Howard CR, Zuckerman AJ et al. Determination of the affinity of antibodies to hepatitis B surface antigen in human sera. Immunol Meth 72:41, 1984

Burnette WN. "Western botting:" Electrophoretic transfer of proteins from sodium dodecyl sulfate-polyacrylamide gels to unmodified nitrocellulose and radiographic detection with antibody and radioiodinated protein A. Anal Biochem 112:195, 1981

Bushnell GW, Louie GV and Brayer GD. High resolution three-dimensional structure of horse heart cytochrome *c*. J Mol Biol 214:585, 1990

Cambier JC, Newell MK, Justement LB et al. Ia binding ligands and cAMP stimulate nuclear translocation of PKC in B lymphocytes. Nature 327:629, 1987

Carbone FR, Fox BS, Schwartz RH et al. The use of hydrophobic, α-helix defined peptides in delineating the T cell determinant for pigeon cytochrome *c*. J Immunol 138:1838, 1987

Coffman RL, Seymour BWP, Lebman DA et al. The role of helper T cell products in mouse B cell differentiation and isotype regulation. Immunol Rev 102:5, 1988

Crestfield AML, Stein WH and Moore S. On the aggregation of bovine pancreatic ribonuclease. Arch Biochem Biophys (Suppl 1):217, 1962

Davies DR, Sheriff S and Padlan EA. Antibody-antigen complexes. J Biol Chem 263:10541, 1988

DeGrado WF, Wasserman ZR and Lear JD. Protein design, a minimalist approach. Science 243:622, 1989

Dorow DS, Shi P-T, Carbone FR et al. Two large immunogenic and antigenic myoglobin peptides and the effects of cyclisation. Mol Immunol 22:1255, 1985

Dyrberg T and Oldstone MBA. Peptides as antigens--importance of orientation. J Exp Med 164:1344, 1986

Dyson HJ, Lerner RA and Wright PE. The physical basis for induction of protein-reactive antipeptide antibodies. Ann Rev Biophys Biophys Chem 17:305, 1988

Engvall E. Enzyme immunoassay ELISA and EMIT. Meth Enzymol 70:419, 1980

Forood B, Feliciano EJ and Nambiar KP. Stabilization of α-helical structures in short peptides via end capping. Proc Natl Acad Sci USA 90:838, 1993

Fricks CE and Hogle JM. Cell-induced conformational change in poliovirus: Externalization of the amino terminus of VP1 is responsible for liposome binding. J Virol 64:1934, 1990

Friguet B, Djavadi-Ohaniance L and Goldberg ME. Some monoclonal antibodies raised with a native protein bind preferentially to the denatured antigen. Mol Immunol 21:673, 1984

Geysen HM, Barteling SJ and Meloen RH. Small peptides induce antibodies with a sequence and structural requirement for binding antigen comparable to antibodies raised against the native protein. Proc Natl Acad Sci USA 82:178:1985

Goodnough MC and Johnson EA. Stabilization of botulinum toxin type A during lyophilization. App Environ Microbiol 58:3426, 1992

Herrington DA, Clyde DF, Losonsky G et al. Safety and immunogenicity in man of a synthetic peptide malaria vaccine against Plasmodium falciparum sporozoites. Nature 328:257, 1987

Hollander Z and Katchalski-Katzir E. Use of monoclonal antibodies to detect conformational alterations in lactate dehydrogenase isoenzyme 5 on heat denaturation and on adsorption to polystyrene plates. Mol Immunol 23:927, 1986

Javaherian K, Langlois AJ, McDanal C et al. Principal neutralizing domain of the human immunodeficiency virus type 1 envelope protein. Proc Natl Acad Sci USA 86:6768, 1989

Jemmerson R and Paterson Y. Mapping epitopes on a protein antigen by the proteolysis of antigen-antibody complexes. Science 232:1001, 1986

Jemmerson R. Antigenicity and native structure of globular proteins: Low frequency of peptide reactive antibodies. Proc Natl Acad Sci USA 84:9180, 1987

Jemmerson R and Hutchinson RM. Fine manipulation of antibody affinity for synthetic epitopes by altering peptide structure: Antibody binding to looped peptides. Eur J Immunol 20:579, 1990

Jemmerson R, Johnson JG, Burrell E et al. A monoclonal antibody specific for a cytochrome *c* T cell stimulatory peptide inhibits T cell responses and affects the way the peptide associates with antigen-presenting cells. Eur J Immunol 21:143, 1991

Jemmerson R and Johnson JG. Different functional boundaries for the major antigenic region of two cytochromes *c*. Proc Natl Acad Sci USA 88:4428, 1991

Kaumaya PTP, Van Buskirk AM, Goldberg E et al. Design and immunological properties of topographic immunogenic determinants of a protein antigen (LDH-C4) as vaccines. J Biol Chem 267:6338, 1992

Lanzavecchia A. Antigen-specific interaction between T and B cells. Nature 314:537, 1985

Laver WG, Air GM, Webster RG et al. Epitopes on protein antigens: Misconceptions and realities. Cell 61:553, 1990

Liu Y and Linsley PS. Costimulation of T cell growth. Curr Opin Immunol 4:265, 1992

Marcovina SM, Adolphson JL, Parlavecchia M et al. Effects of lyophilization of serum on the measurement of apolipoprotein A-1 and B. Clin Chem 36: 366, 1992

Margoliash E, Nisonoff A and Reichlin M. Immunologic activity of cytochrome *c*. I. Precipitating antibodies to monomeric vertebrate cytochromes *c*. J Biol Chem 245:931, 1970

Millan JL. Molecular cloning and sequence analysis of human placental alkaline phosphatase. J Biol Chem 261:3112, 1986

Miyajima A, Kitamura T, Harada N et al. Cytokine receptors and signal transduction. Ann Rev Immunol 10:295, 1992

Myers CD. Role of B cell antigen processing and presentation in the humoral immune response. FASEB J 5:2547, 1991

Nakanishi H, Chrusciel RA, Shen R et al. Peptide mimetics of the thrombin-bound structure of fibrinopeptide A. Proc Natl Acad Sci USA 89:1705, 1992

Newman MA, Mainhart CR, Mallett CP et al. Patterns of antibody specificity during the BALB/c immune response to hen eggwhite lysozyme. J Immunol 149:3260, 1992

Niman HL, Houghten RA, Walker LE et al. Generation of protein-reactive antibodies by short peptides is an event of high frequency: Implications for the structural basis of immune recognition. Proc Natl Acad Sci USA 80:4949, 1983

Nisonoff A, Reichlin M and Margoliash E. Immunologic activity of cytochrome *c*. II. Localization of a major antigenic determinant of human cytochrome *c*. J Biol Chem 245:940, 1970

Noelle RJ, Ledbetter JA and Aruffo A. CD40 and its ligand, an essential ligand-receptor pair for thymus dependent B-cell activation. Immunol Today 13:431, 1992

Paterson Y, Rumsey SM, Benedetti E et al. Sensitivity of polypeptide conformation to geometry. Theoretical and conformation analysis of oligomers of α-aminoisobutyric acid. J Am Chem Soc 103:2947, 1981

Pernis B, Forni L and Amante L. Immunoglobulins as cell receptors. Ann NY Acad Sci USA 190:420, 1971

Raff MC, Sternberg M, Taylor RB et al. Immunoglobulin determinants on the surface of mouse lymphoid cells. Nature 225:553, 1970

Reth M. Antigen receptors on B lymphocytes. Ann Rev Immunol 10:97, 1992

Rothbard JB and Gefter ML. Interactions between immunogenic peptides and MHC proteins. Ann Rev Immunol 9:527, 1991

Rothbard JB. Synthetic peptides as vaccines. Biotechnology 20:451, 1992

Rudensky AY, Preston-Hurlburt P, Hong S-C et al. Sequence analysis of peptides bound to MHC class II molecules. Nature 353:622, 1991

Satterthwait AC, Chiang L-C, Arrhenius T et al. The conformational restriction of synthetic vaccines for malaria. Bull World Health Org 68(Suppl):17, 1990

Satterthwait AC, Arrhenius T, Hagopian RA et al. Conformational restriction of peptidyl immunogens with covalent replacements for the hydrogen bond. Vaccine 6:99, 1988

Schulz GE. A critical evaluation of methods for prediction of protein secondary structures. Ann Rev Biophys Biophys Chem 17:1, 1988

Schulze-Gahmen U, Klenk H-D and Beyreuther K. Immunogenicity of loop-structured short synthetic peptides mimicking the antigenic site A of influenza virus hemagglutinin. Eur J Biochem 159: 283, 1986

Schwab C, Twardek A, Lo TP et al. Mapping antibody binding sites on cytochrome *c* with synthetic peptides: Are results representative of the antigenic structure of proteins? Protein Sci 2:175, 1993

Schwartz RH. T-lymphocyte recognition of antigen in association with gene products of the major histocompatibility complex. Ann Rev Immunol 3:237, 1985

Scibienski RJ. Denaturation of lysozyme by Freund's complete adjuvant. J Immunol 111:114, 1973

Sela M. Antigenicity: Some molecular aspects. Science 166:1365, 1969

Shimonkevitz R, Kappler J, Marrack P et al. Antigen recognition by H-2 restricted T cells. I. Cell-free antigen processing. J Exp Med 158:303, 1983

Spangler BD. Binding to native proteins by antipeptide monoclonal antibodies. J Immunol 146:1591, 1991

Spouge JL, Guy HR, Cornette JL et al. Strong conformational propensities enhance T cell antigenicity. J Immunol 138:204, 1987

Springer TA. Adhesion receptors of the immune system. Nature 346:425, 1990

Stanfield RL, Fieser TM, Lerner RA et al. Crystal structures of an antibody to a peptide and its complex with peptide antigen at 2.8 Å. Science 248:712, 1990

Szakal AK, Kosco MH and Tew JG. Microanatomy of lymphoid tissue during humoral immune responses: Structure function relationships. Ann Rev Immunol 7:91, 1989

Tainer JA, Getzoff ED, Alexander H et al. The reactivity of antipeptide antibodies is a function of the atomic mobility of sites in a protein. Nature 312:127, 1984

Tainer JA, Getzoff ED, Paterson Y et al. The atomic mobility component of protein antigenicity. Ann Rev Immunol 3:5012, 1985

Takano T and Dickerson RE. Conformation change of cytochrome *c*. II. Ferricytochrome *c* refinement at 1.8 Å and comparison with the ferrocytochrome structure. J Mol Biol 153:95, 1981

Townsend MW and DeLuca PP. Use of lyoprotectants in the freeze-drying of a model protein, ribonuclease A. J Parent Sci Tech 42:190, 1988

Walter G. Production and use of antibodies against synthetic peptides. J Immunol Meth 88:149, 1986

Westhof E, Altschuh D, Moras D et al. Correlation between segmental mobility and the location of antigenic determinants in proteins. Nature 311:123, 1984

Wetlaufer DB. Folding of protein fragments. Adv Protein Chem 34:61, 1981

Wilson IA, Haft DH, Getzoff ED et al. Identical short peptide sequences in unrelated proteins can have different conformations: A testing ground for theories of immune recognition. Proc Natl Acad Sci USA 82:5255, 1985

Winslow GM, Scherer MT, Kappler JW et al. Detection and biochemical characterization of the mouse mammary tumor virus 7 superantigen (Mls-la). Cell 71:719, 1992

Wright PE, Dyson HJ and Lerner RA. Conformation of peptide fragments of proteins in aqueous solution: Implications for initiation of protein folding. Biochemistry 27:7167, 1988

Chapter 17

THE POTENCY OF PEPTIDES IN IMMUNE RESPONSES AND IN IMMUNE RESPONSE ANALYSIS: SINGLE AMINO ACID SPECIFICITY

Netty D. Zegers, Eric Claassen and Wim J.A. Boersma

TABLE OF CONTENTS

I. INTRODUCTION

Synthetic peptides are nowadays very potent tools for a broad spectrum of applications in biological research. Anderer and Schlumberger (1965) used a synthetic peptide derived from the C-terminal end of the tobacco mosaic virus coat protein for raising native protein-specific and virus neutralizing antibodies. In the 1970s, more investigators elicited antibodies against identified antigenic epitopes from proteins with synthetic peptides (Arnon et al., 1969, 1971; Fearny et al., 1971; Langbeheim et al., 1976). In that decade, it was a common belief that the production of antisera with peptides against native proteins required the prior identification of antigenic determinants. After demonstration of the possibility to produce antibodies against peptides which were chosen based on knowledge only of the nucleotide sequence from which the amino acid sequence was derived (Walter et al., 1980), an ever-faster moving production of antibodies using synthetic peptides started. By then, synthetic peptides also became favorite tools for vast fields of research, such as mapping proteins for antigenic determinants and T cell determinants, ligand specificity of numerous cellular and nuclear receptors, probes for determining protein structure, immunomodulation, vaccines, autoimmunity, biochemical mechanisms and more. Peptides are used for investigation of peptide-MHC interaction. Sequence analogs can mimic or inhibit protein function. Dramatic changes with peptide analogs were observed in agonist-antagonist character.

Many inheritable diseases are the result of a mutant protein, often with substitution of only a single amino acid, which may lead to dramatic changes in protein structure and function. Tracking carriers with such protein disorders may be a matter of life and death. Virus mutants can be screened with variant-specific monoclonal antibodies (MAbs). Synthetic peptides are indispensable for the production of MAbs with single amino acid specificity. These MAbs are important tools for detection of protein variants. In addition, synthetic peptides are also very useful for induction of antibodies which can discriminate between individual members of a protein family sharing a high degree of homology, e.g., hormone receptors. These antibodies may contribute to the investigation of regulatory mechanisms.

In this chapter, we will discuss the single amino acid specificity of antibodies and the use of peptides as probes to prove single amino acid specificity in biochemical reactions, illustrated with examples from the literature and from work of our own group. A detailed description of the generation of MAbs specific for one amino acid will be given in Section VI. Peptide sequences will be given in a single letter code for amino acids, otherwise the three letter code for amino acids will be used.

II. DETECTION OF VIRUS MUTANTS

Retroviruses show a large mutation frequency, which occurs during reversed transcription or replication.

0-8493-8967-4/95/$0.00+$.50

Mutations are the result of various factors. Reversed transcriptase lacks exonucleolytic (proofreading) activity and the error rate is 10^{-4} per generation. DNA polymerase from the host works more accurately with an error rate of 10^{-5}. The error rate for polymerase II is not known, but it probably will not be different from reversed transcriptase. Major rearrangements of sequences, a high frequency of genetic recombination together with the ability of oncogene capture of the virus genome, contribute further to the high mutation rate. The human immunodeficiency virus (HIV), a retrovirus, is a striking example. Proteins from HIV show extensive variation among isolates, which are expressed in the primary amino acid sequence. These variants may have different biological activity (replication rate, cell tropism, and syncytia induction), which may be related to sequence and to the structural differences in the surface proteins. The third variable domain (V3 domain) of HIV-1 external glycoprotein gp120 evokes a major fraction of neutralizing antibodies during a natural infection. Several studies using MAbs have confirmed the role of the V3 domain in viral infectivity. The V3 domain is therefore considered to be important for candidate vaccines. Polyclonal and monoclonal antibodies directed against preselected epitopes of the V3 domain may be useful reagents to characterize structure-function relationships of viral proteins. Such antibodies may be used to probe epitopes that can be included in candidate vaccines for induction of neutralizing antibodies. In addition, variant-specific antibodies may have valuable diagnostic and therapeutic application.

We used synthetic peptides to generate rabbit polyclonal antibodies (PAbs) and murine MAbs against relatively conserved sequences from IIIB and MN variants of HIV-1 (Laman et al., 1992). The two variants both contain the immunodominant sequence GPGRAF. In the flanking sequences some differences exist: a few amino acid substitutions and one deletion of two amino acids (IIIB variant:IRIQRGPG-RAFVTIG; MN variant: KRIHIGPGRAFYTTK). A IIIB derived peptide overlapping the tip of the V3 domain, IRIQRGPGRAFVTIG, was used to generate MAbs. Two MAbs neutralized IIIB but not MN and inhibited syncytium formation induced by IIIB. The MAbs showed high affinity for gp120 with K_ds of 6.8 $*10^{-11}$ M and 1.6 $*$ 10^{-10} M respectively. PEPSCAN analysis mapped the binding site to the sequence IRIQRGPGR. In addition, two non-neutralizing MAbs were raised with peptides derived from the N- and C-terminal side of the loop. A similar MN derived peptide was used to produce a rabbit PAb. This PAb inhibited syncytium formation induced by HIV-1 IIIB and four field isolates. These findings showed that variant-specific and HIV-1 neutralizing antibodies can be obtained with synthetic peptides and that high affinity anti-peptide MAbs can be generated. Moreover, the PAbs generated through single peptide immunization crossreacted with four different viral strains (field isolates) which indicate major implications for vaccine development.

Another application of variant-specific synthetic peptides was developed in our group (Laman et al., 1991). Peptides derived from the neutralizing epitope of gp41 of HIV-1 (residues 586–608) were conjugated to the enzyme alkaline phosphatase. These enzyme-peptide conjugates may be used for detection of the epitope specificity of antibody forming cells in tissue biopsies, e.g., lymph nodes, from HIV infected individuals. Cryostat sections of lymph nodes were incubated with these conjugates. Enzyme-peptide conjugates bound by antigen-specific antibodies present in antibody forming cells stained with an appropriate substrate. In addition, the antibody subclass isotype (IgG1, IgG2, IgG3, IgM, IgE or IgA) and the specificity of the antibody forming cell could simultaneously be demonstrated in double staining experiments using peptide-alkaline phosphatase conjugates (using a blue substrate) together with isotype-specific antibody-peroxidase conjugates (using a red substrate) incubation. Violet stained (blue + red) cells thus produce peptide-specific antibodies with the isotype under study. Blue stained cells produce peptide-specific antibody with a different isotype, and red cells visualize antibodies directed against different antigens. This method allows correlation of *in vivo* function of B cells with lymph node pathology, clinical stage of the disease and serological data after HIV infection.

III. IMMUNODETECTION OF GENETIC VARIANTS OF PROTEINS

Nowadays more than 4000 inherited disorders are known. New molecular techniques have accelerated the discovery of the molecular defects of the disease. Partly thanks to the Human Genome Project, the genes causing these diseases are tracked. In the last decade, the genes for cystic fibrosis (Rommens et al., 1989; Riordan et al., 1989; Kerem et al., 1989), neurofibromatosis (Wallace et al., 1990), Duchenne's muscular dystrophy (Monaco et al., 1986), Huntington's disease (The Huntington's Research Group, 1993) and others have been located. Some of the disorders are listed in Table 1. (For an exhaustive review, see Cooper and Schmidtke, 1991, 1992.) The majority of the inherited disorders are caused by a mutation in the DNA sequence, by deletion of a single base or of part of the DNA sequence, or by alternative splicing. Point mutations may lead to a single amino acid substitution or to an aberrant protein product if the mutation is a nonsense mutation. Even single amino acid substitutions in the protein product may have dramatic

TABLE 1
Some inherited disorders

Disorder	Protein	Defect	Location	Incidence	References
Hemophilia A	(Factor VIII)	Deletion	Several exons	1:10000 males	Furie and Furie, 1988
		Nonsense mutations	Several exons		Furie and Furie, 1988
Hemophilia B	(Factor IX)	Deletion	Several exons	1:30000 males	Furie and Furie, 1988
		Point mutation	$Arg^{145} \rightarrow His$		
		Point mutation	$Arg^{1} \rightarrow Ser$		
		Point mutation	$Arg^{4} \rightarrow Gln$		
		Point mutation	$Asp^{47} \rightarrow Gly$		
Hemoglobinopathies (thalassemias)	Hemoglobin			1:600 certain ethnic groups	
Familial hyperlipoproteinemia	Apo-E	Point mutation	$Arg^{158} \rightarrow Cys$	1:500	Weisgraber et al., 1982
Inherited emphysema	α_1-Antitrypsin	Point mutation	$Glu^{342} \rightarrow Lys$	1:3500	Bollen et al., 1983; Carrell et al., 1982
Severe combined immunodeficiency	Adenosine deaminase	Deletion		Rare	Valerio et al., 1983; Orkin et al., 1983
		Point mutation	$Arg^{101} \rightarrow Gln$		Bonthron et al., 1985
		Point mutation	$Leu^{304} \rightarrow Arg$		Valerio et al., 1986
Duchenne muscular dystrophy	Dystrophin complex	Deletion		1:3500 (males)	Monaco et al., 1986; Koenig et al., 1987
					Ervasti et al., 1990
Becker muscular dystrophy				1:35000	
Cystic fibrosis	CFTR protein	3 Base deletion	ΔPhe^{508}	1:2500 (Caucasians)	Rommens et al., 1989; Riordan et al., 1989
					Kerem et al., 1989
		Point mutation	$Gly^{542} \rightarrow X$		Ng et al., 1991
		Point mutation	$Gly^{551} \rightarrow Asp$		
		Point mutation	$Arg^{553} \rightarrow X$		
		Point mutation	$Asn^{1305} \rightarrow Lys$		
		Point mutations	Several, rare		Dean et al., 1990
Neurofibromatosis	NF1 gene product	Inserts	Several exons	1:3500	Wallace et al., 1990
	(chromosome 17)	Deletions	Several exons		Viskochil et al., 1990
		Point mutations	Leu→Pro; Arg→stop		Cawthon et al., 1990
Huntington's disease	IT15 gene product	Inserts	Multiple Gln	1:10000	Gusella et al., 1983; Huntington's Res. Group, 1993
Phenylketonuria	Phenylalanine hydroxylase	Point mutations, deletions, inserts		1:10000	Woo et al., 1983; Dilella et al., 1986a, 1986b

effects, e.g., the $Glu^{342} \rightarrow Lys$ substitution in α_1-antitrypsin predisposes the variant homozygous carriers for risk of developing lung emphysema at a relatively young age or the $Arg^{158} \rightarrow Cys$ substitution in apolipoprotein E (Apo-E) which leads to hyperlipoproteinemia. Especially for these two examples, preventive lifestyle measures like no smoking and avoiding areas with smog (α_1-antitrypsin) or a low fat diet (Apo-E) delay onset and severity of the disease.

With development of lifestyle advices and new therapeutic approaches, e.g., replacement therapy and gene therapy, reliable and simple diagnostic assays for tracking variant carriers at risk become very important. Though polymerase chain reaction assays are now broadly applied for screening mutated genes, they are rather expensive, time consuming and not fully reliable. For some disorders, diagnosis early in pregnancy is required, especially when a severe disease develops very early in life. But for other inherited diseases, which can be restored or delayed by therapy or certain lifestyle measures, postnatal screening may be chosen. In that case, diagnostic assays based on antibodies are simpler and cheaper. For specific diseases, these variant protein products may even be detected in serum by monoclonal antibodies specific for the variant amino acid region of the protein. The monoclonal antibodies required can be obtained by immunizing animals with synthetic peptides which contain the variant region. They have been developed in our laboratory for Apo-E variants (Gerritse et al., 1992) and for α_1-antitrypsin variants (Zegers et al., 1991). A detailed description of raising variant-specific antibodies for α_1-antitrypsin will be given in Section VI.

Though most valuable for single amino acid substitutions, the synthetic peptide approach can be applied for deletions also, especially when a single amino acid has been deleted (rare), but also when a part of the sequence has been deleted. A nice demonstration thereof has been described by Ris-Stalpers et al. (1990). An aberrant form of the human androgen receptor, missing 41 amino acids in the N-terminal region caused by a point mutation abolishing normal RNA splicing at the exon4/intron4 boundary leading to a disturbed male sexual differentiation and development in an individual with a 46,XY karyotype but a female habitus, could not be detected with a monoclonal antibody specific for the N-terminal region of the human androgen receptor.

Another demonstration of the powerful technique of the synthetic peptide approach was the elicitation of antibodies specific for tumor-specific *bcr-abl* joining regions in chronic myeloid leukemia (CML) as well as in acute lymphoblastic leukemias (ALL). The reciprocal translocation between chromosome 9 and chromosome 22 as observed in CML and ALL results in a 22q⁻ chromosome, the so-called Philadelphia chromosome. The translocation events create on the Philadelphia chromosome a fusion between two genes: *bcr* and *abl*. Depending on the localization of the breakpoint in the *bcr* gene, different chimeric *bcr-abl* genes are generated, each encoding their own tumor-specific protein. The amino acid sequences at the point of the junction represent unique tumor-specific determinants. Since the fusion of the *bcr* and *abl* genes occurs within a coding triplet, a newly generated amino acid emerges. For example, the b2a2 fusion protein has the sequence INKEE*ALQRP in the junctional region and the b3a2 fusion protein has the sequence KQSSK*ALQRP in the junctional region. The first four residues are from the second (b2) or third (b3) exon of the *bcr* gene, respectively, and the last five residues are from the second exon (a2) of the *abl* gene. These sequences are not unique for the individual with cells with a Philadelphia chromosome, since they are also present in cells expressing the normal *bcr* and *abl* gene encoded proteins. The residues marked with an asterisk (E and K), however, are newly formed. The junctional regions are the aimed determinants for raising specific diagnostic antibodies. Detection of these tumor-specific proteins has a strong clinical significance. We therefore immunized rabbits and mice with peptides spanning these junctional regions. Antibodies thus raised were site specific and could discriminate, using a protein kinase cell assay, between the different Philadelphia proteins in various CML and ALL cell lines and in blood samples of patients (van Denderen et al., 1989, 1990, 1992, 1994).

IV. VARIATION OF AMINO ACID SUBSTITUTION

In this section we will describe examples of investigations with synthetic peptides for which substitution of single residues has been applied. The reader is also directed to the chapters 2 and 3 on PEPSCAN methods. Peptide sequence variation is widely employed in various biochemical research topics.

IV.1. Modulation of ligand function

A synthetic peptide derived from hen egg-white lysozyme (HEL) (1–18) and an analogous peptide in which Phe at position 3 is substituted by Tyr are both immunogenic in (C57Bl/10 x DBA/2)F1 mice. In the parent C57Bl/6 mice, genetically non-responders to HEL, the analogous peptide itself (Tyr^3) induces anti-peptide antibodies that also bind to the wild-type peptide (Phe^3), whereas the wild-type peptide itself is not immunogenic. Thus, a single amino acid substitution of aromatic amino acids — Phe is a non-polar and Tyr is a polar residue — in a peptide may convert the peptide from silent to immunogenic or virtually a nonresponder mouse strain into a responder one (Sette et al., 1986).

Amino acid substitution in a synthetic peptide substrate may convert selective activity from agonistic to antagonistic or vice versa, or may convert the substrate susceptible for another receptor, as will be clear from the following examples. Substance P (RPKPQQFFGLM-(NH_2)) is an undecapeptide that acts as a neurotransmitter. It is part of a family known as tachykinins. Three main types of tachykinin receptors have been suggested to exist in the periphery: SP-P, SP-E and SP-N receptors. The hexapeptide [pGlu^{6}]SP$_{6-11}$ is equipotent with the undecapeptide in SP-P systems. Substitution of Pro for Gly^{9} in [pGlu^{6}]SP$_{6-11}$ discriminates between two receptor subtypes in the same tissue preparation (Laufer et al., 1986).

Replacement of Pro^{7} in the bradykinin sequence with D-Phe is the essential change used to convert kinin analogs into antagonists. The mechanism likely depends, at least in part, on the aromatic and basic character of the peptide (Lawrence et al., 1989). The impact of aromatic and basic amino acids was also demonstrated with analogs of bombesin (Saeed et al., 1989). His^{12} was replaced by several residues. Replacement of His^{12} by Phe resulted in an agonist with a 100-fold decrease in activity. Substitution by D-Phe yielded an antagonist, but with a 10^{4}-fold decrease of affinity. Substitution by a larger and more hydrophobic residue, D-β-naphtylalanine, produced an analog with complete loss of antagonist activity. On the other hand, substitution by a more basic hydrophillic group but with similar size to D-Phe, D-pyridylalanine, or D-Arg^{12} (basic residue) generated very weak agonists. D-Trp^{12} converted the analog inactive, and D-Tyr^{12} rendered a very weak antagonist, 100-fold less potent than the D-Phe^{12} analog.

Wild-type and mutant synthetic peptides conjugates were used as signal peptides of the SV-40 large T antigen to assay their binding to proteins of rat liver nuclei on Western blots. Proteins of 140 and 55 kD were exclusively recognized by wild-type peptide conjugates and less efficiently by conjugates of a mutant peptide which differed by a single amino acid from the wild-type (Wolff et al., 1988; Finlay et al., 1989; Meier and Blobel, 1990). Specific amino acid sequences have been suggested to serve as nuclear transport signals. Most of these sequences consist of a stretch of basic amino acids preceded by a Pro or Gly. Replacement of Lys^{128} of the SV-40 large T antigen by any amino acid but Arg completely abolished its nuclear targeting ability and conferred cytoplasmic localization. The stretch of basic amino acids (Lys or Arg) was thus found to be a prerequisite for the transport signal sequence.

A peptide (ASQKRPSQRHG) derived from myelin basic protein (AA 1–11), which is involved in the induction of experimental autoimmune encephalomyelitis (EAE), with a single amino acid substitution (Lys^{4} $\rightarrow$ Ala^{4}) appeared to show completely different characteristics than the normal peptide. The mutant peptide (Ala^{4}) binds to MHC class II molecules and stimulates encephalitogenic T cells *in vitro* better than the normal peptide. It is nonimmunogenic and non-encephalitogenic *in vivo* in (PL/JxSJL)F1 mice and prevents EAE when administered before, at the time of immunization with the normal peptide or near the time of disease onset (Smilek et al., 1991). Thus, a change of charge (from positive to neutral) at position 4 in this peptide likely confers a different character to the peptide: from disease inductive to disease suppressive.

Actin polymerization can be inhibited by small synthetic peptides LKHAET and LKKTET. Substitution of the third position (His or Lys, both positively charged amino acids) by a Glu (a negatively charged polar amino acid) modulated the action from G-actin to F-actin (Vancompernolle et al., 1992).

IV.2. Analysis of protein structure

Synthetic peptides were used to investigate the structure of two adjacent zinc finger domains of the yeast transcription factor ADR1. These peptides readily bound zinc in a tetrahedral structure, while the mutant peptide containing a deletion for a single amino acid did not. Cys^{134}, Cys^{137}, His^{150} and His^{155} are part of the metal binding site in the second zinc finger ^{130}KPYPCGLCNRCFTRRDLLIRHAQKIHSGNL^{159}. The residues between Cys^{137} and His^{150} form a loop. Deletion of one residue, Asn^{138}, prevents the folding of the peptide into a zinc finger domain. Ligand spacing is thus a major factor for zinc finger folding and has been revealed by use of a peptide analog (Párraga et al., 1990).

The influence of the flanking amino acid sequence on the O-glycosylation of a single threonine was investigated *in vitro* by examining a series of 52 related peptide substrates (O'Connell et al., 1992). Each residue of the parent peptide (derived from the human Von Willebrand factor) was substituted with different amino acids. Substitution of any amino acid tested at position +3, -3 and -2 relative to the threonine markedly decreased O-glycosylation, as did the presence of a charged residue at position -1. The substitutions of amino acids at the other positions of the peptide substrate had little effect.

Single amino acid variation was applied to investigate sequence-structure and structure-function relationships in proteins by various investigators. An elegant study was carried out with peptide analogs from myohemerythrin (Getzoff et al., 1987). The antibody binding determinants of myohemerythrin (MHr) were mapped using 113 overlapping hexapeptide homologs of the MHr sequence. The roles of individual side chains in each determinant were characterized by measuring the reactivities of antisera with sets of 120 peptides that included all possible analogs (replacement nets) differing in sequence from the parent peptide by replacement of a single amino acid residue. Such a replacement net study made it possible to assign each residue of a peptide to one of the following four categories: essential, selected, partially replaceable or generally replaceable. It was shown that each antigenic site contained one or more highly exposed and accessible critical side chains and one or more largely buried and inaccessible critical side chains within the native protein structure. Antibody binding was analyzed combined with three-dimensional crystallographic structure determination. With these techniques, it was found that initial binding of an antibody to solvent-exposed critical amino acid residues may promote local side chain displacements and thereby allow the participation of other, previously buried, critical residues.

Single amino acid deletions at position 19, 20, 21 or 22 in the C-terminal part of the α-helical region (residues 9–22) of salmon calcitonin did not greatly affect either the conformational (measured by circular dichroism) or the biological properties of this protein (Epand et al., 1988). The amphipathic character of this region was kept despite single amino acid deletions.

Ni et al. (1989) used nuclear magnetic resonance (NMR) to study the structural basis for the bleeding disorder caused by the single mutation of Gly^{12} to Val^{12} in the Aα-chain of human fibrinogen. Val^{12} was found to disrupt the type II β-turn involving Glu^{11} and Gly^{12} in normal protein. The positions of Gly^{13} and Gly^{14} were also displaced in this mutant protein. This altered geometry presumably affects the positioning of Arg^{16}–Gly^{17} bond in the active site of thrombin. As a result, the cleavage of the Arg^{16}–Gly^{17} peptide bond in the mutant peptide by thrombin is much slower than the cleavage of the same bond in normal peptides.

IV.3. Optimalization of ligand sequences

Pharmacological industries need to improve and to optimize therapeutic agents. Especially for peptides, used as substrates, improvement of their affinity for receptor binding is rather easy. Each residue is subsequently replaced by one of the coded or non-coded amino acids, and the mutant peptides are assayed for their biological activity and are compared with the original peptide and previous versions of mutant peptides.

For aprotinin, the Kunitz trypsin inhibitor from bovine mast cells, a series of homologs have been prepared by use of peptide-chemical procedures to modulate the inhibitory specificity. Lys^{15} at the P1 position of its reactive site was replaced by any other amino acid. Especially, Val^{15} aprotinin was the most powerful inhibitor in the series for leucocyte elastase, having a dissociation constant which ranks among the smallest values reported for protein inhibitors of this proteinase. Thus, this mutant of aprotinin may be beneficial in the treatment of human diseases which are associated with the destructive potential of leukocyte elastase. Aprotinin itself is an excellent inhibitor of trypsin due to the positively charged side chain of its P1, but a relatively weak inhibitor for chymotrypsin and shows no detectable inhibition with leukocyte elastase. This investigation demonstrated the individual contribution of a single residue to the inhibition of a particular target proteinase and enabled specific inhibitors to be designed (Tschesche et al., 1987).

Six analogs, single amino acid modifications, of the highly δ-opoid receptor selective, conformationally restricted, cyclic peptide [D-Pen2, D-Pen5] enkephalin were evaluated for opoid activity in rat brain receptor binding and mouse vas deferens (MDV) smooth muscle assays. The enkephalin peptide, Tyr-D-Pen-Gly-Phe-D-Pen, contains two D-Pen residues (penicillamine, containing a thiol group in the side chain) which are cyclized via a disulfide bond. The N-terminus was modified by substitution of Tyr by *N,N*-diallyltyrosine and resulted in a very low potency, low affinity agonist despite apparent overall conformational similarity with the original peptide. The low affinity could probably be attributed to a local effect of the diallyl substitution. Modification of the carboxy terminal group from a carboxylic acid to a carboxamide resulted in reduction in δ-receptor selectivity due to enhanced μ-receptor affinity. Substitution of Gly^3 with sarcosine or substitution of Phe^4 with D-Phe or phenylglycine all lead to drastic losses of binding affinity and MDV potency. Particularly noteworthy was the high δ-receptor affinity, potency and selectivity for the analog in which Gly^3 was replaced by α-aminoisobutyric acid. This analog displayed a quite different conformation than the original peptide (Haaseth et al., 1990). This example demonstrates the effects found by replacement of amino acids by non-coded amino acids.

A peptide, which was 2000-fold more reactive with a transmission blocking monoclonal antibody against *Plasmodium falciparum Pfs25* than the peptide derived from the 25 kD surface protein, could be obtained by subsequent amino acid substitution (van

Amerongen et al., 1992). The reactivity of the MAb with the peptide sequence LDTSNPVKT (derived from the surface protein of *P. falciparum*) was low. The improved peptide with the sequence FDDTDPIKK resulted from the amino acid variation.

Besides the examples mentioned above, also other investigations on substrate selectivity and optimalization have been frequently described. Only the subjects of these studies will be mentioned here. Lottenberg et al. (1986) described the amino acid variations at different positions P1–P4 of peptide substrates for factor Xa. Ho et al. (1988) studied the effects of substitutions in β-endorphin on analgesic potency. An increase of antagonistic activity of parathyroid hormone analog was found by replacement studies by Goldman et al. (1988). Prorok and Lawrence (1989) studied the influence of amino acid insertions to peptide substrates for the cAMP-dependant protein kinase.

V. ANTIBODY SPECIFICITY

Poly- and monoclonal anti-peptide antibodies with a unique fine specificity for even single amino acids are applicable to distinguish members of a protein family with high homology or even to discriminate between variant proteins differing in only a single point mutation. Here we review some results from the literature.

A monoclonal antibody against hemoglobin S was produced with a synthetic peptide derived from the N-terminal β-chain of hemoglobin S, where the single amino acid difference between hemoglobins A and S occurs. Hemoglobin A contains the sequence VHLTPEEV and the S variant contains the sequence VHLTPVEK. The negatively charged glutamine has been replaced by the neutral valine, and two residues downstream a second change in charge has occurred: valine has been replaced by the positively charged lysine. With this MAb, one heterozygotic AS cell in a mixture with one million AA cells could be detected. (Jensen et al., 1985).

In human cells, the *ras* protein family consists of several members including normal (proto-oncogene) and mutant (oncogene) forms. The mutant forms which result from somatic mutations of the normal *ras* genes appear to be responsible for the loss of normal growth control. The mutant form typically contains a single amino acid change at position 12. Antibodies were raised against the *ras* gene product p21. Anti-p21 Ser^{12} (Ser is polar, non-charged) was raised with synthetic peptides and recognized only v-Ki-*ras* protein, but not v-Ha-*ras* protein (containing Arg^{12}, polar, positive charge) nor the normal *ras* protein containing Gly (neutral) at position 12. Micro-injection of antibodies specific for Ser^{12} of the oncogenic v-Ki-*ras* protein into cells transformed by this protein caused a transient reversion of the cells to a normal phenotype (Feramisco et al., 1985).

Also Clark et al. (1985) raised polyclonal anti-peptide antibodies against the *ras* gene product p21. These antibodies distinguished the normal p21, containing glycine at amino acid position 12, from the oncogenic form with serine at position 12. This remarkable specificity was achieved by affinity purification. The polyclonal antibodies were adsorbed to a column to which a Gly^{12}-containing peptide was attached and the flow-through was subsequently adsorbed to the Ser^{12}-containing peptide column.

Schoofs et al. (1988) raised antibodies against a synthetic peptide corresponding to the carboxy terminal 24 amino acids (^{305}CPKYVKQNTLKLATGMR-NVPEKQT^{328}) of the heavy chain of the hemagglutinin molecule of influenza virus A/X-31 (H3). This peptide represents a region of the hemagglutinin molecule which is buried in the molecule. When intact virus or protein was used as the immunogen, no antibodies against this region were found. The peptide appeared to possess three antigenic sites, identified with polyclonal mouse anti-peptide sera in a PEPSCAN. Subsequently, two monoclonal antibodies were selected with the 305–328 peptide. One MAb was found to be specific for the most immunodominant epitope of the peptide ^{314}LKLAT^{318}. In a "replacement set" PEPSCAN, it was shown that all five amino acids are essential for the interaction and most probably are contact residues. In addition, no amino acid substitutions were allowed at either site of the epitope. The MAb was highly specific for the H3 virus subtype and did not bind to the H2 subtype. Subtype H3 differs from subtype H2 at position 315 (Lys vs Val), which is included in the ^{314}LKLAT^{318} epitope. A second MAb recognized the ^{322}NVPEKQT^{328} epitope of which N^{322}, E^{325} and Q^{327} using PEPSCAN technology were implicated as contact residues. This study elegantly shows that for influenza virus subtypes, in which the subtype-specific residues are buried in the protein or alternatively are not immunogenic, subtype-specific monoclonal antibodies can only be developed with the help of synthetic peptides.

Antibodies specific for the VP1 protein type O foot-and-mouth disease virus were raised with a synthetic peptide comprising residues 141–160. This region contains a highly conserved RGD (145–147) sequence which contributes to neutralizing antibody binding sites. Anti-peptide antibodies neutralized a wider range of type O variant isolates than the anti-virion serum. Elongation of this peptide at the amino terminus reduced the number of O-variant strains which could be neutralized by the anti-peptide sera. Selected substitutions of a single amino acid at position 148 altered the specificity of the raised antibodies. In particular, the Leu→Ser substitution induced antibodies which neutralized both type O (Leu^{148}) and type A (Ser^{148}) virus-equally (Parry et al., 1989).

Polyclonal site-specific antibodies were raised to synthetic peptides with a length of 12 to 13 amino acids representing HLA-DQw7 and -DQw8 allelic sequences, differing only by one amino acid in position 57 being Asp and Ala, respectively. This position is the least conserved one. The difference investigated is substantial, being a negatively charged Asp and a neutral Ala. These peptide-specific PAbs distinguished single amino acid substitutions in HLA-DQ β-chain alleles associated with insulin-dependant diabetes (Atar et al., 1989).

Immunological discrimination between human apolipoprotein E2 (Arg 158→Cys) and E3 isoforms differing in amino acids was described (Gerritse et al., 1992). A monoclonal antibody was selected after using a panel of several variant-specific peptides. The peptide sequences used were homologous to human Apo-E2 and in all but one amino acid homologous to the murine Apo-E2. The mouse immune system appeared to be tolerant to most of the selected sequences. Immunization with only one of the peptides (AA 154–172) evoked an anti-protein-specific response despite a low predicted antigenicity index. This study clearly showed the utmost frontiers of the peptide approach.

The foregoing examples all demonstrate the possibility and usefulness of raising variant-specific antibodies with the restriction that the substitution has a major impact on the protein, in general implying a change of charge. These examples have been summarized to indicate the limits of the presently available peptide technology. In addition, another example will be extensively described in the next section including all important aspects to be taken into consideration when generating site directed monoclonal antibodies which can discriminate between proteins differing in a single amino acid residue.

VI. RAISING MONOCLONAL ANTIBODIES SPECIFIC FOR GENETIC VARIANTS OF α_1-ANTITRYPSIN

Human α_1-antitrypsin is a serum protease inhibitor which inhibits mainly neutrophil elastase in the lungs. Nowadays more than 70 different isoforms of the protease inhibitor are known, most of them being point mutations in the DNA sequence leading to amino acid substitutions. Severe α_1-antitrypsin deficiency results from a single amino acid substitution (Glu^{342}→Lys) in the molecule which causes α_1-antitrypsin accumulation in the liver where it is synthesized. The change of charge from negative to positive in the Z molecule induces an alteration in the three-dimensional structure of the Z variant protein which, as a consequence, also showed a reduced activity.

Homozygous carriers with the Z type substitution (Glu^{342}→Lys) are at risk to develop lung emphysema at a relative young age. Also, heterozygous carriers with the SZ phenotype may be at risk. Early detection of such carriers offers the possibility to prescribe lifestyle measures, which delay the disease, or to start with replacement therapy. Presently applied methods for detection of variants are not suitable for routine screening since they are difficult to interpret. Simple assays based on monoclonal antibodies, raised using synthetic peptides, which can distinguish between the Z and S variant and the normal M protein, have been developed.

For this purpose, we synthesized several peptides (Table 2) with different length and various overlapping sequences derived from the M and the Z α_1-antitrypsin. Peptides mostly contain only a B cell determinant and are therefore not immunogenic themselves. Hence, peptides were coupled to keyhole limpet hemocyanin (KLH) as a carrier to provide T cell determinants with the bifunctional heterogenous coupling agent m-maleimidobenzoyl-N-hydroxysuccinimide ester (MBS) via a Cys added at one of the termini or via their functional amino or carboxyl groups with glutaraldehyde (GA) (Zegers et al., 1990) or a carbodiimide, (EDC: 1-ethyl-3-(3-dimethylaminopropyl)carbodiimide) (Deen et al., 1990). Balb/C mice were immunized i.p. or s.c. with the peptide-KLH conjugates or with free peptides both given with an adjuvant.

We earlier showed that also short peptides of about ten amino acids may induce native protein-specific antibodies (van Denderen et al., 1989, 1990). Short peptides (10–12 amino acids) (SPEK1A, SPEK-1B and SPEK3) derived from the 342 region of α_1-antitrypsin in general did not elicit antibodies cross-reactive with the native protein. Based on our present experience and also on that of other investigators (Partidos and Steward, 1992), longer peptides of about 15 to 25 amino acids readily lead to protein-specific antibodies of higher affinity. Elongated peptides (SPEK11 to SPEK15) derived from α_1-antitrypsin with various positions of the substituted amino acid were more successful in raising α_1-antitrypsin antibodies.

SPEK11 (derived from the M variant) is almost completely homologous with the mouse α_1-antitrypsin. A Thr (mouse) in stead of a Lys (human) at position 343 at the carboxy terminus of the peptide is the only difference. Immunization of Balb/C mice with SPEK11 induced only weak α_1-antitrypsin-specific antibody responses. The complementary Z derived peptide SPEK15, with two residues different from the mouse sequence, induced stronger α_1-antitrypsin-specific responses. The polyclonal antibodies were not able, however, to discriminate between the M and the Z

TABLE 2
Peptides used to elicit anti α_1-antitrypsin variant-specific antibodies

SPEK1A	*CVLTIDKKGTE	342 position in Z-variant (AA 337–346)
SPEK1B	*CVLTIDEKGTEA	342 position in M-variant (AA 337–347)
SPEK3	*CHKAVLTIDKK	342 position in Z-variant (AA 334–343)
SPEK11	KLSKAVHKAVLTIDEKC*	342 position in M-variant (AA 328–343)
SPEK12	*CKLSKAVHKAVLTIDKK	342 position in Z-variant (AA 328–343)
SPEK13	KAVLTIDKKGTEAAGAC*	342 position in Z-variant (AA 335–350)
SPEK14	KAVLTIDEKGTEAAGAC*	342 position in M-variant (AA 335–350)
SPEK15	KLSKAVHKAVLTIDKKC*	342 position in Z-variant (AA 328–343)
MOUSE	KAVLTIDETGTEAAAV	Homologous sequence of mouse α_1-antitrypsin

Variant-specific residues have been underlined and in redline. Terminal cysteine residues with * superscript are added to the α_1-antitrypsin sequence for thiol coupling possibility. Mouse-specific residues have been underlined (Latimer et al., 1990).

type. SPEK12, which was conjugated to a carrier protein with the distal end compared to SPEK15, also did not result in Z-specific antibodies.

Subsequently, we synthesized peptides SPEK13 and SPEK14 with the substitutions in the middle of the peptides and in addition two extra residues different from the mouse sequence at the carboxy terminus. Immunization, either i.p. or s.c., with KLH-MBS conjugates of the peptides induced polyclonal antibodies specific for α_1-antitrypsin (Table 3). Free peptides only induced a peptide-specific response after s.c. immunization, but failed to induce protein-specific responses. KLH-MBS-SPEK14 even raised antibodies specific for the M variant which did not recognize the Z variant; KLH-MBS-SPEK13 induced polyclonal antibodies which also recognize the M variant, but with lower specificity or affinity than the Z variant.

Conjugates made by coupling with EDC or with glutaraldehyde failed to induce peptide- or protein-specific antibodies. These coupling reagents make use of amino groups (EDC or glutaraldehyde) or carboxyl groups (EDC). Peptides SPEK13, SPEK14 and SPEK15 possess several residues with these functional groups (Lys, Asp, Glu and end groups). The peptides probably will be conjugated to the carrier at more than one site. This probably will modify the specific epitope to such an extent that antibody recognition of the free peptides or of the native protein will be hampered.

Variant-specific monoclonal antibodies were selected after immunization of mice with KLH-MBS-SPEK13 or KLH-MBS-SPEK14. The reaction pattern of the MAbs is shown in Table 4. Three classes of MAbs were obtained. MAbs F50.2.1, F50.3.1 and F50.4.1 specifically recognize only the Z variant of α_1-antitrypsin. MAb F46.4.1 recognizes all variants of α_1-antitrypsin, and MAb F43.8.1 recognizes all variants except the Z variant. With the combination of these MAbs, it is possible to discriminate between hetero- or homozygous Z carriers in different assays.

The selected MAbs have a high affinity for α_1-antitrypsin in ELISA systems in which the protein is coated and also for α_1-antitrypsin variants in immunoblots after isoelectric focusing (Zegers et al., 1991). Subsequently, a sandwich ELISA was designed with a polyclonal antibody to capture different variants of α_1-antitrypsin combined with our selected variant-specific MAbs for rapid and facile discrimination between homo- or heterozygous Z carriers. However, commercially available polyclonal antibodies interfered with the binding of our variant-specific MAbs. We

TABLE 3
Antibody responses after immunization with various conjugates via different routes

		ELISA Coating		
Immunization	**Route**	**Free SP**	α_1**AT(M)**	α_1**AT(Z)**
Free SPEK 13	i.p.	-	-	-
KLH-MBS-SPEK 13	i.p.	+++	+	+
KLH-EDC-SPEK 13	i.p.	-	-	-
KLH-GA-SPEK 13	i.p.	-	-	-
Free SPEK 13	s.c.	++	+	-
KLH-MBS-SPEK 13	s.c.	++	-	+
Free SPEK 14	i.p.	-	-	-
KLH-MBS-SPEK 14	i.p.	+++	++	-
KLH-EDC-SPEK 14	i.p.	-	-	-
KLH-GA-SPEK 14	i.p.	+	-	-
Free SPEK 14	s.c.	++	-	-
KLH-MBS-SPEK 14	s.c.	+++	++	-
Free SPEK 15	i.p.	-	-	-
KLH-MBS-SPEK 15	i.p.	+++	++	+
KLH-EDC-SPEK 15	i.p.	-	-	-
KLH-GA-SPEK 15	i.p.	-	-	-
Free SPEK 15	s.c.	+	-	-
KLH-MBS-SPEK 15	s.c.	+++	+	+

TABLE 4
Specificity of anti-human α_1-antitrypsin monoclonal antibodies

		ELISA			
MAb	**Isotype**	**SPEK13**	**SPEK14**	α_1**-AT(Z)**	α_1**-AT(M)**
F43.8.1	IgG1 κ	-	+	-	+
F46.4.1	IgG1 κ	+	+	+	+
F50.2.1	IgG2b κ	+	-	+	-
F50.3.1	IgG1 κ	+	-	+	-
F50.4.1	IgG2a κ	+	-	+	-

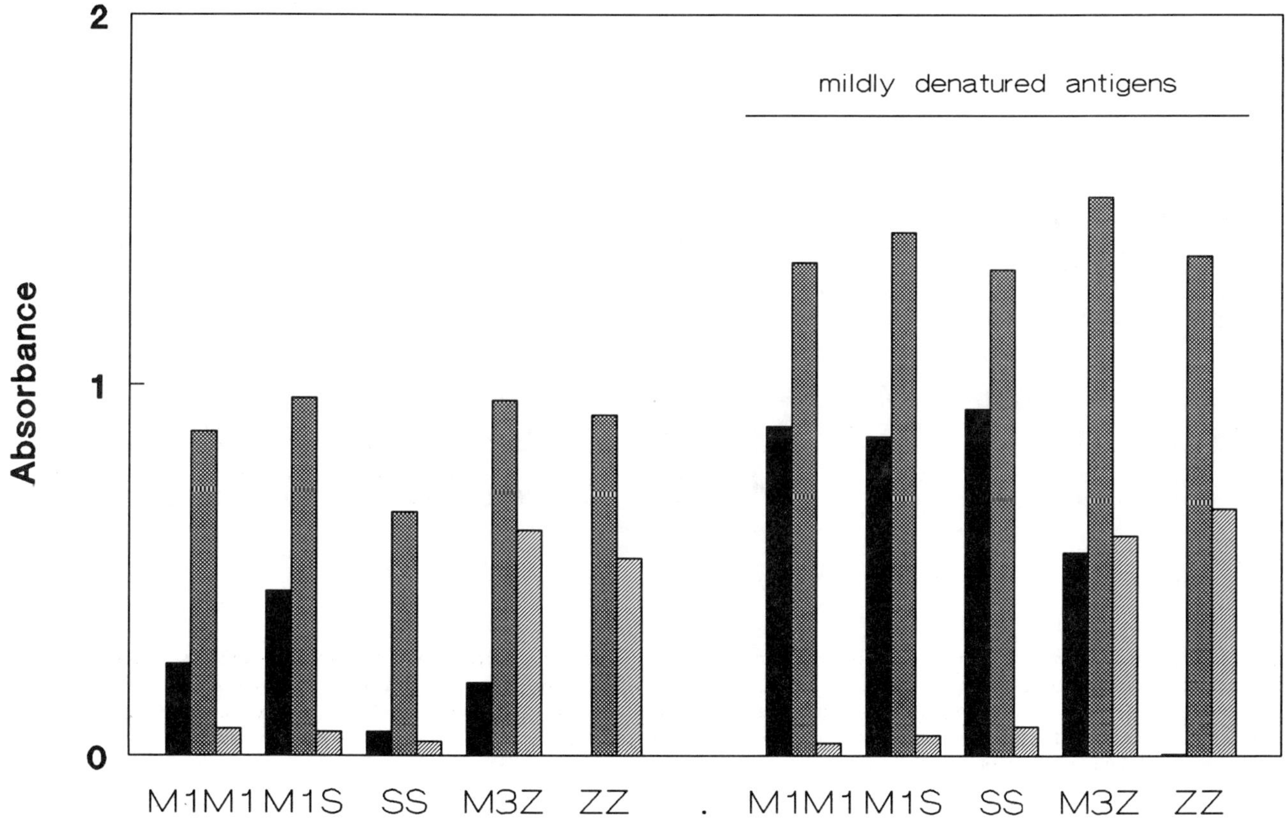

Fig. 1. Comparison of native and denatured α_1-antitrypsin variant-specific detection with a capture ELISA. Human plasma samples with known α_1-antitrypsin phenotype were diluted 1:50. Samples were denatured by heating at 56°C for 30 minutes (right panel) or used in native form (left panel). Variants were captured by variant-specific MAbs: non-Z-specific F43.4.1 (solid bars), all variant-specific F46.4.1 (cross-hatched bars) or Z-specific F50.4.1 (F50.2.1 was used for the right panel) (striped bars). Captured antigens were detected with a polyclonal rabbit serum W9/04 specific for α_1-antitrypsin and a second antibody specific for rabbit Ig conjugated to alkaline phosphatase. Reactivity was measured after 45 minutes for the right panel and after 90 minutes for the left panel.

therefore raised polyclonal antibodies in rabbits using an immune complex of α_1-antitrypsin and polyclonal rabbit Abs specific for α_1-antitrypsin (Sigma, St. Louis, MO). After absorbing these polyclonal antibodies to human serum albumin, they were suitable for use as capture antibodies in a sandwich ELISA (Zegers et al., 1993).

MAb F43.8.1 had a very low affinity for native α_1-antitrypsin in the sandwich ELISA, and it was not possible to detect all offered samples. The MAbs were selected in an ELISA with antigen coated to the surface of a plate. The MAbs therefore readily recognized, to some extent, denatured antigen coated to a plastic surface or coated onto nitrocellulose. In addition, MAbs F46.4.1 and F50.2.1 also recognized α_1-antitrypsin in solution, but MAb F43.8.1 had a much lower affinity for α_1-antitrypsin in solution. When the diluted human plasma samples were mildly denatured by heating at 56°C for 30 minutes, MAb F43.8.1. could detect non-Z α_1-antitrypsin in all samples (Fig. 1). Also MAbs F46.4.1 and F50.2.1 could still readily detect the captured and denatured α_1-antitrypsin. A dipstick method, based on the sandwich ELISA, was also quite suitable to distinguish homo- or heterozygous Z carriers (Fig. 2).

Another set of peptides (Table 5) was synthesized encompassing the substitution of the S variant of α_1-antitrypsin (Glu 264→ Val). Mice were immunized with free peptides or with KLH conjugates made with MBS. With short peptides SP138 and SP139, no peptide- or protein-specific antibody response was found. Elongation of SP138 with three residues at either end, SP136, introduced solubility problems, and therefore the conjugation procedure was not effective. The homologous SP137 (M variant) was soluble and KLH-MBS-SP137 induced peptide- and α_1-antitrypsin (M)-specific antibodies.

Elongation of SP138 to SP190 with three residues only at the N-terminal side resulted in a soluble peptide. SP190 was coupled to KLH or N-terminally

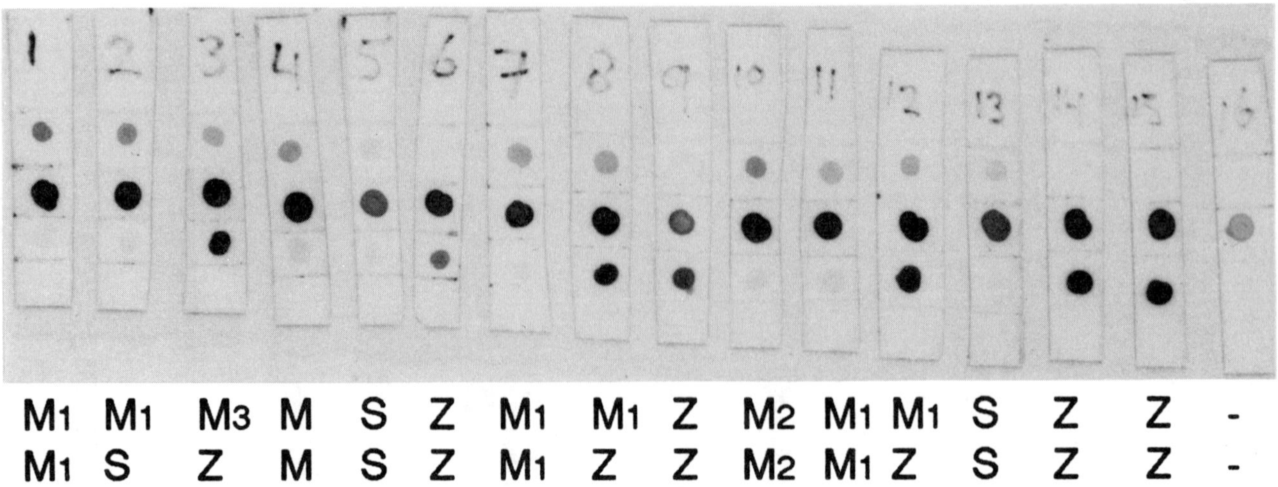

Fig. 2. α_1-Antitrypsin variant-specific detection with a capture dipstick. Human plasma samples with known α_1-antitrypsin phenotype were diluted 1:50. Samples were denatured by heating at 56°C for 30 minutes. Variants were captured by variant-specific MAbs spotted onto nitrocellulose (1 μl of a 2 mg.ml^{-1} solution): non-Z-specific F43.4.1 (upper dots), all variant-specific F46.4.1 (middle dots) or Z-specific F50.2.1 (lower dots). Captured antigens were detected with MAb F91.3.2 specific for α_1-antitrypsin conjugated to β-galactosidase. X-Gal was used as substrate to yield a clear blue spot.

extended with a well-known T cell epitope (Hackett et al., 1985) termed SP191. Both KLH-MBS-SP190 and free SP191 or KLH-MBS-SP191 raised α_1-antitrypsin-specific antibodies. These polyclonal antibodies recognized, however, both the M and the S variant. Monoclonal antibodies resulting from these immunizations also could not discriminate between the S and the M variant. Inhibition ELISAs with peptides SP136, SP137, SP138, SP139 and SP190 indicated that the dominant MAb binding site probably is the DEGK sequence. This finding corresponds with a computer prediction algorithm program (Jameson and Wolf, 1988; Wolf et al., 1988) in which the sequence DEGK was found to be a stretch with a high antigenicity index (Table 6) as compared to the rest of the sequence. Especially for the S variant, the sequence LQHLVNE has a much more lower antigenicity index compared to the homologous sequence LQHLENE of the M variant.

SP256, with the deletion of the immunodominant sequence DEGK, was conjugated to KLH with MBS and injected in mice. We also synthesized the corresponding multiple antigen peptide (Tam, 1988), MAP255, 8 linear peptide sequences with a N-terminal extension of the T cell determinant branched to a core of 7 lysine residues. MAP255 was also used for immunization. In ELISA, a peptide-specific response was found, but not an α_1-antitrypsin-specific response. We may conclude from these results that the sequence LQHLVNELTHDI has a low intrinsic immunogenicity and is too short to mimic the structure of native α_1-antitrypsin.

The experiments described in this section nicely demonstrate the attractiveness and the pitfalls of the peptide strategy for generating variant-specific antibodies. In these cases, one is constrained to a limited possibility of peptide sequence choice. To raise Z-specific antibodies, it was needed to carefully select the peptide sequences. Several overlapping peptides had to be tested. It proved to be essential to locate the substituted amino acid in the middle of the peptide (SPEK13). Also for this region the predicted immunogenicity corresponds with the findings. The residues surrounding the variant-specific amino acids have a higher predicted antigenicity than the preceding sequence (Table 7). If this part of the sequence is situated, however, adjacent to the coupling site, then the steric hindrance probably prevails over the immunogenicity. SPEK15 apparently is a less antigenic sequence. Also the mode of coupling of the peptide to a carrier protein is of importance. Only with conjugates made with MBS were α_1-antitrypsin-specific antibodies produced.

Sometimes peptides are not soluble in aqueous buffers and in DMF or DMSO which cause troubles with conjugation procedures. To overcome this prob-

TABLE 5
Peptides used to elicit S or M variant-specific antibodies

Mouse	DDGKMQHLEQTLSKEL	Homologous mouse α_1-antitrypsin sequence
SP136	DEGKLQHLVNELTHDIC*	S variant AA 256–271
SP137	DEGKLQHLENELTHDIC*	M variant AA 256–271
SP138	KLQHLVNELTC*	S variant AA 259–268
SP139	KLQHLENELTC*	M variant AA 259–268
SP190	C*DEGKLQHLVNELT	S variant AA 256–268
SP191	FERFEIFPKE-DEGKLQHLVNELTC*	T cell epitope + S variant AA 256–268
MAP190	(DEGKLQHLVNELT)$_8$-core	S variant AA 256–268
MAP191	(FERFEIFPKE-DEGKLQHLVNELT)$_8$-core	T cell epitope + S variant AA 256–268
MAP250	(DEGKLQHLVNELTHDI)$_8$-core	S variant AA 256–271
MAP251	(FERFEIFPKE-DEGKLQHLVNELTHDI)$_8$-core	T cell epitope + S variant AA 256–271
MAP255	(FERFEIFPKE-LQHLVNELTHDI)$_8$-core	T cell epitope + S variant AA 260–271
SP256	LQHLVNELTHDIC*	S variant AA 260–271

Variant-specific residues are underlined. Terminal cysteine residues with * superscript are added to the α_1-antitrypsin sequence for thiol coupling possibility. MAP peptides are multiple antigen peptides (8 peptide sequences per molecule) branched to a core of 7 lysine residues according to Tam et al., 1988. The T cell epitope used for N-terminal extension of peptides was derived from influenza virus hemagglutinin (Hackett et al., 1985). Mouse-specific residues have been underlined (Latimer et al., 1990).

lem, linear extension of the peptide with a T cell epitope or synthesis of a branched peptide may be advantageous. For some applications with MAbs generated with synthetic peptides, it is necessary to make the binding site of the protein for the antibody more accessible by denaturing the protein. Direct selection of MAbs in a sandwich ELISA instead of only selection on PVC surface coated protein will probably yield MAbs which have a higher affinity for the native protein in solution. Denaturing the protein may not be necessary. Until now, we did not succeed in generating S variant-specific antibodies, mainly due to the low antigenicity of the variant-specific domain and the immunodominant preceding DEGK sequence.

Experiments based on these considerations led to the production of variant-specific monoclonal antibodies which can discriminate between proteins with only one amino acid different in various immunoassays.

VII. CONCLUSION

Synthetic peptides have proven to be essential means for investigational purposes in a numerous number of disciplines. Single amino acid specificity is encountered Variant-specific residues are underlined. Terminal cysteine residues with * superscript are added to the α_1-antitrypsin sequence for thiol coupling possibility. MAP peptides are multiple antigen peptides (8 peptide sequences per molecule) branched to a core of 7 lysine residues according to Tam et al., 1988. The T cell epitope used for N-terminal extension of peptides was derived from influenza virus hemagglutinin (Hackett et al., 1985). Mouse-specific residues have been underlined (Latimer et al., 1990).
with a large number of functionally diverse biological molecules. With well chosen sequences in synthetic peptides, the specificity of these molecules can be investigated and the function of ligands can be mimicked or inhibited.

MAbs which are single amino acid specific in general can only be produced with the aid of synthetic peptides. These MAbs are extremely useful for early, fast, reliable and cheap detection of variant proteins, e.g., genetic variants or virus mutants. In order to generate variant-specific MAbs, a number of aspects should be carefully taken into account, such as peptide length, sequence, orientation relative to a carrier, coupling mode procedures, carrier, B and T cell epitope selection, immunization scheme and formulation and finally selection methods. In the example described, important aspects were highlighted. This example can serve as a guideline for production of other variant-specific MAbs in the future.

ABBREVIATIONS

AA	Amino acid
ALL	Acute lymphoblastic leukemia
Apo-E	Apolipoprotein E
α_1AT	α_1-Antitrypsin
CML	Chronic myeloid leukemia
EAE	Experimental autoimmune encephalomyelitis
EDC	1-Ethyl-3-(3-dimethylaminopropyl)carbodiimide
ELISA	Enzyme-linked immunosorbent assay
HEL	Hen egg-white lysozyme
HIV	Human immunodeficiency virus
i.p.	Intraperitoneal
i.v.	Intravenous
KLH	Keyhole limpet hemocyanin
MAb	Monoclonal antibody
MBS	m-Maleimidobenzoyl-N-hydroxysuccinimide ester
MHC	Major histocompatibility complex
MHr	Myohemerythrin
PAb	Polyclonal antibody

VIII. REFERENCES

van Amerongen A, Beckers PJ, Plasman HH et al. Peptides reactive with a transmission-blocking monoclonal antibody against Plasmodium falciparum Pfs25: 2000-fold affinity increase by PEPSCAN-based amino acid substitutions. Pept Res 5:269, 1992

Anderer FA and Schlumberger HD. Properties of different artificial antigens immunologically related to tobacco mosaic virus. Biochim Biophys Acta 97:503, 1965

Arnon R and Sela M. Antibodies to unique region in lysozyme provoked by a synthetic antigen conjugate. Proc Natl Acad Sci USA 62:163, 1969

Arnon R, Maron E, Sela M and Anfinson CB. Antibodies reactive with lysozyme elicited by a completely synthetic antigen. Proc Natl Acad Sci USA 68:1450, 1971

Atar D, Dyrberg T, Michelsen B et al. Site-specific antibodies distinguish single amino acid substitutions in position 57 in HLA-DQ beta-chain alleles associated with insulin-dependent diabetes. J Immunol 143:533, 1989

Bollen A, Herzog A, Cravador A et al. Cloning and expression in E. coli of full-length cDNA coding for human α_1-antitrypsin. DNA 2:255, 1983

Bonthron DT, Markham AF, Ginsburg D et al. Identification of a point mutation in the adenosine deaminase gene responsible for immunodeficiency. J Clin Invest 76:894, 1985

Carrell RW, Jeppsson JO, Laurell CB et al. Structure and variation of human α_1-antitrypsin. Nature 298:329, 1982

Cawthon RM, Weiss R, Xu G et al. A major segment of the neurofibromatosis type 1 gene: cDNA sequence, genomic structure and point mutations. Cell 62:193, 1990

Clark R, Wong G, Arnheim N et al. Antibodies specific for amino acid 12 of the *ras* oncogene product inhibit GTP binding. Proc Natl Acad Sci USA 82:5280, 1985

Cooper DN and Schmidtke J. Diagnosis of genetic disease using recombinant DNA. Third edition. Hum Genet 87:519, 1991

Cooper DN and Schmidtke J. Molecular genetic approaches to the analysis and diagnosis of human inherited disease: An overview. Ann Med 24:29, 1992

Dean M, White MB, Amos J et al. Multiple mutations in highly conserved residues are found in mildly affected cystic fibrosis patients. Cell 61:863, 1990

Deen C, Claassen E, Gerritse K et al. A novel carbodiimide coupling method for synthetic peptides enhanced antipeptide antibody responses. J Immunol Meth 129:119, 1990

van Denderen J, Hermans A, Meeuwsen T et al. Antibody recognition of the tumor-specific *bcr-abl* joining region in chronic myeloid leukemia. J Exp Med 169:87, 1989

van Denderen J, van der Plas D, Meeuwsen T et al. Immunologic characterization of the tumor specific *bcr-abl* junction in Philadelphia chromosome-positive acute lymphoblastic leukemia. Blood 76:136, 1990

van Denderen J, ten Hacken P, Berendes P et al. Antibody recognition of the tumor-specific b3-a2 junction of the *bcr-abl* chimeric proteins in Philadelphia-chromosome-positive leukemias. Leukemia 6:1107, 1992

TABLE 6
Antigenicity index for α_1-antitrypsin M and S variant

Residue number	Amino acid	M	S
250	A	-0.600	-0.600
251	I	-0.600	-0.600
252	F	-0.600	-0.600
253	F	-0.600	-0.600
254	L	-0.600	-0.600
255	P	0.250	0.250
256	D	1.300	1.300
257	E	1.300	1.300
258	G	1.300	1.300
259	K	0.900	0.900
260	L	0.750	0.600
261	Q	0.750	-0.300
262	H	0.750	0.300
263	L	0.750	0.300
264	E/V	0.900	0.300
265	N	0.900	-0.300
266	E	0.900	-0.300
267	L	0.750	0.750
268	T	0.600	0.600
269	H	-0.300	-0.300
270	D	-0.600	-0.600
271	I	0.300	0.300
272	I	-0.300	-0.300
273	T	-0.600	-0.600
274	K	-0.600	-0.600
275	F	0.450	0.450
276	L	0.900	0.900
277	E	0.900	0.900

The antigenicity index was calculated according to Jameson and Wolf (1988).

TABLE 7
Antigenicity index for α_1-antitrypsin M and Z variant

Residue number	**Amino acid**	**M**	**Z**
330	S	-0.450	-0.450
331	K	0.450	0.450
332	A	0.450	0.450
333	V	0.600	0.600
334	H	-0.300	-0.300
335	K	-0.600	-0.600
336	A	-0.600	-0.600
337	V	-0.600	-0.600
338	L	-0.600	-0.600
339	T	-0.600	-0.600
340	I	0.000	0.000
341	D	0.900	0.900
342	E/K	1.100	0.900
343	K	1.100	0.900
344	G	1.100	0.900
345	T	0.900	0.900
346	E	0.750	0.750
347	A	-0.300	-0.300
348	A	-0.300	-0.300
349	G	-0.600	-0.600
350	A	-0.600	-0.600
351	M	-0.600	-0.600
352	F	-0.600	-0.600
353	L	-0.600	-0.600
354	E	-0.600	-0.600
355	A	-0.600	-0.600
356	I	-0.600	-0.600
357	P	-0.600	-0.600

The antigenicity index was calculated according to Jameson and Wolf (1988).

van Denderen J, ten Hacken P, Berendes P et al. Recognition of the ALL-specific BCR-ALL junction in P190$^{bcr\text{-}abl}$ by monoclonal antibody ER-FP1. Leukemia in press, 1994

Dilella AG, Kwok SC, Ledley FD et al. Molecular structure and polymorphic map of the human phenylalanine hydroxylase gene. Biochemistry 25:743, 1986a

Dilella AG, Marvit J, Lidsky AS et al. Tight linkage between a slicing mutation and a specific DNA haplotype in phenylketonuria. Nature 322:799, 1986b

Epand RM, Epand RF, Stafford AR et al. Deletion sequences of salmon calcitonin that retain the essential biological and conformational features of the intact molecule. J Med Chem 31:1595, 1988

Ervasti JM, Ohlendieck K, Kahl SD et al. Deficiency of a glycoprotein component of the dystrophin complex in dystrophic muscle. Nature 345:315, 1990

Fearny FJ, Leung CY, Young JD et al. The specificity of antibodies to a peptide determinant of the tobacco mosaic virus protein induced by immunization with the peptide conjugate. Biochim Biophys Acta 243:509, 1971

Feramisco JR, Clark R, Wong G et al. Transient reversion of *ras* oncogene-induced cell transformation by antibodies specific for amino acid 12 of *ras* protein. Nature 314:639, 1985

Finlay DR, Newmeyer DD, Hartl PM et al. Nuclear transport *in vitro*. J Cell Sci Suppl 11:225, 1989

Furie B and Furie BC. The molecular basis of blood coagulation. Cell 53:505, 1988

Gerritse K, De Knijff P, van Ierssel G et al. Immunological discrimination between the human apolipoprotein E2 (Arg158→Cys) and E3 isoforms. J lipid Res 33:273, 1992

Getzoff ED, Geysen HM, Rodda SJ et al. Mechanisms of antibody binding to a protein. Science 235:1191, 1987

Goldman ME, McKee RL, Caulfield MP et al. A new highly potent parathyroidhormoneantagonist:[D-Trp12,Tyr34]bPTH-(7-34)-NH2. Endocrinology 123:2597, 1988

Gusella JF, Wexler NS, Conneally PM et al. A polymorphic DNA marker genetically linked to Huntington's disease. Nature 306:234, 1983

Haaseth RC, Sobczyk-Kojiro K, Medzihradsky F et al. Single residue modifications of the delta opioid receptor selective peptide, [D-Pen2,D-Pen5]-enkephalin (DPDPE). Correlation of pharmacological effects with structural and conformational features. Int J Pept Protein Res 36:139, 1990

Hackett CJ, Hurwitz JL, Dietszgold B et al. A synthetic decapeptide of influenza virus hemagglutinin elicits helper T cells with the same fine recognition specificities as occur in response to whole virus. J Immunol 135:1391, 1985

Ho CL, Ko JL, and Li CH. Beta-endorphin. Receptor binding and peripheral opioid activities of [Gln8]-, [Trp27]-, and [Tyr31]-analogs. Int J Pept Protein Res 32:74, 1988

The Huntington's Disease Collaborative Research Group. A novel gene containing a trinucleotide repeat that is expanded and unstable on Huntington's disease chromosomes. Cell 72:971, 1993

Jameson BA and Wolf H. Predicting antigenicity from protein primary structure: A new algorithm for the prediction of antigenic sites. Comp Appl Biosc 4:181, 1988

Jensen RH, Vanderlaan M, Grabske RJ et al. Monoclonal antibodies specific for sickle cell hemoglobin. Hemoglobin 9:349, 1985

Kerem B, Rommens JM, Buchanan JA et al. Identification of the cystic fibrosis gene: Genetic analysis. Science 245:1073, 1989

Koenig M, Hoffman EP, Bertelson CJ et al. Complete cloning of the Duchenne muscular dystrophy (DMD) cDNA and preliminary genomic organization of the DMD gene in normal and affected individuals. Cell 50:509, 1987

Laman JD, Rácz P, Tenner Rácz K et al. Immunocytochemical determination of antigen and epitope specificity of HIV-1-specific B cells in lymph-node biopsies from HIV-1 infected individuals. AIDS 5:255, 1991

Laman JD, Schellekens MM, Abacioglu YH et al. Variant-specific monoclonal and group-specific polyclonal human immunodeficiency virus type 1 neutralizing antibodies raised with synthetic peptides from the gp120 third variable domain. J Virol 66:1823, 1992

Langbeheim H, Arnon R and Sela M. Antiviral effect on MS-2 coliphage obtained with a synthetic antigen. Proc Natl Acad Sci USA 73:4636, 1976

Latimer JJ, Berger FG and Baumann H. Highly conserved upstream regions of the α_1-antitrypsin gene in two mouse species govern liver-specific expression by different mechanisms. Mol Cell Biol 10:760, 1990

Laufer R, Gilon C, Chorev M et al. [pGlu6,Pro9]SP6-11, a selective agonist for the substance P P-receptor subtype. J Med Chem 29:1284, 1986

Lawrence ID, Warner JA, Cohan VL et al. Induction of histamine release from human skin mast cells by bradykinin analogs. Biochem Pharmacol 38:227, 1989

Lottenberg R, Hall JA, Pautler E et al. The action of factor Xa on peptide p-nitroanilide substrates: Substrate selectivity and examination of hydrolysis with different reaction conditions. Biochim Biophys Acta 874:326, 1986

Meier UT and Blobel G. A nuclear localization signal binding protein in the nucleolus. J Cell Biol 111:2235, 1990

Monaco AP, Neve RL, Colletti-Feener C et al. Isolation of candidate cDNAs for portions of the Duchenne muscular dystrophy gene. Nature 323:646, 1986

Ng ISL, Pace R, Richard MV et al. Methods for analysis of multiple cystic fibrosis mutations. Hum Genet 87:613, 1991

Ni F, Konishi Y, Bullock LD et al. High-resolution NMR studies of fibrinogenlike peptides in solution: Structural basis for the bleeding disorder caused by a single mutation of Gly(12) to Val(12) in the A alpha chain of human fibrinogen Rouen. Biochemistry 28:3106, 1989

O'Connell BC, Hagen FK and Tabak LA. The influence of flanking sequence on the O-glycosylation of threonine *in vitro*. J Biol Chem 267:25010, 1992

Orkin SH, Daddona PE, Shewach DS et al. Molecular cloning of human adenosine deaminase gene sequences. J Biol Chem 258:12753, 1983

Párraga G, Horvath S, Hood L et al. Spectroscopic studies of wild-type and mutant "zinc finger" peptides: Determinants of domain folding and structure. Proc Natl Acad Sci USA 87:137, 1990

Parry NR, Ouldridge EJ, Barnett PV et al. Serological prospects for peptide vaccines against foot-and-mouth disease virus. J Gen Virol 70:2919, 1989

Partidos CD and Steward MW. The effects of a flanking sequence on the immune response to a B and a T cell epitope from the fusion protein of measles virus. J Gen Virol 73:1987, 1992

Prorok M and Lawrence DS. Multiple arginine residues contribute to the increased efficacy of peptide substrates for the cAMP-dependent protein kinase. Biochem Biophys Res Commun 165:368, 1989

Riordan JR, Rommens JM, Kerem B-S et al. Identification of the cystic fibrosis gene: Cloning and characterization of the complementary DNA. Science 245:1066, 1989

Ris-Stalpers C, Kuiper GGJM, Faber PW et al. Aberrant splicing of androgen receptor mRNA results in synthesis of a non-funtional receptor protein in a patient with androgen insensitivity. Proc Natl Acad Sci USA 87:7866, 1990

Rommens JM, Iannuzzi MC, Kerem B-S et al. Identification of the cystic fibrosis gene: Chromosome walking and jumping. Science 245:1059, 1989

Saeed ZA, Huang SC, Coy DH et al. Effect of substitutions in position 12 of bombesin on antagonist activity. Peptides 10:597, 1989

Schoofs PG, Geysen HM, Jackson DC et al. Epitopes of an influenza viral peptide recognized by antibody at single amino acid resolution. J Immunol 140:611, 1988

Sette A, Colizzi V, Appella E et al. Analysis of lysozyme-specific immune responses by synthetic peptides. I. Characterization of antibody and T cell-mediated responses to the N-terminal peptide of hen egg-white lysozyme. Eur J Immunol 16:1, 1986

Smilek DE, Wraith DC, Hodgkinson S et al. A single amino acid change in a myelin basic protein peptide confers the capacity to prevent rather than induce experimental autoimmune encephalomyelitis. Proc Natl Acad Sci USA 88:9633, 1991

Tam JP. Synthetic peptide vaccine design. Synthesis and properties of a high-density multiple antigen peptide system. Proc Natl Acad Sci USA 85:5409, 1988

Tschesche H, Beckmann J, Mehlich A et al. Semisynthetic engineering of proteinase inhibitor homologues. Biochim Biophys Acta 913:97, 1987

Valerio D, Duyvesteyn MGC, Meera-Khan P et al. Isolation of cDNA clones for human adenosine deaminase. Gene 25:231, 1983

Valerio D, Dekker BMM, Duyvesteyn MGC et al. One adenosine deaminase allele in a patient with severe combined immunodeficiency contains a point mutation abolishing enzyme activity. EMBO J 5:113, 1986

Vancompernolle K, Goethals M, Huet C et al. G- to F-actin modulation by a single amino acid substitution in the actin binding site of actobindin and thymosin beta 4. EMBO J 11:4739, 1992

Viskochil D, Buchberg AM, Xu G et al. Deletions and a translocation interrupt a cloned gene at the neurofibromatosis type 1 locus. Cell 62:187, 1990

Wallace MR, Marchuk DA, Andersen LB et al. Type 1 neurofibromatosis gene: Identification of a large transcript disrupted in three NF1 patients. Science 249:181, 1990

Walter G, Scheidtmann K-H, Carbone A et al. Antibodies specific for the carboxy- and amino-terminal regions of simian virus 40 large tumor antigen. Proc Natl Acad Sci USA 77:5197, 1980

Weisgraber KH, Innerarity TL and Mahley RW. Abnormal lipoprotein receptor-binding activity of the human E apoprotein due to Cysteine-Arginine interchange at a single site. J Biol Chem 257:2518, 1982

Wolf H, Modrow S, Motz M et al. An integrated family of amino acid sequence analysis programs. Comput Appl Biosci 4:187, 1988

Wolff B, Willingham MC and Hanover JA. Nuclear protein import: Specificity for transport across the nuclear pore. Exp Cell Res 178:318, 1988

Woo SLC, Lidsky AS, Güttler F et al. Cloned human phenylalanine hydroxylase gene allows prenatal diagnosis and carrier detection of classical phenylketonuria. Nature 306:151, 1983

Zegers ND, Deen C, Gerritse K et al. An improved conjugation method for unidirectional covalent coupling of synthetic peptides to proteins using glutaraldehyde in a dialysis method. J Immunol Meth 130:195, 1990

Zegers ND, Claassen E, Gerritse K et al. Detection of genetic variants of human α_1-antitrypsin with site specific monoclonal antibodies. Clin Chem 37:1606, 1991

Zegers ND, van Holten C, Claassen E et al. Generation and selection of monoclonal antibodies to variants of α_1-antitrypsin specific for one single amino acid using synthetic peptides: Importance of conjugation of synthetic determinants. In: Peptides 1992, eds Schneider CH and Eberle AN, ESCOM Science Publishers, Leiden, The Netherlands. p838, 1993

Chapter 18

PEPTIDE VACCINES

Ruth Arnon and Michael Sela

TABLE OF CONTENTS

I. INTRODUCTION

The development of vaccines has been one of the most important achievements in immunology and medicine to date. It was initiated almost 200 years ago by the famous trial of Jenner, who inoculated a small boy with cowpox, thereby achieving preventive immunity against the fatal smallpox, a process which eventually led to the eradication of this disease. The major breakthrough, occurring just over 100 years ago, was Pasteur's development of a rabies vaccine, a direct result of his pioneering systematic experiments on attenuation of microorganisms, which paved the way for the development of a whole series of vaccines based on viral attenuation. The presently existing vaccines, which consist of killed or live attenuated disease-causing microbial agents or their isolated components, have definitely diminished the incidence, morbidity and mortality of a large number of infectious diseases, including major killers such as diphtheria and polio.

Notwithstanding these tremendous achievements, there have been several crucial drawbacks incurred by the current procedures for vaccine preparation. For example, (1) the difficulty in preparation of sufficient material for vaccine production in the case of some viruses which cannot be cultivated *in vitro*, such as HBV or AIDS virus, and especially in the case of parasites. (2) Safety considerations in the difficulty of ascertaining adequate killing or attenuation of the vaccine preparation, and the hazard which may be caused by exposure of both the vaccinees and those involved in vaccine production. This consideration is of particular consequences in case of fatal incurable diseases such as AIDS. (3) The genetic variations in viruses, or the frequent recurring variations in the antigenic components of parasites, which result in the evolution of new strains with different serological specificity, for which continuous development of new vaccines is obligatory.

For these reasons and others, new approaches are being considered for vaccine development in the future. These include (1) use of recombinant DNA technology for the production of the relevant microbial protein antigens in bacterial, yeast, plant or animal cells for vaccine preparation, (2) use of recombinant DNA techniques for production of live vaccines by introducing the relevant gene(s) into the genome of vaccinia virus or *Salmonella* mutants, (3) application of anti-idiotype antibodies which mimic the antigenic structure of the relevant epitope, and thus could induce protective immunity, and (4) the approach to be described in this chapter — the utilization of synthetic peptides which constitute the relevant protective epitopes of viruses, bacterial toxins or parasites for eliciting a neutralizing immune response toward the disease-causing agent.

Early studies in our laboratory have demonstrated that synthetic antigens containing an immunoreactive region of a protein can give rise to a specific, and often conformation-dependent, immune response toward the intact native protein (Arnon et al., 1971). When the protein in question is a component of a virus, e.g., the coat protein of MS-2 coliphage, the

0-8493-8967-4/95/$0.00+$.50

antibodies induced by a synthetic fragment were capable of neutralizing the viability of the phage (Langbeheim et al., 1976). These findings paved the way for the use of synthetic peptides as the basis for vaccine design (Arnon, 1972, 1986; Sela, 1974; Steward and Howard, 1987). As summarized in several recent reviews (Jolivet et al., 1990; Arnon and Sela, 1992; Arnon and Van Regenmortel, 1992; Ogasawara et al., 1992), this approach has been employed for many microbial antigens. In the following we will focus mainly on research from our own laboratory, which relates to influenza virus and the bacterial toxins of cholera and shiga, but we will mention also results, particularly recent data, from other laboratories to illustrate the wide range application of this concept.

The use of synthetic materials in the field of vaccination against viruses may be envisaged to have several advantages over the use of intact, attenuated, or killed viruses. First, theoretically, it should be possible to attach several peptides, representing the relevant portions of different viruses, to the same carrier molecule. Thus, each adequately designed synthetic antigen will be multivalent and capable to replace several narrowly specific vaccines. Second, this approach might eliminate immunization against many irrelevant antigenic determinants of the virus or of irrelevant proteins which contaminate the viral preparation used for vaccination. Thus, frequently occurring and undesired sidereactions may be avoided. Third, for many vaccines, the addition of an adjuvant is required in order to initiate an immune response, and some of these adjuvants are not suitable for human use. In a synthetic macromolecule this problem of adjuvanticity may find its solution by introduction of certain groups which enhance antigenicity. Antigens so designed may possess adjuvant-like properties and may prove to be less hazardous.

Another advantage of the synthetic approach is the possibility of designing the molecule according to the purpose desired. For example, it is known that the immune response is genetically controlled, a control which is partly dependent on the structure of the macromolecular carrier, and in many cases is linked to the main histocompatibility antigen of the host. It may thus be foreseen that the efficiency of synthetic materials for different individuals may depend on the macromolecular carrier used for its preparation. A detailed study of this correlation might enable the selection of the most efficient vaccine to be used according to tissue typing. Furthermore, induction of *in vivo* antiviral cytotoxic T cell response is often MHC class I dependent, specific toward one or more epitopes. Hence, a cocktail of viral peptides may be used as antiviral T cell vaccines (Aichele et al., 1990).

II. CARRIERS AND ADJUVANTS FOR SYNTHETIC PEPTIDE VACCINES

Before considering the use of synthetic peptides as potential immunogens, it should be realized that a prerequisite for the induction of immunity is an interaction between T and B cells recognizing two distinct determinants (epitopes) of a single antigenic moiety. T-helper cells in general do not recognize native protein antigens, but only the partially degraded molecule, and the T cell epitopes are usually defined as sequential peptides. B cell epitopes are either sequential or conformational and are often discontinuous. A synthetic vaccine should contain both B and T cell epitopes. This situation may be achieved by a special design of the synthetic peptide or alternatively when the peptide is bound to a carrier. Carriers and adjuvants are therefore believed to be essential for endowing synthetic peptides vaccines with adequate efficacy, and efforts are being devoted to their development.

Often the carrier itself acts as an adjuvant, as in the case of liposomes, membranous vesicles formed by the dispersion of phospholipids in aqueous solution. Frisch et al. (1991) examined the immunogenicity of a hexapeptide from the C-terminus of the histone H3 when associated with liposomes. This peptide was not immunogenic if administered without carrier and adjuvant. When encapsulated in large vesicles it was ineffective as well, but covalent coupling of this peptide to the surface of large vesicles yielded both an IgM and IgG response, though of short duration. Furthermore, when covalently linked to the surface of small unilamellar vesicles containing as adjuvant monophosphoryl lipid A, a relatively long-lasting response with memory cell induction was observed. The conclusion from these studies is that for short synthetic peptides liposomes can provide a substitute for a carrier protein, but the vesicle must contain an adjuvant to obtain an efficient immune response (Garcon and Six, 1991).

Schild et al. (1991) have developed a synthetic lipopeptide vaccine, tripalmitoyl-S-glycerylcysteinyl-seryl-serine (P3CSS) coupled to various peptides of influenza virus. They found that immunization with synthetic peptide alone failed to induce virus-specific CTL *in vivo*. However, priming with a synthetic lipopeptide vaccine was successful and was comparable to the priming efficiency seen with the infectious virus. Goodman-Snitkoff et al. (1991) have described a technique to stimulate an immune response to small nonimmunogenic peptides. They covalently linked a peptide to phosphatidylethanolamine (PE), then complexed the peptide-PE conjugate with additional phospholipid and cholesterol. This complex was found highly immunogenic in mice without additional adjuvants.

Lussow et al. (1991) found that heat-shock proteins can act as carrier molecules when conjugated to synthetic peptides without requiring additional adjuvants. A number of integral membrane proteins (Imps) isolated from *Escherichia coli*, when covalently conjugated to antigenic peptides, significantly enhanced the immune response to the peptide as compared to injection of the peptide alone. The antibody response could not be significantly increased by the addition of incomplete Freund's adjuvant.

Redmond et al. (1991) found that the inner capsid protein of bovine rotavirus can be made to self-assemble *in vitro* and form spherical particles which have an inherent capacity to target to cells of the immune system. These spheres can therefore act as novel immunological carriers when coupled to peptides. Every epitope they tested gave an excellent immune response without the use of adjuvants.

One method for the preparation of a completely synthetic product containing both antigen and carrier is the multiple antigen peptide (MAP) system developed by Tam (1988). The surface of these macromolecules consists of multiple clusters of the antigenic epitope with a small oligolysine core at the center. The MAP system has the advantage of being completely chemically defined, containing either multiple copies of a single peptide epitope or different antigenic peptides which can be attached to the same macromolecule.

III. CHOLERA TOXIN

The toxin of *Vibrio cholerae* is composed of two subunits, A and B. Subunit A activates adenylate cyclase, which triggers the biological activity, whereas subunit B is responsible for binding to cell receptors. The B subunit expresses most of the immunopotent determinants, and antibodies against it are capable of neutralizing the biological activity of the intact toxin. In view of its immunodominant role, the B subunit was the obvious candidate for attempting the use of synthetic peptides for immunization. We have synthesized nine peptides, corresponding to various regions of the B subunit, and coupled them to tetanus toxoid (TT). All the conjugates elicited antibodies against the respective homologous peptides, most of which reacted also, to different extents, with the intact B subunit and with native cholera toxin (Jacob et al., 1983).

Of most interest among these peptides were CTP1 (residues 8–20) and CTP3 (residues 50–64). Antisera against these two peptides exerted significant inhibition of the biological activity of cholera toxin. The toxic effect of CT can be demonstrated either *in vivo* by skin vascular permeation and by fluid accumulation in ligated small intestinal loops or *in vitro*, at the biochemical level, by the induction of adenylate cyclase. The inhibitory effect of the antipeptide sera was manifested in all these assays, reaching inhibition values of 60–70% (Jacob et al., 1983, 1984a).

It is noteworthy that the synthetic peptides as such do not cause any toxic or other biological side effects, which is a crucial consideration in the evaluation of their possible suitability for vaccination.

The interest in cholera toxin is accentuated by its relationship with toxins of other microorganisms leading to enterotoxic infections and particularly with that of *E. coli*. The heat-labile toxin (LT) of pathogenic strains of *E. coli* is the causative agent of diarrhea in many tropical countries and, due to its wide spread, probably presents a more serious health problem than cholera. In view of the high level of sequence homology between the B subunits of the LT and CT and the immunological cross-reactivity between the two toxins, with the existence of both shared and specific antigenic determinants, we have investigated the involvement of our synthetic peptides in the antigenic similarity. We demonstrated that, indeed, the antiserum elicited by CTP3 (residues 50–64) is highly reactive with the LT. This is not surprising, since this region is completely identical in the two toxins. The antiserum against CTP1 (residues 8–20) was also cross-reactive, although to a much lower extent. These cross-reactions were observed with toxins obtained from multiple strains of both human and porcine *E. coli*. Moreover, antisera to both CTP1 and CTP3, which are inhibitory toward CT, were found equally effective in neutralizing the biological activity of the *E. coli* LT (Jacob et al., 1984b). This indicates that a vaccine based on synthetic peptides may evolve into a general vaccine against the coli-cholera family of diarrheal diseases.

IV. SHIGA TOXIN

Vaccination against *Shigella* species, to prevent shigellosis, presents a number of problems, and as a consequence no efficient shigella vaccine is available as of yet. Pathogenesis of these infections is dependent on the epithelial cell invasive properties of the organism and, hence, the most recent attempts toward developing shigella vaccines have been based on the expression of virulence-associated membrane antigens. The various species of *Shigella* are also known to produce a protein toxin which experimentally reproduces three major features of the infection (Keusch and Jacewicz, 1975). The role of the toxin in the pathogenesis of *Shigella* is still controversial. However, the detection of neutralizing antibodies against this toxin in sera of convalescent shigellosis patients implies at least partial involvement.

Shiga toxin, isolated from *S. dysenteriae* 1 strains, is one of the most potent of the lethal microbial toxins. It was initially classified as a neurotoxin due to the limb paralysis and death it caused when parenterally administered to sensitive animals. In addition, it is cytotoxic to certain tissue culture cells, and also exerts enterotoxic activity when applied to intestinal ileal loops *in vivo*. This toxin has been purified and has been shown to have a similar structure to that of cholera toxin, consisting of A and B subunits. Antibodies raised against the B subunit, which comprises an oligomer of a B chain, the molecular weight of which is 7716 Da (Seidah et al., 1986), were shown to neutralize the cytotoxic effects of the toxin toward the HeLa cell monolayers and also to inhibit the binding of labelled toxin to these cells (Donohue-Rolfe et al., 1984). In view of this, we have evaluated the B chain of Shiga toxin as a candidate for applying the synthetic vaccine approach.

Computerized plots of hydrophilicity and surface residues identified the amino and carboxy terminal regions of the B chain as the most probable antigenic regions. Hence, several peptides of these regions were synthesized and either conjugated to macromolecular carriers or polymerized. Rabbit antisera raised against all the conjugates and polymers were highly reactive with the respective homologous peptides, and those manifesting the highest titer were further evaluated for their reactivity with the intact toxin. These experiments revealed that all the peptides showed a high extent of cross-reactivity with the Shiga toxin demonstrated by both ELISA and immunoblotting tests (Harari et al., 1988).

More significantly, the antisera had a considerable neutralizing capacity toward all three biological effects of the toxin. (1) The anti-peptide antibodies at a 1:100 dilution led to highly significant inhibition of HeLa cell cytotoxicity, reaching values of 70% inhibition, almost equal to that effected by antibodies against whole Shiga toxin in a similar dilution. (2) Anti-peptide antisera neutralized the enterotoxic activity as measured by fluid secretion into ligated intestinal loops of rats. The antisera against the amino terminal peptide 5–18, and against the carboxy terminal peptide 54–67, showed the highest activity, exhibiting essentially the same level of neutralization (≈80%) achieved with the anti-Shiga holotoxin. (3) Similar results were obtained when the effect of the anti-peptide sera was evaluated on the neurotoxic activity. In these experiments, mice exposed to a certain dose of Shiga toxin usually died within 7 days. However, prior incubation of the toxin with the various anti-peptide sera reduced its lethal effect by ≈60%.

Not only were antisera induced by the various peptides capable of neutralizing the various activities of the toxin, but active immunization of mice and rats with the peptide conjugates led to the development of partial protection against the detrimental effects of the Shiga toxin. Thus, nonimmunized mice succumbed to 96% mortality 7 days after they had been exposed to the lethal dose (2 LD_{50}) of the toxin. In contrast, in mice immunized with the conjugates of the various peptides and exposed to the same dose of toxin, the mortality was much lower, in some cases only 20%, with 80% long-term survival. Furthermore, in rats, immunization with the conjugates led to a marked inhibition of fluid secretion into ligated ileal loops after exposure to the toxin, as compared to the effect in nonimmunized rats.

In shigellosis, as in all other enteric diseases, the intestinal area is the site of the first encounter with the infectious agent. The gut is also the organ into which the Shiga toxin is secreted and where it exerts its detrimental effect. Hence, the notion of developing local immunity by intragastric immunization, which would lead to neutralization of the toxic effect *in situ*, is a very attractive one. Local immunity in the gut is usually assessed in animal models, either rabbits or rats, by determination of the IgA level. In the case of *Shigella*, attempts have been made in the direction of oral immunization using the live bacteria. Such vaccination with live *S. flexneri* noninvasive strains indeed led to primary and IgA memory responses (Keren et al., 1985). In view of our results concerning the immunogenicity and protective potential of the above-mentioned Shigatoxin peptides, we tested whether they could also induce immunity in rats. Indeed, we were able to show that oral immunization with most of the peptide conjugates by intragastric feeding, following priming by intradermal injection in complete Freund's adjuvant, induced elevation of specific IgA antibodies in the sera of immunized rats several days after bile duct ligation. Furthermore, the immunized rats demonstrated significant protection against intestinal fluid secretion induced by injection of Shiga toxin into ligated ileal loops (Arnon et al., 1991). These results indicate, therefore, that synthetic peptides may serve, as such or in a conjugated form, for induction of both systemic and local immunity.

The finding that both termini of the molecule contribute to the neutralizing immune capacity and local immunity can be explained by assuming an appropriate folding of the polypeptide chain of the B subunit. According to the amino acid sequence of the B chain, there are two cysteine residues at both termini of the molecule, namely at positions 4 and 57, which are linked by a disulfide bridge (Seidah et al., 1986). This would lead to proximity between the N-terminal and the C-terminal regions in the three-dimensional structure of the molecule. Furthermore, in the intact toxin, namely the complex between the A chain and the oligomer of the B chain, this hydrophilic region is probably surface exposed and hence constitutes the immunodominant part of the toxin. These results may, therefore, be of importance for the rational design of peptide vaccines against shigellosis.

V. OTHER BACTERIA

Besides the possibility that the synthetic approach could be used to induce neutralizing immunity against bacterial toxin, it could be considered also as a means to reduce bacterial infection. It should be noted that despite the tremendous success of antibiotics and the conventional vaccines against bacterial diseases, in certain instances these measures are not sufficient and an alternative approach for vaccination, such as the use of synthetic peptides, is being explored. A case in point is *Listeria monocytogenes*, a Gram-positive bacterium, which grows in the cytoplasm of eukaryotic cells, and in an immunocompromised person it can lead to severe disease. Pamer et al. (1991) identified a dominant class I MHC-restricted *L. monocytogenes* epitope, by synthesis of 11 nonapeptides based on a peptide-binding motif described recently. This peptide was recognized very efficiently by the CTL clone B9 and represents the first MHC class I-restricted epitope identified in bacteria. The identification of bacterial CTL epitopes has been hampered by the large number and diversity of bacterially expressed proteins, which now may be possible by using this or other peptide-binding motifs.

Another example is *Neisseria gonorrhea*. The development of a vaccine against gonorrhea remains a major problem in the control of this sexually transmitted disease. Heckels et al. (1990) have used a monoclonal antibody to identify a protective highly conserved but nonimmunogenic epitope of the outer membrane protein PIB of *N. gonorrhoea* using a series of overlapping synthetic peptides spanning the deduced sequence of the protein. The identified peptide, comprising the sequence 193–204, was used for immunization of rabbits in either the free form or as a conjugate with KLH. The resulting antisera, in the presence of human complement, were bactericidal for both a homologous and a heterologous strain of the bacteria, indicating that synthetic peptides can be used to induce a protective broad-specificity immune response against epitopes of gonorrhoea PI which are normally nonimmunogenic.

Synthetic peptides served also for the preparation of a vaccine against rheumatogenic streptococci. Already in 1981, Beachey et al. identified specific epitopes of *Streptococcus pyogenes* M protein, of both type-specific and cross-reactive nature, which are related to the virulence. Accordingly, they synthesized several peptide fragments of 14–35 amino acid residues, which led to efficient protective response against several types of group A streptococci. More recently (Beachey et al., 1988), after localizing both protective and cardiac cross-reactive epitopes, they prepared a multivalent vaccine by synthesizing protective epitopes of several serotypes in tandem. These hybrid peptides evoke protective immune response without evoking tissue cross-reactive immunity. Bessen and Fischetti (1990) synthesized peptides which correspond to sequences completely contained within the conserved C repeats of streptococcus M6 protein. When coupled to cholera toxin B subunit and administered orally or intranasally, this peptide led to a significant protection of mice against pharyngeal infection by multiple serotypes of streptococci.

VI. PARASITES

In the case of parasites, where it is difficult to grow the organisms in culture and to prepare sufficient material for production of conventional vaccines, synthetic vaccines are particularly relevant. Studies on synthetic peptide vaccines against parasitic diseases have focused mostly on malaria. The circumsporozoite protein, the major surface antigen of malaria sporozoites, contains a region of multiple tandem amino acid repeats. In *P. falciparum* this region is a tetrapeptide, ASN-ALA-ASN-PRO (NANP), which is repeated up to about 40 times (Zavala et al., 1983). This sequence has been considered, therefore, as the basis for the development of a subunit vaccine against *P. falciparum* malaria. However, both synthetic and recombinant NANP peptides have been shown to be immunogenic only in H-2^b mice. Pessi et al. (1991) recently showed that when the NANP sequence was incorporated into a multiple antigen peptide (MAP) all except one mouse strain tested mounted an antipeptide antibody response.

Saul et al. (1992) were interested in an antigen of the merozoite stage of malaria. They selected three octapeptides from two invariable domains in the N and C terminus of the merozoite surface AG 2 (MSA2) of *P. falciparum*. When conjugated to diphtheria toxoid, these peptides elicited anti-MSA2 antibodies, which also bound the MSA2 homolog from *P. chabaudi*. The mice vaccinated with these conjugates and then challenged with a lethal inoculum of *P. chabaudi* showed a high level of protection and most of them survived. These results suggest that the conserved regions of MSA2, when presented in a suitable immunogenic form, could become a basis for a malaria vaccine that circumvents some of the problems of antigenic diversity.

In the system of schistosomiasis, the 28 kDa protein of *S. mansoni* has received considerable attention as a vaccine candidate. A synthetic peptide derived from the sequence 115–131 has been shown to contain both T and B cell recognition sites. Wolowczuk et al. (1991) synthesized an octameric peptide from this region which was immunogenic in rats, mice and baboons. Immunized Fischer rats were able to mediate platelet-, macrophage-, and eosinophil-dependent cytotoxicity toward schistosomula. Furthermore, rats immunized with this peptide were partially protected against a challenge infection with *S. mansoni* cercariae with a concomitant increase in the level of anti-28 kDa IgG and IgE being observed.

VII. INFLUENZA VIRUS

The currently available influenza vaccines consist of either attenuated or inactivated viral particles. Their effectiveness is only partial, mainly due to the frequent antigenic variation of the external glycoproteins of the virus (Laver, 1979), with each new strain presenting a new challenge to the host immune system.

The influenza virus provides a very suitable model for studying the synthetic approach to vaccination for several reasons: (a) detailed information is available on the structure and function of this virus, as well as on its serological specificities and genetic variations; (b) various reliable assays of the virus are available for evaluating the effect of the immune response on the different viral functions; and (c) sufficient information is available on the amino acid sequence of its major protein components, as well as on their three-dimensional structure and immunochemical properties, to allow the synthesis of peptide fragments that might carry some of its antigenic determinants.

The influenza virus contains two major membranal proteins — hemagglutinin and neuraminidase. The hemagglutinin (HA) occurs as trimeric spikes projecting from the viral proteolipid envelope. It is quantitatively the most important glycoprotein in the viral surface, undergoes constant variations and is the major antigenic protein against which neutralizing antibodies are directed. In view of these properties, it was considered a good candidate for studying the synthetic approach to vaccination.

We have studied several synthetic peptides of the HA molecule, all of which proved immunoreactive, leading to both humoral and cellular immunity (Muller et al., 1982; Shapira et al., 1985a). The peptide with the highest efficacy consisted of 18 amino acid residues corresponding to the sequence 91–108 of the HA molecule. This region, which is common to most H3 strains, was computer predicted to be immunologically reactive and was deliberately chosen to be part of a conserved sequence.

Indeed, a conjugate of this peptide with tetanus toxoid elicited, in both rabbits and mice, antibodies that reacted in a radioimmunoassay with the synthetic peptide, as well as with the intact influenza virus of several strains of type A. These antibodies were capable of inhibiting the capacity of the HA of the relevant strains to agglutinate chicken red blood cells. They also interfered with the *in vitro* growth of the virus in tissue culture, causing up to 60% reduction in viral plaque formation. Furthermore, mice immunized with the peptide-toxoid conjugate were partially protected against further challenge infection with the virus.

The peptide 91–108 does not overlap with any of the four antigenic determinants of the native HA proposed by Wiley et al. (1981) according to the crystallographic data. It is adjacent, however, to the antigenic site D in the three-dimensional structure of the molecule. This could provide an explanation for the partial protective effect achieved by immunization with this peptide. This region is in the interior of the HA trimer spike and is thus probably a "hidden" epitope in the native virus. However, in the infective state of the virus particle, as a result of the "opening" of the trimer, this region is exposed. This enables the antibodies raised by the synthetic peptide to react with the virus and impair its infectivity.

The results reported hitherto were achieved by immunization in complete Freund's adjuvant (CFA) for augmenting the immune reactivity. This adjuvant, which consists of a water-in-oil emulsion containing killed mycobacteria, is a very effective adjuvant, evoking high-level and long-lasting immunity. However, it is not suitable for human use, since it induces local reactions and granulomas, inflammation and fever, due to both the low metabolizable oil and the mycobacteria. To explore the possibility of replacing the CFA by a less harmful substance, we did some experiments with the synthetic adjuvant MDP (N-acetylmuramyl-L-alanyl-D-isoglutamine) (Ellouz et al., 1974). This adjuvant was previously shown to be efficient in enhancing the immunogenicity of a synthetic branched polymeric antigen (Mozes et al., 1980), as well as of conjugates containing peptide fragments of diphtheria toxin (Audibert et al., 1982), of bacteriophage MS-2 (Arnon et al., 1980), and of cholera toxin (Jacob et al., 1986).

We have employed MDP in combination with the 91–108 peptide of hemagglutinin for induction of an anti-influenza response. Our findings revealed that this adjuvant was similar to CFA in the induction of antibodies specific toward the peptide (Shapira et al., 1985b). As for the cross-reactivity with the intact virus, although a marked difference was observed between the titer of antibodies raised in the presence of CFA and the MDP, both adjuvants were capable of inducing protective immunity against a viral challenge. MDP was most efficient when coupled covalently to the conjugate of the 91–108 peptide. In that form, it led to the highest protection level against *in vivo* viral challenge, even higher than that induced in the presence of CFA. It should be emphasized that this conjugate of the synthetic peptide and MDP with the tetanus toxoid is water soluble and was administered in a physiological aqueous solution and hence constitutes a synthetic vaccine with built-in adjuvanticity, which could be suitable for use in humans.

Henrickson et al. (1991) investigated the possibility that linear epitopes on the hemagglutinin-neuraminidase (HN) surface glycoprotein of the human parainfluenza virus type 3 (PIV-3) might induce neutralizing antibodies. However, none out of 37 synthetic peptides, representing 64% of the HN complex, reacted with either neutralizing monoclonal antibodies or high-titer human serum samples. They concluded that the neutralizing epitopes of PIV-3 are conformational and therefore cannot be imitated by linear peptides.

VIII. ADDITIONAL STUDIES ON VIRUSES

The last few years have seen considerable progress in the investigation and development of peptides leading to an antiviral immune response, which could serve as adequate candidates for the production of synthetic vaccines. This is demonstrated by the following examples.

Kast et al. (1991) identified a single peptide from the outside of the Sendai virus nucleoprotein (NP) which is recognized by B6 mice CTL. Subcutaneous immunization of mice with the free peptide led to an NP peptide-specific CTL memory response, as well as an *in vivo* response against the Sendai virus. Consequently, mice immunized with this peptide were protected against lethal virus challenge. This work demonstrates that even free peptide can be effective as an antiviral T cell vaccine.

Snijders et al. (1991) identified linear epitopes on the E2 membrane protein of Semliki Forest virus (SFV) with vaccine potential. Mice immunized with the peptide 240–255 of this protein were protected against viral challenge. This protection correlated well with SFV-specific antibody titer, suggesting antibody-mediated protection.

Dion et al. (1990) synthesized four peptides predicted to be surface-accessible regions of the major viral envelope glycoprotein gp52 of the murine mammary virus. These peptides were coupled to a carrier and tested in protection of Balb/c mice against live viral challenge. One of these peptides (EP-3) significantly reduced the frequency of early onset tumors, and 67% of the test animals were tumor free for the entire observation period. This is in contrast to immunization with the intact gp52 which provided only marginal protection. Schultz et al. (1991) injected subcutaneously a T cell epitope from the nucleoprotein of lymphocytic choriomeningitis virus (LCMV) as a free synthetic peptide. This immunization led to the induction of a specific anti-LCMV CTL response *in vivo*. Protection was demonstrated by inhibition of LCMV replication in mice spleens, which correlated with CD8+ LCMV-specific CTL.

Foot-and-mouth disease virus (FMDV) was one of the earliest viruses in which a synthetic peptide approach was attempted. The major viral protein of antigenic significance is VP1, which is a weak immunogen. But conjugates of a synthetic peptide corresponding to its 141–160 sequence induced antibodies which cross-reacted with the virus and neutralized its infectivity (Bittle et al., 1982). Furthermore, Brown (1988) reported that when this peptide was presented as a linear dimer or tetramer, as little as 40 mg peptide led to protection of pigs against challenge infection. In recent studies (Francis et al., 1991), the multiple antigenic peptide (MAP) approach was employed using the same peptide of the VP1 protein. Monomer, tetramer and octamer structures of this peptide were examined, with the tetrameric structure giving an optimal response. The levels of neutralizing antibody obtained compared favorably with those obtained when the FMDV peptide was attached to various carrier proteins. Steward et al. (1991) have immunized cattle with this synthetic "vaccine" and compared the affinity of the antibodies for the peptide and for the whole virus in sera from protected and nonprotected animals. They found that the affinities of serum antibodies for both the peptide and the virus in protected cattle were significantly higher than those in non-protected animals. In view of this correlation, they recommend that antibody affinity should be one of the parameters considered in the assessment of the efficacy of synthetic immunogens in general.

Hepatitis B virus (HBV), which cannot be grown in culture for the production of a conventional vaccine, has been the target for an extensive study using synthetic peptides. Its envelope proteins contain three major antigenic domains designated S, Pre S2 and Pre S1. Early studies showed that the sequences 117–137 and 139–147 of the S protein may lead to a protective effect. Neurath et al. (1989) showed that the sequence 12–47 of the Pre S1 also induces virus-neutralizing antibodies. Milich et al. (1988) demonstrated that a complex synthetic immunogen, in which the sequence C120–140 serves as a carrier for the sequence 133–140 of the Pre S2 region, induced both T-helper cells and B cells against the native viral envelope. Kumar et al. (1992) were also interested in the T cell response toward synthetic peptides of the S region. They showed that lymphocytes showing cell proliferation on challenge with HBsAg are induced by a cysteine-rich 24 amino acid synthetic peptide derived from HBsAg, which undergoes spontaneous self-oligomerization. The response against the oligomerized peptide was much more significant than that against the intact HBsAg and had a broader restriction profile in various murine strains.

IX. HIV VIRUS

In view of the immense impact of AIDS in society, a major effort is being devoted to the development of a vaccine against this disease. This system will be dealt herewith separately from other viruses since most of the current work on synthetic peptide viral vaccines has focused on the human immunodeficiency virus (HIV), the putative cause of the AIDS disease. Presently, there is no vaccine available toward this disease, and live attenuated vaccines are probably not appropriate in this case, both because of the high risk in reverting to virulence and because HIV infection leads to deleterious immune responses. Hence, the synthetic approach seems especially attractive for this system. An ideal peptide vaccine would consist of only those epitopes known to be protective while avoiding the risk of inducing HIV infection. It is of interest that the significant applicability of this research prompted not only research institutes, but also many companies to develop synthetic peptide vaccines to AIDS (Spalding, 1992).

The V3 loop of the envelope protein, gp 120, is the principal neutralizing determinant (PND) of HIV, and an immunodominant cytotoxic T cell epitope has also been identified in this region (Takakashi et al., 1988). Synthetic peptides based on sequences within this region were indeed found to induce isolate-restricted neutralizing antibody. Hence, immunization with the PND was thought to induce only subtype-specific immunity (Nardelli et al., 1992). However, Javaherian et al. (1990) immunized animals with a 13 amino acid residue peptide from this region and observed that sera from these animals were cross-neutralizing for several HIV isolates. A smaller hexapeptide (GPGRAF) from this region, when used to immunize rabbits, was also capable of neutralizing divergent isolates.

Holley et al. (1991) analyzed the PND sequences of gp 120 from 245 isolates of HIV type 1. Their results suggest that peptide cocktails of limited size may be feasible vaccine candidates. Nardelli et al. (1992) also used the MAP system to present multiple copies of the PND of the divergent HIV isolates IIB, RF and MN and tested its efficacy in three species of animals. The MAP used was either synthesized in a monoepitope configuration, containing four repeats of each individual peptide, or synthesized with a diepitope configuration, adding to the C-terminus of the PND peptides a conserved sequence known to be a Th cell epitope of gp 120. The monoepitope construct yielded a species-dependent antibody response, whereas the diepitope MAPS were immunogenic in all three animal species. Also the antibody titers elicited by the diepitopes were significantly higher than those raised by immunization with the monoepitope MAPs. These results show that a synthetic peptide MAP antigen with a chemically defined structure and without the use of a protein carrier may be potentially useful for the design of an HIV vaccine.

It is generally thought that antiviral CTLs constitute an essential component of a successful immune response to viral infection. Hart et al. (1991) used various hybrid synthetic peptides to construct a carrier-free synthetic immunogen containing only those HIV epitopes necessary to induce a protective immune response. The complete hybrid consisted of (from N- to C-terminus) the first 12 amino acid residues of the gp 41 fusion domain; a T cell epitope of gp 120; a gp 120 B cell epitope from the V3 loop region; and extending from this region, an additional five or six residue segment which comprises a CTL epitope. They found that a carrier-free hybrid peptide, containing either the T1 or F HIV CTL epitope, had the ability to induce $CD8^+$ MHC class I-restricted CTLs *in vivo* in mice.

Takahashi et al. (1992) examined in detail the specificity of CTL recognition of an immunodominant determinant in the hypervariable portion of HIV gp 160. Three mouse CTL lines were examined, but only one elicited CTLs which showed extensive cross-reactivity between isolates. Residue 325 of gp 160 played a critical role in the specificity of the CTL response. Hence, CTLs with different specificities were generated by restimulating of gp 160-primed cells with peptides containing substitutions at position 325. This indicates the potential of synthetic peptide-based vaccines to induce broad specificity CTL toward HIV.

X. CONCLUSIONS

As clearly demonstrated above, there has been considerable effort in the past few years toward developing synthetic peptide vaccines for diverse groups of organisms. Some general conclusions are that a successful synthetic vaccine will probably consist of a cocktail of peptides to overcome the problem of hypervariable sequences and MHC restriction. These vaccines will probably contain both B and T cell epitopes and possibly a CTL-inducing region. Carrier molecules and adjuvants may be required, but these too will probably be synthetic, probably in a combined structure with the peptides that will include built-in adjuvanticity. As is obvious, however, from the rapid progress in this field in recent years, the prospect of constructing successful synthetic peptide vaccines grows as our knowledge in all areas of immunology increases, and it might become a reality in the not too distant future.

ABBREVIATIONS

CFA	Complete Freund's adjuvant
CT	Cholera toxin
CTL	Cytotoxic T lymphocyte
ELISA	Enzyme-linked immunosorbent assay
FMDV	Foot-and-mouth disease virus
HA	Hemagglutinin
HBV	Hepatitis B virus
HIV	Human immunodeficiency virus
KLH	Keyhole limpet hemocyanin
LCMV	Lymphocytic choriomeningitis virus
LT	Heat-labile toxin
MAP	Multiple antigen peptide
MHC	Major histocompatibility complex
MSA2	Merozoite surface antigen 2
MDP	N-Acetylmuramyl-L-alanyl-D-isoglutamine
NP	Nucleoprotein
PND	Principal neutralizing determinant
SFV	Semliki Forest virus
TT	Tetanus toxoid

XI. REFERENCES

Aichele P, Hengattner H, Zingernagel RM et al. Antiviral cytotoxic T-cell response induced by *in vivo* priming with a free synthetic peptide. J Exp Med 171:1815, 1990

Arnon R, Maron E, Sela M et al. Antibodies reactive with native lysozyme elicited by a completely synthetic antigen. Proc Natl Acad Sci USA 68:1450, 1971

Arnon R. Synthetic vaccine - dream or reality? In: Kohen A and Klinberg AM ed: Immunity of Viral and Rickettsial Diseases. Plenum Press, New York, p 209, 1972

Arnon R, Sela M, Parant M et al. Antiviral response elicited by a completely synthetic antigen with built-in adjuvanticity. Proc Natl Acad Sci USA 77:6769, 1980

Arnon R. Synthetic peptides as the basis for future vaccines. TIBS 11:521, 1986

Arnon R, Harari I and Keusch G. Synthetic peptides as the basis for anti-Shiga vaccines. In: Woodrow G and Levine M ed: New Generation Vaccines. Marcel Dekker, New York, p 688, 1991

Arnon R and Sela M. Synthetic approaches to vaccines for infectious and autoimmune disease. Vaccine 10:991, 1992

Arnon R and Van Regenmortel MHV. Structural basis for antigen specificity and design of new vaccines. FASEB J 6:3265, 1992

Audibert F, Jolivert M, Chedid L et al. Successful immunization with a totally synthetic diphtheria vaccine. Proc Natl Acad Sci USA 79:5042, 1982

Beachey EH, Seyer JM, Dale JB et al. Type-specific protective immunity evoked by synthetic peptide of Streptococcus pyogenes M protein. Nature 292:457, 1981

Beachey EH, Bronze M, Dale JB et al. Protective and autoimmune epitopes of streptococcal M proteins. Vaccine 6:192, 1988

Bessen D and Fischetti VA. Synthetic peptide vaccine against mucosal colonization by group A streptococci. I. Protection against a heterologous M serotype with shared C repeat region epitopes. J Immunol 145:1251, 1990

Bittle JL, Houghton RA, Alexander H et al. Protection against foot-and-mouth disease by immunization with a chemically synthesized peptide predicted from the viral nucleotide sequence. Nature 298:30, 1982

Brown F. Use of peptides for immunization against foot-and-mouth disease. Vaccine 6:180, 1988

Croft S, Walsh J, Lloyd W et al. TraT: A powerful carrier molecule for the stimulation of immune responses to protein and peptide antigens. J Immunol 146:793, 1991

Dion AS, Knittel JJ and Morneweck ST. Virus envelope-based peptide vaccines against virus-induced mammary tumors. Virology 179:474, 1990

Donohue-Rolfe AG, Keusch GT, Elson C et al. Pathogenesis of Shigella diarrhea. IX. Simplified high yield publication of Shigella toxin and characterization of subunit composition and function by the use of subunit-specific monoclonal and polyclonal antibodies. J Exp Med 160:1767, 1984

Ellouz F, Adam A, Ciorbaru R et al. Minimal structural requirements for adjuvant activity of bacterial peptidoglycan derivatives. Biochim Biophys Res Commun 59:1317, 1974

Francis MJ, Hastings GZ, Brown F et al. Immunological valuation of the multiple antigen peptide (MAP) system using the major immunogenic site of foot-and-mouth disease virus. Immunology 73:249, 1991

Frisch B, Muller S, Briand JP et al. Parameters affecting the immunogenicity of a liposome-associated synthetic hexapeptide antigen. Eur J Immunol 21:185, 1991

Garcon NMJ and Six HR. Universal vaccine carrier: Liposomes that provide T-dependent help to weak antigens. J Immunol 146:3697, 1991

Goodman-Snitkoff G, Good MF, Berzofsky JA et al. Role of intrastructural/intermolecular help in immunization with peptide-phospholipid complexes. J Immunol 147:410, 1991

Harari I, Donohue-Rolfe A, Keusch G et al. Synthetic peptides of Shiga toxin B-subunit induce antibodies which neutralize the biological activities of the toxin. Infect Immun 56:1618, 1988

Hart MK, Weinhold KJ, Scearce RM et al. Priming of anti-human immunodeficiency virus (HIV) $CD8^+$ cytotoxic T cells *in vivo* by carrier-free HIV synthetic peptides. Proc Nat Acad Sci USA 88:9448, 1991

Heckels JE, Vinji M and Tinsley CR. Vaccination against gonorrhea: The potential protective effect of immunization with a synthetic peptide containing a conserved epitope of gonococcal outer membrane protein 1B. Vaccine 8:225, 1990

Henrickson KJ, Kingsbury DW, van Wyke Coelingh KL et al. Neutralizing epitopes of human parainfluenza virus type 3 are conformational and cannot be imitated by synthetic peptides. Vaccine 9:243, 1991

Holley LH, Goudsmit J and Karplus M. Prediction of optimal peptide mixtures to induce broadly neutralizing antibodies to human immunodeficiency virus type 1. Proc Natl Acad Sci USA 88:6800, 1991

Jacob CO, Sela M and Arnon R. Antibodies against synthetic peptides of the B subunit of cholera toxin: Cross-reaction and neutralization of the toxin. Proc Natl Acad Sci USA 80:7611, 1983

Jacob CO, Sela M, Pines M et al. Adenylate cyclase activation by cholera toxin as well as its activity are inhibited with antibodies against related synthetic peptides. Proc Natl Acad Sci USA 8:7893, 1984a

Jacob CO, Pines M and Arnon R. Neutralization of heat labile toxin of E. coli by antibodies to synthetic peptides derived from B subunit of cholera toxin. EMBO J 3:2889, 1984b

Jacob CO, Arnon R and Sela M. Anti-cholera response elicited by a completely synthetic antigen with built-in adjuvanticity administered in aqueous solution. Immunol Lett 14:43, 1986

Javaherian K, Langlois AJ, LaRosa GJ et al. Broadly neutralizing antibodies elicited by the hypervariable neutralizing determinant of HIV-1. Science 250:1590, 1990

Jolivet M, Lise L, Grass-Masse H et al. Polyvalent synthetic vaccines: Relationship between T epitopes and immunogenicity. Vaccine 8:35, 1990

Kast WM, Roux L, Curren J et al. Protection against lethal Sendai virus infection by *in vivo* priming of virus-specific cytotoxic T lymphocytes with a free synthetic peptide. Proc Natl Acad Sci USA 88:2283, 1991

Keren DF, McDonald RA, Scott PJ et al. Effect of antigen form on immunoglobulin. A memory response of intestinal secretion to *Shigella flexneri*. Infect Immun 47:123, 1985

Keusch GT and Jacewicz M. Pathogenesis of Shigella diarrhea. Relationship of Shigella enterotoxin, neurotoxin, and cytotoxin. J Infect Dis 131:533, 1975

Kumar A, Kaul S, Manivet V et al. Comparison of immune response to a native viral antigen and a synthetic peptide derived from it: Implications for vaccine development. Vaccine 10:814, 1992

Langbeheim H, Arnon R and Sela M. Antiviral effect on MS-2 coliphage obtained with a synthetic antigen. Proc Natl Acad Sci USA 73:4636, 1976

Laver WG and Air GM. Structure and Variation in Influenza Virus. Elsevier, Holland, 1979

Lussow AR, Barrios C, Van Embden J et al. Mycobacterial heat-shock proteins as carrier molecules. Eur J Immunol 21:2297, 1991

Milich OR, Hughs JL, McLachlan A et al. Hepatitis B synthetic immunogen comprised of nucleocapsid T-cell sites and an envelope B-cell epitope. Proc Natl Acad Sci USA 85:1610, 1988

Mozes E, Sela M and Chedid L. Efficient genetically controlled formation of antibody to a synthetic antigen poly ($_L$Tyr, $_L$Glu)-poly (DLAla)-poly ($_L$Lys) covalently bound to a synthetic adjuvant (N-acetylmuramyl-L-alanyl-D-isoglutamine). Proc Natl Acad Sci USA 77:4933, 1980

Muller M, Shapira M and Arnon R. Anti-influenza response achieved by immunization with a synthetic antigen. Proc Natl Acad Sci USA 79:569, 1982

Nardelli B, Lu Y-A, Shiu DR et al. A chemically defined synthetic vaccine model for HIV-1. J Immunol 148:914, 1992

Neurath RB, Seto B and Strick N. Antibodies to synthetic peptides from the pre S1 region of the Hepatitis B virus (HBV) envelope (env) protein are virus-neutralizing and protective. Vaccine 7:234, 1989

Ogasawara K, Naruse H, Itoh Y et al. A strategy for making synthetic peptide vaccines. Proc Natl Acad Sci USA 89:8995, 1992

Pamer EG, Harty JT and Bevan MJ. Precise prediction of a dominant class I MHC-restricted epitope of Listeria monocytogenes. Nature 353:852, 1991

Pessi A, Valmori D, Migliorni P et al. Lack of H-2 restriction of the Plasmodium falciparum (NANP) sequence as multiple antigen peptide. Eur J Immunol 21:2273, 1991

Redmond MJ, Ohmann HB, Hughes HP et al. Rotavirus particles function as immunological carriers for the delivery of peptides from infectious agents and endogenous proteins. Mol Immunol 28:269, 1991

Saul A, Lord R, Jones GL et al. Protective immunization with invariant peptides of the *Plasmodium falciparum* antigen MSA2. J Immunol 148:208, 1992

Schild H, Deres K, Wiesmuller K-H et al. Efficiency of peptides and lipopetides for in-vivo priming of virus-specific cytotoxic T cells. Eur J Immunol 21:2649, 1991

Schulz M, Zinkernagel RM and Hengartner H. Peptide-induced antiviral protection by cytotoxic T cells. Proc Natl Acad Sci USA 88:991, 1991

Seidah NG, Donohue-Rolfe A, Lazure C et al. Complete amino acid sequence of Shigella toxin B-chain. J Biol Chem 261:13928, 1986

Sela M. Vaccins synthetiques - un rêve ou une réalité? Bull Inst Pasteur 72:73, 1974

Shapira M, Misulovin Z and Arnon R. Specificity and cross-reactivity of synthetic peptides derived from a major antigenic site of influenza hemagglutinin. Mol Immunol 22:23, 1985a

Shapira M, Jolivet M and Arnon R. Synthetic vaccine against influenza with built-in adjuvanticity. Int J Immunopharmac 7:719, 1985b

Snijders A, Benaissa-Trouw BJ, Oosterlaken TAM et al. Identification of linear epitopes on Semliki Forest virus E2 membrane and their effectiveness as a synthetic peptide vaccine. J Gen Virology 72:557, 1991

Spalding BJ. In hot pursuit of an HIV vaccine. Biotechnology 10:24, 1992

Steward MW, Stanley CM, Dimarchi R et al. High-affinity antibody induced by immunization with a synthetic peptide is associated with protection of cattle against foot-and-mouth disease. Immunology 72:99, 1991

Steward MW and Howard CR. Synthetic peptides: A next generation of vaccines? Immunology Today 8:51, 1987

Takahashi H, Cohen J, Hosmalin A et al. An immunodominant epitope of the human immunodeficiency virus envelope glycoprotein gp160 recognized by class I major histocompatibility complex molecule-restricted murine cytotoxic T lymphocytes. Proc Natl Acad Sci USA 85:3105, 1988

Takahashi H, Nakagawa Y, Pendleton CD et al. Induction of broadly cross-reactive cytotoxic T cells recognizing an HIV-1 envelope determinant. Science 255:333, 1992

Tam JP. Synthetic peptide vaccine design: Synthesis and properties of a high-density multiple antigenic peptide system. Proc Natl Acad Sci USA 85:5409, 1988

Townsend ARM and Skehel JJ. The influenza A virus nucleoprotein gene controls the induction of both subtype specific and crossreactive T cells. J Exp Med 160:552, 1984

Townsend ARM, Rothbard J, Gotch FM et al. The epitopes of influenza nucleoprotein recognized by cytotoxic T lymphocytes can be defined with short synthetic peptides. Cell 44:959, 1989

Wiley DC, Wilson IA and Skehel JJ. Structural identifications of the antibody binding sites of Hong Kong influenza hemagglutinin and their involvement in antigenic variation. Nature 289:373, 1981

Wolowczuk I, Auriault C, Bossus M et al. Antigenicity and immunogenicity of a multiple peptidic construction of the *Schistosoma mansoni* SM-28 GST antigen in rat, mouse and monkey. 1. Partial protection of Fischer rat after active immunization. J Immunol 146:1987, 1991

Zavala F, Cochrane AH, Nardin EH et al. Circumsporozoite proteins of malaria parasites contain a single immunodominant region with two or more identical epitopes. J Exp Med 157:1947, 1983

Chapter 19

CYTOTOXIC T LYMPHOCYTE EPITOPE BASED PEPTIDE VACCINES

W. Martin Kast, Remco M.P. Brandt and Cornelis J.M. Melief

TABLE OF CONTENTS

I. SUMMARY

Peptides are the only components of pathogens or tumor antigens that can be recognized by the cellular arm of the immune system, i.e., the T cell system. These peptides are presented to cytotoxic T lymphocytes (CTL) in the groove on top of major histocompatibility complex (MHC) class I molecules at the cell membrane.

Normally, peptides derived from processed (foreign) proteins are loaded in the cell's endoplasmic reticulum into empty MHC class I molecules and subsequently transported to the surface of the cells. In this chapter it is argued that there is no absolute need for MHC class I molecules to be loaded with peptides via this biosynthetic processing route, but that loading can be achieved by the addition of proper peptides to the outer surface of cells after which these peptides either fill up MHC class I molecules that arrive at the cell surface empty or exchange with peptides that are loaded endogenously in these MHC class I molecules. This presentation of the peptides of choice by the MHC class I molecules bypasses the need, or for instance, of infecting the cells with a virus to effectively induce a CTL response against that virus.

Therefore, the strategy of immunization with a T cell epitope based peptide vaccine might be regarded as a "new generation vaccine".

In this chapter, two examples of the possibilities to successfully use CTL epitope based vaccines for the prevention of a lethal Sendai virus infection and of human papillomavirus induced tumor outgrowth in mice will be given and placed in the context of recent literature.

II. INTRODUCTION

Cytotoxic T lymphocytes (CTL) are of the utmost importance in the resistance against many viral infections (Kast and Melief 1991a) and (virus induced) tumors (Melief and Kast 1991, 1992; Melief 1992). These CTL recognize short peptides from endogenous antigens bound in the presenting groove of major histocompatibility complex (MHC) class I molecules (Fig. 1) (Townsend et al., 1986; Bjorkman et al., 1987a; Van Bleek and Nathenson, 1990; Monaco et al., 1992).

0-8493-8967-4/95/$0.00+$.50

Molecular details of the interaction between peptides and MHC class I molecules have been revealed in studies on the three-dimensional structure of the peptide binding groove occupied either by heterogeneous mixtures of naturally processed peptides or by single peptide epitopes (Bjorkman et al., 1987b; Garrett et al., 1989; Madden et al., 1991; Saper et al., 1991; F
remont et al., 1992; Matsumura et al., 1992; Zhang et al., 1992).

Sequence analysis of naturally processed peptides bound to MHC class I molecules indicated that the majority of these peptides have a restricted size of 9 ± 1 amino acids and bind to pockets in the antigen presenting groove via key amino acids occupying one of the residue positions 2–5 and the C-terminus of the peptides.

These pocket binding amino acids are referred to as anchor residues and are generally allele specific (Falk et al., 1991; Rötzschke and Falk 1991; Falk and Rötzschke 1993; Rammensee et al.,1993). Recently, it became apparent that in addition to the role of the anchor positions also non-anchor positions can also contribute to peptide binding to MHC molecules (Ruppert et al., 1993).

Peptides that are presented in the groove of MHC class I molecules are derived from proteins which are present in the cytosol of cells. These proteins are fragmented by proteolytic systems such as proteasomes, and their peptide fragments are subsequently transported via "transport associated proteins" (TAP) into the endoplasmic reticulum of cells where the peptides bind to newly synthesized MHC class I molecules. TAP independent transport has also been documented (Zhou et al., 1993 a,b; Ossevoort et al., 1993). Finally, these peptide-loaded MHC class I molecules are transported to the cell surface where they can be recognized by the T cell receptor of CTL. This scheme of events has recently been reviewed in a molecular model (Monaco, 1992).

However, an alternative possibility to present a peptide in the groove of an MHC class I molecule would be to load these molecules on the outer surface of the cells by the addition of the exogenous peptides. The feasibility of such an approach is illustrated by the fact that the MHC class I molecules on TAP deficient cells, which basically transport empty MHC class I molecules to the cell surface, can be loaded efficiently with exogenously added peptides of the right configuration (De Bruijn et al., 1991, 1992; Feltkamp et al., 1993; Houbiers et al., 1993; Kast et al., 1993a; Nijman et al., 1993a,b).

It is likely, however, that also on normal cells some empty MHC class I molecules would appear at the cell surface that subsequently could be filled with exogenously added peptides of choice or that alternati-

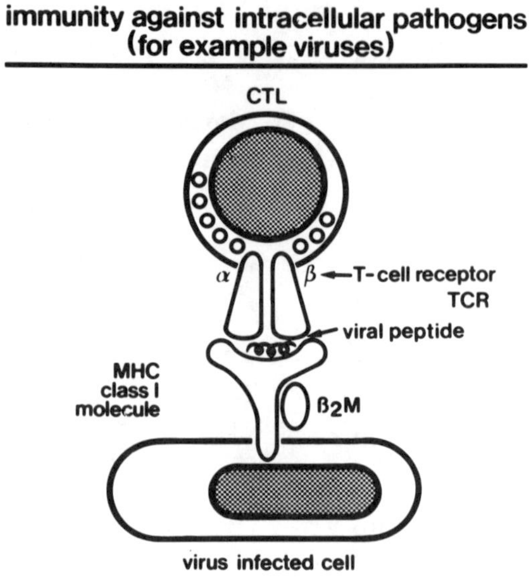

Fig.1.

vely some exchange would take place with endogenous peptides that arrive at the cell surface in MHC class I molecules.

III. *IN VIVO* POTENCY OF CTL

The biologic potency of CTL, specifically recognizing these peptides, is well illustrated by adoptive transfer of CTL to clear viral infections or eradicate tumor cells. Two examples are given here from our own research.

The first one is adoptive transfer experiments with CTL to eradicate human adenovirus induced tumors in mice. Large established and aggressively growing tumors were completely eradicated following transfer of cloned, *in vitro* grown human adenovirus-specific CTL, provided that interleukin-2 on which most CTL depend was administered (Fig. 2) (Kast et al., 1989). The second example is the cure of a lethal pneumonia caused by Sendai virus in mice, again after the adoptive transfer of virus-specific CTL and either interleukin-2 or T helper cells that produce IL-2 (Fig. 3) (Kast et al., 1986).

Although both examples clearly show the efficacy of CTL transfer, clinical application of this technology would be hampered by the very laborious, costly and time-consuming generation of such CTL clones. Because of MHC polymorphism, adoptively transferred CTL are required to be patient derived and thus have to be custom made for each patient to avoid allogenic rejection of CTL. Therefore, the idea arose several years ago to activate these antigen-specific CTL by vaccination with the antigen-specific peptides recognized by these CTL.

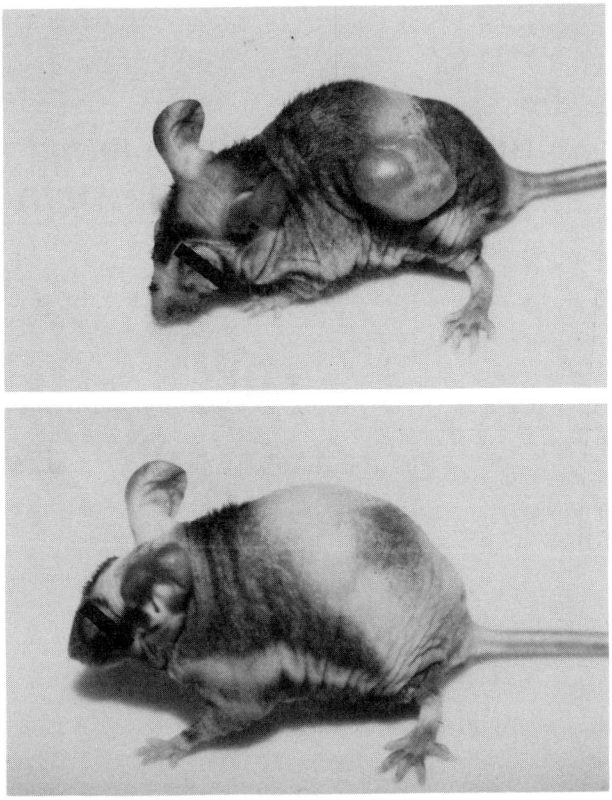

Fig.2. Pictures of a nude mouse bearing an Adenovirus induced tumor before (2A) and after (2B) treatment with cloned Adenovirus-specific CTL and IL-2. (please note the protection of the privacy of this mouse).

IV. IDENTIFICATION OF EPITOPES RECOGNIZED BY CTL

The prerequisite for this alternative strategy was the identification of the nature of peptides recognized by CTL. Several approaches to this question have been explored, such as restricted protein degradation yielding smaller fragments of the protein that could still sensitize target cells for killing by CTL. However, almost all published CTL epitopes identified this way up to around 1990 are longer than the actual active peptides. Another strategy for identifying CTL epitopes was then developed. This strategy made use of synthesizing multiple peptides on pins allowing to scan an entire protein sequence encompassed by hundreds of peptides (PEPSCAN) (Geysen et al., 1984, 1985; Van der Zee et al., 1989), followed by subsequent testing whether these peptides were recognized by CTL. Despite the fact that this method suffered from clear disadvantages, including undetectable quantities and unknown concentrations of peptides and the inconvenience of always synthesizing an extra C-terminal aspartic acid, it is remarkable that this strategy yielded several well-defined CTL epitopes such as the 10 amino acid (AA) long H-2D^b restricted human adenovirus type 5 E1a epitope (AA 234–243), (Kast et al., 1989; Kast and Melief 1991b) and the 9 AA long H-2K^b restricted Sendai virus nuclear protein epitope (AA 324–332), (Kast et al., 1991c; Schumacher et al., 1991).

Even screening with 12-mer peptides (actuallly 13 when the C-terminal aspartic acid is counted) did not prevent these epitope identifications. Although these 13-mer peptides actually did not bind to the MHC class I molecules, smaller byproducts of the proper binding size were present (Schumacher et al., 1991). Therefore, this indirectly led to identification of peptides of the optimal size, most likely corresponding to the major species of naturally processed peptide (Van Bleek and Nathenson, 1990).

New and better strategies have now been developed involving the use of correctly sized peptides in overlapping sets that were generated in quantifiable amounts (Kast et al., 1993a; Feltkamp et al., 1993) (see lab protocol 1). Also, strategies which make use of the predictive capacity of peptide MHC binding motifs have been succesful (Houbiers et al., 1993; Nijman et al., 1993 a,b).

V. *IN VIVO* EFFICACY OF CTL EPITOPE BASED PEPTIDE VACCINES AGAINST VIRUS INFECTIONS

The first attempts to use CTL epitopes identified this way in anti-viral vaccination experiments were successful. Both in a lymphocytic choriomeningitis model (Schulz et al., 1991) and a Sendai virus model (Kast et al., 1991c, 1993b), CTL epitope based peptide vaccination led to a nearly complete protection from a later challenge with an otherwise lethal dose of virus. In these vaccinations, 100 μg of peptide was emulsified in incomplete Freund's adjuvant (IFA) and injected subcutaneously. The adjuvant was essential. Since the classical IFA most likely will not be allowed in a clinical setting, recently some alternative adjuvants have been tested in the Sendai virus model (Kast et al., 1993b). The non-ionic block copolymer, Titermax, was active, while others like aluminium hydroxide were not. Also, a new and clinically allowed adjuvant was successfully applied with CTL epitope based peptide vaccines (Feltkamp et al., unpublished observations). Alternatives for the use of adjuvants in peptide vaccination are being explored at the moment, such as the coupling of peptides to lipophilic molecules (Deres et al., 1989; Schild et al., 1991; Nardelli and Tamp, 1993), and these studies have indicated an efficient priming of virus-specific CTL, but no protection yet.

The concentration of the MHC class I binding peptide in IFA could be lowered to about 1 μg per mouse to still induce the same high level of protection (Kast et al., 1993b). The protective effects were decrea-

mice suffering from lethal pneumonia caused by Sendai virus
(= para-influenza virus)
cough!
cough!
no treatment
inoculation with Sendai virus specific cytotoxic- and helper-T lymphocyte clones
whoopee!
rigor mortis
W.M. Kast and C.J.M. Melief - University Hospital, Leiden -

Fig. 3.

sed in a dose dependent manner if concentrations lower than 1 μg per mouse were used (Kast et al., 1993b).

VI. *IN VITRO* IMMUNIZATION WITH CTL EPITOPE BASED PEPTIDES

Interestingly, identical titration curves on a molar basis were obtained when protective effects were compared between immunizations with peptides of exactly the right 9 AA length or peptides that were extended at both ends with 3 or 4 AA. The protective effects of the longer 16 AA peptides therefore cannot be explained by contamination with peptides of optimal size for binding. Rather an *in vivo* processing mechanism must be involved in the generation of the proper 9 AA size peptide out of the 16 AA peptide (Kast et al., 1993b). The possibility of using peptides that are slightly longer than the real epitope could have advantages for the strategy of peptide vaccination, since different peptides capable of binding different MHC molecules could overlap, resulting in the possibility that 1 longer peptide could be used for two or more MHC phenotypes.

Interestingly, if peptides of the right or of longer size were compared for their ability to induce primary *in vitro* CTL responses, thus without *in vivo* immunization, a clear difference between the induction capacity of peptides of the proper size versus longer peptides would be apparent in that optimally binding peptides (9 AA) are up to 1000-fold more potent in primary CTL induction compared to, for example, a peptide of 16 AA encompassing the 9 AA sequence (De Bruijn et al., 1991). Apparently, the processing

TABLE 1
Frequency of HLA-A antigens°

Antigen	Caucasoid	Negroid	Frequency (%) Japanese	Chinese	Hispanics
A*0101	29	10	1	9	10
A*0201	46	30	42	54	43
A*0301	21	16	1	7	15
A*1101	10	4	20	33	7
A*2401	17	9	58	33	27

°Calculated from Central Data Committee 1991

mechanism of peptides which exists *in vivo* is not present in *in vitro* cell cultures.

The possibility for inducing primary CTL *in vitro* is a powerful tool for screening peptides which are known to bind to a particular MHC molecule for their immunogenicity (De Bruijn et al., 1991, 1992; Kast et al., 1993a) (see lab protocols 3,4). Especially in human studies, screening for peptide immunogenicity *in vivo* would otherwise be difficult.

VII. *IN VIVO* EFFICACY OF CTL EPITOPE BASED PEPTIDE VACCINES AGAINST VIRUS INDUCED TUMORS

Besides the apparent protective capacity of peptides against viral infections, we recently demonstrated that CTL epitope based peptide vaccination was also very effective in the protection against human papillomavirus type 16 (HPV 16) induced tumors in mice (Feltkamp et al., 1993). Although HPV16 is a human virus, it is tumorigenic in mice. These tumors contain HPV16 DNA coding for proteins in which CTL epitopes can be present. Indeed, CTL epitopes were identified and mice were subsequently immunized with one of these CTL epitope based peptides. Four weeks after one booster immunization, these mice were challenged with a tumorigenic dose of an HPV16 induced tumor.

Only 7% of the mice developed a late tumor, whereas in the sham immunized control mice approximately 80% developed an early progressive and lethal tumor. This illustrates once more the remarkable efficacy of CTL epitope based peptide vaccines.

Interestingly, mice could also be protected by immunization with irradiated HPV16 induced tumor cells. However, the CTL that were induced in such mice were directed against another yet unidentified peptide than the one used for the peptide immunization mentioned above (Feltkamp et al., unpublished observations).

By definition, the already characterized peptide is therefore immunosubdominant. However, after forcing the immune system to react against such a subdominant peptide by introducing the peptide in isolation, the subsequently induced CTL response was sufficient to reach almost complete protection. In situations in which tumors apparently have already escaped from the immune reaction directed against immunodominant epitopes, the use of immunizations with subdominant CTL epitope based peptide vaccines might have an added value over strategies that would induce immunity against immunodominant epitopes.

VIII. IDENTIFICATION OF CLINICALLY RELEVANT CTL EPITOPE BASED PEPTIDES

HPV16 is involved in the induction of cervical carcinoma in women, and HPV DNA is found in nearly all cervical carcinoma biopsies (Resnick et al., 1990; Van den Brule et al., 1991). It is obvious that CTL epitopes of HPV 16 now need to be identified for peptide based vaccines in human beings. Despite the polymorphism of the human MHC system (HLA), peptide binding studies with only relatively few HLA alleles, i.e., the HLA types A*0101, A*0201, A*0301, A*1101 and A*2401, would suffice to select candidate HPV epitopes for a large proportion of individuals in human Caucasoid, Oriental and Hispanic populations. Obviously, Negro populations require a larger set of alleles (see Table 1).

HPV 16 peptides capable of binding to HLA-A*0201 molecules have recently been identified with a cellular assay (Kast et al., 1993a) (see lab protocol 2). This assay makes use of human processing defective T2 cells. These cells express empty and unstable HLA-A*0201 molecules at their cell surface due to a defect in the TAP genes. These molecules can be stabilized by exogenously added HLA-A*0201 binding peptides. The stabilization of the HLA-A*0201 molecules results in a higher cell surface expression which can be visualized on a FACS. Recently, we have also used that method to identify several new HLA-A*0201 restricted epitopes from antigens such as p53 and influenza virus (Houbiers et al., 1993; Nijman et al., 1993a,b). A new molecular peptide-MHC binding assay that allows the identification of HPV16 peptides for the other four HLA-A alleles (see Table 1) has been performed recently (Kast et al., 1994). In this assay, peptides are tested for competition of binding to purified HLA class I molecules at room temperature in the presence of ß2 microglobulin with radiolabeled, optimal-sized peptides (see labprotocol 5). Those five sets of HLA-A binding peptides of HPV16 will form the basis of a CTL epitope based peptide vaccine for cervical carcinoma that is now in preparation.

IX. ADVANTAGES OF CTL EPITOPE BASED PEPTIDE VACCINES

Many different vaccine strategies are being developed at this moment. Of these strategies, peptide vaccination has distinct advantages over live vector vaccination, tumor vaccination or protein vaccination. Besides being cheap and safe, it would be possible to direct the immune system toward a dominant or even subdominant epitope, avoiding possible suppressor epitopes (Krzych et al., 1988) and giving the virus or tumor cell fewer possibilities to escape immunity by creating mutant sequences. The potency of vaccination with a subdominant CTL epitope was recently demonstrated (Feltkamp et al., 1993). Peptide vaccination would also bypass the problem of exogenous proteins present in vaccines not being processed in the appropriate manner for MHC class I molecules, a potential problem with proteins used for immunizations. Another advantage of peptide based vaccines is their remarkable versatility. For example, CTL epitopes could be mixed or coupled to promiscuous helper T cell epitopes of various origin or combined with lymphokines to increase immunological help.

Further research on such issues will clarify the potential role, strategies and importance of these CTL epitope based peptide vaccines in the near future.

X. LABPROTOCOLS

X.1. Peptide-MHC binding assay for H-2^b molecules using mouse RMA-S cells (refs 1,2)

1) Culturing of RMA-S cells:

RMA-S cells (H-2^b haplotype;[3,4]) are cultured in Iscove's modified Dulbecco's Medium with glutamax I (IMDM: Gibco BRL, Paisley, Scotland) supplemented with 100 IU penicillin (Brocades Pharma, Leiderdorp, The Netherlands), 100 μg/ml kanamycin (Sigma, St. Louis, MO, cat.no. K0254), 20 μM ß mercapto-ethanol (Merck, Darmstadt, Germany, cat.no.15433; = complete medium) and 8 % fetal calf serum (FCS: Hyclone Lab., Inc., Logan, Utah). The cells are cultured in a concentration of approximately 5 x 10^4/ml for 3 days in an incubator at 37°C, 5% CO_2 and humidified air.

2) Cold incubation of RMA-S cells:

The cultured RMA-S cells are washed extensively (at least three times) in serum-free complete IMDM and put on a concentration of 2 x10^6/ml in complete IMDM containing 2% human pooled serum (HPS: should be tested for lack of suppression activity in mixed lymphocyte cultures). Incubate the cells for 36 h at 26°C, 5% CO_2 and humidified air.

RMA-S cells, due to a defect in the transporter molecules associated with antigen processing (TAP), express unstable empty MHC molecules at the cell surface. These empty MHC molecules can be stabilized by either exogenously adding binding peptides or culturing at 26°C. Both procedures result in an increased MHC expression (5,6).

3) Peptide incubation of RMA-S cells:

The cold incubated RMA-S cells are washed two times in serum-free complete IMDM and put at a concentration of 4 x10^6 cells/ml.

The wells of a 96-well-V-bottom plate (Greiner, Frickenhausen, Germany, cat.no.651101) are filled with 12 μl peptide at a concentration of 500 μg/ml and with 50 μl RMA-S cells. Each well contains 2 x 10^5 RMA-S cells in the presence of 100 μg/ml peptide. Gently rock the plate for 2 min and subsequently incubate for 4 h at 26°C .

4) FACS experiment:

After peptide incubation, add 75 μl "PBA"(= PBS supplemented with 0.5% BSA [bovine serum albumine 35%, Sigma, St. Louis, MO, cat no. A7409] and 0.02 % sodiumazide [NaN_3 10%, Merck, Darmstadt, Germany, cat.no. 822335]) to the wells and centrifugate for 3 min at 1200 rpm. Remove the supernatant and loose up the pellet. Repeat this washing step with 100 μl PBA. Add 25 μl saturated amounts of first antibody[@] to the pellet and incubate for 30 min at 4°C in the dark.

Wash the cells twice with 100 μl PBA as described above.
Add 25 μl saturated amounts of Fitc conjugated goat anti-mouse IgG antibody (Boehringer Mannheim, Mannheim, Germany, cat.no. 605240) to each well and incubate for 30 min at 4°C in the dark. Wash these cells once with 100 μl PBA and transfer the cells to "Facs tubes" (Micronic, Lelystad, The Netherlands, cat.no.226-C2) in a total volume of 100 μl PBA#. Fluorescence of viable T2 cells is measured at 488 nm on a FACScan flow cytometer (Becton Dickinson, Franklin Lakes, NJ).

@ First antibody:
a) anti-H2-D^b antibody: clone 28.14.8S, IgG2a subclass (7).
b) anti-H2-K^b antibody: clone B8.3.24, IgM subclass (8).

The cells should be fixated with 0.5% paraformaldehyde in PBA if the samples are not measured the same day.

Each experiment should include:
a) positive control: K^b/D^b binding peptide (S9: FAPGNYPAL; [6,9]); D^b binding peptide (A10: SGPSNTPPEI; [10,11]).
b) negative control: 12 μl PBA instead of peptide and no first antibody.
c) blanc: no peptide and no antibodies.
d) background control: no peptide, only the first and second antibody.

A peptide is considered as a binding peptide when the fluorescence index (FI) ≥ 0.5 (1,2).
The FI is calculated as the mean fluorescence (MF) of each peptide subtracted and divided by the MF of the background control.

5) References

1. Feltkamp, M.C.W., Smits, H.L., Vierboom, M.P.M., Minnaar, R.P., de Jongh, B.M., Drijfhout, J.W., ter Schegget, J., Melief, C.J.M. and Kast, W.M. (1993) Eur. J. Immunol. 23, 2242
2. Feltkamp, M.C.W., Vierboom, M.P.M., Toes, R.E.M., Ossendorp, F., ter Schegget, J., Melief, C.J.M. and Kast, W.M. submitted 1995
3. Ljunggren, H.-G. and Kärre, K. (1985) J.Exp.Med. 162, 1745
4. Kärre, K., Ljunggren, H.-G., Piontek, G. and Kiessling, R. (1986) Nature 319, 675
5. Ljunggren, H.-G., Stam, N.J., Öhlen, C., Neefjes, J.J., Höglund, P., Heemels, M.-T., Bastin, J., Schumacher, T.N.M., Townsend, A., Kärre, K. and Ploegh, H.L. (1990) Nature 346, 476
6. Schumacher, T.N.M., De Bruijn, M.L.H., Vernie, L.N., Kast, W.M., Melief, C.J.M, Neefjes, J.J. and Ploegh, H.L. (1991) Nature 350, 703
7. Ozato, K. and Sachs, D.H. (1981) J.Immunol. 126, 317
8. Köhler, G., Fisher-Lindahl, K. and Heusser, C. (1981) The Immune System, vol.2, Karger, Basel, 202
9. Kast, W.M., Roux, L., Blom, H.J.J., Voordouw, A.C., Meloen, R.H., Kolakofsky, D. and Melief, C.J.M. (1991) Proc. Natl. Acad. Sci. 88, 2283
10. Kast, W.M., Offringa, R., Peters, P.J., Voordouw, A.C., Meloen, R.H., Van der Eb, A.J. and Melief, C.J.M. (1989) Cell 59, 603
11. Kast, W.M. and Melief, C.J.M. (1991) Int. J. Cancer, supplement 6, 90

X.2. Peptide-MHC binding assay for HLA-A*0201 molecules using human T2 cells (1,2)

1) Culturing of T2 cells:

T2 cells (HLA-A*0201+) are cultured in Iscove's modified Dulbecco's Medium with glutamax I (IMDM: Gibco BRL, Paisley, Scotland) supplemented with 100 IU penicillin (Brocades Pharma, Leiderdorp, The Netherlands), 100 μg/ml kanamycin (Sigma, St. Louis, MO, cat.no. K0254; = complete medium) and 8% fetal calf serum (FCS: Hyclone Lab, Inc., Logan, Utah). The cells are cultured in a concentration of approximately 2,5 x 10^5 /ml for 3 days in an incubator at 37°C, 5% CO_2 and humidified air.

2) Peptide incubation of T2:

The cultured T2 cells are washed two times in serum-free complete IMDM with 20 μM ß mercapto-ethanol (Merck, Darmstadt, Germany, cat.no. 15433) and put on a concentration of 2 x10^6 cells/ml.
The wells of a 96-well-V-bottom plate (Greiner, Frickenhausen, Germany, cat.no. 651101) are filled with 10 μl peptide at a concentration of 500 μg/ml and with 40 μl T2 cells. Each well contains 8 x 10^4 T2 cells in the presence of 100 μg/ml peptide. Gently rock the plate for 2 min and subsequently incubate for 16 h at 37°C, 5% CO_2 and humidified air@.

@ In order to enhance the binding capacity of the peptides, human ß2 microglobulin (ß2-m: Sigma, St. Louis, MO, cat.no. M4890) can be added at a concentration of 10 μg/ml during the overnight incubation (1). ß2-m should be dissolved in serum-free IMDM and stored at -20°C at a concentration of 1 mg/ml.

3) FACS experiment:

After peptide incubation, add 75 μl "PBA" (= PBS supplemented with 0.5% BSA [bovine serum albumine 35%, Sigma, St. Louis, MO, cat.no. A7409] and 0.02% sodiumazide [NaN_3 10 %, Merck, Darmstadt, Germany, cat.no. 822335]) to the wells and centrifugate for 3 min at 1200 rpm. Remove supernatant and loose up the pellet. Repeat this washing step with 100 μl PBA.

Add 50 μl saturated amounts of BB7.2 monoclonal antibody (anti HLA-A2, IgG2b subclass, [3]) to the pellet and incubate for 30 min at 4°C in the dark. Wash the cells twice with 100 μl PBA as described above.

Add 25 μl saturated amounts of Fitc conjugated goat anti-mouse IgG antibody (Boehringer Mannheim, Mannheim, Germany, cat.no. 605240) to each well and incubate for 30 min at 4°C in the dark. Wash these cells once with 100 μl PBA and transfer the cells to "Facs tubes" (Micronic, Lelystad, The Netherlands, cat.no.226-C2) in a total volume of 100 μl PBA#.

Fluorescence of viable T2 cells was measured at 488 nm on a FACScan flow cytometer (Becton Dickinson, Franklin Lakes, NJ).

The cells should be fixated with 0.5% paraformaldehyde in PBA if the samples are not measured the same day.

Each experiment should include:
a) positive control: A2 binding peptide (Influenza virus matrix 1 58–66: GILGFVFTL, [4]).
b) negative control: 10 μl PBA instead of peptide and no BB7.2.
c) blanc: no peptide and no antibodies.
d) background control: no peptide, only the first and second antibody.

A peptide is considered as a binding peptide when the fluorescence index (FI) ≥ 0.5 (5).
The FI is calculated as the mean fluorescence (MF) of each peptide subtracted and divided by the MF of the background control.

4) References

1. Nijman, H.W., Houbiers, J.G.A, Vierboom, M.P.M., van der Burg, S.J., Drijfhout, J.W., D'Amaro, J., Kenemans, P., Melief, C.J.M. and Kast, W.M. (1993) Eur.J.Immunol. 23, 1215
2. Kast, W.M., Brandt, R.M.P., Drijfhout, J.W. and Melief, C.J.M. (1993) J. Immunotherapy 14, 115
3. Parham, P. and Brodsky, F.M. (1981) Human Immunol. 3, 277
4. Nijman, H.W., Houbiers, J.G.A., van der Burg, S.H., Vierboom, M.P.M., Kenemans, P., Kast, W.M. and Melief, C.J.M. (1993) J.Immunotherapy 14, 121
5. Drijfhout, J.W., Brandt, R.M.P., D'Amaro, J., Kast, W.M. and Melief, C.J.M. (1994) Human Immunol. in press

X.3. *In vitro* induction of mouse H-2^b restricted primary CTL responses using the processing defective cell line RMA-S (1)

The expression of MHC class I on RMA-S (2,3) is increased by incubating RMA-S at 26°C (4) and/or in medium containing an MHC class I binding peptide. RMA-S will present this peptide bound to MHC class I in high amounts and is therefore a good antigen presenting cell (APC) for cytotoxic T lymphocytes (CTL). In the CTL response inducing method described later, the RMA-S is used as APC and nylon wool passed (so T cell enriched) spleen cells from unprimed mice are used as responder cells.

METHOD

1) Peptide loading of MHC class I on RMA-S cells:

a) RMA-S cells at a concentration of 2 x 10^6 cells/ml are incubated for 40 h at 26°C in a T 75 flask (Becton Dickinson, Franklin Lakes, NJ, cat.no. 3024) in IMDM (Iscove's Modified Dulbecco's Medium: Gibco BRL, Paisley, Scotland, cat.no. 21980-032; = complete medium) containing 2% human pooled serum (HPS: should be tested for lack of suppression activity in mixed lymphocyte cultures) with L-glutamine (2 mM, ICN Biochemicals Inc., Costa Meisa, CA cat.no. 15-801-55), 100 IU/ml penicillin (Brocades Pharma, Leiderdorp, The Netherlands), 100 μg/ml kanamycin (Sigma, St. Louis, MO, cat.no. K0245, 20 μM ß mercapto-ethanol (Merck, Darmstadt, Germany, cat.no.15433).

b) These RMA-S cells incubated at 26°C are spun down and subsequently washed three times in serum-free complete IMDM. Subsequently, these cells are resuspended in 1 ml serum-free complete IMDM for 4 h at 26°C in the presence of 50 μg/ml synthetic peptide at a concentration of approximately 30 x 10^6 cells/ml.

2) Mitomycin-C treatment of peptide-loaded RMA-S cells (APC):

Mitomycin-C was dissolved in distilled water at a final concentration of 1 mg/ml. The peptide-loaded RMA-S cells are spun down and thereafter are treated at a concentration of 20 x 10^6 cells/ml for 1 h at 37°C with 50 μg/ml mitomycin-C in serum-free complete IMDM (Gibco BRL, Paisley, Scotland, cat.no. 21980-032). Subsequently, the RMA-S cells were washed five times in complete IMDM.

3) Primary CTL response induction:

Approximately 2.5 x 10^5 nylon wool (Robbins Scientific Corporation, Sunnyvale, CA, cat.no.1078-02-0) passed spleen cells from unprimed C57Bl/6 mice and 62.5 x 10^3 mitomycin-C treated, peptide-loaded RMA-S cells are seeded in 100 μl complete IMDM medium [L-glutamine (2 mM, ICN Biochemicals Inc., Costa Meisa, CA, cat.no. 15-801-55), 100 IU/ml penicillin (Brocades Pharma, Leiderdorp, The Netherlands), 100 μg/ml kanamycin (Sigma, St. Louis, MO, cat.no.K0245) containing 8% fetal calf serum (Hyclone Lab. Inc., Hogan, Utah) per well of a 96-well U-bottom plate (Costar, Cambridge, MA, cat.no. 3799). The cultures are maintained for 5–6 days at 37°C in an incubator with 5% CO_2 and humidified air.

4) Cytotoxicity assay:

On day 5–6, the bulk cultures are tested for peptide specificity in a cytotoxicity assay using EL-4, RMA, RMA-S with and without peptide as target cells.

5) Cloning by limiting dilution:

Five or six days after incubation of nylon wool-passed spleen cells and peptide-loaded RMA-S cells the effector cells are isolated with a density gradient (Lymphoprep: Cedarlane Lab., Hornby, Ontario, Canada, cat.no. CL 5035) and washed in complete IMDM with 8% FCS. The primary bulk is cloned by limiting dilution. One hundred viable cells from the primary bulk are seeded in each well of a 96-well U-bottom plate (Costar, Cambridge, MA, cat.no. 3799) with 100 μl complete IMDM with 8% FCS. In other 96-well U-bottom plates, 10, 1 or 0.3 cells per well are seeded. To all these wells, rat cell supernatant as lymphokine source (corresponding with 50 IU/ml IL-2;[5,6]) and 5 x 10^4 peptide-loaded 2.5 Gy irradiated lipopolysaccharide induced B cell blasts as stimulator cells are added. LPS blasts are obtained after incubating 100 x 10^6 C57Bl/6 spleen cells for 5 days at 37°C, 5% CO_2 in humidified air with 30 μg/ml LPS (LPS-B, Bacto Lab., Difco, Detroit, MI, cat.no. 3923-10-6) in 20 ml complete IMDM supplemented with 8% FCS. Viable LPS blasts are incubated for 1 h at 26°C with the selected peptide at 25 μg/ml in 5 ml complete IMDM supplemented with 2% HPS. These cells are washed two times before adding to the wells.

6) Expand clones:

When growth and specificity of the CTL clones are established, they are restimulated weekly with peptide-loaded (irradiated) spleen cells (five-fold excess of stimulator cells over CTL) and rat cell supernatant. To this end, spleen cells are incubated for 4 h at 37°C with 50 μg/ml peptide at a concentration of 50 x 10^6 cells/ml and subsequently washed.

7) References

1. De Bruijn, M.L.H., Schumacher, T.N.M., Nieland, J.D., Poegh, H.L., Kast, W.M. and Melief, C.J.M. (1991) Eur. J. Immunol. 21, 2963
2. Ljunggren, H.-G. and Kärre, K. (1985) J.Exp.Med. 162, 1745
3. Kärre, K., Ljunggren, H.-G., Piontek, G. and Kiessling, R. (1986) Nature 319, 675
4. Ljunggren, H.-G., Stam, N.J., Öhlen, C., Neefjes, J.J. Höglund, P., Heemels, M.-T., Bastin, J., Schumacher, T.N.M., Townsend, A., Kärre, K. and Ploegh, H.L. (1990) Nature 346,476
5. Kast, W.M., de Waal, L.P. and Melief, C.J.M. (1984) J. Exp. Med. 160, 1752
6. Kast, W.M., Offringa, R., Peters, P.J., Voordouw, A.C., Meloen, R.H., van der Eb, A.J. and Melief, C.J.M. (1989) Cell 59, 603

X.4. *In vitro* induction of human HLA-A*0201 restricted primary CTL responses using the processing defective cell line 174CEM.T2 (1-3)

The expression of HLA-A*0201 on 174CEM.T2 (T2) (4) is increased by incubating T2 cells in medium containing the relevant peptide. T2 will present the relevant peptide bound to HLA-A*0201 in high amounts and is therefore a good antigen presenting cell (APC) for cytotoxic T lymphocytes. In the response inducing method described below, T2 is used as APC and post-Ficoll mononuclear cells of an HLA-A*0201 positive donor are used as responder cells.

METHOD

*1) Peptide loading of HLA-A*0201 on T2 cells:*

T2 cells at a concentration of 2 x 10^6 cells/ml are incubated for 13 h at 37°C in a T 25 flask (Becton Dickinson, Franklin Lakes, NJ, cat.no. 3013) in serum-free IMDM (Iscove's Modified Dulbecco's Medium: Gibco BRL, Paisley, Scotland, cat.no. 21980-032) with L-glutamine (2 mM, ICN biochemicals Inc., Costa Meisa, CA, cat.no. 15-801-55), 100 IU/ml penicillin (Brocades Pharma, Leiderdorp, The Netherlands), 100 μg/ml kanamycin (Sigma, St. Louis, MO, cat.no. K0245; = complete medium) and the synthetic peptide at a concentration of 80 μg/ml.

2) Mitomycin-C treatment and irradiation of peptide-loaded T2 (APC):

Mitomycin-C was dissolved in distilled water at a final concentration of 1 mg/ml. The peptide-loaded T2 cells are spun down and subsequently treated at a concentration of 20 x 10^6 cells/ml for 1 h at 37°C with 50 μg/ml mitomycin-C in serum-free complete RPMI (Gibco BRL, Paisley, Scotland, cat.no. 21980-032). Subsequently, the T2 cells are washed three times in complete RPMI and γ-irradiated with 20 Gray (Gy). Hereafter, the cells are washed once in complete RPMI.

3) Responder cells:

Responder cells are mononuclear peripheral blood lymphocytes (PBL) of a HLA-A*0201 subtyped donor. The PBL are separated from a buffy coat with a density gradient (Ficoll lymphoprep, Nycomed Pharma, Oslo, Norway, cat.no. 105033) and washed two times in complete RPMI with 15% HPS. After separation and washing, the PBL are resuspended in complete RPMI medium containing 15% human pooled serum (HPS should be tested for lack of suppression activity in mixed lymphocyte cultures).

4) *Incubation of primary immune responses:*
Approximately 4 x 10^5 PBL and 1 x 10^5 mitomycin-C treated and irradiated, peptide-loaded T2 cells are seeded in 100 μl complete RPMI medium [L-glutamine (2 mM, ICN Biochemicals Inc., Costa Meisa, CA, cat.no. 15-801-55), 100 IU/ml penicillin (Brocades Pharma, Leiderdorp, The Netherlands), 100 μg/ml kanamycin (Sigma, St. Louis, MO, cat.no. K0245)] containing 15% human pooled serum (HPS should be tested for lack of suppression activity in mixed lymphocyte cultures) per well of a 96-well U-bottom plate (Costar, Cambridge, MA, cat.no. 3799). The cultures are maintained for 7 days at 37°C in an incubator with 5% CO_2 and humidified air.

5) *Restimulation (day 7):*
EBV transformed B lymphocytes of the HLA-A*0201 donor (autologous) are washed in serum-free complete RPMI and put on a concentration of 2 x 10^6 cells/ml. These EBV are incubated for 2.5 h at 37°C in a 50 ml tube with the selected peptide in a concentration of 100 μg/ml. Hereafter, the cells are washed two times in complete RPMI and γ-irradiated (100 Gy) followed by another washing step.
Seven days after incubation of PBL and peptide-loaded T2 cells (headers 1–4), the responder cells are restimulated with peptide-loaded autologous EBV. For this purpose all cells and medium out of the 96-well plate are harvested. Viable primary bulk cells are isolated with a density gradient and washed in complete RPMI. In each well of a 96-well U-bottom plate 5 x 10^4 of the primary bulk cells in 50 μl complete, 15% HPS containing RPMI medium are seeded and 50 μl of complete, 15% HPS containing RPMI medium with 2 x 10^4 autologous, irradiated (30 Gy) PBL and 5 x 10^4 autologous, peptide-loaded EBV are added.
The cultures are maintained for 7 days at 37°C in an incubator with 5% CO_2 and humidified air.

6) *Restimulation (day 14):*
Fourteen days after incubation of PBL and peptide-loaded T2 cells (headers 1–4) the responder cells are restimulated with the selected peptide. To do so the procedure under header 5 is repeated.

7) *CD4 depletion of primary CTL bulk:*
Twenty-one days after incubation of PBL and peptide-loaded T2 cells, the responder cells are isolated with a density gradient and washed in complete RPMI with 15% HPS. These viable cells are depleted for $CD4^+$ cells. To do so, CD4 coated beads (Dynal, Oslo, Norway, cat.no.111.06) are washed twice with depletion medium (complete RPMI with 1% fetal calf serum [Hyclone Lab. Inc., Logan, CA]). To a 10 ml round-bottom tube, viable cells at a concentration of 20 x 10^6 cells/ml and beads are added in a ratio of cells:beads = 1:20. Subsequently the cells are incubated for 45 min at 4°C (stirr regularly). After the incubation is completed, depletion medium is added to a total volume of 5 ml and the tube is held against the magnet MPC-1 (Dynal, Oslo, Norway, cat.no. 120.01) for 1 min. The non-adherent cells are harvested and washed twice in depletion medium. This CD4 depleted primary bulk is cloned by limiting dilution.

8) *Cloning by limiting dilution:*
One hundred viable cells from the CD4 depleted primary bulk are seeded in each well of a 96-well U-bottom plate (Costar, Cambridge, MA, cat.no. 3799) with 50 μl complete RPMI medium containing 15% HPS. In other 96-well U-bottom plates, 10, 1 or 0.3 cells per well are seeded. To all these wells 5 x 10^4 pooled and irradiated (30 Gy) PBL of four different donors and 1 x 10^4 irradiated (100 Gy) of a pool of three different HLA-A*0201$^+$ EBV transformed B cells are added in 50 μl complete RPMI medium with 15% HPS, human recombinant IL-2 at a concentration of 120 IU/ml (=20 Cetus Units: Eurocetus, Amsterdam, The Netherlands) and 2x Leuco-agglutinin (PHA-L: Sigma, St. Louis, MO, cat.no. L2769)[@].

[@] Dissolve 1 mg Leuco-A in 10 ml serum-free complete RPMI (0.1 mg/ml = 100x).

9) *Expand clones (from day 14 on):*
Add per well, in a final volume of 100 μl
- 2 x 10^4 viable cells
- 5 x 10^4 irradiated PBL-pool (as header 8)
- 10^4 irradiated, peptide-loaded EBV-pool (as header 5)
- 6 IU recombinant IL-2
- 1x Leuco-A

10) References

1. Houbiers, J.G.A., Nijman, H.W., van der Burg, S.H., Drijfhout, J.W., Kenemans, P., van de Velde, C.J.H., Brand, A., Momburg, F., Kast, W.M. and Melief, C.J.M. (1993) Eur.J. Immunol. 23, 2072
2. Kast, W.M., Brandt, R.M.P., Drijfhout, J.W. and Melief, C.J.M. (1993) J.Immunotherapy 14, 115
3. Nijman, H.W., Houbiers, J.G.A., van der Burg, S.H., Vierboom, M.P.M., Kenemans, P., Kast, W.M. and Melief, C.J.M. (1993) J.Immunotherapy 14, 121
4. Salter, R.D. and Cresswell, P. (1986) EMBO J. 5, 943

X.5. Quantitative peptide-MHC binding assay for HLA-A*0201 molecules using purified class I molecules (1–3)

This method utilizes purified class I molecules and radiolabeled synthetic probes. Peptides tested for their binding to the class I molecules compete for this binding with specific radiolabeled standard peptides.

*1) Iodination of the HLA-A*0201 standard peptide:*

The standard peptide for HLA-A*0201 bindings assays, the HBc 18–27 F->Y6 sequence FLPSDYFPSV, is iodinated (NaI^{125}: NEN,cat.no.NEZ-033A) by use of the chloramine T method (4). Peptides are usually tested at one or two high doses and the 50% inhibitory concentration (IC_{50}) of peptides yielding positive inhibition is determined in subsequent experiments, in which four to six further dilutions are tested. To allow comparison of the data obtained in different experiments, a relative binding figure is calculated for the peptide by dividing the IC_{50} of the positive control for inhibition of unlabeled probe peptide by the IC_{50} for each tested peptide. The value of the IC_{50} for the standard HLA-A*0201 peptide is 5 nM. Each competitor peptide should be tested in at least two independent experiments. Each peptide is arbitrarily categorized as a high, weak, or negative binder when it falls into the following categories: < 500 nM, 500 to 5000 nM, and > 5000 nM, respectively.

*2) Isolation and purification of HLA-A*0201:*

The homozygous HLA-A*0201+ human EBV transformed B cell line JY is used as a source of A*0201 molecules (1–3).

Approximately 10^{11} JY cells are lysed in Nonidet P-40 (NP40: Fluka Biochemika, Buchs, Switzerland, art.no. 74385) and are filtered through 0.45 μm filter and purified by affinity chromatography. The cell lysate is passed over two pre-columns of inactivated Sepharose CL4B (Pharmacia, Roosendaal, The Netherlands, cat.no. 17096303) and protein A-Sepharose (Sigma, St. Louis, MO, cat.no. P3391). Class I molecules are then purified from the cell lysate by depletion of HLA-B and HLA-C molecules by repeated passage over protein A Sepharose beads conjugated with the anti HLA-B and HLA-C antibody B1.23.2 (5). Subsequently, the anti-HLA-A/B/C antibody W6/32 (anti-human HLA class I α-chain: IgG2a subclass; [6]) is used to capture HLA-A*0201 molecules. Molecules are eluted with pH 11.5 diethylamine/1% n-octyl glucoside (Sigma, St. Louis, MO, cat.no. D3131 and 06004) neutralized with 1 mM Tris pH 6.8, and concentrated by ultrafiltration on Amicon 30 kD cartridge, and stored at 4°C. Protein purity, concentration and effectiveness of depletion steps are monitored by SDS-polyacrylamide gel electrophoresis.

3) Peptide-MHC binding assay:

A quantitative assay for HLA-A*0201 binding peptides is based on the inhibition of binding of a radiolabeled standard peptide to detergent solubilized MHC molecules (1–3).

For binding assays, MHC amounts that resulted in binding of 15% of the radiolabeled synthetic probe (normally in the 10 to 50 nM range) is incubated in 0.05 % NP40/PBS with approximately 5 nM of radiolabeled HBc 18–27 F->Y6 peptide and titrated amounts of unlabeled competitor peptide to be tested (ranging from 10 μM to 1 nM) in the presence of 1 μM human ß2 microglobulin (ß2-m : Sigma, St. Louis, MO, cat.no. M4890) and a mixture of protease inhibitors (final concentration 1 mM PMSF (Sigma, St. Louis, MO, cat.no. D7626), 1.3 mM 1.10 phenanthroline (Sigma, St. Louis, MO, cat.no. P9375), 73 μM pepstatin A (Sigma, St. Louis, MO, cat.no. P4265), 8 mM EDTA (Sigma, St. Louis, MO, cat.no. E5134), 200 μM N-α-p-tosyl-L-lysine chloromethyl ketone (Sigma, St. Louis, MO, cat.no. T7254)). After 2 days at 23°C, the percentage of MHC bound radioactivity is measured by size exclusion chromatography on a TSK 2000 gel filtration and the 50% inhibitory dose is calculated for each peptide (7).

4) References

1. Ruppert, J., Sidney, J., Celis, R., Kubo, R.T., Grey, H.M. and Sette, A. (1993) Cell 74, 929
2. Kast, W.M., Brandt, R.M.P., Sidney, J., Drijfhout, J.W., Kubo, R.T., Grey, H.M., Melief, C.J.M. and Sette, A. (1994) J.Immunol. 152, 3904
3. Sette, A., Sidney, J., del Guercio, M.-F., Southwood, S., Ruppert, J., Dahlberg, C., Grey, H.M. and Kubo, R.T. (1994) Molec. Immunol. 31, 813
4. Buus, S., Sette, A., Colón, S.M., Miles, C. and Grey, H.M. (1987) Science 235, 1352
5. Rebai, N. and Malissen, B. (1983) Tissue Antigens 22, 107
6. Barnstable, C.J., Bodmer, W.F., Brown, G., Galfre, G., Milstein, C., Williams, A.F. and Briegler, A. (1978) Cell 14, 9

7. Sette, A., Southwood, S., O'Sullivan, D., Gaeta, F.C.A., Sidney, J. and Grey, H.M. (1992) J.Immunol. 148, 844

ACKNOWLEDGEMENTS

We would like to thank Drs. Stephen Schoenberger and Joe D'Amaro for critically reading the manuscript. This review is partly based on the results of studies funded by the Netherlands Organisation for Scientific Research, the Dutch Cancer Society and the National Institute of Health of the U.S.A., Bethesda, MD. W.Martin Kast is a senior fellow of the Royal Netherlands Academy of Arts and Sciences.

ABBREVATIONS

AA	Amino acid
CTL	Cytotoxic T lymphocyte
FACS	Fluorescence activated cell sorter
HLA	Human leukocyte antigen
HPV	Human papillomavirus
IFA	Incomplete Freund's adjuvant
MHC	Major histocompatibility complex
TAP	Transporter associated with antigen processing

XI. REFERENCES

Bjorkman PJ, Saper MA, Samraoui B et al. The foreign antigen binding site and T cell recognition regions of class I histocompatibility antigens. Nature 329:512, 1987a

Bjorkman PJ, Saper MA, Samraoui B et al. Structure of the human class I histocompatibility antigen, HLA-A2. Nature 329:506, 1987b

Central Data Analysis Committee. Allele frequencies. Section 6.3 Splits Combined (Five Loci). In: The data book of the 11th International Histocompatibility Workshop. Yokohama, Japan, p 807, 1991

De Bruijn MLH, Schumacher TNM, Nieland JD et al. Peptide loading of empty MHC molecules on RMA-S cells allows induction of primary cytotoxic T lymphocyte responses. Eur J Immunol 21:2963, 1991

De Bruijn MLH, Nieland JD, Schumacher TNM et al. Mechanisms of induction of primary virus specific cytotoxic T lymphocyte responses. Eur J Immunol 22:3013, 1992

Deres K, Schild H, Wiesmüller K-H et al. *In vivo* priming of virus-specific cytotoxic T lymphocytes with synthetic lipopeptide vaccine. Nature 342:561, 1989

Falk K and Rötzschke O. Consensus motifs and peptide ligands of MHC class I molecules. Sem Immunol 5:81, 1993

Falk K, Rötzschke O, Stevanovic S et al. Allele-specific motifs revealed by sequencing of self-peptides eluted from MHC molecules. Nature 351:290, 1991

Feltkamp MCW, Smits HL, Vierboom MPM et al. Vaccination with cytotoxic T cell epitope-containing peptide protects against a tumor induced by human papillomavirus type 16-transformed cells. Eur J Immunol 23:2242, 1993

Fremont DH, Matsumura M, Stura EA et al. Crystal structures of two viral peptides in complex with murine class I $H\text{-}2K^b$. Science 257:99, 1992

Garrett TP, Saper MA, Bjorkman PJ et al. Specificity pockets for the side chains of peptide antigens in HLA-Aw68. Nature 342:692, 1989

Geysen HM, Meloen RH and Barteling SJ. Use of peptide synthesis to probe viral antigens for epitopes to a resolution of a single amino acid. Proc Natl Acad Sci USA 81:3998, 1984

Geysen HM, Barteling SJ and Meloen RH. Small peptides induce antibodies with a sequnece and structural requirement for binding antigen comparable to antibodies raised against the native protein. Proc Natl Acad Sci USA 82:178, 1985

Houbiers JGA, Nijman HW, Van der Burg SH et al. *In vitro* induction of human cytotoxic T lymphocyte responses against peptides of mutant and wild type p53. Eur J Immunol 23:2072, 1993

Kast WM, Bronkhorst AM, De Waal LP et al. Cooperation between cytotoxic and helper T lymphocytes in protection against lethal Sendai virus infection. Protection by T cells is MHC-restricted and MHC-regulated, a model for MHC-disease associations. J Exp Med 164:723, 1986

Kast WM, Offringa R, Peters PJ et al. Eradication of adenovirus E1-induced tumors by E1A specific cytotoxic T lymphocytes. Cell 59:603, 1989

Kast WM and Melief CJM. *In vivo* efficacy of virus derived peptides and virus-specific cytotoxic T lymphocytes. Immunol Lett 30:229, 1991a

Kast WM and Melief CJM. Fine peptide specificity of cytotoxic T lymphocytes directed against adenovirus induced tumors and peptide-MHC binding. Int J Cancer Sup 6:90, 1991b

Kast WM, Roux LM, Curren J et al. Protection against lethal Sendai virus infection by *in vivo* priming of virus-specific cytotoxic T lymphocytes with a free synthetic peptide. Proc Natl Acad Sci USA 88:2283, 1991c

Kast WM, Brandt RMP, Drijfhout JW et al. HLA-A2.1 restricted candidate CTL epitopes of human papillomavirus type 16E6 and E7 proteins identified by using the processing defective human cell line T2. J Immunotherapy 14:115, 1993a

Kast WM, Brandt RMP and Melief CJM. Strict peptide length is not required for the induction of CTL mediated antiviral protection by peptide vaccination. Eur J Immunol 23:1189, 1993b

Kast WM, Brandt RMP, Sidney J et al. The role of HLA-A motifs in identification of potential CTL epitopes in human papillomavirus type 16 E6 and E7 proteins. J Immunol 152:3904, 1994.

Krzych U, Fowler AV, Miller A et al. Induction of helper and suppressor T cells by nonoverlapping determinants on the large protein antigen beta-galactosidase. FASEB J 2:141, 1988

Madden DR, Gorga JC, Strominger JL et al. The structure of HLA-B27 reveals nonamer self-peptides bound in an extended conformation. Nature 353:321, 1991

Matsumura M, Fremont DH, Peterson PA et al. Emerging principles for the recognition of peptide antigens by MHC class I molecules. Science 257:927, 1992

Melief CJM and Kast WM. T cell immunotherapy of cancer. Res Immunol 142:425, 1991

Melief CJM. Tumor eradication by adoptive transfer of cytotoxic T lymphocytes. Adv Cancer Res 58:143, 1992

Melief CJM and Kast WM. Lessons from T cell responses to virus-induced tumours for cancer eradication in general. Cancer Surveys 13:81, 1992

Monaco JJ. A molecular model of MHC class I-restricted antigen processing. Immunol Today 13:173, 1992

Nardelli B and Tamp JP. Cellular immune responses induced by *in vivo* priming with a lipid-conjugated multimeric antigen peptide. Immunology 79:355, 1993

Nijman HW, Houbiers JGA, Vierboom MPM et al. Identification of peptide sequences that potentially trigger HLA-A2.1 restricted cytotoxic T lymphocytes. Eur J Immunol 23:1215, 1993a

Nijman, HW, Houbiers JGA, Van der Burg SH et al. Characterization of cytotoxic T lymphocyte epitopes of a self protein, p53, and a non-self protein, influenza matrix: Relationship between MHC peptide binding affinity and immune responsiveness to peptides. J Immunotherapy 14:121, 1993b

Ossevoort MA, Sijts AJAM, Van Veen KJH et al. Differential effect of transporter TAP2 gene introduction into RMA-S cells on viral antigen processing. Eur J Immunol 23:3082, 1993

Rammensee H-G, Falk K and Rötzschke O. Peptides naturally presented by MHC class I molecules. Ann Rev Immunol 11:213, 1993

Resnick RM, Cornelissen MT, Wright DK et al. Detection and typing of human papillomavirus in archival cervical cancer specimens by DNA amplification with consensus primers. J Natl Cancer Inst 82:1477, 1990

Rötzschke O and Falk K. Naturally-occurring peptide antigens derived from the MHC class I-restricted processing pathway. Immunol Today 12:447, 1991

Ruppert J, Sidney J, Celis E et al. Prominent role of secondary anchor residues in peptide binding to HLA-A2.1 molecules. Cell 74:929, 1993

Saper MA, Bjorkman PJ and Wiley DC. Refined structure of the human histocompatibility antigen HLA-A2 at 2.6Å resolution. J Mol Biol 219:277, 1991

Schild H, Deres K, Wiesmüller K-H et al. Efficiency of peptides and lipopeptides for *in vivo* priming of virus-specific cytotoxic T cells. Eur J Immunol 21:2649, 1991

Schulz M, Zinkernagel RM and Hengartner H. Peptide-induced antiviral protection by cytotoxic T cells. Proc Natl Acad Sci USA 88:991, 1991

Schumacher TNM, De Bruijn MLH, Vernie LN et al. Peptide selection by MHC class I molecules. Nature 350:703, 1991

Townsend ARM, Rothbard J, Gotch FM et al. The epitopes of influenza nucleoprotein recognized by cytotoxic T lymphocytes can be defined with short synthetic peptides. Cell 44:959, 1986

Van der Zee R, Van Eden W, Meloen RH et al. Efficient mapping and characterization of a T cell epitope by the simultaneous synthesis of multiple peptides. Eur J Immunol 19:43, 1989

Van Bleek GM and Nathenson SG. Isolation of an endogenously processed immunodominant viral peptide from the H-$2K^b$ molecule. Nature 348:213, 1990

Van den Brule AJC, Walboomers JMM, Du Maine M et al. Difference in prevalence of human papillomavirus genotypes in cytomorphologically normal cervical smears is associated with a history of cervical intraepithelial neoplasia. Int J Cancer 48:404, 1991

Zhang W, Young ACM, Imarai M et al. Crystal structure of the major histocompatibility complex class I H-2^d molecule containing a singel viral peptide: Implications for peptide binding and T cell receptor recognition. Proc Natl Acad Sci USA 89:8403, 1992

Zhou X, Glas R, Liu T et al. Antigen processing mutant T2 cells present viral antigen restricted through H-$2K^b$. Eur J Immunol 23:1802, 1993a

Zhou X, Glas R, Momburg F et al. TAP2-defective RMA-S cells present Sendai virus antigen to cytotoxic T lymphocytes. Eur J Immunol 23:1796, 1993b

Chapter 20

PEPTIDE APPLICATIONS IN AUTOIMMUNE DISEASE RESEARCH

Koen Gerritse

TABLE OF CONTENTS

I. AUTOIMMUNITY AND TOLERANCE

Autoimmunity is characterized by the failure of the immune system to discriminate between self and nonself. In a normal functioning organism, the immune system is quite capable of recognizing and processing a foreign antigen, leaving the host cells and tissues unaffected. However, sometimes the immune system fails to maintain this so called self tolerance, resulting in destruction of cells and tissues. Although many distinctive cells of the immune system are involved in autoimmunity, only B and T cells are involved in the actual recognition and discrimination between self and nonself antigens.

I.1. Autoreactive cells and MHC

T cells can functionally be divided in cytotoxic T cells (Tc) and T helper (Th) cells. The Tc cells are involved in the lysis of target cells. Th cells activate and provide help to B cells and other T cells to proliferate and differentiate. The T cells recognize protein antigens as processed peptides. The processing involves the ingestion of antigens by antigen presenting cells, like B cells, dendritic cells and macrophages; a subsequent controlled proteolytic fragmentation; and the association of the immunogenic peptides with major histocompatibility complex (MHC) proteins. The complex of immunogenic peptide and MHC will be expressed on the membrane of the antigen presenting cells. Subsequently, the immunogenic peptides in the context of MHC proteins will be recognized by the T cell receptor (TCR).

The MHC proteins are the products of a particular region of the genome. In man, the MHC region is the human lymphocyte antigen (HLA) gene cluster located on chromosome 6. The MHC region in the mouse is the H-2 gene cluster located on chromosome 17. The MHC proteins play a role in interactions

0-8493-8967-4/95/$0.00+$.50

between lymphocytes and in interactions between lymphocytes and antigen presenting cells. Three types of MHC molecules, class I, II and III, have been identified in both mice and man. Processed antigens in association with MHC class I products are primarily recognized by cytotoxic T cells. Processed antigens associated with MHC class II products are recognized by T helper cells, whereas MHC class III genes are encoding for components of the complement system, a complex group of serum proteins which mediate a variety of inflammatory reactions. Several autoimmune diseases are associated with specific MHC haplotypes. Heterozygote HLA-DR3, HLA-DR4 individuals have an increased risk of developing insulin dependent diabetis melitis (for review, see Hitman, 1989). Multiple sclerosis and rheumatoid arthritis have been shown to be associated with HLA haplotypes DR2, B7 (Ota et al., 1990; Jingwu et al., 1990; Pette et al., 1990) and DR3, DR4 (Gao et al., 1988), respectively. These observations indicate that specific MHC molecules are involved in the presentation of particular autoantigens epitopes.

MHC class I and II molecules themselves do not have the ability to discriminate between self derived peptides and peptides originating from foreign antigens. As a consequence, both self and nonself peptides will be bound to the MHC molecules and presented to T cell receptors. However, in a normal functioning organism, no response will be evoked against self antigens, in other words the immune system is tolerant to self antigens. Several mechanisms for induction of T cell tolerance can be distinguished. In an early stage of development, the presentation of self antigens in the thymus to T cells leads to the elimination of the T cells which recognize the self antigens. This process is called thymic tolerance or clonal deletion. However not all self antigens are transported to the thymus or are transported only in very low concentrations, especially self antigens originating from immune privileged sites like the central nervous system (CNS) or testis. For these antigens, an additional tolerance induction system, i.e., clonal anergy, is operational. T cells, which are anergic to autoantigens, will not be activated when autoantigens are presented to them, and as a consequence autoimmunity will not occur. The precise mechanisms by which the T cells are deleted or anergized remains to be revealed.

B cells are characterized by their ability to produce and secrete antibodies. Antibodies are bifunctional molecules. The Fab fragment of the antibody recognizes the antigen, whereas the Fc portion is capable to interact with Fc receptors. Fc receptors are expressed on the surface of several cell types of the immune system. Antibodies recognize self or nonself antigens which are in solution or expressed on cell surfaces. The antibody binding to antigens may be an initial step in the ultimate elimination of toxins or viruses. Antibodies can also act as mediators between antigens and effector cells which express Fc receptors on their membrane surface. Neutrophils and most specialized antigen presenting cells express Fc receptors on their membranes, which allow them to phagocytose antigen-antibody complexes for intracellular degradation. Furthermore, antibodies can sensitize target cells for attack of cytolytic cells which express Fc receptors. The mechanism of this antibody mediated immune defence is called the antibody dependent cellular cytotoxicity (ADCC). Several cells, like macrophages, eosinophils, cytotoxic T cells, natural killer cells and killer cells, can attack target cells making use of the ADCC mechanism.

Whether clonal deletion or anergy play a role in the discrimination of self and nonself antigen by B cells, like these mechanisms play a role in self and nonself antigen recognition by T cells, is unclear. The recognition and binding of antigens with immunoglobulins, which are expressed on the membrane of the B cells, generate an initial intrinsic activation signal. The strength of the initial activation signal is one of the factors which is of influence on the process of discrimination between self and nonself antigens. There is evidence that not only a high antigen receptor occupancy has a negative influence on the strength of the initial activation signal, but a low receptor occupancy as well (Goodnow, 1992). A high receptor occupancy can be caused by the recognition of self proteins present in high concentrations, like several serum proteins, whereas low antigen receptor occupancy can be caused by recognition of self proteins, which are transported in low concentrations, like self antigens originating from immune privileged sites. Not only a negative influenced intrinsic signal causes B cell inactivation or deletion, but a positive influenced intrinsic cellular signal as well. Receptor linkage leads to a very potent intrinsic cellular signal, as was shown for lysozyme, which is expected to crosslink the membrane receptors extensively. Subsequently, the potent intrinsic signal is leading to the elimination of lysozyme reactive B cells (Hartley et al., 1991). These findings suggest that potential autoreactive B cells, from which the antigen receptors are crosslinked by self antigens, could be eliminated in the same way.

If the B cell is not eliminated or inactivated due to the strength of the intracellular signal, the antigens will be internalized, processed and subsequently presented in the context of MHC class II molecules to Th cells. In turn, the recognition of the processed antigens in association with MHC class II molecules by the Th cell receptor activates Th cells. The activated Th cells will produce and secrete cytokines and will express the gp39 membrane activation marker (Armitage et al., 1992; Noelle et al., 1992). The binding of the gp39 marker to its B cell ligand CD40 is required for an optimal stimulation of B cells to proliferate and differentiate (Paulie et al., 1989). As a consequence, clonal deletion or anergy of Th cells will prevent the ultimate activation of autoreactive B cells as well.

I.2. Molecular mimicry

The autoantigen itself is not always required for the activation of autoreactive cells. It is possible that autoreactive cells are activated by crossreactive epitopes of pathogenic agents like bacteria and viruses. This phenomenon, sharing of antigenic determinants, is called molecular mimicry. By using computer analysis to compare virus and bacterial protein sequences to human protein sequences, many sequence homologies have been identified. Subsequently, by making use of peptides corresponding to these shared epitopes, immunological crossreactivity has been identified.

It was revealed that a 160 kD surface protein of *Trypanosoma cruzi* shares epitopes with a 48 kD protein present in mammalian axonal and myenteric plexus cells. Recombinant peptides from various regions of the 160 kD surface protein of *T. cruzi* were used to inhibit the binding of polyclonal anti-160 kD surface protein antibodies to nerve cells. A peptide of 48 amino acids, derived from expression of base pairs 611–761 of the DNA sequence of the 160 kD surface protein of *T. cruzi*, completely inhibited the anti-160 kD surface protein antibody binding to nerve cells. Other peptides, not originating from the region 611–761 of the DNA sequence of the 160 kD surface protein, did not inhibit the binding of anti-160 kD surface protein antibodies to nerve cells at all. Subsequently, three peptides, which encompass the 48 amino acid peptide from the 160 kD surface protein of *T. cruzi*, were synthesized and used to inhibit the specific binding of the anti-160 kD surface protein antibodies. A 12 amino acid peptide corresponding to DNA bases 615–651, from the 160 kD surface protein of *T. cruzi*, completely inhibited the antibody binding to nerve cells. The other two synthetic peptides did not inhibit the specific binding of the antibodies to the nerve cells (Voorhis et al., 1991). Molecular mimicry by antigens of type 1 streptococcal M protein was investigated by making use of monoclonal antibodies raised with synthetic peptides corresponding to the amino terminal site of type 1 streptococcal M protein. One of the monoclonal antibodies reacted with mesangial cells of renal glomeruli and myocardium. A monoclonal antibody, which was directed against amino acid sequence 13–19 of streptococcal M protein, crossreacted with a 43 kD mesangial/myocardial protein in Western blot (Kraus et al., 1990). These results show that type 1 streptococcal M protein contains at least one crossreactive epitope with the 43 kD protein of human glomeruli and myocardium cells.

Besides many bacterial structural proteins, there are indications that, due to molecular mimicry, especially a group of autologous bacterial proteins, the so-called heat shock proteins (HSP) or stress proteins could play a role in the induction of autoimmune diseases as well. Upon bacterial infection, bacterial stress proteins will be processed by antigen presenting cells and subsequently presented as peptides to γ/δ receptor bearing T cell, which will proliferate and differentiate. Besides the defense mechanism of these γ/δ receptor bearing T cells, natural antibodies against HSP epitopes represent a parallel B cell defense mechanism. Because bacterial and host stress proteins have a sequence homology over 40%, there is a constant risk of autoimmune disease due to molecular mimicry. A considerable percentage of T cells, isolated from the synovial fluid of patients with rheumatoid arthritis, which are reactive with the 65 kD bacterial HSP, comprises T cells which are reactive with the human 60 kD HSP as well. It was hypothesized that these T cells must be specific for shared epitopes on both 65 kD bacterial HSP and human 60 kD HSP. This was investigated by screening mononuclear cells isolated from two juvenile rheumatoid arthritis patients on reactivity with peptides corresponding to human 60 kD HSP. T cells from both patients recognized the human 60 kD HSP peptide sequences 241–255 (Quayle et al., 1992). The human 60 kD peptide sequences 241–255 appear to be highly homologous with an amino acid sequence of the mycobacterial 65 kD HSP. These results suggest that bacterial HSP epitopes could play a role in the breakdown of tolerance in patients with rheumatoid arthritis.

These findings indicate that epitopes of nonself antigens, which are homologous to self proteins, may drive autoreactive B or T cell clones to expand and cause autoimmune pathogenesis. Therefore, molecular mimicry is considered to be an important mechanism which plays a role in the induction of human autoimmune diseases.

When the mechanisms of elimination or inactivation of self reactive cells are somehow overridden, the recognition of autoantigens, or crossreactive epitopes of foreign antigens, leads to the activation of self reactive T and B cells. However, the presence of activated autoantigen reactive T and B cells or antibodies *per se* does not imply the organism will automatically develop an autoimmune disease. One of the reasons the autoimmune response will not occur is the lack of an appropriate amount of self antigen, or antigen with crossreactive epitopes to stimulate the self reactive cells continuously. Furthermore, the presence of autoantigen reactive T and B cells and antibodies does not necessarily implicate that these cells or the autoantibodies are involved in the induction or enhancement of the autoimmune disease.

Especially, the role of autoantibodies in the process of several autoimmune diseases remains unclear. Myasthenia gravis is the only autoimmune disease in which the direct involvement of anti-acetylcholine antibodies as pathogenic effector molecules was established (Fuji and Lindstrom, 1988). However, within almost the complete spectrum of autoimmune diseases studied, ranging from organ specific autoimmune diseases like Hashimoto's thyroiditis (Delves and Roitt, 1984) and multiple sclerosis to nonorgan specific autoimmune diseases like systemic lupus erythematosus (Lehman et al., 1984), autoantibodies have been detected. Although these findings make the production of autoantibodies one of the hallmarks of autoimmunity, it is not known whether these autoantibodies are an epi-phenomenon or are directly involved in the pathogenesis of autoimmune diseases. In addition, the fact that autoantibodies can easily be detected by ELISA, not only in the serum of persons with autoimmune diseases but in the serum of healthy individuals as well, makes the role of autoreactive antibodies and B cells in the process of autoimmunity even more controversial.

I.3. Experimental autoimmune diseases

The study of autoimmune diseases using animal models provides an important and useful way of learning more about basic mechanisms used by the immune system to discriminate between self and foreign antigens. Although there are many differences between the experimental animal diseases and their human autoimmune counterparts, the experimental animal models are generally the only tool to provide knowledge concerning the fundamental mechanisms involved in disease induction or enhancement.

There are several inbred laboratory animals known which occasionally spontaneously develop autoimmune diseases, like the spontaneous diabetis in rats (Chao et al., 1989) and the lupus-prone mice (Tsao et al., 1990), which closely mimic their human counterparts. In addition, much work has been done to generate experimental animal models for a variety of autoimmune diseases. The principle of experimental autoimmune disease induction is based on interference with the immune system in such way that the natural occurring tolerance is overruled. Autoimmune diseases can be induced in two distinctive procedures. First, autoimmune disease induction can be accomplished by the adoptive transfer of autoreactive cells to naive recipients. Autoreactive cells can be isolated from animals with autoimmune diseases or can be isolated after *in vitro* stimulation with autoantigen. Second, autoimmune diseases can be induced by the artificial presentation of an excess of autoantigens to the immune system. The introduction of autoantigens can be accomplished in two ways. Autoantigens can be introduced indirectly by the inoculation of viruses which causes tissue damage. Subsequently, due to tissue damage, autoantigens will be released and presented to the immune system, and autoantigen reactive cells will be activated. Inoculation of animals with Theiler's murine virus (Dal Canto and Lipton, 1975), Semliki Forest virus (Suckling et al., 1978) or murine coronavirus JHM (Weiner, 1973) induces experimental autoimmune diseases which have several similarities with human autoimmune demyelination. However, the most commonly used method for the introduction of autoantigens is the direct administration of autoantigens to animals by immunization. Because a normal functioning animal will be tolerized to the administered autoantigens, the use of adjuvant is a requirement.

II. PEPTIDES AND AUTOIMMUNITY

Peptides, syntheticly produced, produced by enzymatic degradation of the native proteins or cloned in micro-organisms, have been proven to be very important tools in autoimmune disease research. Synthetic peptides, analogous to autoantigen sequences, can be used to identify the epitopes of autoreactive T and B cells (Table 1). Subsequently, knowledge concerning the epitope specificity and the availability of specific autoreactive cell lines or clones makes it possible to study the autoreactive T and B cell repertoire diversity in both human or experimental autoimmune diseases. However, a disadvantage of the peptide approach to identify the epitopes of autoreactive cells is that only sequential epitopes and no conformational epitopes of autoreactive B cells can be detected. In addition, by applying synthetically produced T and B cell epitopes for induction of autoimmune diseases in animals, several pathogenic epitopes have been identified. The use of synthetic peptides for induction of autoimmune diseases has many advantages as compared to the use of intact autoantigens for induction of autoimmune diseases. Induction of autoimmune diseases with synthetic peptides leads in general to a higher disease incidence related to the disease incidence as observed after autoimmune disease induction with intact autoantigen. In addition, MHC restriction for several pathogenic epitopes has been revealed by application of peptides for autoimmune disease induction in animal strains with different MHC haplotypes. Furthermore, the peptide induced autoimmune diseases are usually more strictly defined as compared to the autoimmune diseases induced with intact autoantigen. Because the pathogenic effector cells in the peptide induced models are reactive with a limited number of epitopes only, the detection and isolation of these pathogenic effector cells is uncomplicated. The identification and isolation of the pathogenic cells has

TABLE 1
The use of synthetic peptides to reveal the epitope specificity of immune reactive cells in autoimmune diseases

Disease	Autoantigen peptides analogue to	B or T cell epitope	References
Autoimmune chronic active hepatitis	Hepatitis C virus	B cell	Lenzi et al., 1991
Interstitial nephritis	Tubular basement membrane	T cell	Neilson et al., 1991
Autoimmune thyroid disease	Thyroid peroxidase	B cell	Banga et al., 1991
	Thyroglobulin	B cell	Henry et al., 1992
Uveitis	Interphotoreceptor retinoid Binding protein	B cell	Hoeksema et al., 1990
Systemic rheumatic diseases	Ribonucleoprotein complex	T cell	O'Brien et al., 1990
Sjogren's syndrome	La/SS-B ribonucleoprotein	B cell	McNeilage, 1992
Dilated cardiomyopathy	Adenylate cyclase β receptor	B cell	Limas et al., 1992
Idiopathic thrombocytopenic purpura	Platelet fibronectin receptor	B cell	De Souza et al., 1992
Atherosclerosis	Elastin	B cell	Gminski et al., 1991
Primary biliary cirrhosis	Dihydrolipoamide acetyltransferase	B cell	Tuaillon et al., 1992b
Myasthenia gravis	Acetylcholine receptor	B cell	Ashizawa et al., 1992
Guillain Barré	Po and P2 myelin proteins	T cell	Khalili-Shirazi et al., 1992
Hashimoto's thyroiditis	Thyroid peroxidase	T cell	Kawakami et al., 1992
Multiple sclerosis	Myelin basic protein	T cell	
	Proteolipid protein	T cell	Trotter et al., 1991

allowed further analysis of the mechanisms of the autoimmune response. A more recent application of peptides in autoimmune disease research is the use of peptides as modulators in autoimmune responses. Furthermore, in many studies, synthetic peptides were used to generate monoclonal antibodies, which in turn were used to study function, catabolism and structure of self proteins or to detect metabolic products of degenerated self proteins.

In the next section, the role of autoimmune T and B cells in the pathogenesis of peptide induced experimental autoimmune diseases will be described. As a model for peptide induced autoimmune diseases in general, peptide induced experimental autoimmune encephalomyelitis (EAE) will be described extensively. In addition, several other important peptide induced autoimmune diseases will be highlighted and a variety of peptide applications in both human and animal autoimmune disease research will be paraphrased.

II.1. Multiple sclerosis and experimental autoimmune encephalomyelitis

Multiple sclerosis (MS) is characterized by the breakdown of the myelin sheaths which surround the axons, resulting in microscopic lesions (plaques). The breakdown of the myelin sheaths exclusively affects the CNS. Due to the impaired conductivity of the demyelinated axons, clinical characteristics like visibility and respiration problems, paralysis and incontinence will be manifest. Since MS is accompanied by an inflammatory response of immune competent cells like T cells, B cells and macrophages (Traugott et al., 1983; Weller, 1985), the disease is considered to be immune mediated.

II.1.a. T cell reactivity in MS

T cell sensitization to various myelin components may be important to the pathogenesis of MS. Extensive, partly conflicting, data have been presented concerning the presence, distribution and function of T cells specific for myelin components in peripheral blood and cerebrospinal fluid (CSF), which seems to fluctuate with the clinical course of MS (Freedman and Antel, 1988). As compared to other putative antigenic components of the myelin sheaths, myelin basic protein (MBP) is the most extensively investigated protein for its involvement in MS. MBP specific T cells have been detected in the sera and cerebrospinal fluid of both MS patients and patients with other neurological diseases (Ota et al., 1990; Gorcy et al., 1983). Overlapping synthetic peptides of MBP were used to identify the antigenic determinants of human T cells. Some determinants, MBP amino acid sequence 5–25, 35–47, 65–75 and 81–100, were recognized by T cells derived from more than one individual (Burns et al., 1991). Because of the epitope specificity agreement of several T cells isolated from separate individuals, it was suggested that these epitopes are particularly immunogenic in humans. Jingwu et al. (1990) used MBP fragments derived by peptic cleavage and synthetic peptides to reveal the epitope specificity of 17 MBP reactive T cell clones generated from four MS patients. The T cell clones showed a striking similarity in reactivity with the C-terminal MBP peptide 149–171. Almost the same MBP epitope, sequence 154–172 was recognized by a majority of MBP reactive T cell clones derived from both MS patients and healthy subjects (Martin et al., 1990). The MBP sequence 149–171 overlaps the encephalitogenic epitope in rhesus monkey (Karkhanis et al., 1975). However, the association between the MBP epitope 149–171 and encephalitogenicity in humans remains to be revealed.

Myelin proteolipid protein (PLP) can induce a T cell mediated chronic relapsing autoimmune encephalomyelitis in Hartley guinea pig (Yoshimura et al., 1985), Lewis rat (Yamamura et al., 1986) and rabbits (Veen et al., 1986). Therefore, PLP is considered to be an important autoantigen to the pathogenesis of MS as well. T cells reactive with PLP have been isolated from patients with early relapsing remitting MS (Trotter et al., 1991). PLP reactive T and B cells were also detected in blood and CSF from control patients, but these cells were detected in much lower numbers as compared to MS patients (Sun et al., 1991). No difference in reactivity to PLP peptide 139–151 was measured between T cells from peripheral blood from MS patients compared to normal subjects and patients with other neurological diseases.

II.1.b. B cell reactivity in MS

Although the role of antibodies and B cells specific for myelin components in MS is less well documented, there is circumstantial evidence of antibody and antibody forming cell (AFC) involvement in the process of demyelination. Persistent oligoclonal IgG synthesis in the CSF (Walsh and Tourtelotte, 1983) and a high antibody specificity index (Felgenhauer and Reiber, 1992) are indicative for a sustained B cell activation within the CNS of MS patients. In both CSF and sera of MS patients, antibodies directed against components of brain tissue (Henneberg et al., 1991) and a variety of putative CNS antigens like cerebellar soluble lectin (Zanetta et al., 1990a, 1990b), glycolipids (Endo et al., 1984; Hirsch and Parks, 1976), myelin basic protein (Link et al., 1990), proteolipid protein (Sun et al., 1991) and myelin associated glycoprotein (Baig et al., 1991) were detected. Warren and Catz (1986) detected antibodies in MS patients directed against myelin components, particurlaly during exacerbations of clinical signs. In contrast, this correlation was not observed by Link and co-workers (Link et al., 1990). Gorcy reported the detection of elevated frequencies of anti-MBP antibodies in CSF of MS patients and patients with other neurological diseases as compared to healthy controls (Gorcy et al., 1983). Others were not able to demonstrate a statistical difference in anti-MBP antibody titers in CSF (Chou et al., 1983) nor in the frequencies of anti-MBP AFCs in blood (Jingwu et al., 1991) of MS patients and healthy controls. In our laboratory, we have detected MBP specific antibody forming cells in brain tissue sections of MS patients. Although MBP specific antibodies or antibody forming cells can be detected in the sera or cerebrospinal fluid of both MS patient and controls, MBP specific antibody forming cells in CNS tissues can be detected in MS patients only (Gerritse et al., 1994).

II.1.c. Experimental autoimmune encephalomyelitis

EAE is considered to be the best available animal model for MS. The most generally used method for EAE induction is the subcutane immunization of susceptible primates and rodents with whole CNS homogenate or with separated and purified autologous or heterologous CNS protein components like MBP or PLP emulsified in Freunds complete adjuvant (Paterson, 1976). The precise mechanisms which lead to the damage of myelin sheaths remain to be investigated. However, it is hypothesized that the autoantigens in the inoculum will be presented to T cells, which will proliferate and differentiate. A part of these autoantigen primed T cells, and probably antigen specific B cells as well, will enter the CNS. A second contact of these immune competent T and B cells with CNS autoantigens, to which they were primed, will activate the T and B cells again and an immune response against the antigens will be the result. Another frequently used method of EAE induction is the adoptive transfer of myelin reactive cells isolated from animals immunized with myelin components (Mokhtarian et al., 1984; Veen et al., 1989). More recently, it became possible to induce EAE with sequences corresponding to amino acid sequences from myelin sheath proteins (Table 2). Although the myelin sheath consists of several protein components, EAE can be induced by the immunization with amino acid sequences of MBP and PLP only.

Based on the course of appearance of clinical signs, two types of EAE can be distinguished: first, acute EAE which is characterized by only one single period in which clinical signs are manifest; and second, chronic relapsing-remitting EAE (CR-EAE) which is characterized by two or even more periods with clinical or neuropathological signs (Raine, 1985; Polman et al., 1988). The course of appearance of clinical signs in CR-EAE resembles the appearance of clinical signs in MS more closely as compared to the course of appearance of clinical signs in acute EAE (Alvord, 1985).

Not all animal strains are equally susceptible to EAE induction. Dunkin Hartley guinea pigs, Lewis rat and SJL/J mice are susceptible to induction of EAE. However, Brown Norway rats and BALB/C mice are resistant to EAE induction. In general, the differences in susceptibility to EAE induction are attributed to MHC class II restriction. On the other hand, quantitative changes in local MHC class II expression might be of influence on the susceptibility of disease induction as well. Animals which are highly susceptible to EAE induction have an increased expression of MHC class II molecules in the CNS (de Groot et al., 1991). Furthermore, there are some indications that a decreased susceptibility to EAE induction is caused by a disturbed antigen recognition of T cell receptors (Zamvill and Steinmann, 1990).

II.1.d. Peptide applications in EAE mice

The encephalitogenicity and MHC restriction of myelin protein epitopes were determined by using proteolytic fragments and synthetic peptides, analogous to sequences of myelin proteins, for immunization purposes in several strains of mice. SJL/J (H-2^s) and B10.T(6R) (H2^q) mice develop EAE in response to the MBP C-terminal fragment 89–169 of guinea pig or rat MBP (Pettineli et al., 1982). To identify the encephalitogenic epitope(s) within this long C-terminal MBP sequence 89–169, several partly overlapping peptides were synthesized for EAE induction in SJL/J (H-2^s) mice. Three overlapping MBP peptide sequences 91–103, 89–101 and 84–102 appear to be encephalitogenic although with a very low incidence. MBP synthetic peptides containing MBP sequence 89–101 appear to induce chronic EAE in B10RIII (H-2^r) mice as well (Jansson et al., 1991). The EAE incidence in SJL/J (H-2^s) mice was remarkably improved when the same peptides were used to obtain peptide reactive T cells, which in turn were used for EAE induction by adoptive transfer to naive recipients (Su and Sriram, 1992). A clinical manifestation of EAE in SJL/J (H-2^s) mice was observed after immunization with the MBP peptides corresponding to amino acid sequences 87–114, 87–98, 91–104, 88–98 and 89–98 (Kono et al., 1988). These results implied that at least one encephalitogenic epitope on the C-terminal fragment 89–169 is located around amino acid sequence 87–104 of the MBP molecule. Synthetic peptides analogous to the long C-terminal MBP sequence were used by Bourdette and co-workers (1989) to select T cell lines from MBP immunized SJL/J (H-2^s) mice. Selected lines proliferated in response to the 89–169 fragment of MBP and synthetic peptides 87–99, 90–99 and 91–99. These T cell lines induce a chronic relapsing form of EAE when transferred to naive mice. The reported susceptibility of SJL/J (H-2^s) mice to MBP sequences, when used for

TABLE 2
Encephalitogenic amino acid sequences for EAE induction

Myelin basic protein sequences:			
Species:	**Haplotype:**	**Residues:**	**References:**
Mouse:			
PL/J	H-2^{u}	1–37	Fritz et al., 1985
		1–11	Zamville et al., 1986
SJL/J	H-2^{s}	84–102	Su and Sriram, 1992
		87–114	Kono et al., 1988
		87–98	Kono et al., 1988
		88–98	Kono et al., 1988
		89–169	Pettinelli et al., 1982
		89–104	Su and Sriram, 1992
		89–98	Kono et al., 1988
		91–103	Su and Sriram, 1992
		91–104	Kono et al., 1988
SWR	H-2^{q}	89–101	Jansson et al., 1991
NFR/N	H-2^{q}	89–101	Jansson et al., 1991
B10RIII	H-2^{r}	89–169	Pettinelli et al., 1982
		89–101	Jansson et al., 1991
Rat:			
Lewis	RT-11	1–88	Levine et al., 1990
		45–88	Levine et al., 1990
		63–88	Levine et al., 1990
LOU/M	RT-1w	110–129	Hashim et al., 1991
Buffalo		87–99	Jones et al., 1992
Proteolipid protein sequences:			
Species:	**Haplotype:**	**Residues:**	**References:**
Mouse:			
PL/J	H-2^{u}	35–47	Zamville et al., 1988
		43–64	Whitham et al., 1991a
SJL/J	H-2^{s}	139–151	Tuohy et al., 1989
		141–150	Kuchroo et al., 1991
		141–149	Kuchroo et al., 1991
SWR	H-2^{q}	103–116	Sobel et al., 1991

immunization purposes to induce EAE, could not be confirmed in our laboratory using MBP sequences 22–37, 45–64, 70–89, 90–108 and 123–141. A relapsing-remitting form of EAE can be induced in SJL/J (H-2^{s}) mice, either by active immunization with the major encephalitogenic PLP peptide sequence 139–151 (Tuohy et al., 1989; Sobel et al., 1990; McRae et al., 1992) or following adoptive transfer of PLP peptide sequence 139–151 specific T cells (McRae et al., 1992; Whitham et al., 1991b). In our laboratory, the encephalitogenicity of the 139–151 epitope was confirmed in SJL/J (H-2^{s}) mice. However, we did not observe the relapsing-remitting course as was described by McRae and colleagues. In our experiments, the induced EAE appears to be monophasic. Immunization with PLP peptides sequence 141–150 and 141–149 induces EAE as well, although with a lower incidence as compared to the EAE incidence obtained after immunization with PLP peptide sequence 139–151 (Kuchroo et al., 1991).

Although it is possible to induce EAE with separate proteins or even peptide sequences, it does not imply that these proteins or sequences are the specific antigens or epitopes responsible for the EAE induction with spinal cord homogenate. To determine the encephalitogenic proteins and epitopes in the spinal cord homogenate, the *in vitro* proliferative responses to myelin sheath proteins and MBP and PLP peptides of lymphoid cells isolated from animals with spinal cord homogenate induced EAE were revealed. Lymph node cells only responded with PLP peptide 139–151, not with PLP peptide sequence 141–152 nor with MBP or MBP synthetic peptides. Mononuclear cells, isolated from the CNS of animals with spinal cord induced EAE, reacted preferentially with PLP synthetic peptides. In addition, PLP peptide 139–151 specific cells from spinal cord homogenate immunized mice could transfer EAE in naive recipients. These data suggest that the PLP sequence is the major encephalitogenic epitope in spinal cord immunized SJL/J (H-2^s) mice (Whitham et al., 1991b).

It was revealed that the intravenous administration of spinal cord homogenate or MBP coupled to splenocytes inhibits the development of both active and adoptive EAE in SJL/J mice. Furthermore, the administration of MBP peptide sequence 84–104, which contains the major encephalitogenic epitopes for SJL/J mice, inhibited completely the peptide specific T cell response. However, the induction of EAE by passive transfer of encephalitogenic T cells to naive recipients was only partly inhibited after MBP peptides coupled to splenocytes were administered. The usage of other MBP peptides, sequence 91–104 and sequence 17-27, did not influence the disease induction by passive transfer of T cells at all. This suggests that there are other MBP epitopes besides the encephalitogenic MBP epitopes within sequence 84–104, which contribute to the suppression of EAE (Tan et al., 1992).

Although the SJL/J mice is the most generally used mouse strains for EAE induction, EAE can be induced in other mice strains as well. It was revealed that the SWR and NFR/N mice, both with the MHC H-2^q haplotype, are susceptible to EAE induction with the MBP peptide sequence 89–101. By making use of truncated synthetic peptides, the minimal encephalitogenic epitope of MBP for SWR (H-2^q) mice was localized to amino acids 87–99 (Cross et al., 1991). The encephalitogenic PLP epitope for SWR (H-2^q) mice is sequence 103–116 (Tuohy et al., 1988; Sobel et al., 1991). This sequence is not encephalitogenic in PL/J (H-2^u) mice. In addition, the SJL/J (H-2^s) encephalitogenic PLP epitope sequence 139–151 is not encephalitogenic for the SWR (H-2^q) mice. Both encephalitogenic epitopes for SWR (H-2^q) and SJL/J (H-2^s) mice are not encephalitogenic in PL/J (H-2^u) mice. One of the encephalitogenic MBP epitopes described for the PL/J (H-2^u) mouse is the N-terminal peptide 1–37 (Fritz et al., 1985). PL/J (H-2^u) mice also develop EAE after immunization with the α-amino acetylated synthetic peptide analogous to MBP sequence 1–11, which appears to be MHC class II restricted (Zamvill et al., 1986). EAE induction in the PL/J (H-2^u) mouse was also observed after immunization with PLP peptide sequence 43–64 (Whitham et al., 1991a).

Lamont and co-workers (1990) showed that some peptides analogous to one of the T cell epitopes of ovalbumin could inhibit the T cell activation when co-administrated with the antigen in the same adjuvant emulsion. These observations implicated that the administration of peptide analogs as an MHC antagonist could be used in the treatment of MHC associated autoimmune diseases like EAE. This hypothesis was investigated by Wraith and co-workers (1990). They showed that two peptides were able to compete with the autoantigen for binding to the MHC molecule and could inhibit the induction in (H-2^u) mice.

II.1.e. Peptide applications in EAE rat

Due to a higher EAE incidence in rats as compared to mice, the rat is the preferably used animal species for disease induction with complete spinal cord homogenate or with individual myelin components. Various encephalitogenic epitopes have been described for EAE induction in the Lewis rat. Two major encephalitic T cell regions of guinea pig or rat MBP have been defined: sequences 72–89 and 87–99 (Offner et al., 1987, 1989). EAE induction in Lewis rat is even possible without the aid of adjuvant by immunizing with MBP fragments sequence 1–88, 45–88 and 63–88, although there are some disease incidence differences, depending on the sequence used (Levine et al., 1990). One of the major encephalitogenic T cell regions in Lewis rat, MBP sequence 87–99, is encephalitogenic in Buffalo rats as well (Jones et al., 1992). The LOU/M rat, which is resistant to EAE induction with guinea pig MBP in CFA, is susceptible to EAE induction by a synthetic peptide corresponding to sequence 110–129 of human MBP. However, EAE can only be induced with this peptide when pertussis toxin or cyclophosphamide is administered simultaneously (Hashim et al., 1991).

Synthetic peptides were used to reveal the major MBP and PLP epitopes recognized by B cells isolated from EAE Lewis rat. A panel of 27 monoclonal antibodies selected for their reactivity with native PLP were tested in a solid phase assay on their reactivity on synthetic peptides analogous to certain sequences of PLP to reveal their epitope specificity. Eleven monoclonal antibodies reacted with the C-terminal peptide. A pentapeptide sequence 272–276 from this region is sufficient enough for antibody binding. The terminal phenylalanine was found particularly important for antibody binding. Furthermore, deletion, modification and replacement of this residue markedly reduced the antibody-peptide interaction. Nine other monoclonal antibodies reacted with another B cell epitope of PLP, namely with sequence 209–217 (Yamamura et al., 1991). Seven monoclonal antibodies did not recognize any of the peptides used in this study and their specificity remains to be identified. Perhaps they react with one of the sequences which was not included in the study. On the other hand, the non-reactivity of the monoclonal antibodies could be explained by presuming that these monoclonal antibodies are directed against conformational epitopes.

The process of recovery in EAE in Lewis rat is characterized by an increasing diversity of T cell clones directed against nonencephalitogenic epitopes of MBP. It was suggested that these clones especially could play a role in the process of recovery. Of particular interest in this context is the non-encephalitogenic T cell clone C4 which is directed against MBP sequence 55–69. Passive transfer of clone C4 could induce protection against both active and passive EAE. T cell clone C4 was found to express Vβ 8.6 in its TCR. Residues 39–59 of this TCR Vβ 8.6 chain were found to be crossreactive with the corresponding sequence of the Vβ chain 8.2, which is known to induce protective anti-idiotypic T cells and antibodies. The protective characteristics of sequence 55–69 are believed to induce indirectly anti-TCR T cells, which are in turn responsible for the observed protection (Offner et al., 1991a). However, to reveal whether or not it is possible to treat an already established disease, synthetic peptides corresponding to the Vβ 8 chain of the TCR were administered to rats with clinical signs of EAE. The treatment with synthetic Vβ 8 TCR chain peptide sequence 39–59 reduced disease severity. In addition, the animals recovered faster as compared to untreated animals (Offner et al., 1991b). It was suggested that the natural occurring population of anti-Vβ 8 T and B cells will be boosted due to the injection of Vβ 8 TCR synthetic peptide. It also implicates that the TCR Vβ 39–59 sequence is a natural idiotype involved in EAE recovery. Human TCR peptides may be as effective in enhancing natural regulation of autoreactive T and B cells that share common V genes.

The immunomodulating capacity of pentapeptides were evaluated on the development of EAE in both a high susceptible strain of rat (Lewis) and a low susceptible strain of rat (Wistar). Two opioid pentapeptides, methionine-enkephalin and leucine-enkephalin, were used in the experiments of Jankovic and Maric (1987). After EAE was induced by immunization with guinea pig spinal cord homogenate, animals received daily either a high dose or a low dose of these two peptides. The administration of the high dose methionine-enkephalin significantly reduced the neurological signs and intensity of histological lesions in both brain and spinal cord as well. In addition, the loss of body weight which usually accompanies the EAE was also prevented with a high dose of methionine-enkephalin. In contrast, the low dose administration of the methionine-enkephalin potentiated the neurological and histopathological features of the disease, whereas the leucine-enkephalin administration had no effect at all in both rat strains independent of the dose administered.

Encephalitogenic peptides, which induce EAE in rat when emulsified in CFA, were targeted to B cells for presentation to T cells by administration of the peptides which were covalently coupled to anti-IgD antibodies. This form of antigen presentation favors the antibody production and downregulates the T cell response against the encephalitogenic epitope. Lewis rats injected with the encephalitogenic peptide were protected against EAE when injected with the peptide in its encephalitogenic form (emulsified in CFA) at the same time. Spleen cells from the protected animals, after a short period of peptide stimulation *in vitro*, were able to induce EAE when passively transferred to naive recipients (Day et al., 1992). These results suggest that a cell mediated response can be inhibited by targeting the specific antigen to B cells without inducing a T cell anergy. The targeting of autoantigen peptides to B cells is a potential approach for preventing autoimmune diseases which are essentially cell mediated.

Encephalitogenic T cells from EAE rat make use of common germline T cell receptor genes, despite the differences in epitope specificity and MHC restriction. Furthermore, in the Lewis rat there are some idiotypic determinants of the β chain of the T cell receptor which are conserved among encephalitogenic T cells (Burns et al., 1989; Chluba et al., 1989). A stronger restriction of TCR gene usage was found among encephalitogenic cells isolated from the spinal cord as compared to encephalitogenic cells isolated from lymph nodes (Offner et al., 1992). Identical observations were done concerning the T cell receptor gene usage of encephalitogenic T cells in the EAE mouse model (Acha-Orbea et al., 1988), although not consistently (Su and Sriram, 1992). This V gene bias raised the possibility of TCR peptide therapy using V gene peptides. Howell et al. (1989) used these findings to prevent the induction of EAE by immunizing Lewis rat with synthetic peptide analogous to an idiotypic determinant of the β chain VDJ region and the Jα chain of the T cell receptor which is conserved among the known encephalitogenic T cells prior to challenge of MBP. Various effects were observed dependent on the selected sequence for immunization. The effects vary from slight decreased severity and duration of disease to delayed onset of disease or even complete protection (Howell et al., 1989; Stevens et al., 1992). In contrast to these results, others found that immunization of Lewis rat with such peptides in complete Freunds adjuvant prior to induction of EAE with MBP leads to responses ranging from profound disease enhancement to lack of disease (Desquenne-Clark et al., 1991; Sun, 1992). Although the obtained results making use of T cell receptor vaccines are not consistent, the findings of Howell and colleagues (1989) suggest that the utility of TCR peptide vaccination for modulating the activity of autoreactive T cells represents a possible therapeutic approach for T cell mediated pathogenesis. However, vaccination with T cell receptor sequences is only successful if there is a restricted TCR gene usage for the disease inducing T cells. Although not all reports are consistent in this respect, it seems that there is no restricted TCR gene usage of disease inducing MBP reactive T cells among MS patients. In certain reports, it was described that there is perhaps a preferential usage of TCR genes in only one patient, but there is a different usage between patients. Vaccination with peptide mixtures which cover a variety of TCR sequences is still an option, although there is a great probability that one will be vaccinated against nondisease inducing T cells.

Tolerance against EAE can also be induced by intravenous injection of guinea pig MBP peptide sequence 68–86 coupled to splenocytes, in Lewis rat, which received MBP specific T cells 2 days prior to the splenocyte injection. In contrast, the usage of guinea pig MBP peptide sequence 68–84 had no inhibitory effect on disease development after passive transfer of MBP reactive T cells (Pope et al., 1992).

The protocols used for EAE induction, with either MBP or PLP peptides, differ in many aspects like antigen quantity, route of administration, number of injections, use of adjuvant, etc. These many differences between the several protocols do not permit one to point out a consensus protocol. However, some general remarks can be made that one should take into consideration when using peptides for EAE induction.

- The susceptibility to EAE is reduced significantly in old mice (Endoh et al., 1990).
- The induction of EAE in mice is critically dependent on the concentration of pertussis vaccine (Bernard and Carnegie, 1975).
- The quantity of peptides used for EAE induction varies from 100 to 400 μg per inoculum.
- The EAE incidence and severity of clinical and pathological signs is dependent on the sequence used for immunization and the strain of animals.
- Most encephalitogenic epitopes described are MHC class II restricted (Table 2).

II.2. Uveitis and experimental autoimmune uveoretinitis

Experimental autoimmune uveoretinitis (EAU) is a highly organ specific inflammatory disease of the uveal tract and retina of the eye. EAU is a well recognized model for human uveitis. Although the pathogenesis of human uveitis is still unclear, it was suggested that molecular mimicry could play a role in the induction of the disease (Shinohara et al., 1991). There are several indications that at least two retinal autoantigens, S-antigen (S-Ag) a 45 kD protein present in photoreceptor cells and interphotoreceptor retinoid-binding protein (IRBP), are involved in the pathogenesis of both human and animal uveitis. EAU can be induced in susceptible animals by active immunization with human or bovine S-Ag or IRBP or by passive transfer of bovine S-Ag activated T cells. Using synthetic peptides analogous to sequences of the putative autoantigens, for EAU induction purposes, several uveitogenic epitopes have been identified.

II.2.a. S-antigen

A synthetic peptide, corresponding to sequence 303–322 of bovine S-Ag, is capable of inducing EAU in Lewis rats, monkeys and Hartley guinea pigs. The induction of EAU with bovine S-Ag peptide sequence 303–322 in Lewis rat seems to be MHC restricted, since EAU induction with this peptide is not possible in rats with a non-Lewis genetic background (Hirose et al., 1990). Another uveitogenic region for Lewis rats of the retinal bovine S-antigen (sequence 339–352) was identified and characterized by Merryman et al. (1991). Oral administration of the bovine retinal S-Ag peptide sequence 303–322 to Lewis rat, prior to EAU induction with the same peptide in complete Freunds adjuvant, leads to EAU suppression. It was revealed that both clinical and histological characteristics of EAU were suppressed in a dose dependent manner. Since oral administration of a control peptide had no suppressive effect on EAU induction at all, it was concluded that suppression of the clinical and histological signs was peptide sequence 303–322 specific. Furthermore, spleen or lymph node T cells isolated from Lewis rat, to which the uveitogenic peptide was administered orally, could induce tolerance when transferred passively to naive recipients (Singh et al., 1992). This implicates that the oral administration of EAU inducing peptides initiates a peptide specific cell population which is suppressive for EAU. EAU induction in Hartley guinea pigs can also be performed following immuni-zation with synthetic peptides corresponding to bovine retinal S-antigen peptide sequences 281–302, 221–240 and 241–260, although there are differences in strength of the inflammatory response depending on the sequence used for EAU induction. Bovine retinal S-antigen peptide sequences 281–302 and 303–320 induce a moderate inflammatory response in the retina. In contrast, the immunization with retinal S-antigen peptide sequences 221–240 and 241–160 induce only a mild inflammatory response (Singh et al., 1989). To investigate the role of retinal S-antigen peptide sequence 302–322 in human uveitis, peripheral mononuclear cells of 21 patients with uveitis were tested for proliferation upon stimulation with bovine S-Ag peptide sequence 303–322. Mononuclear cells of six patients proliferated upon stimulation with intact S-Ag, whereas the mononuclear cells of four of these six patients also reacted specifically upon stimulation with the bovine S-Ag 303–322 peptide (Hirose et al., 1990). Because the amino acid sequences 303–322 of bovine and human S-Ag are identical, it was suggested that the S-Ag amino acid sequence 303–322 might be involved in the pathogenesis of human uveitis as well.

II.2.b. Interphotoreceptor retinoid-binding protein

EAU can be induced in Lewis rats by administration of high doses of bovine IRBP synthetic peptide sequence 1091–1115. Small amounts of the peptide induces a non pathological cellular immune response that can be measured by a lymphocyte proliferation assay. It has been hypothesized that the unusual functional dissociation between uveitogenic and immunogenic properties of the 23 amino acid peptide is attributed to different sites of the peptide. One site is stimulative to the lymphocytes, which can induce uveoretinitis, whereas the other site, the immunodominant site, stimulates lymphocytes to proliferate in culture. However, the relevance of this hypothesis remains to be determined (Kotake et al., 1991a).

EAU in Lewis rat can also be induced with IRBP peptide sequence 1169–1191. The shortest peptide within this sequence which is found to be immunologically active is IRBP peptide sequence 1182–1190. To determine the contribution of the individual amino acids of IRBP peptide sequence 1182–1190 on the capacity of the peptide to induce EAU and to stimulate IRBP or IRBP peptide sequence 1169–1191 sensitized lymphocytes *in vitro*, peptides analogous to bovine IRBP peptide sequence 1181–1191 were synthesized. In each analog peptide, only one residue was substituted with alanine. The analogs designated A(1184), A(1186) and A(1187) exhibited similar activities as IRBP peptide sequence 1181–1191. The other analog peptides showed reduced activities in various patterns. It was not possible to induce EAU with analog peptide A(1182), nor was it possible with this analog peptide to stimulate IRBP or IRBP peptide sequence 1169–191 sensitized lymphocytes *in vitro*. Peptide analogs A(1185), A(1190) and A(1183) appeared to be weakly uveitogenic, and relatively high concentrations of these peptides were needed for the *in vitro* stimulation of IRBP or IRBP peptide sequence 1169–1191 sensitized lymphocytes. Peptide analogs A(1188) and A(1189) appeared to be uveitogenic, but they were not able to stimulate sensitized lymphocytes *in vitro*. Also, these peptide analogs were able to inhibit the interaction between peptide 1181–1191 and IRBP peptide sequence 1181–1191 specific cells. It was suggested that residues 1188 and 1189 play a role in the interaction between the IRBP peptide and TCR, whereas residues 1182 and 1190 and perhaps residues 1183 and 1185 are important for the binding of the peptide to the MHC (Kotake et al., 1991b).

II.3. Experimental autoimmune neuritis

Experimental autoimmune neuritis (EAN), an animal model of human Guillain-Barré syndrome, is an autoimmune inflammatory demyelinating disease of the peripheral nervous system. The disease is monophasic, followed by complete recovery. The exacerbation is characterized by paralysis due to impaired conductivity of demyelinated axons in the peripheral nervous system. EAN can be induced in several animals by immunization with the myelin P_2 protein (Brostoff et al., 1972) and can be transferred to naive Lewis rat by P_2 reactive T cell clones derived from Lewis rat immunized with P_2 protein. More recently, peptides were used also for EAN induction (Olee et al., 1989). Immunization with P_2 peptide sequence 53–78 appeared to be successful in inducing severe clinical and histological EAN in Lewis rat (Rostami and Gregorian, 1991). Furthermore, T cell lines reactive with the P_2 synthetic peptide sequence 53–78, isolated from lymph nodes and spleen from animals immunized with P_2 synthetic peptide sequence 53–78, can induce severe clinical and histological signs of EAN upon adoptive transfer in naive syngeneic recipients.

II.4. Experimental autoimmune thyroiditis

Experimental autoimmune thyroiditis serves as an animal model for human autoimmune thyroid diseases, such as Grave's disease, Hashimoto's thyroiditis and idiopathic myxoedema. Thyroglobulin is known as a T cell dependent autoantigen in both human and experimental thyroid autoimmune diseases. The epitope specificity of two distinct, murine thyroglobulin autoreactive T cell hybridomas, CH9 and ADA2, was revealed by using of synthetic peptides. It was demonstrated that both clones recognize an epitope containing a thyroxine at residue position 2553 in human thyroglobulin. Using sets of overlapping 5-mer and 12-mer peptides around the thyroxine at position 2553, it was revealed that a 9-mer peptide, beginning at asparagine at position 2551, was the best recognized peptide by both CH9 and ADA2 T cell clones. The thyroxine residue 2553 appears to be a critical residue since replacement of residue 2553 with any of the natural occurring amino acids resulted in peptides which could not stimulate the thyroglobulin autoreactive T cells *in vitro* at all (Champion et al., 1991). Experimental autoimmune thyroiditis can be induced in CBA/J mice with a 40 amino acid peptide corresponding to a highly hydrophobic fragment of human thyroglobulin. After immunization of CBA/J mice with this peptide, the same lymphocytic infiltrates in the thyroid glands were generated as compared to the infiltrates generated with native porcine thyroglobulin. Antibodies crossreactive with both human and porcine thyroglobulin were detected in sera of porcine thyroglobulin immunized mice. In contrast, no crossreactive antibodies against human- or porcine thyroglobulin could be detected in mice immunized with the 40 amino acid thyroglobuline peptide. The autoantibody response against the peptide was significantly increased in peptide immunized mice as compared to the anti-peptide response in intact porcine thyroglobulin immunized mice (Texier et al., 1992).

Another putative thyroid autoantigen which may be involved in autoimmune thyroiditis is thyroid peroxidase, a glycoprotein enzyme which catalyzes the iodination of thyroglobulin and the coupling of iodinated tyrosines. Porcine thyroid peroxidase peptide sequence 774–788 is thyroiditogenetic in C57Bl/6 mice and can stimulate lymph node cells isolated from C57Bl/6 mice immunized with porcine thyroid peroxidase (Kotani et al., 1992).

Although the significance of autoantibodies in the pathogenesis of autoimmune thyroid diseases remains unclear, autoantibodies against human thyroid peroxidase have been detected in the sera from patients with autoimmune thyroid disease. Sera from 61 patients with autoimmune thyroid diseases were tested on a panel of peptides corresponding to human thyroid peroxidase. None of the peptides used was recognized by all of the sera tested. The antibody response to human thyroid peroxidase appears to be highly heterogeneous in human thyroid autoimmune diseases (Zanelli et al., 1992).

II.5. Experimental autoimmune oophoritis

Human autoimmune oophoritis is an ovarian autoimmune disease, resulting in premature damage of the ovary. Immunization of B6AF1 mice with a peptide corresponding to a 15 amino acid sequence of ZP3, the sperm-binding component of the *zona pellucida* that surrounds oocytes, induces an autoimmune oophoritis, which serves as the counterpart of the human disease. The experimental disease is considered to be T cell mediated. Oophoritis can be induced by adoptive transfer of $CD4^+$ T cells, derived from animals with peptide induced oophoritis, to naive recipients. Furthermore, truncated peptides, as large as 8 amino acids, were still oophoritogenic, but did not elicit an anti-peptide antibody response (Rhim et al., 1992). The ZP3 proteins of mice and human show an amino acid sequence homology of 67%. Therefore, it is possible that the peptides which were used for induction of experimental autoimmune oophoritis are the target epitopes in the human disease as well. The identification of the target antigen could facilitate the understanding of the pathogenesis of the human autoimmune disease.

II.6. Myasthenia gravis and experimental autoimmune myasthenia gravis

II.6.a. Myasthenia gravis

The pathogenesis of the autoimmune disease myasthenia gravis is characterized by an antibody mediated response against acetylcholine receptors (Achr) in the neuromuscular junction. The autoantibody response is T cell dependent. Many experiments were performed to reveal the epitope specificity of the T cells involved in the ultimate stimulation of B cells which produce anti-Achr antibodies. Making use of a panel of synthetic peptides corresponding to the complete sequence of the δ subunit of the acetylcholine receptor, the epitope specificity of Achr specific T cell lines, isolated from the blood of myasthenia gravis patients, was revealed. The peptides which were most frequently recognized by the T cells were situated within sequence 121–290 of the Achr δ subunit (Protti et al., 1991). In addition, it was shown that several Achr specific T cells of some myasthenia gravis patients could not be activated by the panel of peptides corresponding to the δ subunit Achr sequence at all. The cells which could be stimulated with synthetic peptides corresponding to the δ subunit of the Achr could not be activated by synthetic peptides corresponding to the complete sequence of α or γ subunits of the Achr. In contrast, when only $CD4^+$ T cells were studied *in vitro*, by the removal of the $CD8^+$ cells from total peripheral blood of myasthenia gravis patients, four epitopes were identified which were recognized most frequently, i.e., Achr α subunit sequences 48–67, 101–137, 293–337 and 308–437 (Manfredi et al., 1992).

II.6.b. Experimental autoimmune myasthenia gravis

Experimental autoimmune myasthenia gravis (EAMG) can be induced in C57BL/6 mice by injection of purified *Torpedo nicotinic* Achr. The response can be measured by T cell proliferation, antibody production and the development of muscle weakness, resembling the most prominent characteristic of human myasthenia gravis. The susceptibility to EAMG is influenced by MHC locus products. The bm12 mutation, due to several amino acid substitutions in the I-Ab molecule in C57BL/6 mice, which is originally highly susceptible to EAMG induction, yields a mutant strain (bm12) which is resistant to EAMG induction. The consequences of the bm12 mutation on the $CD4^+$ T cell response to the *Torpedo nicotinic* Achr α subunit were investigated by making use of synthetic peptides. Overlapping peptides, corresponding to the complete sequence of the *Torpedo nicotinic* Achr α subunit, were used to identify the $CD4^+$ T cell epitopes. $CD4^+$ T cells from C57BL/6 mice recognize *Torpedo nicotinic* Achr α subunit peptide sequences 150–169, 181–200 and 360–378. $CD4^+$ T cells from mice with the bm12 mutation in the I-A molecule did not react with any of the synthetic peptides used. In contrast, $CD4^+$ T cells isolated from mice with the I-A bm12 mutation, injected with the *Torpedo nicotinic* Achr α subunit peptides, responded strongly to Achr peptide sequence 360–378, to a lesser extent to Achr peptide residues 181–200 and not to Achr peptide sequence 150–169. However, only $CD4^+$ T cells sensitized against the Achr peptide sequence 181–200 showed crossreactivity with the Achr. It was tested whether or not the lack of response of the $CD4^+$ T cells to Achr α subunit sequence 150–169 and the low response to Achr α subunit sequence 181–200 was due to the inability of the antigen presenting cells to present the synthetic T cell epitopes to T helper cells. Both antigen presenting cells with the bm12 mutation and the nonmutated C57BL/6 antigen presenting cells were used to present the T cell epitope peptides to specifically sensitized $CD4^+$ T cells from C57BL/6 mice. Antigen presenting cells with and without the bm12 mutation were able to present all T cell epitopes used to the sensitized $CD4^+$ T cells. It was concluded that the induced resistance to EAMG may be due to the change in the T cell receptor epitope repertoire of the $CD4^+$ Th cells and not to an inability of the antigen presenting cells, carrying the I-A molecule with the bm12 mutation to present the T cell epitopes to the T cells (Bellone et al., 1991, 1992).

Similar experiments were performed by Infante et al. (1991). The fine specificity of T cell clones isolated from the bm12 strain and the C57BL/6 strain of mice was determined by making use of synthetic peptides. Cells from the C57BL/6 strain reacted predominantly with Achr α subunit peptide sequence 146–162. Cross-reactivity was observed with Achr α subunit peptide sequence 111–126 by comparing the sequences of the two peptides and measuring the T cell responses to truncated peptides, the major T cell epitope was revealed to be Achr α subunit peptide sequence 148–152. Immunization of C57BL/6 mice with both peptides separately revealed T cell responses to both peptides. In contrast, immunization with the two peptides separately in the bm12 strain revealed only T cell responses to peptide sequence 111–126. Because the bm12 strain failed to develop a T cell response to the Achr α subunit peptide sequence 148–152, it was postulated that the 148–152 epitope is the disease related epitope.

II.7. Rheumatoid arthritis

Rheumatoid arthritis is characterized by the presence of rheumatoid factors, (e.g., circulating autoantibodies) directed against a variety of autoantigens. Synthetic peptides were used to reveal the fine specificity of human rheumatoid factors reactive with histones. The histone sequences most frequently recognized by rheumatoid factors were histone 1 peptide sequences 1–16 and 204–218, histone 2A peptide sequence 1–20 and 65–85 and histone 3 peptide sequence 1–21 (Tuaillon et al., 1992a). Other antibody epitopes have been identified by making use of peptides corresponding to the CB-11 sequence of bovine type II collagen. Sera from collagen II immunized rats were screened for reactivity with the peptides. Twenty-one different antibody binding sites were recognized. Some of the epitopes which were recognized by the rat antibodies are identical to sequences from human type II collagen (Worthington et al., 1991).

Synthetic peptides derived from the 65 kD mycobacterial protein were used to inhibit the antigen presentation in rheumatoid arthritis by blocking the peptide/MHC binding *in vitro*. Cross competition was observed between 65 kD mycobacterial protein peptides sequences 65–85 and 412–426 when antigen presenting cells were cultured in the presence of stimulator peptides and T cell clones. This peptide approach permits the identification of MHC specific peptide agonists (Mendez-Samperio and Jiminez-Zamudio, 1991).

II.8. Chronic immune thrombocytopenic purpura

Immune thrombocytopenic purpura is an autoimmune disorder due to autoantibodies directed to platelet glycoprotein GPIIIa. Binding of these anti-platelet antibodies results ultimately in the lysis of the platelets. To reveal the antigenic epitopes of these antibodies, plasma of immune thrombocytopenic purpura patients was tested for reactivity with nine synthetic peptides corresponding to different regions of glycoprotein IIIa. Antibodies in 5 out of 13 plasmas tested bound with peptide sequence 721–744 and with peptide sequence 742–762. Both peptides are part of the carboxy terminal site of the protein. The carboxy terminal site of the peptide is presumed to be a cytoplasmatic domain. It was concluded that the two mentioned sequences of the GPIIIa protein are important sequences in the development of anti-platelet autoantibodies. However, it is not known whether these antibodies play a role in the induction of the disease or if they occur as a secondary phenomenon during the course of the platelet lysis (Fujisawa et al., 1991). Besides the molecular mimicry of bacterial epitopes, there are some indications that molecular mimicry of viral epitopes contributes to the induction of autoimmunity in rheumatoid arthritis patients as well. A synthetic peptide corresponding to the glycine/alanine repeat sequence of Epstein-Barr virus nuclear antigen-1 is recognized by antibodies in the sera of patients with rheumatoid arthritis. It was shown that the antibody titers in the sera of patients with rheumatoid arthritis were raised fourfold as compared to the antibody titers of healthy controls. Furthermore, it was revealed that purified anti-peptide antibodies crossreacted with human epidermal keratin, denaturated collagen type II and actin (Baboonian et al., 1991).

III. CONCLUSIVE REMARKS

The availability and use of peptides, synthesized, expressed by clones or generated enzymematically, in autoimmune research has directed progress in understanding the underlying mechanisms of both cellular and humoral immune responses in autoimmune diseases. By making use of peptides, many pathogenic epitopes of autoantigen reactive cells or antigen epitopes of pathogenic microorganisms, which causes a crossreactive autoimmune response upon infection, have been revealed. The identification of disease inducing epitopes on autoantigens may lead to therapeutic approaches to establish tolerance or anergy to these epitopes. Several experiments have shown that peptides corresponding to TCR sequences can be used as vaccines in experimental autoimmune disease. The use of peptides may even have potential clinical applicability for treatment of human autoimmune diseases. However, the use of human TCR peptides can only be effective in enhancing natural regulation of autoimmune T cells if these T cells share the use of the same TCR genes. The applicability of TCR peptides for interference in immune responses of human autoimmune diseases is now under study.

In addition, the use of peptides makes it possible to induce a variety of experimental autoimmune diseases which mimic the human counterpart. The simplicity of the peptide induced models makes it possible to study the pathogenic processes rather easily. However, although the peptide induced models are more strictly defined and simplified as compared to the protein and tissue homogenate induced experimental diseases, the peptide induced models are only a small reflection of the human disease. In other words, the peptide approach will mimic only a small part of the immune responses which normally occur in autoimmune diseases. As a consequence, many other processes, perhaps pathogenic relevant processes, and the influence of these processes on each other will not be mimicked by the administration of just one single epitope. The administration of peptide cocktails is a possible approach to induce experimental diseases which have more resemblance with the human disease.

Only linear peptide sequences can be synthesized and as a consequence only these peptides can be used for induction of autoimmune diseases or other applications. The conformational epitopes, epitopes which are recognized due to their specific three-dimensional structure, cannot be synthesized and subsequently cannot be studied. If more advanced peptide-synthesize or modification methods become available, the conformational synthetic peptide epitopes will be used for autoimmune diseases research as well.

ABBREVIATIONS

Achr	Acetylcholine receptor
ADDC	Antibody dependent cellular cytotoxicity
AFC	Antibody forming cell
CNS	Central nervous system
CFA	Complete Freunds adjuvant
CR	Chronic relapsing-remitting
CSF	Cerebrospinal fluid
EAE	Experimental allergic encephalomyelitis
EAN	Experimental autoimmune neuritis
EAU	Experimental autoimmune uveitis
EAMG	Experimental autoimmune myasthenia gravis
HLA	Human lymphocyte antigen
HSP	Heat shock protein
IRBP	Interphotoreceptor retinoid-binding protein
MBP	Myelin basic protein
MHC	Major histocompatibility complex
MS	Multiple sclerosis
PLP	Proteolipid protein
S-Ag	S antigen
TCR	T cell receptor
Tc	T cytotoxic
Th	T helper

IV. REFERENCES

Acha-Orbea H, Mitchell DJ, Timmermann et al. Limited heterogeneity of T-cell receptors from lymphocytes mediating autoimmune encephalomyelitis allows specific immune intervention. Cell 54:263, 1988

Alvord EC. Disseminated encephalomyelitis: Its variation in form and their relationship to other diseases of the nervous system. In: Vinken PJ, Bruyn GW and Klawans HL.Eds. Handbook of

Clinical Neurology. Vol 3, Rev Ser 3, Ed. Koetsier JC: Demyelinating Diseases. Elsevier, Amsterdam 467, 1985

Armitage RJ, Fanslow WC, Strockbine L et al. Molecular characterization of murine ligand for CD40. Nature 357:80, 1992

Ashizawa T, Ruan KH, Jinnai K and Atassi MZ. Profile of the regions on the alpha-chain of human acetylcholine receptor recognized by autoantibodies in myasthenia gravis. Mol Immunol 29:1507, 1992

Baboonian C, Venables PJ, Williams DG et al. Cross reaction of antibodies to a glycine/alanine repeat sequence of Epstein-Barr virus nuclear antigen-1 with collagen, cytokeratin, and actin. Ann Rheum Dis 50:722, 1991

Baig S, Olsson T, Yu-Ping J et al. Multiple Sclerosis: Cells secreting antibodies against myelin-associated glycoprotein are present in cerebrospinal fluid. Scand J Immunol 33:73, 1991

Banga JP, Barnett PS, Ewins DL et al. Mapping of autoantigenic epitopes on recombinant thyroid peroxidase fragments using the polymerase chain reaction. Autoimmunity 6:257, 1990

Bellone M, Ostlie N, Lei SJ et al. The I-Abm12 mutation, which confers resistance to experimental myasthenia gravis, drastically affects the epitope repertoire of murine $CD4^+$ cells sensitized to nicotinic acetylcholine receptor. J Immunol 147:1484, 1991

Bellone M, Ostlie N, Lei S et al. T helper function of $CD4^+$ cells specific for defined epitopes on the acetylcholine receptor in congenic mouse strains. J Autoimmun 5:27, 1992

Bernard CCA and Carnegie PR. Experimental autoimmune encephalomyelitis in mice: Immunologic response to mouse spinal cord and myelin basic proteins. J Immunol 114:1537, 1975

Bourdette DN, Vandenbark AA, Hashim G et al. T cell lines selected with synthetic peptides are highly encephalitogenic in SJL/J mice. J Neuroimmunol 22:255, 1989

Brostoff S, Burnett P, Lampert P et al. Isolation and characterization of a protein from sciatic nerve myelin responsible for experimental allergic neuritis. Nature 235:210, 1972

Burns FR, Li X, Shen N et al. Both rat and mouse T-cell receptors specific for the encephalitogenic determinant of myelin basic protein use similar $V\alpha$ and $V\beta$ chain genes even though the major histocompatibility complex and encephalitogenic determinants being recognized are different. J Exp Med 169:27, 1989

Burns J, Littlefield K, Gomez C et al. Assessment of antigenic determinants for the human T-cell response against myelin basic protein using overlapping synthetic peptides. J Neuroimmunol 31:105, 1991

Champion BR, Page KR, Parish N et al. Identification of a thyroxine-containing self-epitope of thyroglobulin which triggers thyroid autoreactive T-cells. J Exp Med 174:363, 1991

Chluba J, Steeg C, Becker A et al. T-cell receptor β chain usage in myelin basic protein specific rat T lymphocytes. Eur J Immunol 19:279, 1989

Chao NJ, Timmerman L, McDevitt HO et al. Molecular characterization of MHC class II antigens (b domain) in the diabetes-prone and -resistant rat. Immunogenetics 29:231, 1989

Chou CHJ, Tourtelotte WW and Kibler RF. Failure to detect antibodies to myelin basic protein or peptide fragments of myelin basic protein in CSF of patients with MS. Neurology 33:24, 1983

Cross AH, Hashim GA and Raine CR. Adoptive transfer of experimental allergic encephalomyelitis and localization of the encephalitogenic epitope in the SWR mouse. J Neuroimmunol 31:59, 1991

Dal Canto MC and Lipton HL. Primarily demyelination in Theiler's virus infection. An ultrastructural study. Lab Invest 33:626, 1975

Day MJ, Tse AG, Puklavec M et al. Targeting autoantigen to B-cells prevents the induction of a cell mediated autoimmune disease in rats. J Exp Med 175:655, 1992

Delves PJ and Roitt IM. Idiotypic determinants on human thyroglobulin autoantibodies derived from the serum of Hashimoto patients and EB virus transformed cell lines. Clin Exp Immunol 57:33, 1984

De Souza SJ, Sabbaga J, D'Amico E et al. Anti-platelet autoantibodies from patient recognize an epitope in GpIIb/IIa deduced by complementary hydropathy. Immunology 75:17, 1992

Desquenne-Clark L, Esch TR, Otvos L et al. T-cell receptor peptide immunization leads to enhanced and chronic experimental allergic encephalomyelitis. Proc Natl Acad Sci USA 88:7219, 1991

Endo T, Stewart SS, Kundu SK et al. Antibodies to glycosphingolipids in patients with Multiple Sclerosis. Ann NY Acad Sci 436:213, 1984

Endoh M, Rapoport SI and Tabira T. Studies of experimental allergic encephalomyelitis in old mice. J Neuroimmunol 29:21, 1990

Felgenhauer K and Reiber H. The diagnostic significance of antibody specificity indices in multiple sclerosis and herpes virus induced diseases of the central nervous system. Clin Investig 70:28, 1992

Freedman MS and Antel JP. Immunoregulatory circuits in MS: Is there a "short"? Ann Neurol 24:183, 1988

Fritz RB, Skeen MJ, Chou CHJ et al. Major histocompatibility complex-linked control of the murine immune response to myelin basic protein. J Immunol 134:2328, 1985

Fuji Y and Lindstrom J. Specificity of the T cell immune response to acetylcholine receptor in experimental autoimmune Myasthenia Gravis response to subunits and synthetic peptides. J Immunol 140:1830, 1988

Fujisawa K, O'Toole THE, Tani P et al. Autoantibodies to the presumptive cytoplasmic domain of platelet glycoprotein IIIa in patients with chronic immune thrombocytopenic purpura. Blood 77:2207, 1991

Gao XJ, Ball EJ, Dombrausky L et al. Class II human leucocyte antigen genes and T cell receptor polymorphisms in patients with rheumatoid arthritis. Am J Med 85:14, 1988

Gerritse K, Deen C, Fasbender M et al. The involvement of specific anti myelin basic protein antibody forming cells in multiple sclerosis immunopathology. J Neuroimmunol 49:153, 1994

Gminski J, Drozdz M, Ulfig-Maslanka R et al. Evaluation of elastin metabolism in children from families with high risk of artherosclerosis. Artherosclerosis 91:185, 1991

Goodnow CC. Transgenic mice and analysis of B-cell tolerance. Ann Rev Immunol 10:489, 1992

Gorcy MK, Wroblewska Z, Pleasure D et al. CSF antibodies to myelin basic protein and oligodendrocytes in multiple sclerosis and other neurological diseases. Acta Neurol Scand 7:338, 1983

De Groot CJA, Van de Pal RHM, Lopes-Cardozo M et al. Interferon-gamma induced Ia antigen expression on cultured neuroglial cells and brain macrophages from rat spinal cord and cerebrum. Intern J Neurosci 56:53, 1991

Hartley SB, Crosbie J, Brink R et al. Elimination from peripheral lymphoid tissues of self-reactive B lymphocytes recognizing membrane bound antigens. Nature 353:765, 1991

Hashim G, Vandenbark AA, Gold DP et al. T-cell lines specific for an immunodominant epitope of human basic protein define an encephalitogenic determinant for experimental autoimmune encephalomyelitis-resistant LOU/M rats. J Immunol 146:515, 1991

Henneberg A, Mayle DM and Kornhuber HH. Antibodies to brain tissue in sera of patients with chronic progressive multiple sclerosis. J Neuroimmunol 34:223, 1991

Henry M, Zanelli E, Peiechaczyk M, Pau B et al. A major thyroglobulin epitope defined with monoclonal antibodies is mainly recognized by human autoantibodies. Eur J Immunol 22:315, 1992

Hirose S, Donoso LA, Shinohara T et al. Lymphocyte responses to peptide M and to retinal S-antigen in uveitis patients. Jpn J Opthalmol 34:298, 1990

Hirsch HE and Parks ME. Serological reactions against glycolipid-sensitized liposomes in multiple sclerosis. Nature 264:785, 1976

Hitman GA. The major histocompatibility complex and insulin dependent (type I) diabetes. Autoimmunity 4:119, 1989

Hoeksema R, Hwan SB, Rothova A et al. Serum antibody response to human and bovine IRBP in uveitis. Curr Eye Res 9:1177, 1990

Howell MD, Winter ST, Olee T et al. Vaccination against experimental allergic encephalomyelitis with T-cell receptor peptides. Science 246:668, 1989

Infante AJ, Thompson PA, Krolick KA et al. Determinant selection in murine experimental autoimmune myasthenia gravis. Effect of the bm12 mutation on T-cell recognition of acetylcholine receptor epitopes. J Immunol 146:2977, 1991

Jankovic BD and Maric D. Enkephalins and autoimmunity: Differential effect of methionine-enkephalin on experimental allergic encephalomyelitis in Wistar and Lewis rat. J Neurosci Res 18:88, 1987

Jansson L, Olsson T, Höjeberg B et al. Chronic experimental autoimmune encephalomyelitis induced by the 89–101 myelin basic protein peptide in B!) RIII (H-2) mice. Eur J Immunol 216:93, 1991

Jingwu Z, Chou CH, Hashim G et al. Preferential peptide specificity and HLA restriction of myelin basic protein-specific T cell clones derived from MS patients. Cell Immunol 129:189, 1990

Jingwu Z, Chin Y, Henderikx P et al. Antibodies to myelin basic protein and measles virus in multiple sclerosis: Precursor frequency analysis of the antibody producing B-cells. Autoimmunity 11:27, 1991

Jones RE, Bourdette D, Offner H et al. The synthetic peptide 87–99 peptide of myelin basic protein is encephalitogenic in Buffalo rats. J Neuroimmunol 37:203, 1992

Karkhanis YD, Carlo DJ, Brostoff SW et al. Allergic encephalomyelitis. Isolation of an encephalitogenic peptide active in the monkey. J Biol Chem 5:1718, 1975

Kawakami Y, Fisfalen ME and De Groot LJ. Proliferative responses of peripheral blood mononuclear cells from patients with autoimmune thyroid diseases to synthetic peptide epitopes of human thyroid peroxidase. Autoimmunity 13:17, 1992

Khalili-Shirazi A, Highes RA, Brostoff SW et al. T cell responses to myelin proteins in Guillain-Barre syndrome. J Neurol Sci 111:200, 1992

Kono DH, Urban JL, Horvath SJ et al. Two minor determinants of myelin basic protein induce experimental allergic encephalomyelitis in SJL/J mice. J Exp Med 168:213, 1988

Kotake S, Redmond TM, Wiggert B et al. Unusual immunologic properties of the uveitogenic interphotoreceptor retinoid-binding protein-derived peptide R23. Invest Opthalmol Vis Sci 32:2058, 1991a

Kotake S, Smet de MD, Wiggert B et al. Analysis of the pivotal residues of the immunodominant and highly uveitogenic determinant of interphotoreceptor retinoid-binding protein. J Immunol 146:2995, 1991b

Kotani T, Umeki K, Yagihashi S et al. Identification of thyroiditogenic epitope on porcine thyroid peroxidase for C57BL/6 mice. J Immunol 148:2084, 1992

Kraus W, Dale JB and Beachey EH. Identification of an epitope of type 1 streptococcal M protein that is shared with an 43-kDa protein of human myocardium and renal glomeruli. J Immunol 145:4089, 1990

Kuchroo VK, Sobel RA, Yamamura T et al. Induction experimental allergic encephalomyelitis by myelin proteolipid-protein specific T-cell clones and synthetic peptides. Pathobiology 59:305, 1991

Lamont AG, Sette A and Grey HM. Inhibition of antigen presentation *in vitro* and *in vivo* by MHC antagonist peptides. Int Rev Immunol 6:49, 1990

Lehman TJ, Hanson V, Zvaifler N et al. Antibodies to nonhistone nuclear antigens and antilymphocyte antibodies among children and adults with systemic lupus erythematosus and their relatives. J Rheumatol 11:644, 1984

Lenzi M, Johnson PJ, McFarlane IG et al. Antibodies to hepatitis C virus in autoimmune liver disease: Evidence for geographical heterogeneity. Lancet 338:277, 1991

Levine S, Saltzman A and Deibler GE. Peptides of myelin basic protein are encephalitogenic in rats without the aid of emulsions. PSEBM 195:325, 1990

Limas CJ, Goldenberg IF and Limas C. Assessment of immune modulation of beta-adrenergic pathways in human dilated cardiomyopathy: Influence of methodologic factors. Am Heart J 123:96, 1992

Link H, Baig S, Olsson O et al. Persistent anti-myelin basic protein IgG antibody response in multiple sclerosis cerebrospinal fluid. J Neuroimmunol 28:237, 1990

Manfredi AA, Protti MP, Wu XD et al. $CD4^+$ T-epitope repertoire on the human acetylcholine receptor alpha subunit in severe myasthenia gravis: A study with synthetic peptides. Neurology 42:1092, 1992

Martin R, Jaraquemada D, Flerlage M et al. Fine specificity and HLA restriction of myelin basic protein-specific cytotoxic T cell lines from multiple sclerosis patients and healthy individuals. J Immunol 145:540, 1990

McNeilage LJ, Umapathysivam K, Macmillan E et al. Definition of a discontinuous immunodominant epitope at the NH_2 terminus of the La/SS-B ribonucleoprotein autoantigen. J Clin Invest 89:1652, 1992

McRae BL, Kennedy MK, Tan LJ et al. Induction of active and adoptive relapsing experimental autoimmune encephalomyelitis (EAE) using an encephalitogenic epitope of proteolipid protein. J Neuroimmunology 38:229, 1992

Mendez-Samperio P and Jiminez-Zamudio L. Peptide competition at the level of MHC binding sites using T cell clones from a rheumatoid arthritis patient. J Autoimmun 4:795, 1991

Merryman CF, Donoso LA, Zhang XM et al. Characterization of a new, potent, immunopathogenic epitope in S-antigen that elicits T-cells expressing V beta 8 and V alpha 2-like genes. J Immunol 146:75, 1991

Mokhtarian F, McFarlin DE and Raine CS. Adoptive transfer of myelin basic protein-sensitized T cells produces chronic relapsing demyelinating disease in mice. Nature 309:356, 1984

Neilson EG, Sun MJ, Kelly CJ et al. Molecular characterization of a major nephritogenic domain in the autoantigen of anti-tubular basement membrane disease. Pro Natl Acad Sci USA 88:2006, 1991

Noelle RJ, Roy M, Shepherd DM et al. A 39-kDa protein on activated helper T cells binds CD40 and transduces the signal for cognate activation of B cells. Proc Natl Acad Sci USA 89:6550, 1992

O'Brien RM, Cram DS, Coppel RL et al. T-cell epitopes on the 70 kD protein of the (U1) RNP complex in autoimmune rheumatological disorders. J Autoimmun 3:747, 1990

Offner H, Hashim GA and Vandenbark AA. Response of rat encephalitogenic T lymphocytes lines to a synthetic peptides of myelin basic protein. J Neurosci Res 17:344, 1987

Offner H, Hashim GA, Celnik B et al. T-cell determinants of myelin basic protein include a unique encephalitogenic I-E restricted epitope for Lewis rats. J Exp Med 170:335, 1989

Offner H, Vainiene M, Gold DP et al. Protection against experimental encephalomyelitis. Idiotypic autoregulation induced by a nonencephalitogenic T-cell clone expressing a cross-reactive T-cell receptor V gene. J Immunol 146:4165, 1991a

Offner H, Hashim GA and Vandenbark AA. T-cell receptor therapy triggers autoregulation of experimental encephalomyelitis. Science 251:430, 1991b

Offner H, Vainiene M, Gold DP et al. Characterization of the immune response to a secondary encephalitogenic epitope of basic protein in Lewis rats. I. T-cell receptor peptide regulation of T-cell clones expressing cross-reactive Vβ genes. J Immunol 148:1706, 1992

Olee T, Powers JM and Brostoff SW. A T-cell epitope for experimental allergic neuritis. J Neuroimmunol 19:167, 1988

Olee T, Weise M, Powers M et al. A T-cell epitope for experimental allergic neuritis is an amphipatic α helical structure. J Neuroimmunol 21:235, 1989

Ota K, Matsui M, Milford EL et al. T-cell recognition of an immuno-dominant myelin basic protein epitope in multiple sclerosis. Nature 346:183, 1990

Paterson PY. Experimental autoimmune (allergic) encephalomyelitis: Induction, pathogenesis, and suppression. In: Miescher PA and Muller-Eberhard HJ (Eds.), Textbook of Immunopathology. Vol. 1, Gruene and Straton, New York, 179, 1976

Paulie S, Rosen A, Ehlin-Hendriksson B et al. The human B lymphocyte and carcinoma antigen, CDw40, is a phosphoroprotein involved in growth signal transduction. J Immunol 142:590, 1989

Pette M, Fujita K, Wilkinson D et al. Myelin autoreactivity in multiple sclerosis: Recognition of myelin basic protein in the context of HLA-DR2 products by T lymphocytes of multiple sclerosis patients and healthy controls. Proc Natl Acad Sci USA 87:7968, 1990

Pettinelli CB, Fritz RB, Chou CHJ et al. Encephalitogenic activity of guinea pig myelin basic protein in the SJL mouse. J Immunol 129:1209, 1982

Polman CH, Matthaei I, de Groot CJA et al. Low-dose cyclosporine A induces relapsing remitting experimental allergic encephalomyelitis in the Lewis rat. J Neuroimmunol 17:209, 1988

Pope L, Paterson PY and Miller SD. Antigen specific inhibition of the adoptive transfer of experimental autoimmune encephalomyelitis in Lewis rats. J Neuroimmunol 37:177, 1992

Protti MP, Manfredi AA, Wu XD et al. Myasthenia gravis. T epitopes on the delta subunit of human muscle acetylcholine receptor. J Immunol 146:2253, 1991

Quayle AJ, Wilson KB, Li SG et al. Peptide recognition, T cell receptor usage and HLA restriction elements of human heat shock protein (hsp) 60 and mycobacterial 65-kDa hsp-reactive T cell clones from rheumatoid fluid. Eur J Immunol 22:1315, 1992

Raine CS. Experimental allergic encephalomyelitis and experimental allergic neuritis. In: Vinken PJ, Bruyn GW and Klawans HL. Eds. Handbook of Clinical Neurology Vol 3, Rev. Ser. 3, Ed. Koetsier JC: Demyelinating diseases. Elsevier, Amsterdam, 429, 1985

Rhim SH, Millar SE, Robey F et al. Autoimmune disease of the ovary induced by a ZP3 peptide from the mouse zona pellucida. J Clin Invest 89:28, 1992

Rostami AM and Gregorian SK. Peptide 53–78 of myelin P_2 protein is a T-cell epitope for the induction of experimental autoimmune neuritis. Cell Immunol 132:433, 1991

Shinohara T, Singh VK, Yamaki K et al. Molecular mimicry may play a role in autoimmune uveitis. Prog Clin Biol Res 362:163, 1991

Singh VK, Donoso LA, Yamaki K et al. Uveitogenic sites in bovine S-antigen. Autoimmunity 3:177, 1989

Singh VK, Kalra HK, Yamaki K et al. Suppression of experimental autoimmune uveitis in rats by the oral administration of the uveitopathogenic S-antigen fragment or a cross-reactive homologous peptide. Cell Immunol 139:81, 1992

Sobel RA, Tuohy VK, Lu Z et al. Acute experimental allergic encephalomyelitis in SJL/J mice induced by a synthetic peptide of myelin proteolipid protein. J Neuropathol Exp Neurol 49:468, 1990

Sobel RA, Tuohy VK and Lees MB. Parental MHC molecule haplotype expression in SJL/J x SWR)F1 mice with acute experimental allergic encephalomyelitis induced with two different synthetic peptides of myelin proteolipid protein. J Immunol 146:543, 1991

Stevens DB, Karpus WJ, Gould KE et al. Studies of V beta 8 T cell receptor peptide treatment in experimental autoimmune encephalomyelitis. J Neuroimmunol 37:123, 1992

Su XM and Sriram S. Analysis of TCR Vβ gene usage and encephalitogenicity of myelin basic protein peptide p91–103 reactive T-cell clones in SJL mice: Lack of evidence for V gene hypothesis. Cell Immunol 141:485, 1992

Suckling AJ, Pathak S, Jagelman S et al. Virus associated demyelination. A model using avirulent Semliki Forest virus infection of mice. J Neurologic Sci 36:147, 1978

Sun JB, Olsson T, Wang WZ et al. Autoreactive T and B cells responding to myelin proteolipid protein in multiple sclerosis and controls. Eur J Immunol 21:1461, 1991

Sun D. Synthetic peptides representing sequence 39 to 59 of rat V beta 8 TCR fail to elicit regulatory T cells reactive with V beta 8 TCR on rat encephalitogenic T cells. Cell Immunol 141:200, 1992

Tan LJ, Kennedy MK and Miller SD. Regulation effector stages of experimental autoimmune encephalomyelitis via neuroantigen-specific tolerance induction. J Immunol 148:2748, 1992

Texier B, Bedin C, Tang H et al. Characterization and sequencing of a 40-aminoacid peptide from human thyroglobulin inducing experimental autoimmune thyroiditis. J Immunol 148:3405, 1992

Traugott U, Reinherz EL and Raine CS. Multiple Sclerosis. Distribution of T-cells, T-cell subsets and Ia-positive macrophages in lesions of different ages. J Neuroimmunol 4:201, 1983

Trotter JL, Hickey WF, Veen van der RC et al. Peripheral blood mononuclear cells from multiple sclerosis patients recognize myelin proteolipid protein and selected peptides. J Neuroimmunology 33:55, 1991

Tsao BP, Ebling FM, Roman C et al. Structural characteristics of the variable regions of immunoglobulin genes encoding a pathogenic autoantibody in murine lupus. J Clin Invest 85:530, 1990

Tuaillon N, Martin T, Knapp AM et al. Double reactivity of monoclonal and polyclonal rheumatoid factors for IgG and histones: Mapping of binding sites by means of histone synthetic peptides and anti-Id antibodies. J Autoimmun 5:1, 1992a

Tuaillon N, Andre C, Briand JP et al. A lipoyl synthetic octadecapeptide of dihydrolipoamide acetyltransferase specifically recognized by anti-M2 autoantibodies in primary biliary cirrhosis. J Immunol 148:445, 1992b

Tuohy VK, Lu Z, Sobel RA et al. A synthetic peptide from myelin proteolipid protein induces experimental allergic encephalomyelitis. J Immunol 141:1226, 1988

Tuohy VK, Lu Z, Sobel RA et al. Identification of an encephalitogenic determinant of myelin proteolipid protein for SJL/J mice. J Immunol 142:1523, 1989

Veen van der RC, Sobel RA and Lees MB. Chronic experimental allergic encephalomyelitis and antibody responses in rabbits immunized with bovine proteolipid apoprotein. J Neuroimmunol 11:321, 1986

Veen van der RC, Trotter JL, Clark HB et al. The adoptive transfer of chronic relapsing experimental allergic encephalomyelitis with lymph node cells sensitized to myelin proteolipid protein. J Neuroimmunol 21:183, 1989

Voorhis van WC, Schlekewy L and Trong HL. Molecular mimicry by Trypanosoma cruzi: The F1-160 epitope that mimics mammalian nerve can be mapped to a 12- aminoacid peptide. Proc Natl Acad Sci USA 88:5993, 1991

Walsh MJ and Tourtelotte WW. In: Hallpike JF, Adams CWM and Tourtelotte WW Eds. Multiple Sclerosis. : Chapman & Hall, London, 275, 1983

Warren KG and Catz J. Diagnostic value of cerebrospinal fluid anti-myelin basic protein in patients with multiple sclerosis. Ann Neurol 20:20, 1986

Weiner LP. Pathogenesis of demyelination induced by a mouse hepatitis virus (JHM virus). Arch Neurol 28:298, 1973

Weller PO. Pathology of Multiple sclerosis. In: Mathews WB, Acheson ED, Batchelor JR et al. Eds. McAlpine's Multiple Sclerosis, Churchill Livingstone, Edinburgh, 301, 1985

Whitham RH, Jones RE, Hashim GA et al. Location of new encephalitogenic epitope (residues 43 to 64) in proteolipid protein that induces relapsing experimental autoimmune encephalomyelitis in PL/J and (SJL x PL) F1 mice. J Immunol 147:3803, 1991a

Whitham RH, Bourdette DN, Hashim GA et al. Lymphocytes from SJL/J mice immunized with spinal cord respond selectively to a peptide of proteolipid protein and transfer relapsing demyelinating experimental autoimmune encephalomyelitis. J Immunol 146:101, 1991b

Worthington J, Brass A and Morgan K. Identification of antibody epitopes within the CB-11 peptide of type II colagen. I: Detection of antibody binding sites by epitope scanning. Autoimmunity 10:201, 1991

Wraith DC, Smilek DE, Mitchell DJ et al. T-cell recognition in experimental autoimmune encephalomyelitis: Prospects for immune intervention with synthetic peptides. Int Rev Immunol 6:37, 1990

Yamamura T, Namikawa T, Endo M et al. Experimental allergic encephalomyelitis induced by proteolipid apoprotein in Lewis rat. J Neuroimmunol 12:143, 1986

Yamamura T, Konola JT, Wekerle H et al. Monoclonal antibodies against myelin proteolipid protein: Identification and characterization of two major determinants. J Neurochem 57:1671, 1991

Yoshimura T, Kunishita T, Sakai K et al. Chronic experimental allergic encephalomyelitis in guinea pigs induced by proteolipid protein. J Neurol Sci 69:47, 1985

Zamvill SS, Mitchell DJ, More AC et al. T-cell epitope of the autoantigen myelin basic protein that includes encephalomyelitis. Nature 324:258, 1986

Zamvill SS, Mitchell DJ, Powell MB et al. Multiple discrete encephalitogenic epitopes of the autoantigen myelin basic protein include a determinant for I-E class II restricted T-cells. J Exp Med 158:1181, 1988

Zamvill SS and Steinmann L. The T lymphocyte in experimental allergic encephalomyelitis. Ann Rev Immunol 8:579, 1990

Zanelli E, Henry M and Malthiery Y. Use of recombinant epitopes to study the heterogeneous nature of the autoantibodies against thyroid peroxidase in autoimmune thyroid disease. Clin Exp Immunol 87:80, 1992

Zanetta JP, Warter JM, Lehmann S et al. Presence of antibodies to lectin CSL in the blood of patients with multiple sclerosis. CR Acad Sci III 311:327, 1990a

Zanetta JP, Warter JM, Kuchler S et al. Antibodies to cerebellar soluble lectin CSL in multiple sclerosis. Lancet 335:1482, 1990b

INDEX

A

K

L

M

Q

R

S

T

U

V

W

X